Marine Unmanned Platform Technology

Jintao Wang · Feng Ding · Shangqing Zhu · Ye Lv ·
Bin Wang

Marine Unmanned Platform Technology

Jintao Wang
Department of Electronic Engineering
Tsinghua University
Beijing, China

Feng Ding
Department of Electronic Engineering
Tsinghua University
Beijing, China

Shangqing Zhu
Beijing Municipal Road and Bridge Management and Maintenance Group Co. Ltd.
Beijing, China

Ye Lv
China Railway Information Technology Group Ltd.
Beijing, China

Bin Wang
Shijiazhuang Tiedao University
Shijiazhuang, Hebei, China

ISBN 978-981-96-5609-7 ISBN 978-981-96-5610-3 (eBook)
https://doi.org/10.1007/978-981-96-5610-3

Jointly published with Tsinghua University Press
The print edition is not for sale in China (Mainland). Customers from China (Mainland) please order the print book from: Tsinghua University Press.

Translation from the Chinese Simplified language edition: "海洋无人平台技术" by Jintao Wang et al., © Tsinghua University Press 2023. Published by Tsinghua University Press. All Rights Reserved.

© Tsinghua University Press 2025

This work is subject to copyright. All rights are solely and exclusively licensed by the Publisher, whether the whole or part of the material is concerned, specifically the rights of reprinting, reuse of illustrations, recitation, broadcasting, reproduction on microfilms or in any other physical way, and transmission or information storage and retrieval, electronic adaptation, computer software, or by similar or dissimilar methodology now known or hereafter developed.
The use of general descriptive names, registered names, trademarks, service marks, etc. in this publication does not imply, even in the absence of a specific statement, that such names are exempt from the relevant protective laws and regulations and therefore free for general use.
The publishers, the authors, and the editors are safe to assume that the advice and information in this book are believed to be true and accurate at the date of publication. Neither the publishers nor the authors or the editors give a warranty, express or implied, with respect to the material contained herein or for any errors or omissions that may have been made. The publishers remain neutral with regard to jurisdictional claims in published maps and institutional affiliations.

This Springer imprint is published by the registered company Springer Nature Singapore Pte Ltd.
The registered company address is: 152 Beach Road, #21-01/04 Gateway East, Singapore 189721, Singapore

If disposing of this product, please recycle the paper.

Contents

Part I Introduction

1 Overview of Marine Unmanned Platforms 5
  1.1 Overview of Drones 5
  1.2 Overview of Surface Unmanned Vehicles 8
  References 11

2 Development History of Marine Unmanned Platforms 13
  2.1 Development History of Marine Unmanned Aircraft Platform ... 13
    2.1.1 Origins of Marine UAV Platforms 13
    2.1.2 First Modern Drone 14
    2.1.3 Early Target Drone 16
    2.1.4 Early Unmanned Reconnaissance Aircraft 18
    2.1.5 Long-Range Reconnaissance Drones 18
    2.1.6 First Unmanned Helicopter 19
    2.1.7 The Birth of the Twin-Tailed Truss-Propelled Drone 19
    2.1.8 Maritime UAV Development in the Twenty-First Century .. 21
  2.2 Development History of Marine Unmanned Vessel Platforms 24
    2.2.1 Results of USV Development in the USA 28
    2.2.2 Israeli USV Development Results 29
    2.2.3 Results of USV Development in Other Countries 29
    2.2.4 Overview of Domestic Developments 33
  Bibliography 39

3 Introduction to Marine Unmanned Platform System 41
  3.1 Introduction of Composition 41
    3.1.1 UAS Components 41
    3.1.2 Unmanned Boat System Components 43

	3.2	Key Technologies of Marine Unmanned Platform	47
		3.2.1 Key Technologies for Marine UAV Platforms	47
		3.2.2 Key Technologies for Unmanned Boat Platforms	56
	3.3	Typical Applications of Marine Unmanned Platform	58
		3.3.1 Typical Applications of Marine Drones	58
		3.3.2 Typical Applications of Marine Unmanned Boats	59
	Bibliography		63

Part II Acoustics

4	**Overview of Acoustic Technology**		67
	4.1	History of Acoustic Technology	67
		4.1.1 Overview of the Development of Acoustics	67
		4.1.2 Relationship Between Acoustics and Other Disciplines	68
	4.2	Main Content and Development of Modern Acoustics	70
		4.2.1 Electroacoustics	71
		4.2.2 Ultrasonics	73
		4.2.3 Hydroacoustics	75
	4.3	Development Status of Foreign Acoustic Technology	78
		4.3.1 Acoustic Emission Techniques	78
		4.3.2 Sonar Technology	80
	4.4	Current Status of Development Status of Domestic Acoustic Technology	85
		4.4.1 Sonar Technology	89
	References		90

5	**Principles of Acoustic Technology**		93
	5.1	Acoustic Emission Technology Principle	93
		5.1.1 Basic Concepts	94
		5.1.2 Basic Principles of Acoustic Emission	103
		5.1.3 Acoustic Emission Data Processing	106
		5.1.4 Typical Acoustic Emission Technology Products	113
	5.2	Principle of Sonar Technology	115
		5.2.1 Basic Concepts	116
		5.2.2 Principle of Sonar Technology	122
		5.2.3 Data Processing	132
		5.2.4 Typical Sonar Products	152
	References		159

6	**Acoustic Key Technologies**		161
	6.1	Acoustic Emission Key Technologies	161
		6.1.1 Signal Analysis and Processing Techniques	163
		6.1.2 Acoustic Emission Source Localization Technology	178
		6.1.3 Pattern Recognition of Acoustic Emission Signals	189

	6.2	Key Technologies of Sonar	202
		6.2.1 Beam Forming Technology	202
		6.2.2 Beam-Forming Acoustic Imaging Technology	209
		6.2.3 Multi-beam Sonar Technology	218
		6.2.4 Side-Scan Sonar Technology	238
		6.2.5 Synthetic Aperture Sonar (SAS) Key Technology	242
	References		253
7	**Application of Acoustic Technology in Marine Engineering Inspection**		255
	7.1	Application of Acoustic Emission Technology	255
		7.1.1 Fatigue Crack Signal Identification of Tube Nodes of Offshore Platforms	256
		7.1.2 Corrosion Damage Signal Clustering Recognition of Tension Strand	263
	7.2	Application of Sonar Technology	276
		7.2.1 Detection of Underwater Structure Engineering	276
		7.2.2 Application in Marine Platform Detection	279
		7.2.3 Bridge Underwater Structure Inspection	293
	References		302

Part III Optoelectronics

8	**Overview of Optoelectronic Technology**		307
	8.1	Development History of Optoelectronic Technology	307
		8.1.1 Visible Light Detection	308
		8.1.2 Infrared Detection	309
		8.1.3 Microminiature Imaging Sensors	313
		8.1.4 Imaging Polarisation Detection	313
		8.1.5 Multi-spectral/Hyperspectral Imaging Technology	314
		8.1.6 LIDAR Imaging Technology	315
		8.1.7 Multi-sensor Data Fusion Technology	316
	8.2	Development Status of Foreign Airborne Photoelectric Equipment	316
		8.2.1 Sniper Optical Targeting Pods	317
		8.2.2 Litening Optical Targeting Pods (LITENING)	317
		8.2.3 ATFLIR Optical Targeting Pods	318
		8.2.4 Damocles Electro-Optical Targeting Pod (DOT)	318
		8.2.5 OLS-35 Optronic Radar	319
		8.2.6 T-50 Optronic Integrated System	320
		8.2.7 F-35 Electro-Optical Targeting System	321
	8.3	Development Trend of Domestic Airborne Photoelectric Detection Equipment	322
		8.3.1 Airborne Earth-to-Ground Photoelectric Detection Equipment Form Development Trend	322

		8.3.2	Technology Development Trend of Airborne Ground-to-Ground Photoelectric Detection Equipment	323

	References			325
9	**Principles of Photovoltaics**			**327**
	9.1	Principle of UAV TV Camera and Tracking and Positioning		327
		9.1.1	Overview	327
		9.1.2	CCD Structure and Principles	329
		9.1.3	CCD Classification and Characteristics	336
		9.1.4	Optical Stabilisation Platform Principle and Characteristics	346
		9.1.5	TV Image System Tracking and Localisation	353
	9.2	Principle of Infrared Imaging for UAV		360
		9.2.1	Physical Basis of Infrared	361
		9.2.2	Infrared Imaging Technology	362
		9.2.3	Infrared Detectors	364
		9.2.4	Infrared Detector Cooling	371
		9.2.5	Infrared Thermal Imaging System	376
	9.3	Principle of Laser Scanning Technology		379
		9.3.1	Basic Concept	380
		9.3.2	Technical Principle	381
		9.3.3	Data Processing	384
		9.3.4	Typical Products	391
	Bibliography			392
10	**Optical Key Technology**			**393**
	10.1	Three-Dimensional Laser Scanning Technology		393
	10.2	Ultra-High Resolution Photoelectric Imaging Technology		395
	10.3	Offshore High Humidity, Salt Spray and Temperature Protection Technology		399
	10.4	High-Precision Stable Platform Technology		400
	10.5	High-Precision Target Recognition Tracking and Positioning Technology		401
	10.6	Photoelectric Information Processing Technology		402
	Bibliography			403
11	**Application of Optical Technology in Marine Engineering**			**405**
	11.1	Application of Airborne Marine Lidar in Offshore Engineering		405
		11.1.1	Application of Red Tide and Pollution Monitoring	405
		11.1.2	Marine Sounding Applications	406
		11.1.3	Application of Three-Dimensional Landscape Simulation in Coastal Zone	407
		11.1.4	Underwater Military Target Detection	408
		11.1.5	Fish Detection	412

		11.1.6	Reef Detection and Distress Surveys	414
		11.1.7	Detection of Marine Underwater Resources	415
		11.1.8	Detection of Marine Plankton	415
	11.2	Application of Infrared Detection in Offshore Engineering		416
	Bibliography			421

Part IV Radar

12 Overview of Radar Technology 425
    12.1 What Is Radar? 426
        12.1.1 Functions of Radar 426
        12.1.2 Radar Operating Frequency and Usage Characteristics 429
    12.2 Development History of Radar Technology 433
        12.2.1 The Infancy of Radar Technology 433
        12.2.2 Maturity of Radar Technology 434
        12.2.3 Period of Development of Radar Technology 436
    12.3 Current Status and Trend of Radar Technology Development in Domestic and Overseas Marine Field 437
        12.3.1 Digitalisation of Radar Hardware Platforms 437
        12.3.2 Softwareisation of Radar Capability Implementation 439
        12.3.3 Blurring of Radar Functional Boundaries 440
        12.3.4 Diversification of Radar Information Acquisition Domains 441
    References 443

13 Radar Technology Principles 445
    13.1 Basic Principles of Radar Transmitter 448
    13.2 Fundamentals of Radar Antennas 450
    13.3 Fundamentals of Radar Receivers 451
    13.4 Fundamentals of Radar Signal Processing 454
        13.4.1 Pulse Compression 454
        13.4.2 Moving Target Indication 457
        13.4.3 Moving Target Detection 459
        13.4.4 Pulse Detection 462
    References 466

14 Radar Key Technologies 467
    14.1 Sea Clutter Processing Technology 467
    14.2 Dynamic Platform Beam Control Technology 472
    14.3 Sea Environment Adaptability 474
        14.3.1 Basic Concepts of Material Corrosion 475
        14.3.2 Selection of Metal Materials for Radar Design in Ocean Engineering 476

		14.3.3	Selection of Polymer Materials for Radar Design in Ocean Engineering	477
		14.3.4	Some Considerations for Radar Design in Offshore Engineering	479
	14.4	Electromagnetic Compatibility Control Technology		480
		14.4.1	Significance of Electromagnetic Compatibility Design of Radar in Ocean Engineering	480
		14.4.2	Design of Electromagnetic Compatibility of Radar in Offshore Engineering	481
	14.5	Atmospheric Waveguide Processing Technology		485
	References			489
15	**Application of Radar Technology in Marine Engineering**			491
	15.1	Military Application of Radar in Ocean Engineering		491
		15.1.1	US AN/SPY-1 Series Shipboard Radar	491
		15.1.2	New SMART Three-Coordinate Radar from the Netherlands	492
		15.1.3	German TRS Series Radar	494
	15.2	Navigation Radar		496
		15.2.1	Application Form of Navigation Radar	499
		15.2.2	Domestic and Foreign Navigation Radar Development Power	501
	15.3	Synthetic Aperture Radar		502
	References			504

Part V Communications

16	**Overview of Communications Technology**			509
	16.1	Development History of Maritime Communication Technology		509
	16.2	Development Status of Foreign Maritime Communication Technology		510
		16.2.1	Current Status of Foreign Maritime Shortwave/ultra Shortwave Communications	510
		16.2.2	Current Status of Foreign Digital Microwave Communications	511
		16.2.3	Status of Foreign Maritime Satellite Communications	512
	16.3	Domestic Maritime Communication Technology Development Status		513
	References			515
17	**Principles of Communications Technology**			517
	17.1	Airborne Platform Communications		517
		17.1.1	Basic Principle of Air Platform Communication	518
		17.1.2	Main Features of Air Platform Communication	519

	17.1.3	Airborne Platform Carriers	520
17.2		Surface Platform Communications	524
	17.2.1	Application of Unmanned Boats in Maritime Communication	524
	17.2.2	Application of Buoy in Maritime Communication	525
	17.2.3	Application of Mesh/Ad-Hoc Network in Maritime Communication	526
Bibliography			526

18 Communication Key Technology 529
    18.1 Channel Receiver Realisation Technology 529
        18.1.1 Overview 530
        18.1.2 Mathematical Foundations 533
        18.1.3 Single-Carrier Systems 544
        18.1.4 Multi-carrier Systems 563
        18.1.5 Introduction to Receivers Within the European DVB-T Standard [15, 16] 570
        18.1.6 Introduction to Receivers Within the Chinese DTMB Standard [17–19] 574
    18.2 Modulation and Coding Techniques 578
        18.2.1 Overview 578
        18.2.2 Single-Carrier Modulation Technology 581
        18.2.3 Multi-carrier Modulation 586
        18.2.4 SC-FDE 588
        18.2.5 Channel Coding Techniques 589
    18.3 Channel Characteristics of Wideband Transmission Techniques 590
        18.3.1 Overview 590
    References 592

19 Application of Communication Technology in Marine Engineering 595
    19.1 Wide-Band Multichannel Relay Systems 595
    19.2 Duplex Multichannel Shared Systems 597
    19.3 UAV Relay Communication Systems 598
    19.4 Adaptive Joint C4ISR Nodes 600
    Bibliography 601

Part VI Prospects

20 Future Outlook for Unmanned Maritime Systems 605
    20.1 An Important Form of Future Warfare is the Unmanned Nature of Warfare 605
    20.2 Characteristics of Future Development of Unmanned Equipment 606
    Bibliography 607

Part I
Introduction

The research and application of unmanned platforms have developed rapidly over the past decade or two. Since the use of unmanned platforms in high-risk and complex environments can greatly reduce casualties and is more flexible in terms of design and functionality, research on and application of unmanned platforms have been carried out rapidly, first in the military field. At the same time, research on unmanned platforms for civilian applications is also progressing rapidly.

The unmanned platform is relative to the traditional manned platform, mainly including unmanned ground vessels (UGVs), unmanned aerial vessels (UAVs), unmanned underwater vessels (UUVs) and unmanned surface vessels (USVs), of which the research on unmanned ground vessels, unmanned aerial vessels and unmanned underwater vessels started earlier and developed faster, and has already achieved a lot of results.

Unmanned vessels are relatively new, but they have received more and more attention in recent years. Unmanned vessels are surface vessels that can plan and navigate autonomously according to navigation information and can be remotely operated by shore-based personnel while completing tasks such as environmental information collection and target detection. Unmanned boats can perform information reconnaissance, infiltration and combat tasks in various complex environments and can play a role in clearing sea obstacles for other ships in future sea battles and in establishing safe sea passages, etc. Since the 21st century, western countries have attached great importance to the research and application of surface unmanned boats. The United States, Israel, Japan and other countries have already taken the unmanned boat as one of the most important research fields of the navy. In recent years, China has also paid more and more attention to the research of surface unmanned boats, but the research of unmanned boats is still in the preliminary stage of theoretical exploration, and there are fewer researches on the experiments of real boats. In the future naval war, more and more emphasis on shallow landing operations, sea and air and underwater co-operation operations, low-casualty operations and so on. Since unmanned boats do not require crew members to work on them, there is no danger of casualties even in the face of an enemy with highly effective and lethal weapons, especially in high-risk environments and areas containing high levels of radiation and pollution. Therefore, research on unmanned boats has attracted the attention of more and more countries.

At present, it is of great significance to conduct research on high-speed unmanned craft on the surface of the sea in China.

With the American "Spartan Scout" unmanned boat to join the U.S. military combat sequence and play an active role in local wars, the future of the ship is bound to move towards automation, intelligence, unmanned direction, high-speed surface unmanned boat is the product of this historical trend. Unmanned boat will become an indispensable and important member of the future war, playing an irreplaceable role in seizing the information superiority position, carrying out precise military strikes and fulfilling special combat tasks. At present, western countries have successively carried out research on unmanned boats for various purposes, and in order to comply with this historical trend and catch up with the world's advanced level in the field of military science and technology, the research on intelligent surface unmanned boats in various fields has become an imminent task for China.

In recent years, with the rapid development of information technology, wireless communication, precision navigation and artificial intelligence and their wide application in the military field, intelligent unmanned systems have become an important part of modern warfare, and the world's major military powers have attached great importance to the application of unmanned systems equipment in the military field. In the future, intelligent unmanned equipment will profoundly affect the way of combat and subvert the rules of war, while intelligent unmanned equipment, as a collection of cutting-edge technologies, represents the highest level of a country's scientific and technological strength. War under the condition of information technology has changed the confrontation between man and man, man and machine, and machine and machine into the confrontation between combat systems, in which case, if there is a local or a certain technical advantage, it is very likely to affect the final outcome of the whole war. In China, for reasons of history and geography, as well as real economic interests, the rights and interests of the sea are constantly being violated by Western Powers and neighbouring countries, and local conflicts to safeguard national sovereignty may break out at any time. At present, China's marine military equipment level still lags behind the advanced countries in the West, the intelligent surface unmanned boat into other large and medium-sized surface ships in the combat sequence, to make up for the shortcomings of the level of military equipment at a lower cost, and give full play to the characteristics of unmanned boat movement and flexibility, functional diversity, etc., which will form a local tactical advantage, and achieve the purpose of influencing the local and even the whole conflict. In the conflicts that may break out in the future, the use of surface unmanned boats can quickly set up surface information nodes, establish maritime communication networks, carry out information relay, and also carry out such tasks as target detection, maritime search and rescue, fire strikes, security patrols, logistical support and baiting of target vessels.

Unmanned vessels can also be used in civilian fields, such as cleaning up the sea oil spill and other kinds of rubbish, clean up the surface of the water cyanobacteria, hydrological and meteorological detection, etc., unmanned boat can also be used for sea patrol, reconnaissance, etc., to deal with marine emergencies and the ocean, large lakes and other aspects of the environmental monitoring and disaster early warning, etc., one of the important uses of the sea is the search and rescue, that is, the wind and waves are relatively large, and the search and rescue personnel cannot be reached by the unmanned boat. One of the important uses is search and rescue at sea, i.e. when the wind and waves are strong and the search and rescue personnel cannot reach, the unmanned vessels will rescue the trapped people. In recent years, some domestic scientific research institutes and colleges have carried out research on unmanned boats one after another.

Domestic research institutes have conducted in-depth studies on the system composition and principles of unmanned boats. As the core design idea of unmanned boat, the system structure elaborates its composition principle from the overall perspective, which will be of great significance in guiding the development of unmanned boat moulding products. Meanwhile, in the research of motion control system, the domestic research is mainly limited to the demonstration of control theory, model-based performance research and computer simulation, etc., based on the actual ship, diesel-powered, water-jet propulsion for the propulsion of the control system hardware and software research books are very few. This book explores the domestic and international development of surface unmanned boats by analysing and collating the existing relevant information on unmanned boats at home and abroad and drawing on the technical achievements of other manned and unmanned platforms. Based on the research on the current status of the development and products of surface unmanned boats at home and abroad, and combining with the current domestic and international surface unmanned boat market demand, this book provides a systematic introduction of unmanned surface platforms through collation and analytical studies. After finishing and analysing the research, the unmanned surface platform is systematically introduced, and the main professional and technical fields involved in the platform are discussed in chapters according to the four professional directions of acoustics, optoelectronics, radar and communication. Each chapter is divided into several parts for introduction, according to the overview and domestic and international status, key technologies, typical applications, in an effort to systematically introduce the contents of each speciality, and better help you understand the unmanned platform.

Chapter 1
Overview of Marine Unmanned Platforms

1.1 Overview of Drones

Unmanned aerial vehicle system is a complex system composed of multiple subsystems with unmanned aircraft as the main body, integrating aeronautical technology, information technology, control technology, measurement and control technology, sensing technology, new materials, new energy and other multidisciplinary technologies, and has become a new development direction of aerospace. The development history of UAV can be traced back to the 1920s, due to the technological progress and war demand, UAV has gradually developed into one of the important parts of the weaponry of all countries in the world, especially the developed countries, and unmanned has also increasingly become one of the directions of the future war development, while the UAV is also being developed to the civilisation. At the end of the twentieth century, the development of unmanned aerial vehicles has entered a new era and three waves of development have been formed. At present, all major countries in the world are actively developing unmanned aerial vehicles, despite their different directions and degrees of development, and are expanding them into the civilian sphere while further developing their military applications, with the climax of the development of unmanned aerial vehicles on the horizon.

Unmanned aeroplanes, abbreviated to "UAV", are unmanned aircraft operated by radio remote control equipment and self-contained programme control devices. From the technical point of view, it can be divided into unmanned helicopters, unmanned fixed-wing aircraft, unmanned multi-rotor aircraft, unmanned airships, unmanned parachute aircraft and so on. Unmanned Aircraft System (UAS), abbreviated as "UAS", is a group of equipment that takes UAV as the main body and is equipped with related sub-systems to fulfil specific tasks. UAS generally consists of an unmanned aircraft platform, a measurement, control and information transmission subsystem, a flight control and navigation system, a mission payload, a launch and recovery system, and a ground transport and security system.

The above definition of drones, with its emphasis on the aerodynamic provision of lift to the vehicle, contemplates mainly unmanned aerial vehicles in airspace. However, with the rapid development of drone technology, drones have begun to develop beyond airspace, and near-space drones and space drones have begun to appear. Represented by the near-space unmanned aerial vehicles (UAVs) and airborne UAVs that the United States has been testing, UAVs are moving to a higher, more distant and faster airspace, and the development of UAV technology has given UAVs a new outreach and connotation.

The above definition of drones, with its emphasis on the aerodynamic provision of lift to the vehicle, contemplates mainly unmanned aerial vehicles in airspace. However, with the rapid development of drone technology, drones have begun to develop beyond airspace, and near-space drones and airborne drones have begun to appear. With the U.S. test flights of near-space unmanned aerial vehicles and air and space drones as representatives, unmanned aerial vehicles are moving to higher, farther and faster air and space, and the development of unmanned aerial vehicle technology has given unmanned aerial vehicles a new outreach and connotation.

An unmanned aerial vehicle (UAV) that requires a power unit, is capable of sustained, controlled mission flight in the air or controlled flight in both aerospace and space, is capable of carrying mission payloads of a civil or military nature, and is disposable or reusable. In particular, drones capable of sustained cruise flight in near-space are referred to as near-space drones, and drones with both aircraft and spacecraft flight capabilities are referred to as airborne drones.

UAS can be classified by mass, range and flight altitude into micro UAVs (generally <1 kg in mass), small UAVs (generally <20 kg in mass and <30 km in range), proximity UAVs (capable of 100 km in range), medium-range UAVs (capable of 500 km in range), medium-altitude and long-range UAVs (with a range of 500 km, an endurance of 20 h or more, and a flight altitude of 5000–10,000 m) and high-altitude and long-range UAVs (with a range of 10,000 km, endurance of 20 h or more, and a flight altitude of 15,000–10,000 m). (with a flight altitude of 5000–10,000 m) and high-altitude long-endurance UAVs (with a range of 10,000 km, an endurance of 20 h or more, and a flight altitude of 15,000 m) [1].

Compared with manned aerial vehicles, drones have the advantages of versatility, low cost, good cost-effectiveness, no risk of casualties, strong survivability, good manoeuvrability, ease of use and so on, and they are suitable for carrying out the so-called "3D" missions, which are "boring, dirty and dangerous". It is suitable for carrying out "boring, dirty and dangerous" so-called "3D" missions, can carry out reconnaissance in nuclear and chemically contaminated areas and at the front line of war, and can fly in extremely bad weather, thus playing an extremely important role in modern warfare. The missions performed by drones are categorised as attack-kill and non-attack-kill. The U.S. Navy has assigned the following mission divisions to UAVs.

(1) Above the Horizon: detection, classification, targeting, combat damage assessment.
(2) Amphibious operations support: naval fire support, mines/mines.

1.1 Overview of Drones

(3) Command and control communications/radio.
(4) Intelligence: weather, electronic support measures, signals intelligence, nuclear and biological weapons sensors, surveillance, reconnaissance.
(5) Hybrid operations: attack, deception, psychological warfare, search and rescue operations.
(6) Electronic warfare/countermeasures: electronic countermeasures, electronic support measures, jammers, aerial decoys.

Reconnaissance/surveillance missions are by far the main activity of UAVs, with sensors and data links being the focus of their current development. Target observation is second, and electronic warfare is its third mission. After the success of the reconnaissance/surveillance, target observation and electronic warfare missions, the other missions will eventually be expanded.

Maritime UAV applications are an important part of UAS applications, and 71% of the Earth's area is covered by the oceans, providing more exploitable uses and space for UAVs to function [2]. UAVs are highly manoeuvrable and operate in a space that is virtually unlimited by geography, which gives them a great performance advantage at sea. Maritime UAVs can be divided into shipborne UAVs and land-based UAVs according to the take-off and landing modes. Among them, the shipborne UAV consists of take-off and landing shipborne platform and UAV, the UAV can be realised on the deck of the sailing ship for take-off and recovery, which greatly increases the coverage of the airspace and sea area of the UAV, and the shipborne UAV is shown in Fig. 1.1 [3]. Land-based UAVs achieve take-off and recovery on land and can fly missions at sea.

Due to the narrow space of the ships at sea and the large wind speed on the sea surface, the ships not only have to move forward, but also the deck lifting and sinking, transverse and longitudinal rocking movements are quite violent, which puts forward high requirements for the take-off and landing of the UAVs. At the same time, the combat ships carrying maritime UAVs usually carry many electromagnetic detection and reconnaissance equipments and weapons, etc., which requires that the UAV measurement and control system can adapt to the space of the ship and the

Fig. 1.1 Schematic diagram of the composition of a maritime unmanned aircraft system

complex electromagnetic environment of the outside world, so as to achieve that it will not be interfered with by other signals and will not interfere with the normal work of other equipments.

In terms of military applications, foreign maritime UAVs have been developed earlier and have become relatively systematic, with the representative ones being the RQ-2A Pioneer used by the US military, the Scout UAV and the Orbiter UAV developed by the Israeli military, and the Scanning Eagle UAV developed by the Boeing Company. In addition to the fixed-wing aircraft listed above, the RQ-8 Fire Scout rotary-wing shipborne unmanned aerial vehicle (UAV) is also frequently used in maritime military operations.

For civil use, maritime UAVs can be applied to maritime search and rescue, emergency assistance, maritime resource exploration, meteorological detection, etc. Domestic research institutes have also gradually attached importance to the research and development of maritime UAVs in civil use, such as the U650 UAV developed in China, which can be used for maritime supply delivery and monitoring tasks. The development of UAVs in the marine field has received more and more attention, and in the future, more domestic and foreign R&D organisations will design more models to adapt to different marine application environments.

1.2 Overview of Surface Unmanned Vehicles

Unmanned boat is an abbreviation for surface unmanned boat, which refers to a type of surface small craft that is directly navigated autonomously or remotely controlled to achieve normal navigation, manoeuvring and operations. UAVs can carry out designated missions by carrying various mission payloads. It consists of a platform, a mission payload, a communications system and a manoeuvring system.

(1) Platform: a combination of hull, turbine and electrical equipment.
(2) Mission load: equipment configured to fulfil the assigned mission.
(3) Communication system: wireless transmission system for transmitting various commands, status information, images, video and audio data between the platform and the platform, and the platform and the control station of the mother ship/shore-based/air-based.
(4) Manoeuvring system: located on the platform and mother ship/shore base to identify, process and make decisions on the collected information, so as to realise the platform's autonomous or remote-controlled navigation system.
(5) Platform in situ manual control: manual control on the platform for maintenance, overhaul and emergency.
(6) Remote control: in the mother ship / shore-based use of autopilot on the platform propulsion device and a variety of equipment and systems for remote control of the navigation mode.
(7) Autonomous navigation: the navigation mode of safe navigation in accordance with the target mission without any human intervention.

1.2 Overview of Surface Unmanned Vehicles

Unmanned craft can be divided into five categories as follows.

(1) Category 1: Unmanned craft designed to navigate at a distance of more than 200 n miles from shore with a minimum design meaningful wave height (Hs) of 6 m.
(2) Category 2: Unmanned craft designed to navigate at a distance of not more than 200 n miles from shore with a minimum design meaningful wave height (Hs) of 4 m.
(3) Category 3: Unmanned craft designed to navigate at a distance of not more than 20 n miles from shore with a minimum design meaningful wave height (Hs) of 2 m.
(4) Category 4: Unmanned craft designed to navigate at a distance of not more than 10 n miles from shore with a minimum design meaningful wave height (Hs) of 1 m.
(5) Category 5: Unmanned craft designed to navigate at a distance of not more than 5 n miles from shore with a minimum design meaningful wave height (Hs) of 0.5 m.

Surface unmanned craft are mainly used to carry out dangerous and unsuitable tasks for manned vessels, and once equipped with advanced control systems, sensor systems, communication systems and weapons systems, they can carry out a wide range of war and non-war military tasks, such as reconnaissance, search, detection and demining, search and rescue, navigation and hydro-geological surveys, antisubmarine warfare, anti-special operations and patrols, anti-piracy and anti-terrorist attacks, and so on.

So far, various countries in the world have developed unmanned boats for various purposes (Fig. 1.2). The architecture design of unmanned boats based on different functional requirements is also different. This book focuses on the use of unmanned boats for security and defence missions in offshore waters. As a front-line defence for offshore defence, such unmanned craft can patrol important ports and islands, collect information on offshore waters, monitor and follow up suspicious targets, and, when it is determined that a suspicious target is dangerous, the weapons system carried on board can strike and destroy it. If it carries warning and guidance equipment, it can be used as civilian equipment by the maritime authorities to assist them in completing maritime management and maritime search and rescue tasks. In order to achieve the above tasks, first of all, we must have accurate and reliable navigation and navigation and positioning functions, and at the same time, we also need to have the ability to collect and process information about the surrounding environment and the ability to intelligently intervene in the processing of the target.

(1) The unmanned boat should have accurate and reliable navigation and navigation functions. On the one hand, the unmanned boat should make use of the GPS, gyroscope, electronic chart and other equipments equipped by itself to navigate autonomously along the planned path, and when encountering obstacles, it can set up obstacle avoidance paths independently to bypass the obstacles; on the other hand, it should also be able to accept the remote control order from the

Fig. 1.2 Surface Unmanned Vessel [4]

shore-based controllers to realise the remote-control navigation. Outside the visual range of the controller, the unmanned boat should be able to accurately locate its own position, and accurately measure the heading and the speed, acceleration and angular acceleration of the unmanned boat. When navigating at high speed, it should ensure good hull stability. Unmanned craft need to have a sufficiently long power supply for equipment, which can be achieved by equipping them with high-performance batteries and axle-belt generators. The use of large-capacity fuel tanks and advanced fuel-saving technologies will enable the unmanned craft to have a long cruising time to ensure that the range of the unmanned craft covers the waters under its jurisdiction.

(2) The unmanned boat should have the function of providing real-time on-site environmental information and its own navigation information to the shore-based controller. The performance characteristics of different sensors should be taken into account to optimise the combination and collect environmental information as comprehensively as possible. Due to the special environment at sea, the unmanned boat should be equipped with a special mounting platform to minimise the distortion of the collected information and screen jitter caused by seawater shaking when the sensors are working. The unmanned boat will also accurately return the measured information of its own navigation to the shore operator to facilitate the controller to understand the real-time information of the unmanned boat so as to make the correct control instructions. The format of the information transmitted to the controller includes images, pictures, audio and so on, which are processed and presented to the controller in real time.

(3) Unmanned boats should be equipped with the ability to intervene intelligently to deal with suspicious targets. Non-lethal equipment carried by the unmanned craft can be used to stop offences and eliminate potential threats, or they can be destroyed by direct strikes with the help of weapons systems. The remote-controlled machine guns and small-calibre guns carried by unmanned boats can be used to strike and destroy all kinds of suicide boats and mines placed on the surface of the water, which is low-cost and safe. When civilian unmanned craft are used for maritime management, shore personnel can be alerted to violations of the law by means of the bright lights on board the remotely controlled

unmanned craft, and the unmanned craft can be used to carry out rescue operations at sea in respect of capsized ships and people who have fallen into the water.

Compared with other unmanned equipment, the development of surface unmanned craft has lagged behind, but the degree of autonomy is increasing. The degree of autonomy is the core indicator of the advancement of unmanned systems. According to the degree of autonomy, surface unmanned boats can be divided into three categories: remote-controlled, semi-autonomous and fully autonomous. Since the fully autonomous control mode requires a high degree of intelligence, it is extremely difficult to realise and is still in the research and exploration stage. At present, the surface unmanned boats of various countries mostly adopt semi-autonomous type, and from the main features and functions of the surface unmanned boats that have been put into service or are under research in foreign countries, the fully autonomous surface unmanned boat is the development goal of the surface unmanned boat in the future.

References

1. https://blog.csdn.net/u011326478/article/details/79297485
2. Xingda, Chen. 2018. An Overview of the Application and Development of Maritime Drones. *China Strategic Emerging Industries* 12.
3. https://www.163.com/dy/article/E7OG4PS20511DV4H.html
4. https://m.sohu.com/a/195809245_358040

Chapter 2
Development History of Marine Unmanned Platforms

2.1 Development History of Marine Unmanned Aircraft Platform

2.1.1 Origins of Marine UAV Platforms

The history of drones is, in fact, the history of all aircraft. From the Chinese kite flying gracefully through the air centuries ago to the introduction of the first hot air balloon, unmanned aerial vehicles predate the risky manned vehicles. According to legend, Zhuge Liang (180–234 A.D.), the warlord of Shu Han during the Three Kingdoms period, was one of the early users of drones, lighting oil lamps in paper balloons to heat up the air and then releasing the balloons over enemy camps at night, misleading the enemy into believing that divine power was at work. In modern times, drones are primarily autonomous/remotely operated aerial platforms capable of mimicking manned aircraft manoeuvres. The names of drones have changed many times over the years. Aircraft manufacturers, civil aviation authorities and the military have each given them different names. Aerial torpedoes, radio-controlled vehicles, remotely piloted platforms, remotely controlled vehicles, autonomously controlled platforms, unmanned platforms, pilotless remotely piloted aircraft and airborne unmanned platforms (which have evolved into drones) have all been used to describe these "unmanned on board" flying machines.

In the early days of aviation, the development and testing of unmanned aerial vehicles offered significant advantages, at least to the extent that those who were very hands-on did not have to risk the loss of life or limbs, and in the 1890s, German aviation pioneer Otto Lillingshire used unmanned gliders as a test bed for main-lift wing designs and light aerostructures, and despite some accidents, no injuries occurred and the tests made great progress. Despite a number of accidents during the experiments, there were no injuries and the tests made great progress. Early unmanned aircraft, despite attempts to use the "no man on board" model, were limited in their widespread use by the lack of satisfactory control methods. Aeronautical research soon turned to the use of "test pilots" to fly these ground-breaking vehicles, but the attempts to break through to unmanned glider technology came at a terrible price, and even aviation pioneer Lee Lindsell was killed in a flight experiment in 1896.

In terms of modern drone use, historical drones have often followed a consistent pattern of use—what are today called 3D missions, i.e., dangerous, harsh and boring missions. "Dangerous" means that someone is trying to shoot down the aircraft or that the pilot may face additional operational risks to his or her life; "hostile" means that the mission environment may be contaminated with chemical, biological, radiological, or even nuclear substances that the human body cannot be exposed to; and "boring" means that the mission environment may be contaminated with chemical, biological, radiological, or even nuclear substances that the human body cannot be exposed to. "Dull" refers to repetitive or persistent missions in which the pilot is susceptible to fatigue and stress. Currently, there is also 1D, or deep, which refers to missions beyond the operational radius of manned aircraft, and together are also referred to as 4D. More broadly, the challenges faced by unmanned systems from deep space, deep sea, deep earth and other applications are all missions of a deep nature.

2.1.2 First Modern Drone

In late 1916, the United States began funding Sperry's development of unmanned aerial torpedoes through the Navy. Elmer Sperry devoted his entire team to what was then the most difficult aeronautical endeavour. According to the Navy's contractual instructions, Sperry was to build a small, lightweight aircraft capable of self-launching without a pilot, flying unmanned and guided to a target 1000 yd (914.4 m, 1 yd = 0.9144 m) away, and detonating its warheads close enough to the warship to form an effective strike against the ship (Fig. 2.1). Since the aircraft had only a short history of 13 years since its introduction, even to build an airframe capable of carrying a large warhead, or a large radio with batteries, heavy electric brakes and large mechanical three-axis gyroscopic stabilisers were incredible in themselves, not to mention that it was unimaginable to integrate these primitive technologies to form an effective flight profile.

Sperry appointed his son, Lawrence Sperry, to lead the test flights on Long Island, New York. By the time the United States entered World War I in 1917, the technologies had been fused and testing had begun. Thanks to substantial funding from the U.S. Navy, the project was able to withstand a series of setbacks, including multiple crashes of the Curtiss N-9 aerial torpedo and the complete failure of various components. Everything that could go wrong did, including catapult failures, engine stoppages, and multiple airframe crashes due to stalls, flips, and crosswinds. However, Sperry's team persevered, and on 6 March 1918 the prototype Curtiss was launched unmanned, flew smoothly for 1000 yd, swooped down to its target at a predetermined

2.1 Development History of Marine Unmanned Aircraft Platform

Fig. 2.1 Early drones

time and place, and then recovered and landed successfully, making it the world's first true "drone". "UAV". Thus the UAS was born.

Not to fall behind the Navy, the United States Army invested in a concept for an aerial bomb similar to the aerial torpedo. The Army's efforts further improved Sperry's mechanical gyro-stabilisation technology to a level almost approaching that of the Navy project. Charles Kettering designed a light biplane that incorporated aero-stability features not valued in the manned aircraft programme (such as excessive inverted angle on the main wing), thereby improving the aircraft's roll stability at the expense of precision and some manoeuvrability. Ford Motor Company was commissioned to design a new lightweight V-4 engine with 41 hp (1 hp = 745.7 W) and 151 lb (68.5 kg, 1 lb = 0.4536 kg). The landing gear has a wide wheelbase to minimise ground roll during landing. In order to further reduce costs and to emphasise the expendability of the vehicle, the fuselage is made of cardboard and paper, in addition to the traditional cloth skin. In addition, the vehicle is equipped with a catapult system with non-adjustable full-throttle settings.

The aerial bomb invented by Kettering, named the "bedbug", had very high long-range and high-altitude performance, travelling a distance of 100 miles (160.9 kms, 1 mile = 1609 m) and an altitude of 10,000 ft (3050 m, 1 ft = 0.305 m) in a number of test flights. To demonstrate the effectiveness of its airframe components, Kettering built a model with a pilot's cockpit so that test pilots could fly the aircraft. Unlike the Navy's aerial torpedo, the aerial bomb was the first drone to go into large-scale production (the aerial torpedo never entered service or production). Although produced too late for use in the First World War, it still played an active role in testing for 12–18 months after the war. The aerial bomb was strongly supported by Colonel Henry Harper Arnold, who later became a five-star general in charge of the entire U.S. Army Air Corps during World War II, and the project gained prominence in October 1918 when Secretary of War Newton Baker observed a test flight. After the end of the First World War, 12 Bugs, along with several aerial torpedoes, continued to be flight-tested at Carlstrom Proving Ground, Florida.

2.1.3 Early Target Drone

Surprisingly, most of the world's UAV research efforts after the First World War were not oriented towards weapons platforms (e.g., aerial torpedoes and aerial bombs, etc.), but focused primarily on the technical application of unmanned target aircraft. During the peaceful period between the two world wars (1919–1939), when the impact of aircraft combat capability on ground/sea combat effectiveness began to be recognised, militaries around the world increased their investment in air defence weaponry, which in turn contributed to the demand for physical-like targets. It is against this background that unmanned target aircraft have emerged. Unmanned target aircraft also play a key role in testing air warfare doctrine. The Royal Air Force and the Royal Navy engaged in a heated debate over the ability of aircraft to sink ships. In the early 1920s, General Billy Mitchell of the Army Air Corps sank a trophy German warship and a number of old warship targets, much to the dismay of the United States Navy. Contrarians of these actions argued that a fully manned battleship armed with anti-aircraft anti-aircraft guns could easily shoot down incoming aircraft. The British tested this view by flying unmanned target aircraft over similarly equipped ships. To everyone's amazement, in 1933, unmanned target aircraft flew more than 40 times over Royal Navy warships armed with the latest anti-aircraft guns and were never shot down. Thus, not only did drone technology play a key role in defining the doctrine of air power operations, but it also provided important data for the huge investment made by the United States, Britain and Japan in the development of aircraft carriers, an investment that was justified by the crucial role they played in the ensuing Second World War.

The unmanned target aircraft programme in the United States was largely influenced by the successful development of the Sperry Courier light biplane. There were two versions of this aircraft, manned and unmanned, which could be used militarily as both manned and torpedo-carrying aircraft. The U.S. Army ordered 20 of these aircraft and named them the "Courier Air Torpedo" in 1920. However, in the early 1920s, United States efforts in this area suffered a major setback when Sperry's son, Lawrence Sperry, was tragically killed in an aeroplane accident and the Sperry Aircraft Company withdrew from the current drone design.

As the U.S. Army lost interest in the MAT programme, it turned its attention to unmanned target aircraft, and in 1933, Reginald Denny perfected a radio-controlled aircraft that was only 10ft long and powered by a single-cylinder 8hp engine. Named OQ_19 by the Army and later renamed MQM-33, the flexible, lightweight unmanned aircraft, of which some 48,000 were produced, became the world's favourite unmanned target aircraft throughout the Second World War.

In the late 1930s, the United States Navy returned to the field of unmanned aircraft, the Naval Research Laboratory developed the "Curtiss" N2C-2 unmanned target aircraft. Weighing 2500 lb (1134 kg), with a radial engine and biplane design, this target aircraft played an active role in solving the problem of identifying areas where the Navy's anti-aircraft guns were not powerful enough. Just as the British Air Force's early experience with drones evaded countless shots from well-armed naval ships, the

2.1 Development History of Marine Unmanned Aircraft Platform

Fig. 2.2 Firebee UAV targets

USS Utah failed to shoot down an N2C-2 drone simulating an attack on it. Strangely enough, the U.S. Navy has named these drones NOLO (no live operator onboard), and the U.S. Navy also developed the technology to fly human-controlled drones in the late 1930s under the auspices of the Navy's unmanned target aircraft programme.

In 1951, Rane Aeronautical Corporation developed a derivative, the Firebee Military Target Aircraft (Fig. 2.2), based on the Rane Firebee, which was the first turbojet-powered UAV and holds the record for the most versatile UAV ever flown. It was the first UAV to be propelled by a turbojet engine and has the record of being the most versatile UAV ever flown. First flown in 1955, the Firebee was one of the first jet-propelled UAVs and was used primarily by the USAF for intelligence-gathering missions and radio communications monitoring.

Similarly, in the inter-war period, the Royal Navy attempted to develop unmanned aerial torpedoes and unmanned target aircraft using the same airframe. Several attempts were made to launch them from ships, but these failed. The Royal Aircraft Factory eventually succeeded by combining a long-range gun with a Lynx engine, known as the Larynx, and the RAF followed this up by developing the first practical target aircraft by fitting automated controls to an existing manned aircraft. This was done by converting the Fairleigh Scout-IUF manned aircraft into a gyro-stabilised radio-controlled aircraft, now known as the Queen. A total of five were built, but all crashed on their first flight, except for the last one, which was successful in a sea firing experiment. The next step in the development was to combine the Fairey flight control system with the highly stable de Havilland Moth to assemble what is now known as the Fairey Queen Bee. The Fairey Queen Bee is now known as the Fairey Target Aircraft. This target aircraft proved to be more reliable than the previous Queen. A total of 420 Fairey Queen Bees were ordered by the Royal Air Force. Since then, the letter Q has been used in the name of unmanned aircraft to indicate unmanned operation. The United States military has also adopted this protocol.

In the inter-war period, almost all countries with an aviation industry began to develop drones in one form or another, the main form still being unmanned target aircraft, with the exception of Germany. Inventor Paul Schmidt pioneered the pulse jet engine in 1935. It was a low-cost, easy-to-operate, high-performance thrust device. After inspecting his work, Luftwaffe General Erhard Milch suggested that the new pulse-jet engine be adapted to unmanned aircraft.

2.1.4 Early Unmanned Reconnaissance Aircraft

From the first successful flight of a drone in 1918 until the Second World War, drones were used primarily as target aircraft and weapons delivery platforms. During the ensuing Cold War, drone development rapidly shifted to reconnaissance and decoy missions. This trend has continued to the present day, with nearly 90% of drones involved in various forms of data collection activities in areas such as military, law enforcement and environmental monitoring. The main reason why drones were not used for reconnaissance during the Second World War had more to do with imaging technology and navigation requirements than with the aircraft platform itself. Cameras in the 1940s required relatively more accurate navigation to obtain the required data for the area of interest, but navigation technology then was not comparable to that of today, or even to that of a trained pilot with a map. The end of the war was a major turning point, however, and with the advent of radar mapping, improvements in radio navigation technology, and the use of Loran-type networks and inertial navigation systems, drones finally became autonomous and could fly back and forth between their point of origin and their target area with sufficient accuracy.

The first high-performance unmanned reconnaissance aircraft, later called the GAM-67, was converted from a high-altitude target aircraft, the YQ-1B, with a camera; the turbojet-powered aircraft was originally launched from a B-47 aircraft for the suppression of enemy air defence systems. The proposal to retrofit the camera was cancelled after only 20 aircraft had been converted, mainly because of the short range and high cost.

2.1.5 Long-Range Reconnaissance Drones

The United States Air Force pioneered the development of the first mass-produced long-range, high-speed unmanned aerial vehicle (UAV), designed primarily for reconnaissance missions, but later expanded to include a range of other missions, such as suppression of enemy air defences and weapons delivery. Ryan's I47, later renamed the AQM-34 Firefly, has the longest UAV service record. Based on an early Ryan Aircraft target aircraft from the late 1950s, the aircraft was fitted with a turbojet engine, a low drag wing and fuselage configuration, and could fly at altitudes of over 50,000 ft (15,250 m) and speeds of up to 600 n mile/h (1105 km/h, high subsonic).

Known to operators as the "Bug", the UAV has had a long service life and can fly in a variety of profiles at both high and low altitudes, performing a variety of tasks such as electronic signals intelligence gathering, photo reconnaissance, and transmitting radar decoy signals. During its operational use from the early 1960–2003s, the UAV underwent a number of improvements. A number of unique breakthroughs were applied to the UAV, including aerial launch from the wing mounts of the DC-130 aircraft, and recovery by air parachute from the H-2 ("Hulk") helicopter. Later in

Fig. 2.3 AQM-34 UAV

its service life, the UAV was renamed the AQM-34 and performed a number of important missions, as well as relatively routine missions such as targeting of fighter air-to-air missiles (Fig. 2.3).

2.1.6 First Unmanned Helicopter

The QH-50DASH, a radio-controlled anti-submarine helicopter, introduced by the U.S. Navy in the early 1960s, was the first unmanned helicopter of its kind and the first to take off and land from a ship at sea. This specially constructed counter-rotating rotor aircraft was the first unmanned helicopter and the first to take off and land from a ship at sea. The DASH drone was designed to increase the range of anti-submarine torpedoes. In the early 1960s, a typical destroyer could detect submarines at distances of more than 20 miles (32.2 kms) but could only fire its weapons from a distance of <5 miles (8.05 kms). The small, compact unmanned helicopter could simply fly out to maximum detection range and drop a self-seeking torpedo on a submerged submarine. The QH-50DASH was remotely controlled, with the pilot on the transom manoeuvring the take-off and landing, and then guided by a gyro-stabilised autopilot to a position tracked by the radar of the mother ship. More than 700 were built from the early 1960s to the mid-1970s, ending their service as target towing aircraft for anti-aircraft guns. The UAV was also used by countries such as France and Japan (Fig. 2.4).

2.1.7 The Birth of the Twin-Tailed Truss-Propelled Drone

In the late 1960s, the U.S. Marine Corps successfully developed the Bikini drone, a groundbreaking effort that laid the foundation for one of the most popular unmanned aerial vehicles, and from which came the RQ-7 Shadow drone, the largest UAV produced outside of the hand-launched "The RQ-7 Shadow UAV, the largest UAV produced except for the hand-delivered Big Raven UAV." The most important feature

Fig. 2.4 QH-50DASH unmanned helicopter

of the Bikini airframe is that the camera is mounted in the nose position, giving a virtually unobstructed field of view. This led to a propulsive engine layout, which was further simplified by the adoption of a twin-truss configuration. Although triangular propulsion layouts have also been attempted, as in the case of the classic Skyhawk UAV, this aerodynamic layout makes weight and balance a more challenging design point, as the elevator arms are usually fixed, whereas the twin tail trusses can easily be extended.

In the late 1970s, Israel developed a small tactical battlefield surveillance drone called the Scout, manufactured by Israel Aerospace Industries (IAI), based on the Bikini drone model. It is accompanied by IAI's UAV-A decoy drone and the Mabat drone developed by Ryan. Of these, the decoy drone is designed to be used against air defence missile forces by spoofing the premature activation of radar or even firing missiles at the drone itself. The Mabat is designed to collect radar signals associated with anti-aircraft missiles. The Scout, on the other hand, utilizes the operations of the other two drones, focusing on target information collection and post-fire damage assessment of air defence missile forces. In addition, Scout provides close-up battlefield imagery intelligence to ground manoeuvre unit commanders, the first of its kind for a drone. This approach differs significantly from previous reconnaissance drones, mainly in that the imagery is of more operational/strategic significance and the negatives can be subsequently developed or sent electronically to collection centres for analysis. Based on the development of small-scale computer technology, such "bird's eye view" images can be transmitted in real time to the commanders of mobile units on the ground, directly influencing their decision-making process in commanding small groups of soldiers or even individual tanks.

In the 1980s, Israel developed the twin-tailed miniature Pioneer UAV (Fig. 2.5), based on experience with the Scout and Dog UAVs. The United States Naval Air Systems Command procured this model in bulk and equipped it for service on Iowa-class battleships, using the drone's main task of flying over the target and sending back live video imagery of the relevant position to its own side for use in guiding artillery, assessing the effectiveness of strikes, fortification construction, troop deployments and combat operations, and so on.

Fig. 2.5 Pioneer drone

2.1.8 Maritime UAV Development in the Twenty-First Century

Since the beginning of the twenty-first century, countries have been strengthening their UAV capabilities and improving their UAV spectrum, resulting in a situation of competing development. At present, the main types of maritime drones include wide-area maritime surveillance drones, fixed-wing combat/security drones, vertical take-off and landing drones, small tactical drones and submarine-launched drones.

(1) Wide-area maritime surveillance drones: mature technology, high efficiency of use

Wide-area reconnaissance of maritime targets is indispensable to maritime operations, mainly targeting the deployment, training and manoeuvres of naval bases, surface ships, underwater vehicles and aircraft of the countries concerned in the main strategic directions. In recent years, the United States Navy has permanently deployed two Global Hawk wide-area maritime surveillance drones at the United Arab Emirates airbase to carry out reconnaissance missions. Since the Iranian nuclear crisis, whenever a large naval warship passes through the Strait of Hormuz, a Global Hawk wide-area maritime surveillance drone flies overhead. The drone also closely monitors the movements and deployments of Iranian naval and littoral forces and transmits this information in real time back to United States maritime vessels and ground stations.

In addition to the Global Hawk, the United States Navy is accelerating the development of the next generation of maritime drones is the MQ-4C "Mermaid Poseidon". The UAV has a wingspan of 39.9 m, a length of 14.5 m, a width of 4.6 m, a maximum take-off weight of 14.6 t, a maximum internal load of 1.452 t, a maximum flight altitude of 17,000 m, a maximum endurance of 28 h, and is equipped with active phased-array radar, a multispectral targeting system, an electronic support system, and an automatic identification system and other maritime surveillance sensors. In the future, this aircraft will cooperate with P-8A "Poseidon" manned anti-submarine patrol aircraft to carry out maritime surveillance missions.

(2) Fixed-wing combat/security UAVs: direct participation in combat operations

Fixed-wing combat/security UAVs are directly involved in combat operations and are known as a "sharp sword" in the maritime unmanned combat system. Large fixed-wing UAVs are carrier-based and difficult to develop, represented by the X-47B stealth carrier-based UAV and the MQ-25 Stingray unmanned refuelling aircraft.

X-47B stealth shipborne UAV has participated in the US Navy's "Shipborne Unmanned Aerial Surveillance and Strike" project, the UAV has an endurance of 6 h, two built-in bomb bays and can carry 2t of weapons. Since the completion of the first land flight in February 2011, the UAV has completed the ejection take-off, landing obstruction, night flight, and manned aircraft co-operation and aerial refuelling and other testing projects, the relevant technology has matured. However, due to the UAV range is too short, insufficient ammunition, cannot undertake the expected combat mission, so in the 2017 fiscal year budget, the U.S. Navy cancelled the project, and instead to create a "shipborne aerial refuelling system", the selected UAV is named MQ-25 "Stingray" MQ-25 Stingray.

As an unmanned refuelling aircraft, MQ-25 "Stingray" can carry 15,000 lb (about 6800 kg) fuel, 500n mile (about 920 km) away from the mothership, 4–6 carrier aircraft refuelling operations in the air, so that the carrier aircraft combat radius in the existing basis for expansion of 300–400mile (about 400 km). 400mile (about 480–640 km), lengthening the stagnation time, expanding the scope of activities. In the future, this type of unmanned aircraft will cooperate with a variety of long-range strike weapons, so that the carrier aircraft will have the ability to launch strikes deep inland, or expand the carrier formation's air defence alert, interception and anti-submarine area.

(3) Vertical take-off and landing UAVs: the "best partner" for frigates.

Fixed-wing UAVs need to be carried by flat-deck warships such as aircraft carriers or amphibious assault ships. For frigates with only one helicopter landing platform, vertical take-off and landing UAVs are the best choice.

Among vertical take-off and landing UAVs, unmanned helicopters have a long history of equipment. The first generation of unmanned helicopters can carry 1–2 anti-submarine torpedoes, with sonar and radar for anti-submarine warfare, and is an important aviation anti-submarine force.

Since the twenty-first century, the U.S. Navy has launched the MQ-8 "Fire Scout" unmanned helicopter. This is a large unmanned helicopter, the use of manned helicopters as the basis for research and development, plus the installation of mature equipment, so the development progress is very fast. At present, many types of U.S. warships have taken the naval model of this unmanned helicopter as a "standard", in order to cooperate with the manned helicopters to carry out maritime reconnaissance, target screening, opposite strikes, missile relays, anti-mine and other combat tasks.

In addition to unmanned helicopters, tilt-rotor drones have also developed rapidly in recent years. Tilt-rotor UAVs have both the vertical take-off and landing and hovering capabilities of helicopters and the advantages of fixed-wing aircraft, such

2.1 Development History of Marine Unmanned Aircraft Platform

as faster speed, longer range and higher lift, which give them more tactical advantages than unmanned helicopters and have become the focus of research and development in various countries.

(4) Small tactical drones: strong versatility and wide range of equipment.

Small tactical UAV has strong versatility, can be used on small ships without flight decks, and can even be loaded into submarines with narrow space, mainly carrying out intelligence surveillance/reconnaissance, airway escort, guarding of high-value targets, communication relay and other tasks, and the representative model is the Scanning Eagle UAV. The UAV has a wingspan of only 3.1 m, a length of 1.2 m, a maximum take-off weight of 18 kg, a maximum endurance of 28 h, a cruising speed of 90 km/h, and a maximum flight altitude of 4880 m. The head load compartment can be readily replaced by a variety of loads such as optoelectronic/infrared sensors, biochemical detectors, and laser pointers. Owing to the difficulty of its development and its versatility, the small drone has become the most widely equipped unmanned aerial vehicle in various countries and has frequently appeared on the battlefields of the Middle East in recent years.

(5) Submarine-launched drones: a "multiplier" of submarine warfare.

Submarine-launched drones is not a new topic, as early as the 1940s, some countries have envisaged the launch of unmanned aerial vehicles from nuclear submarines to carry out aerial non-aggressive tasks to achieve the "submarine + fighter" mode of operation, but due to the technical difficulties ultimately given up. Twenty-first century, with the technological advances, a number of countries have carried out research on the In the early twenty-first century, with the progress of technology, some countries have carried out the research on submarine-launched UAV. The most enthusiastic about the development of submarine-launched small drones is Germany, the German Navy developed submarine-launched UAV system can carry three drones, using catapulted from the submarine to launch, through the antenna to receive real-time reconnaissance of the image of the drone. The communication distance between the drone and the submarine is 30 km, beyond which the reconnaissance data can be stored and the images transmitted to the submarine after the normal communication distance is restored. However, these drones are disposable and are mainly used for tactical reconnaissance before special operations.

The U.S. "Cormorant" submarine-launched UAV project is designed to equip nuclear submarines with foldable wings, which can be used like launching an intercontinental missile. The aircraft is mainly used for special operations reconnaissance, but can also carry out attack missions, after the completion of the mission to return to the submarine area splashed down in the sea waiting for recovery. Due to a number of factors, the project was eventually cancelled. However, submarine-launched drones, as a "multiplier" of submarine combat power, are still receiving attention and development by various countries.

2.2 Development History of Marine Unmanned Vessel Platforms

The history of unmanned surface craft dates back to 1898, when the famous inventor Nicolas Truss invented a remotely operated boat called the "Wireless Robot". First used in combat during World War II, they were initially designed in the form of torpedoes to remove mines and obstacles from wave breakers. The U.S. Navy also added radio-controlled steering and mine-clearing rockets to small landing craft for shallow minefield operations in the latter part of the Second World War.

After the end of the Second World War, the surface unmanned boats have been further developed, mainly used for mine clearance and battlefield damage assessment (BDA) and other tasks. With the progress of information technology, automatic control technology, navigation technology and material science, surface unmanned boat technology has also been newly developed, and up to now, more than 10 countries have developed and deployed surface unmanned boats, as shown in Table 2.1.

In the late 1970s, European navies began to develop a new mine countermeasure system, which led to a new phase of research on unmanned boats. At that time, the unmanned boat was designed as a radio-controlled vessel, and the manned mine countermeasures system sailed behind the unmanned boat. The highlight of this design was that it increased the distance between the dangerous waters and the manned platform, which greatly reduced the risk to personnel. In addition, a manned

Table 2.1 Milestones in the development of surface unmanned vehicles

Number	Times	Study	Nations
1	1898	Remotely operated unmanned boat (world's first surface unmanned boat)	USA
2	1960s	Remote-controlled minesweeper	USA
3	1960s	Small remotely operated surface unmanned craft	USSR
4	1970–1980s	Owing to technical constraints, the development of surface unmanned boats has not made a great breakthrough, and they are mainly used in military manoeuvres and as sea targets for artillery firing	USA
5	1990s	Remote-controlled mine-hunting combat prototype	USA
6	2003	Spartan Scout drone	USA
7	2005	stingray surface drone	Israeli
8	2006	Protector unmanned aerial vehicle (UAV)	Israeli
9	2010	piranha unmanned aerial vehicle (UAV)	USA
10	2013	Large unmanned craft, continuous autonomous operation for 48 h	Singapore
11	2014	The drones have the ability to protect their own ships and can launch autonomous attacks using "swarm warfare" and can launch autonomous attacks using "swarm warfare"	USA

2.2 Development History of Marine Unmanned Vessel Platforms

platform is also realised to operate several unmanned dinghies at the same time by remote control, which improves efficiency.

Since the twenty-first century, the surface intelligent unmanned boat has entered a new period of rapid development. Since the "9–11 incident", based on the needs of anti-terrorism at sea, research with intelligence collection, tracking, reconnaissance, demining, anti-submarine, search and rescue, precision strike and other functions of the surface unmanned boat has become the United States as the leader of the Western countries of the important direction of the research. 2001, the U.S. Navy formally put forward the construction of the coastal combatant ship plan. In 2001, the U.S. Navy formally put forward a plan to build a coastal combat ship. In this plan, it is clearly proposed that the surface high-speed unmanned boats and unmanned aircraft, unmanned ground vehicles, unmanned underwater submarines together constitute an unmanned combat system, to complete such tasks as reconnaissance, detection, vigilance, anti-submarine and other tasks in order to collaborate with the special operations.

A typical high-speed surface unmanned craft is the Spartan Scout developed by the United States. It is based on the standard 7 m and 11 m rigid inflatable boats of the US Army, and France and Singapore are jointly involved in the project. "The main design objectives of Spartan Scout are: to enhance early warning capability in maritime space by relying on networked ISR to protect troops from surprise attacks; to validate the performance of sensors and weapons under unmanned conditions; and to minimise manned operations to reduce casualties. So far, from the technology has reached the requirements of unmanned autonomous control, and according to the needs of the mission in a modular way to quickly replace the mission module. The boat has been deployed to the USS Gettysburg cruiser, with other forces, to carry out the "Operation Enduring Freedom" and "Iraqi Freedom Movement" in the Arab region. Operation Enduring Freedom" and "Iraqi Freedom Campaign" in the Arab region.

The Phantom Guardian is another representative high-speed surface unmanned craft developed by the United States Robotic Marine Corporation. It is mainly used to carry out missions such as intelligence gathering, coast guard, logistic supply and climate monitoring. In addition, there are other organisations in the United States that have carried out research on unmanned boats, and have achieved good results. For example, Maggit Defence Systems developed the "water tiger fish", mainly used for the target target ship towing. DRS developed the "sea owl" unmanned boat, as a maritime delivery platform, the implementation of surveillance and reconnaissance tasks. Ltd. and SearoBotics Co. jointly developed the high-speed unmanned boat "Interceptor", which adopts mixed-fuel engine and water-jet propulsion, and can reach a speed of more than 40kn, and can be remotely controlled by radio or autonomously navigated in accordance with a set route. The unmanned boat is mainly used for maritime drug enforcement, anti-piracy patrols, harbour security and other tasks.

As a major scientific and technological country, Israel is also at the forefront of the development of unmanned boats in the world. Because of its special geographical location, Israel's maritime counter-terrorism situation is extremely serious. Against this background, the Israeli government has launched a variety of unmanned boats. Among them, the "Protector" is the most eye-catching. 2003, Rafael Weapon Design Bureau delivered the first batch of "Protector" surface unmanned boats to the Israeli military. In addition to the characteristics of the American Spartan Scout, the boat has an emphasis on stealth. It is mainly used for maritime counter-terrorism operations, force protection and reconnaissance, etc. In December 2006, Israel's Ebilt Systems launched the Silver Marlin unmanned boat. The boat is known as the second generation of unmanned boats, with a length of 10 m, a weight of 4000 kg, a maximum speed of 45 kn, and a maximum range of 500 n miles, and is mainly used for maritime patrol missions. Since then, the company also launched the "stingray" unmanned boat, the maximum speed of 40 kn, endurance time of 8 h. This kind of unmanned boat relative to other unmanned boats is the outstanding characteristics of the hull size is small, stealth ability is strong, is mainly used for electronic reconnaissance, coast guard.

In order to prevent hostile countries from carrying out sudden attacks by sea invasion, the Ministry of Defence of Japan has begun to carry out technological research on surface unmanned boats and unmanned underwater submarines. To this end, in 2008, the Diet adopted a six-year unmanned boat research plan with a total allocation of 6 billion yen. According to the plan, unmanned boats will be deployed in large numbers along Japan's coastline for the purposes of coastal guarding, prevention of invasion by forces of other countries, monitoring and arming of sensitive islands, and detection of dangerous vessels. Against this background, Yamaha has developed the high-performance UMV-H and UMV-O unmanned boats. The UMV-H is 4.4 m long, deep V-shaped, equipped with a 90 kW water jet propeller, and can reach a maximum speed of 40 kn. The UMV-O is mainly used for oceanic and atmospheric changes, and was delivered to the Japan Science and Technology Agency in 2003.

In addition, France, Germany, the United Kingdom, Singapore and other countries have also accelerated their research on unmanned boats. The naval research organisation of the United Kingdom successfully developed the surface unmanned boat "Fenriel" in 2002 by drawing on the "Spartan Scout" of the United States, which was incorporated into the navy and mainly used to carry out various high-risk missions. After being incorporated into the navy, it is mainly used to carry out various high-risk missions, and in 2010, Singapore launched its latest unmanned boat, the Venus, which is mainly used to combat piracy and maritime smuggling.

From the development of advanced unmanned boat can be seen, in the unmanned boat hull material, rigid inflatable boat is the mainstream of foreign unmanned boat. At present, the surface unmanned boat is developing in the direction of intelligence, modularity and standardisation. Intelligent high-speed surface unmanned boats are equipped with a large number of naval forces, which can play a positive role in establishing a fast and safe sea channel, establishing a security alert system, armed protection and precision strikes for large forces.

Surface unmanned boat is a kind of small surface platform with autonomous planning and sailing ability, and can autonomously complete the tasks of environment perception and target detection, and can undertake the tasks of intelligence collection, surveillance and reconnaissance, mine clearance, anti-submarine, precision strike, search and arrest, hydrography and geography survey, anti-terrorism, and

2.2 Development History of Marine Unmanned Vessel Platforms

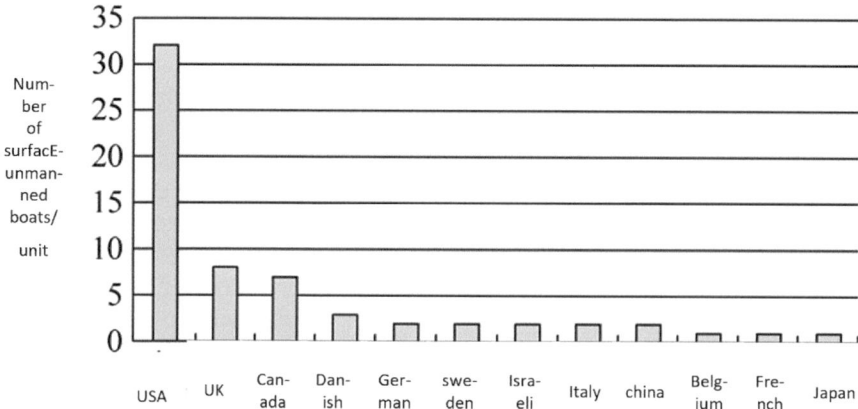

Fig. 2.6 Distribution of major countries producing surface unmanned vehicles (USVs)

relay communication, etc. At present, the navies of many countries are equipped with a large number of surface unmanned boats.

At present, the navy of many countries are actively starting to deploy unmanned surface vehicles, to carry out the development of surface unmanned boat countries and regions, including the United States, Israel, Europe, Japan, etc., but only the United States, Israel, some of the models equipped with troops, as shown in Fig. 2.6. Countries are competing to develop multifunctional surface unmanned boats that integrate anti-mine warfare, anti-submarine warfare, electronic warfare and other capabilities. Although the domestic research on unmanned boats started late, it has been developing fast and gradually transitioned from the conceptual design stage to the practical application stage, but there is still a big gap compared with foreign countries.

In view of various operational needs, especially the need for counter-terrorism operations, national navies have increased the research and development of surface unmanned boats, and with the advancement of guidance technology and control technology in surface unmanned boats, it has ushered in a high-speed development stage, and at present, the main institutions in foreign countries that have developed surface unmanned boats include Northrop Grumman, Raytheon, Lockheed Martin, Dassin Defence Systems Ltd., Meggitt Defence Systems, Liquid Robotics, SeaRobotics Corporation, the United States Naval Research Institute and Israel's Rafael, Elbit Systems, Aerospace Industries, Aeronautical Defence Systems, Aeronautical Defence Systems, as well as Britain's QinetiQ, and so on.

2.2.1 Results of USV Development in the USA

The United States is the world's most powerful naval strength of the country, USV development has been the attention of the U.S. Navy, its development ideas and top-level planning is very clear and clear, in the relevant research and development of the world's leading position (Fig. 2.7). Twenty-first century, the U.S. Navy in the twenty-first century Maritime Power—Navy Vision, put forward in 2015 before the introduction of new unmanned platforms into the future networked combat system. -At the beginning of the twenty-first century, the U.S. Navy proposed in the "twenty-first century Maritime Power—Naval Vision" that new unmanned platforms will be introduced into the future networked combat system by 2015. In July 2007, the U.S. Navy released the "Naval Surface Unmanned Vessel Master Plan" for the first time, which set up seven mission tasks of surface unmanned vessel—Anti-Mine Warfare, Anti-Submarine Warfare, Maritime Security, Anti-Ship Warfare, Support for Special Forces Operations, Electronic Warfare, and Support for Maritime Blockade Operations, pointing out the focus of future surface unmanned boat development and the direction of technological research for the U.S. industry, academia and international partners. Since then, the U.S. military began to co-ordinate the development of unmanned systems across all military branches, and unanimously released the Unmanned Systems Roadmap, which provides an overall plan for the operational requirements of surface unmanned boats, key technology areas, and interconnectivity with other unmanned systems. Among them, the latest version of the Unmanned Systems Roadmap released in December 2013 provides a more detailed description of the technological development priorities and capability requirements of surface unmanned vessels in the next 5 years (near term), 10 years (medium term) and 25 years (long term): the near-term technological development priorities of surface unmanned vessels will focus on the areas of enhanced power systems, communication systems and sensor systems, while the medium- and long-term priorities are the development of high-efficiency autonomous systems, obstacle avoidance algorithms and security and safety systems, and the development of the latest technology and capabilities of surface unmanned vessels. In the medium to long term, the focus will be on the development of efficient autonomous systems, obstacle avoidance algorithms and security architectures, etc. The capability needs of surface unmanned vessels in the near term are to increase autonomy in performing specific tasks in locally controlled areas and to improve networking capabilities, in the medium term to expand the range of operations and increase the number of mission types, and in the long term to realise global autonomous mission performance.

The Unmanned Systems Roadmap identifies the technical challenges faced by surface unmanned craft as including, inter alia, maritime endurance and survivability in harsh environments. It also points out that in order to maximise the potential of unmanned systems, all types of future unmanned systems must achieve seamless interoperability. Over the past few years, the United States Navy has invested more than $500 million in the development of surface unmanned boats, and there are now

2.2 Development History of Marine Unmanned Vessel Platforms

'Spartan Scout Surface Unmanned Vehicle   'Phantom Guardian unmanned aerial vehicle (UAV)

Fig. 2.7 U.S. surface unmanned vehicle developments

a variety of models under development, experimentation and evaluation, as shown in Table 2.2.

2.2.2 Israeli USV Development Results

Israel also attaches great importance to the development of surface unmanned aerial vehicles (UAVs), and in recent years has launched a variety of types of UAVs, as shown in Table 2.3. Based on the fact that Israel possesses rich UAV development technology, it has a unique competitive advantage through the application of UAV technology to the research and development of surface unmanned aerial vehicles. Israel is second only to the United States in terms of the types of surface unmanned aerial vehicles it has developed, its development level is among the most advanced in the world, and a few of its surface unmanned aerial vehicles have been successfully exported overseas.

2.2.3 Results of USV Development in Other Countries

The USVs developed by the United States and Israel mainly serve the state, while the USVs developed by other countries serve more in the civil field, such as shipping, waterway surveying, marine environmental protection, etc., including the United Kingdom, France, Germany, Italy, Sweden, Singapore and other countries are also active in the development of surface unmanned vessels, see Table 2.4.

Table 2.2 U.S. surface unmanned vehicle results

Name	Control method	Performances	Mandate	Maximum load	Manufacturer/institution
Sea owl eagle	Remote controller	Speed 45 kn. Endurance 10 h (speed 12 kn) and 24 h (speed 5 kn)	Minefield reconnaissance, shallow water surveillance, maritime interdiction and securing the perimeter of port terminals	200 kg	Led by the American Institutes for Research in the 1990s, it was the U.S. Navy's first attempt to develop an unmanned surface vessel
Spartan spy	Remote control or semi-automatic	Speed 28 kn (Maximum speed 50 kn in sea state below Class 3); Endurance 8 h (Maximum 48 h); Range 150 n mile (Maximum 1000 n mile)	Environmental surveillance, intelligence reconnaissance, precision strike, anti-mine, anti-submarine	1350–2300 kg	United States Navy defence project, whose advance technology concept demonstration project was initiated in 2002
Ghost defender	Remote control or automatic	Speed over 40 kn. Maximum power 266 hp	Maritime policing and protection	–	American Robotic Ships, Inc
Piranha (fish)	Remote control or automatic	Speed over 40 kn. Maximum power 200 hp	Naval surface targets	–	Meggitt Defence Company, USA
Unmanned long-range hunting vessel	Remote controller	Speed over 10 kn. Maximum range 700 m. Endurance 20–40 h			

(continued)

2.2 Development History of Marine Unmanned Vessel Platforms

Table 2.2 (continued)

Name	Control method	Performances	Mandate	Maximum load	Manufacturer/institution
Depth up to 30 ft or less in deep water	Rapid reconnaissance, detection, classification, identification and accurate positioning of mines (especially anchor and submerged mines); anti-submarine exploration; surface surveillance; coastal intelligence reconnaissance and collection	–	Lockheed Martin presented in the 1990s		
Trimaran watercraft unmanned aerial vehicle (UAV)	Remote controller	Speed 35 kn	Intelligence surveillance, reconnaissance search and mine countermeasures	–	–
Blue knight	Remote control or autonomous	Speed 50 kn (can work normally in class 3 sea state)	Assault	Displacement 129 t	Fully modular design structure allows for the replacement of different weapons and sensors according to battlefield requirements
Sea hunt	Remote controller	Captain 40 m	Deep tracking of enemy submarines	–	United States Department of Defense Advanced Research Projects Agency (DARPA)

Table 2.3 Israel's surface unmanned vehicle achievements

Name	Control method	Performances	Mandate	Maximum load	Manufacturer/institution
Protector	Remote control or autonomous	Speed up to 40 kn; Displacement 4 t	Maritime force protection; intelligence surveillance and reconnaissance; anti-mine warfare; electronic warfare; precision strike; counter-terrorism	1000	Developed jointly by Rafael and Aeronautical Defence Systems in 2003
Starfish	Remote controller	Speed 45 kn. Range 300 n mile	Port/strategic facility protection; coastal patrol; ship and oil rig protection; ISR mission; target marking; jamming and decoying; force protection and extended optical electromagnetic field (EMF)	2500	Israel Aerospace Industries
Stingray	Remote controller	Speed 40 kn. Endurance 8 h	Offshore intelligence reconnaissance and surveillance; and electronic warfare and electronic reconnaissance electronic warfare and electronic reconnaissance	150	Wholly owned by Elbit Systems, Israel

(continued)

2.2 Development History of Marine Unmanned Vessel Platforms

Table 2.3 (continued)

Name	Control method	Performances	Mandate	Maximum load	Manufacturer/institution
Silver marlin	Remote control or autonomous	Maximum speed 45 kn. Maximum range 500 n mile. Endurance 24 h	ISR tasks; force protection/counter-terrorism; anti-ship and anti-mine; search and rescue; harbour and waterway patrols; electronic warfare	2500	Developed by Israel's Elbit Systems, it is a second-generation surface unmanned vehicle
KATANA's new generation of surface unmanned vehicles	Remote control or autonomous	–	Protection of the exclusive economic zone, maritime boundaries, port security, offshore gas rigs and pipelines; shallow water patrols and identification, tracking and classification of electronic warfare targets, both near and far; provision of real-time intelligence imagery and attacking targets on command, etc	–	KATANA is the latest addition to the family of surface unmanned craft produced by Israel Aerospace Industries

2.2.4 Overview of Domestic Developments

At present, China is also following the lead of Japan, the United Kingdom, Germany and other countries in joining the development of unmanned surface craft. In the field of R&D and application of surface unmanned aerial vehicles, although China started late, it is developing rapidly, and has gradually transitioned from the initial conceptual design stage to the practical application stage. China's independent research and development of water surface unmanned boat units including colleges and universities, research institutes and related enterprises, such as Harbin Engineering University, Shanghai University, Shanghai Maritime University, Dalian Maritime University, the Chinese Academy of Sciences, Shenyang Institute of Automation and Shenyang Aerospace Xinguang Company, Zhuhai Yunzhou Intelligent Science and Technology Co., Ltd. and so on, has been successfully applied to a few kinds of water surface unmanned boat control mode, performance parameters and application scenarios are shown in Table 2.5, several typical water surface unmanned boat

Table 2.4 The achievements of unmanned surface vehicles in other countries

Name	Main load	Nations	Manufacturers/institutions	Note
Rapid mobile mine clearance	–	UK	Quinnettique	Demo boat production completed in 2008, testing commenced in 2009
Sentinel	Microwave control chain, day/night high-resolution camera, sonar, radar, optional photoelectric sensors, chemical sensors and environmental sensors	Uk	Quinnettique	It's been sea-tested
Maritime series rapid targeting	Front view camera, microphone/speaker, GPS navigation unit	UK	British Autonomous Watercraft Corporation	The target boat was tested by the British Navy on the high seas in 2008–2009
Tharpur	GPS navigation system, audio (currently VHF), video communication system	Canada	International Submarine Engineering Canada	Relevant testing has been initiated in 2001
ART-STER unmanned target boat	Navigation cameras, side-scan sonar, echo detectors, magnetometers	French	ECA France	The French Navy expects to procure more than 6 ships
Inspector	K-STER extinguishers, side-scan sonobuoys, forward-looking/obstacle avoidance sonobuoys, multi-band echo sounders, magnetometers	French	ECA France	7 delivered to the Navy for testing in December 2006
Rheinlandia	Satellite, radar, optoelectronic and other sensors	German	Rheinland Mantle Defence, Germany	A series of sea trials were conducted in 2009
U-RANDER	Visible/infrared camera, compass, inertial sensors, GPS, forward-looking/side-scan sonar	Italy	Calzoni S.r.l., Italy	Exhibited at "NATO Harbour Protection Experiment—2008"

(continued)

2.2 Development History of Marine Unmanned Vessel Platforms

Table 2.4 (continued)

Name	Main load	Nations	Manufacturers/institutions	Note
SAM 3	Electromagnetic signal effectors, electrically/hydraulically driven acoustic signal effectors and electrical signal effectors	Sweden	Kaukum AB	Comprehensive tests were conducted in August 2008
Gold star	Force protection module: radar, electro-optical sensors, small-calibre remotely operated weapons Anti-submarine warfare: active drogue sonar; and Mine warfare: synthetic aperture radar (SAR), disposable mine neutralisation device (DND) Electronic warfare, maritime surveillance and precision firepower: electronic warfare systems, proximity missile systems	Singapore	Singapore Electronics Technology and France Jointly developed with France	Displayed at the Singapore Airshow in 2010 Integration of Hitrole Naval Remote Weapon Station completed in 2011

control mode, performance parameters and application scenarios. 5, several typical unmanned boat appearance as shown in Fig. 2.8. Unmanned boats have a wide range of application prospects in the field of marine development and construction, water engineering applications.

In 2008, the first high-speed monohull unmanned boat for engineering application in China, "Tianxiang 1", was launched, followed by the first engineering prototype of surface unmanned intelligent measurement platform, "Jinghai" series of USVs, "Haiteng 01" high-speed surface unmanned boat, "Haiyi 1" USV and so on, "Haideng 01" high-speed surface unmanned boat, Haiyi 1 USV and so on. The research time of unmanned boat in civil field is relatively short. Zhuhai Yunzhou Intelligent Technology Co., Ltd. is a fast-developing unmanned boat research and development enterprise in China in recent years, and is also the world's first enterprise to do environmental protection unmanned boat. It has successively developed

Table 2.5 Development achievements of China's surface unmanned vessels

Name	Control method	Performances	Mandate	Manufacturer/institution
"Sky Sign 1"	Autonomous or remote control	Range of hundreds of kilometres; continuous operation for about 20 days at a time	Response to marine emergencies and environmental monitoring and early warning of disasters in oceans, large lakes, etc	"Tianxiang 1 is the first unmanned surface vessel developed by Aisino Xinguang Company in China for engineering purposes, and the first unmanned surface vessel used for gas-phase detection
"Seikai"	Autonomous	Maximum speed 18 kn Maximum range 120 n miles Full load weight 2.3 t	Detection of environmental elements in water bodies, environmental measurements and marine hydrographic measurements	Developed by Shanghai University
"Hai Teng 01"	Autonomous, semi-autonomous, remotely controlled		Maritime cruising, hydrographic surveying, waterborne oil spill control and recovery, search and rescue at sea, wreck exploration and salvage	Developed by Shanghai Maritime University
"Pilot" marine surveying and mapping vessel	–		Environmental monitoring, scientific research and exploration, underwater surveying and mapping, search and rescue, security patrol, maritime emergency response	Developed in September 2014 by Zhuhai Yunzhou Intelligent Technology Co
Unmanned surveying vessel "Yunzhou"	Autonomous or remote control		Submarine topographic surveys, hydrographic surveys, maritime search and rescue	Developed in November 2015 by Zhuhai Yunzhou Intelligent Technology Co

2.2 Development History of Marine Unmanned Vessel Platforms

Elephant in the Sky 1

Songkran

Hai Teng 01

Yunzhou L30 'Lookout' high-speed unmanned boat

First unmanned survey vessel

Yunzhou Intelligent 'Lookout 2' Unmanned Surveillance and Fighting Unmanned Vessel

Fig. 2.8 China's surface unmanned craft development results

Pilot, ESM30, MC120, etc., and its design models cover monohull, catamaran and wave-piercing etc. In August 2016, CIMC Raffles signed a USV co-operation and development agreement with a company in Shanghai, indicating that our country will start to research and develop and produce marine intelligent equipments with independent intellectual property rights and more high-tech content, and will make greater contribution to the country's scientific and technological innovation and the development of the local economy. This also means that the domesticated unmanned boat will soon start a historical precedent. In addition to the above enterprises, some domestic universities and other research institutions have also conducted a lot of research on USV. Professor Yang Songlin of Jiangsu University of Science and Technology has done a lot of research on USV, including: intelligent control technology research, boat type and line design research, optimisation modelling and new optimisation methods, etc. The types of boats studied include monohull skiffs, catamarans, trimarans and five-hulled boats. Harbin Engineering University has conducted research in the field of USV and underwater robots earlier, in which a lot of research has been done in path planning and motion control.

Comprehensive research status at home and abroad, the development of unmanned vessels will present the following five trends.

(1) Structural modularity

Unmanned boats adopt modular structure design, and can be assembled with a variety of "plug-and-play" type mission modules on the basis of basic unmanned boats. At the same time, generalised and standardised platforms, technologies, components and interfaces can effectively reduce the risks and costs of the development and use of surface unmanned boats, lower the difficulty of logistics and maintenance, and enhance the ability of mutual coordination between them and other platforms. Modular design and open architecture enhance functional diversity, accelerate development progress and reduce development costs. This design and development feature of surface unmanned aerial vehicles will be maintained in the future. When the structure is modular, special attention should be paid to ensure the high reliability of the system performance, so that the unmanned boat can be safely recovered after completing the required tasks.

(2) Functional intelligence

At present, most of the unmanned boats in service belong to semi-autonomous type, in order to realise fully autonomous unmanned boats, it is necessary to enhance the self-adaptive level and autonomous decision-making ability of unmanned boats, and the anti-rocking ability to cope with the adverse sea conditions should also be enhanced to increase the level of intelligence of each functional module. Highly intelligent unmanned vessels can reduce the dependence on remote controllers, lower the requirement for communication bandwidth, and enhance the ability to perform missions over the visual distance. Unmanned vessels will certainly develop in the direction of full autonomy.

(3) System networking

On the one hand, the system network of unmanned ships should realise the integrated network control between unmanned ships and mother ships and unmanned ships, so as to enhance the ability of unmanned ships to collaborate with each other and carry out missions; on the other hand, it should realise the integrated network of unmanned combat system. As a future war to complete the information confrontation, the important means of special operations mission, as early as 2001, the U.S. Navy in the Littoral Combat Ship combat system proposed the use of unmanned boats, unmanned submarines and unmanned aircraft constitute the Navy's unmanned combat system to complete such as intelligence collection, anti-submarine, anti-mine, reconnaissance and detection, precision strikes and other combat tasks.

(4) Wide application

Unmanned ships have been widely used in military field, and have played an important role in the process of transforming military structure. In the military, unmanned vessels can carry out mine clearance, anti-submarine warfare, electronic warfare, support for special operations and other military tasks. In recent years, there have been some attempts to apply unmanned boats in civil affairs, for example, the application of unmanned boats for meteorological monitoring. In the near future, unmanned boats can be applied to large-area ocean mapping and water quality monitoring, large-scale search and rescue, in order to enhance the coverage capacity while reducing the labour intensity and operating time, and the application of small unmanned boats to large-scale unmanned boat transition gradually.

(5) Equipment localisation

In the research of unmanned boats, the United States and Israel have been at the forefront of the world. Such as the United States "Spartan Scout" has been deployed to the "Gettysburg" cruiser, and participated in the Arabian Gulf region "Operation Enduring Freedom" and "Operation Iraqi Freedom". Operation Iraqi Freedom in the Arabian Gulf region. The Israeli Protector was delivered to the Israel Defence Forces in 2003 and is in service with the national and Singaporean navies. Since 1 September 2012, the European Union has invested 6.7 million euros to carry out a three-year research on the unmanned cargo ship driving system called "MUNIN", exploring the unmanned ship driving based on the mode of autonomous navigation and shore-based monitoring. Compared with the world's advanced level, China's research on unmanned vessels is still in a backward state, and we need to gradually strengthen the research on unmanned vessels to take the initiative in the field of unmanned driving in the future.

Bibliography

1. https://blog.csdn.net/weixin_45839894/article/details/114020234
2. https://www.163.com/dy/article/C5R4A4F505148ALS.html
3. https://baijiahao.baidu.com/s?id=1602967564004531474&wfr=spider&for=pc
4. https://baike.baidu.com/item/%E5%A4%A9%E8%B1%A1%E4%B8%80%E5%8F%B7
5. https://www.163.com/dy/article/EVJGRCUL0511DV4H.html
6. https://baike.baidu.com/item/%E2%80%9C%E6%B5%B7%E8%85%BE01%E2%80%9D%E5%8F%B7/18703070
7. https://www.163.com/news/article/DRIVEMMO000181KT.html
8. https://news.sina.com.cn/o/2018-12-29/doc-ihqfskcn2489405.shtml

Chapter 3
Introduction to Marine Unmanned Platform System

3.1 Introduction of Composition

3.1.1 UAS Components

The maritime unmanned aircraft system consists of an aircraft system, a shipborne system, a mission payload and an integrated support system. The aircraft system includes the vehicle platform, the propulsion system, the flight control system, the navigation system, the on-board part of the take-off/landing system, and the data-link on-board terminal, etc.; the shipboard system includes the on-board command and control sub-systems (mission control station, take-off and landing control station, and take-off and landing guidance station), the on-board part of the take-off/landing system, the data-link on-board terminal (link station), the intelligence processing system, and the shipboard auxiliary equipments, etc.; the mission payload is the equipment for the UAS to fulfil its combat mission. The mission payload is the equipment for UAS to complete the combat mission, and the mission payload on board mainly consists of photoelectric reconnaissance, SAR imaging, meteorological detection, surveying and mapping, communication relay, technical reconnaissance, electronic countermeasures, and airborne weapons and other equipment. The comprehensive security system is the support guarantee for the UAS to work normally, which mainly includes human resources, training, UAS technical maintenance and other security resources, as well as security equipment such as meteorological detection, communication and airport facilities. Figure 3.1 shows the block diagram of UAS.

In the aircraft system, the on-board portion of the take-off/landing system works in conjunction with the on-board portion of the ship to accomplish the launch and recovery of the UAV. The propulsion system provides the power for the UAV. The airframe system refers to the vehicle platform of the UAV. The navigation system provides navigation and targeting information for the UAS to fulfil its tactical missions through satellite navigation, early warning aircraft guidance, shipboard

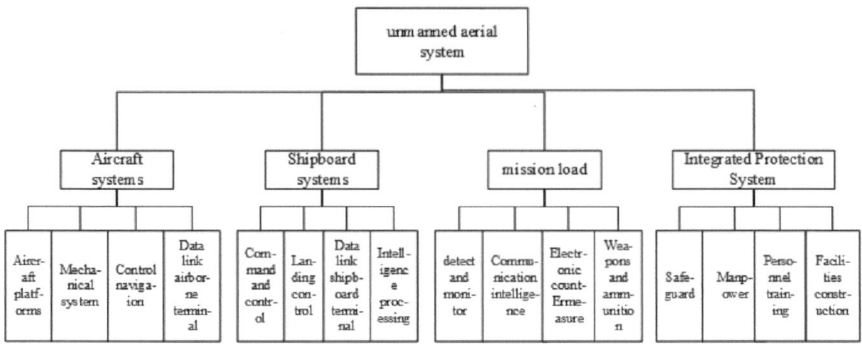

Fig. 3.1 Block diagram of UAS components

guidance, and the UAV's own target discovery and tracking capabilities. The flight control system is the core of the onboard part of the UAV, which monitors, controls and commands other onboard subsystems, accepts the instructions from the shipboard mission control station, coordinates the work of various onboard subsystems, and sends the status of the UAV and other required information to the shipboard command and control subsystem. Under the monitoring and command of the shipboard command and control system, the onboard guidance, navigation and control system controls the UAV to complete the scheduled flight and mission. Therefore, the guidance, navigation and control system is the central controller for coordinating, managing and controlling the various sub-systems of the UAV, and it is also the core of the management and control of the UAV.

In the shipborne system, the shipborne part of the take-off/landing system is an important guarantee to complete the launch and recovery of the UAV. The data link shipboard terminal and the airborne terminal work together to provide communication between the shipboard station and the UAV, realising the monitoring and commanding of the UAV and completing the predetermined tasks. The shipborne command and control sub-system sends control commands through the remote control telemetry data link and receives the status data and mission information transmitted from the UAV, provides the operator with an understanding of the status of the UAV and the battlefield situation through the graphical interface, monitors and commands the operation of the UAV, and provides the operator's intervention capability in the case of accidents or malfunctions of the UAV.

In the shipboard system, the command and control sub-system is in the core position, comprehensively monitoring, controlling and commanding the work of other sub-systems, providing the operator with comprehensive battlefield information and UAV status information, and arranging the various sub-systems to complete the predetermined tasks according to the operator's orders. It can make reasonable dispositions for emergencies and notify the operator in time.

Data link The shipboard terminal and the airborne terminal constitute the remote control telemetry data link of the UAS, which is responsible for the uploading and downloading of commands, data and intelligence information of the UAS. The uplink is the remote control link, which is used to transmit the control of the UAV and the command of the mission load. The downlink is the telemetry link, which is used to transmit the status information of the aircraft, and in addition, there is a downlink telemetry data transmission channel.

3.1.2 Unmanned Boat System Components

According to the physical distribution, the unmanned boat system can be divided into two major subsystems: the unmanned boat subsystem and the shore-based monitoring subsystem. The unmanned boat subsystem consists of the hull of the unmanned boat and all kinds of equipment on board. The shore-based monitoring subsystem is set up on the shore or other surface ships to monitor and control the unmanned boats.

1. Physical structure of unmanned boat subsystems

(1) Hull and auxiliary structural components module

The hull and auxiliary structural components are the carriers of the unmanned boat subsystem. The hull is the carrying platform for all the equipments of the unmanned boat, ensuring the safety and stability of the equipments is the most basic requirement for it. The hull has a great impact on the manoeuvrability, flexibility, range, load capacity and other functions of the UAV. The types of boats available for unmanned boats include trimarans, composite boats, jet skis, rigid inflatable boats, hydrofoils, surface effect boats, and so on. Auxiliary structural components include on-water side structures and support platforms, which are mainly used to install equipment including navigation equipment and weapon systems. Glass fibre, carbon fibre, etc. can be used to manufacture auxiliary structural components.

(2) Motion control module

The motion control module is responsible for regulating and controlling the speed and heading of the unmanned boat. The main equipments include engine, propeller, servo-hydraulic cylinder and so on. Based on the consideration of endurance and energy supply, the engine is a high-performance diesel engine that can be operated by electronic control. The unmanned boat needs to have a high speed to carry out the military mission, and the propeller chooses the water jet propeller with good performance to control the speed of the water jet propeller by controlling the oil intake of the diesel engine, and the rudder function is achieved by changing the angle of the nozzle of the water jet propeller.

(3) Energy module

The energy module consists of a fuel tank that provides energy for the diesel engine and a battery pack that supplies power to the electronic equipment on board. In order to prevent the unmanned boat from running out of energy in the middle of the mission, the fuel tank and the battery pack have the function of real-time display of the remaining energy and automatic alarm when the energy is insufficient. In order to ensure the safe return of the unmanned boat, when the remaining fuel is tight, the energy intelligent management system will automatically switch off the equipment that is not in use for the time being. Since the unmanned boat carries more electronic equipment, power consumption is large, in order to ensure that the unmanned boat uninterrupted power supply during a single mission, the design can be used silver-zinc battery pack. Silver-zinc battery pack is 5–6 times more powerful than ordinary lead-acid battery pack. If the unmanned boat is equipped with other large electronic equipment, the installation of a shaft generator on the diesel engine can be considered. If the unmanned boat is to carry out long-term tasks, solar panels can be considered.

(4) Navigation and collision avoidance module

As the unmanned boat to perform the task, the need for accurate navigation and positioning information, and a variety of navigation methods exist at the same time the advantages and disadvantages, in order to give full play to their respective advantages, the design of a variety of navigation equipment to participate in the combination of navigation methods. The GPS receiver module is used to obtain the latitude and longitude information of the unmanned boat. In order to avoid electromagnetic interference, which affects the signal reception, the GPS receiver should be installed in a position far away from other antennas. Equipped with laser gyroscope to get the acceleration and angular acceleration information of the UAV. Measure the speed and range information by the odometer. Read the water depth information by echo sounder to prevent stranding in shallow water. Equipped with a small navigation radar with ARPA function, this radar can provide radar video and target information to the control and command system on the one hand, and on the other hand, it can track and lock the target, and provide the nearest encounter time and distance by analysing the target's movement trajectory.

(5) Communication module

Considering the distance between the unmanned boat and the shore-based control equipment when it goes out to perform the task, the design adopts a combination of satellite communication, wireless network communication and microwave communication. Microwave communication is mainly used for communication within the line-of-sight range because of its short distance. Wireless communication can be realised in near-shore waters by arranging wireless network and adopting wireless broadband technology, and communication within 50 km can be realised by setting up antennae in unmanned boats and control terminals. For longer range, satellite communication is used. When the unmanned boat is working, the communication mode can be chosen flexibly according to the actual needs.

(6) Environmental information acquisition module

3.1 Introduction of Composition

The environmental information acquisition module is mainly used for the reconnaissance and monitoring of the waters around the unmanned boat. It is designed to install black and white/colour cameras on the unmanned boat to get pictures or videos of the scene, forward-looking infrared sensors to get environmental information at night, laser rangefinders to get the distance to the target, and azimuth indicators to get the relative orientation information. These information acquisition devices are centrally located in the control head. The head is the support equipment for installing and fixing the information acquisition equipment, and it has the function of 360° horizontal movement and a certain range of up and down movement. According to the needs of the collection angle of the collection equipment, under the action of external control signals, can be completed at the specified speed of the required horizontal and vertical movement and to achieve the aperture, focus adjustment and sensor closure and opening and other functions. The information sensing equipment and the head are located in the multi-functional broadcast tower. Based on the purpose of listening, the unmanned boat is designed with a directional audio capture card, using VHF equipment to obtain communication information around the unmanned boat through real-time listening to the VHF communication channel, Automatic Identification System (AIS) can obtain real-time information about the identity of the nearby vessels and the navigation status.

(7) Load platform module

The load platform module mainly consists of a rotating platform and other load equipment. The rotating platform can drive the weapon system carried on it to rotate to the required position. According to the need, the rotating platform is installed in the front part of the unmanned boat. The weapon system uses control signals to destroy dangerous targets. To complement the target strike, the rotating platform is also fitted with a target identification and tracking device to lock on to the target. Non-lethal weapons, including bright lights and loudspeakers, are mounted to provide bright light warnings and shouts to non-lethal hazardous non-compliant vessels.

(8) Command and control module

Command and control module is the brain of the unmanned boat, the information collected by other modules will be summarised here. The core of the command and control module is a PC/104 PC, which is responsible for summarising and processing all the information of the unmanned boat, and transmitting the video, audio, image and other information to the shore-based control unit through the appropriate communication link after compression. At the same time, it is responsible for receiving the control instructions from the shore-based monitoring system, analyzing the corresponding control signals and sending them to the corresponding equipment to realize the required actions and functions. The navigation management system is installed in the PC/104 ICM to control the unmanned boat according to the specified procedures when it is in autonomous navigation mode or when it loses contact with the shore-based monitoring system due to a malfunction. In order to facilitate remote control operation, a console corresponding to the control interface of the shore-based operator is installed on the UAV. Meanwhile, in order to ensure the safety of the unmanned boat, there is an emergency return system inside the navigation management system, when the unmanned boat fails or the operation procedure is disordered, the system will be activated automatically, so that the unmanned boat will return to the shore in time to receive maintenance.

The above is the physical structure design of the unmanned boat subsystem, because of the modular design, as long as the replacement or switching different modules, can perform different tasks, the specific structure layout as shown in Figs. 3.2 and 3.3.

2. Physical structure of shore-based monitoring subsystem

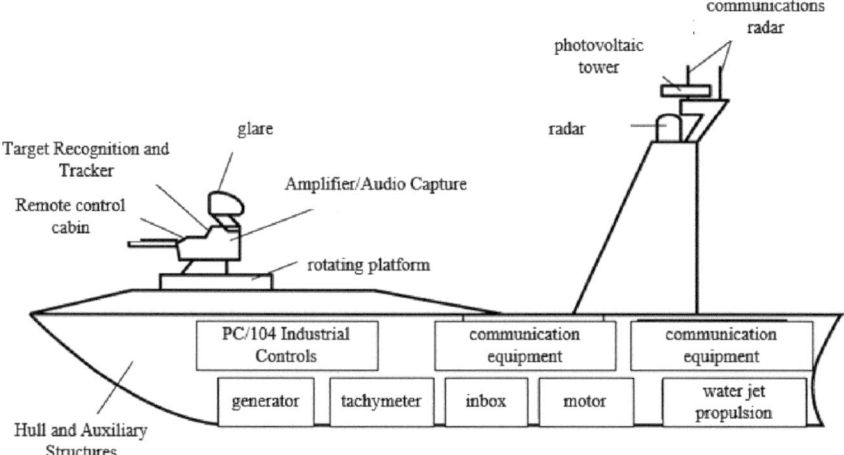

Fig. 3.2 Side view of unmanned boat equipment layout

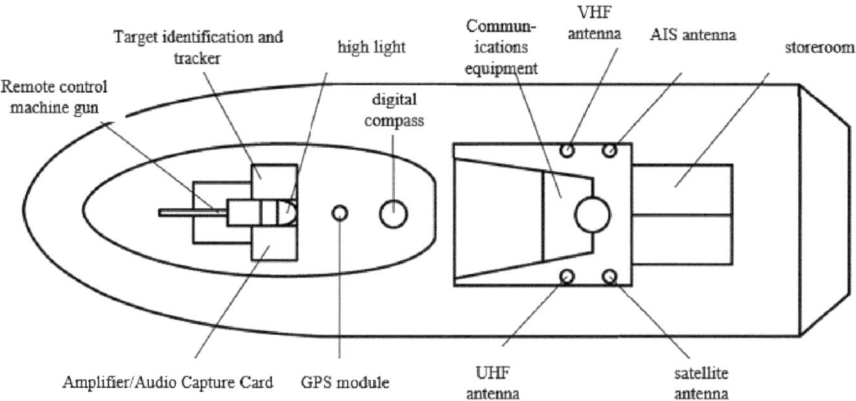

Fig. 3.3 Top view of unmanned boat equipment layout

As the command and control centre of the unmanned boat, the shore-based monitoring module can remotely control the unmanned boat according to the mission requirements. Considering the practical application, the mobile workstation mode is adopted to facilitate flexible and convenient arrangement. In this way, the shore-based monitoring module can be mounted on other ships or placed on the shore. According to the needs of the mission, it can also be connected to the upper level command centre through the network and controlled from the upper level control terminal. The shore-based monitoring system communicates with the unmanned vessel in real time through the selected communication mode, receives data from various information acquisition devices on the unmanned vessel, and provides reference for the operator to give control instructions. After the control instructions are transmitted to the unmanned vessel through the communication equipment, they are decoded and classified by the control unit on the unmanned vessel and sent to the corresponding equipment to realise the required operation.

The shore-based monitoring system consists of several computers connected to form a local area network (LAN), as shown in Fig. 3.4. One of the computers is used to display the video information of the surface target transmitted by the radar, one is used to display the information of the surrounding environment transmitted by various sensors on the unmanned boat, and the other is used to display the comprehensive information of the unmanned boat, including the positional coordinates of the unmanned boat, the heading, the speed, and the remaining fuel, power and other working status information. There is an electronic chart system installed in the computer, which can plan the route of the UAV. At the same time, the computer is also installed with the operation software to control the unmanned boat, through which not only can the unmanned boat carry out the direction and speed control, but also can carry out the remote control operation of all kinds of sensors in the optoelectronic tower, as well as the control of the weapon equipment and strong light on the loading platform, etc. These control instructions are transmitted through the communication link. These control commands are sent to the unmanned craft through the communication link.

3.2 Key Technologies of Marine Unmanned Platform

3.2.1 Key Technologies for Marine UAV Platforms

Ships equipped with maritime drones will be able to enhance their mission capabilities. However, the space of ships at sea is narrow, the wind speed on the sea surface is large, and the ships not only have to move forward, but also the deck lifting and sinking, transverse and longitudinal rocking movements are quite violent. The use of maritime UAVs must also ensure that the normal operation of ships at sea does not produce or less interference. It is precisely because of the harsh environment for the

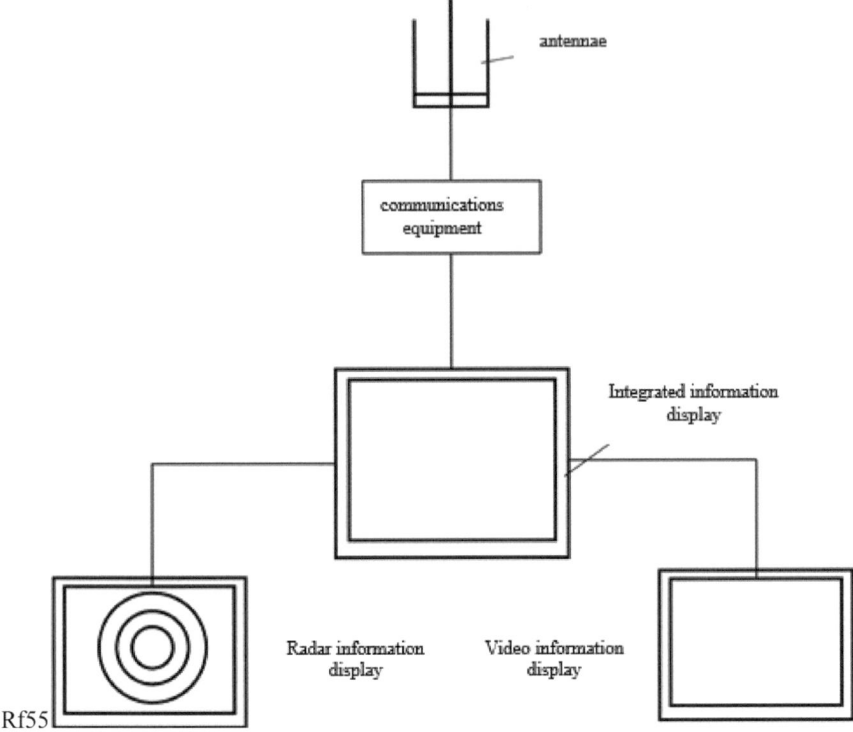

Fig. 3.4 Unmanned vessel shore-based monitoring system composition

use of maritime UAVs that the technical requirements are higher and the development is more difficult, and the key technologies that need to be solved are as follows.

(1) Shipboard recovery technology

For the maritime vertical take-off and landing UAV, it is relatively simple to take off and land on a ship at sea. It can make full use of the ability of hovering, small speed forward flight and side flight, and accurately land on the narrow deck under the visual remote control of the ship's surface operator. In medium to high sea conditions, the UAV can be prevented from skidding and tipping with the help of harpoon-grids or landing nets.

In the case of fixed-wing UAVs at sea, take-off from a frigate generally requires the use of booster rockets or other means to accelerate the fixed-wing UAV. The problem of landing is more complicated, and currently the commonly used methods are 'parachute recovery' and 'crash net recovery'. 'Parachute recovery' means that the UAV will fall to the nearby sea surface by its own parachute, and then be recovered manually by the crew. This method requires the UAV to have enough load and volume to provide space for the parachute bag, and also requires the UAV to be cleaned and repaired on board. The 'crash net recovery' is to manoeuvre the UAV to accurately crash into the blocking net erected at the stern of the ship, and then safely recover the aircraft through the energy-damping system. Crash net recovery technology is suitable for recovering UAVs from small sites or ships, but it requires higher control precision and may cause damage to the aircraft. Overseas maritime fixed-wing unmanned aerial vehicles mostly adopt the crash net recovery technology, such as the 'Pioneer' maritime fixed-wing unmanned aerial vehicle system equipped by the United States Navy.

The main problems of the above two recovery methods are as follows: first, there is a certain amount of damage to the aircraft and even mission equipment, which needs to be repaired or cleaned, and the preparation time for another sortie is longer; second, it is very difficult to use it in medium sea conditions, and it also poses a threat to the safety of the personnel and the ship.

The use of fixed-wing UAVs on large ships, such as large landing ships or aircraft carriers, can be recovered in more ways due to the larger landing site, such as remote-controlled automatic rappelling, rope obstruction, etc. The safe recovery of UAVs on board is also relatively simple.

Therefore, the main development trend of maritime UAVs is to adopt automatic on-board recovery technology and to have automatic on-board take-off and landing capability. Good on-board recovery technology should be able to ensure the normal use of the UAV under 5–6 level sea conditions.

(2) Autonomous flight and control capability

Autonomous flight and control capability is an important symbol of today's UAV towards practicality and top use. It has the ability to perceive the flight environment and make course correction, and it can also transmit its own state and adjust itself within a certain range. Autonomous flight and control capability than visual remote control and programming flight on a new level.

In 2000, the U.S. Navy and Air Force research institutions for the development of fixed-wing unmanned aerial vehicles put forward the definition of 'autonomous operation' (autonomousoperation), which can also be translated as 'autonomous combat', and its connotation is broader than that of autonomous flight. Autonomous flight is broader. In recent years, the United States Department of Defence has adopted the same approach to the autonomous flight and control capability of unmanned aerial vehicles (UAVs) in its report on the medium- and long-term development roadmap for UAVs (systems). The U.S. Army defines the autonomous control capability of UAS as 10 levels, as shown in Table 3.1, of which 'Pioneer' is at the level of 1.5, 'Predator' is at the level of 2, and 'Global Hawk' is at the level of 2.5. Pioneer is at level 1.5, Predator is at level 2, and Global Hawk is at level 2.5.

As can be seen from Table 3.1, the autonomous control capability can be roughly divided into two levels. The first level is for single machine or single system autonomous control capability requirements, level 1–4; the second level is for multi-machine or multi-system autonomous control capability requirements, level 5–10.

Most scholars in China agree with and directly quote the definition of the United States UAV autonomous control capability level. For level 1, it may be better to

Table 3.1 Autonomous control capability level definitions

Level	Definition
10	Fully autonomous swarm
9	Group strategic goals
8	Distributed control
7	Group tactical goals
6	Group tactical replan
5	Group coordination
4	Onboard route re-plan
3	Adapt to failure and flight conditions
2	Real time health diagnosis
1	Remotely guided

use the term 'remotely guided' (remotelyguided). For a single aircraft or a single system, a UAS with primary autonomous flight capability should be able to complete the predetermined flight and combat tasks according to the mission planning; and a UAS with intermediate (≥ 2) or higher autonomous flight capability should be able to carry out route modification and replanning on its own according to the threat judgement and its own status, and successfully complete the given combat tasks.

There are some questionable points in the above classification and definition of autonomous control capability of UAVs. Especially for maritime UAVs, to realise autonomous landing of maritime vertical UAVs on small ships and maritime fixed-wing UAVs on large ships is quite difficult both technically and in terms of use. In order to achieve truly autonomous landing, in addition to the dynamic base problem brought about by the motion of the mother ship, it also involves attitude problems such as the rise and fall of the mother ship's deck, the transverse and longitudinal rocking motion under medium and above sea state. Therefore, the autonomy capability of single system of maritime UAV should be increased between level 2 and level 3 by adding the level of 'Autonomous Landing', i.e., ① Long-range Guidance Control; ② Real-time State Diagnosis and Control; ③ Automatic Following and Landing Control; ④ Adaptation to Failure and Flight Condition; ⑤ On-board Path Re-planning (Threat Determination and Avoidance). In this way, it may be more convenient for the in-depth study of the problems related to maritime unmanned aerial vehicles (UAVs).

For the autonomous control capability of the fleet, the technical connotation of distributed control at level 8 is already reflected in levels 7 or 9, so levels 7, 8 and 9 could be combined into two levels. In this way, the autonomous capability of maritime UAS remains at level 10, but with a different classification and connotation than that of the United States military. The biggest difference is that it highlights the characteristics and difficulties of maritime UAS technology and use.

Level 3, 'Automatic Follow and Landing Control', should also include the capability that if a maritime fixed-wing UAV fails to align with the deck landing point, or fails to hook onto the arresting cord, at the moment of unsuccessful landing

3.2 Key Technologies of Marine Unmanned Platform

(within a fraction of a second), the maritime UAV must be able to take remedial measures autonomously, such as automatically increasing throttle/thrust to resume flight control.

(3) Overall configuration and design

The overall design of a vehicle is mainly determined by its mission and tactical technical indicators, but the overall configuration is not unique, and the designer has a lot of room for play. Especially when some main war technology indexes can't fully correspond to the overall design programme, or it is difficult to consider thoroughly in the design stage, and the technical and usage problems will be fully exposed only after a large number of test flight experiments or even batch trials.

Looking at the development and equipment of maritime UAVs in various countries around the world, there is a common situation of directly transferring shore-based UAVs to ships, which results in some UAVs failing to meet the requirements of the Party A in the experimental test flight stage, and some of them are found to be difficult to achieve the expected combat tasks immediately after being put into use, or the scope of use and functions are obviously reduced. In the author's opinion, the above situation is partly due to the lack of detailed technical specifications of the Party A, and partly due to the deviation of the design concept of the contractors, which only meets the clear quantitative specifications and gives less consideration to quantitative or qualitative specifications of the use environment, reliability and applicability, and so on.

Take the maritime unmanned helicopter as an example, Canada's CL-327 has good performance in land use, but the U.S. Navy believes that it is difficult to meet the requirements of shipboard use after experimental evaluation. At present, most of the unmanned helicopters at home and abroad are designed with the concept of civil or shore-based helicopters to carry out the overall configuration design, and the result is that: although the layout of the whole aircraft is reasonable and the aerodynamic efficiency is high, the use and capability of the helicopter after boarding the ship will be greatly reduced.

Therefore, the development of an unmanned maritime helicopter cannot be based on the aerodynamic general design points of a shore-based/civilian helicopter. A high-performance maritime unmanned helicopter must fully consider the following three capabilities and deal with the related conflicts.

Firstly, wind resistance must be very strong. Compared with the land, the wind on the sea is much stronger. Excellent wind resistance not only ensures the safety of UAV use, but also improves its applicability. This requires a larger paddle load, a larger power reserve, and a smaller fuselage cross-section/side area. This is contradictory to improving aerodynamic efficiency and reducing power and fuel consumption.

Secondly, the anti-skidding and capsizing ability should be strong. This is especially important because of the frequent take-off and landing in medium and above sea conditions. This requires the centre of gravity to be as low as possible and the landing gear to be as widely spaced as possible. This contradicts the requirement to raise the fuselage to accommodate mission loads under the belly and the design concept of reducing structural weight.

Thirdly, overload and shock resistance should be strong. The ship lifting and sinking and transverse rocking motion will greatly increase the impact load of the fuselage structure, and it is necessary to improve the overload and impact resistance of the maritime unmanned helicopter, which leads to an increase in the structural weight of the whole aircraft. Under the same material system, the empty weight of the maritime type is generally significantly higher than that of the shore-based type. This is in contradiction to the lightweight design concept of the shore-based unmanned helicopter.

Take the Fire Scout being developed by Northrop Grumman as an example, the initial prototype was the RQ-8A (shore-based). The U.S. Navy soon found that its operational capability was too low to meet its operational requirements. The contractor immediately changed the design of the MQ-8B, which is almost a completely new configuration: the propellers were changed from three blades to four; the engine was replaced, increasing the power by more than 30%; the mission load was more than doubled, and the maximum take-off weight was also increased by 20%. At the same time, the fuselage appearance and hangers, etc. have also made greater modifications, the purpose is to develop a stronger combat capability, wider scope of application, excellent performance of the sea unmanned helicopters, to meet the U.S. Navy's future operational use of the requirements.

(4) Ship surface automatic/autonomous take-off and landing technology

The most fundamental difference between maritime UAVs and land-based UAVs is the process of UAV take-off and landing, which is extremely complex, and how to guide it to fly according to the required trajectory to achieve automatic take-off and landing is one of the key technologies of maritime UAVs. Typical UAV landing guidance system is the UCARS of SierraNevada Company (Fig. 3.5), which is a kind of take-off and landing guidance system with radar system as the main guidance means of UAVs. The maritime system can detect the range of UAVs through radar to achieve accurate tracking of UAVs, and together with the on-board equipments, the precise landing guidance can be achieved. The UCARS system has developed two versions, V1 and V2, and has been successfully applied to the RQ-8B Fire Scout maritime unmanned helicopter system.

The single-radar UAV landing guidance system is more suitable for rotary-wing UAVs. In order to adapt to the use of fixed-wing UAVs on board ships, the United States began to develop the Joint Precision Approach Landing System (JPALS) based on satellite navigation in May 1996 by Raytheon, which is more suitable for the landing and takeoff of fixed-wing UAVs on ships with flat decks (e.g., aircraft carriers or amphibious assault ships). In 2013, with JPALS support, the United States X-47B completed a carrier-based take-off and landing experiment.

In addition to the United States, the French SADA automatic deck take-off and landing system, the D2AD automatic landing system for unmanned aerial vehicles, the ADS autonomous landing system and the Deckfinder take-off and landing assistance system for the Austrian S-100 unmanned helicopters (Fig. 3.6) have already made it possible for unmanned aerial vehicles to take off and land on naval platforms.

Fig. 3.5 Drone common automatic recovery system (UCARS)

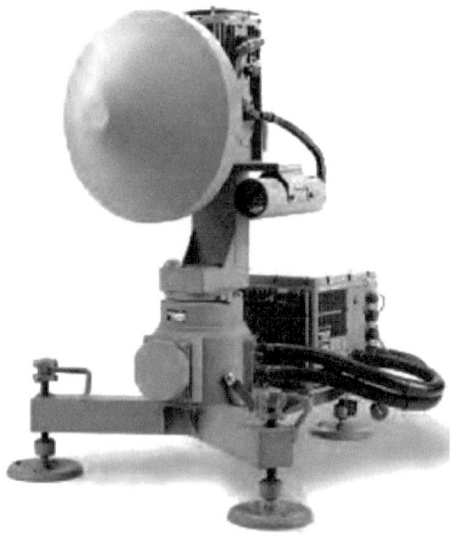

Fig. 3.6 Deckfinder takeoff and landing assist system

As the most fundamental feature of unmanned systems at sea, the maritime UAV platform is the carrier for executing tasks and the basis for realising system functions. The urgent problem that needs to be solved when the UAV is on board is how to take off and land on the deck of a moving ship. At present, the main guidance technologies for landing on ships at home and abroad include GPS guidance, radar guidance, TV guidance, infrared imaging guidance, photoelectricity guidance, etc. The use of a certain guidance method alone cannot meet the requirements for the development of the UAV platform.

Simply using a guidance method cannot meet the requirements of the rapid development of UAV landing guidance, the comparison of different guidance methods is shown in Table 3.2, so it is necessary to study a variety of guidance methods coexisting UAV landing guidance methods, usually the solution for the use of radar, photoelectricity and satellite combination of common guidance methods.

(5) Antenna integration and link transmission technology

Compared with land-based UAS, the measurement and control datalink antenna of maritime UAS needs to be installed on the combat ship. Due to space constraints and movement characteristics on board, the measurement and control antenna, although designed to be as small as possible, may still be unable to meet the installation requirements. In addition, with the development of phased array radar technology at sea, the general measurement and control antenna will cause the expansion of the radar radiation area RCS of the ship, resulting in the exposure of the ship's target, which seriously reduces the probability of survival. Therefore, the measurement and control datalink antenna of maritime UAS needs to be integrated with other mission antennas to reduce the radiation area and deployment requirements.

At the same time, due to the complex electromagnetic environment and confrontation environment at sea, the data link transmission information needs to enhance anti-jamming performance and encryption performance.

(6) New material application and power system technology

Table 3.2 Comparison of different guidance methods

Guidelines	Advantages	Disadvantages
GPS guidance	Highly accurate, easy to use and relatively mature technology	It's completely controlled by the US military and would be totally irrelevant in the event of war
Radar guidance	Capable of searching, intercepting and tracking targets around the clock and not easily intercepted, detected and interfered with by the enemy	Large size of the equipment, limited by the installation space on board
TV guide	Highly resistant to electromagnetic interference and can operate in radio silence	Close proximity and inoperative at night and in inclement weather
Infrared imaging guidance	Strong anti-interference ability, high sensitivity, high resolution, can work all-weather	Limited by positioning accuracy at proximity
Optoelectronic guide	Highly resistant to electromagnetic interference and can operate in radio silence	Proximity of action

3.2 Key Technologies of Marine Unmanned Platform

In recent years, maritime UAV adopts a lot of new materials, new power and other advanced technologies. In terms of new materials, composite materials, such as glass fibre and carbon fibre, are used in large quantities in the airframe mechanism to reduce the weight of the structure and the number of parts; titanium alloy is used in moving parts, such as rotor shafts and propeller hubs, to improve the life of moving parts. The modular design and the extensive use of new materials have reduced the empty weight of maritime UAVs, significantly increased their structural life and improved their maintenance performance.

In terms of power systems, some maritime drones under development use new rotor engines or small turboshaft engines to improve power-to-weight ratios and lifespans, and to reduce vibration and fuel consumption. In order to meet the requirements for safe use on board ships, most of the power systems use heavy fuel oil or are being improved to use heavy fuel oil and adopt full authority digital engine control (FADEC) technology.

At present, most small and medium-sized unmanned helicopters use reciprocating piston engines. The main advantages of these engines are high efficiency and low fuel consumption, but the disadvantages are larger outline size and weight. The German Wanggel invented the triangular rotor engine, referred to as rotor engine. The rotor engine has the advantages of light weight, small volume, high specific power, few parts, smooth operation, good high-speed performance, etc. In the 1980s, Germany and the United Kingdom were in the leading position in the rotor engine, and the United States has increased the rotor engine R&D efforts in recent years, and the level of obvious improvement. In the U.S. military unmanned aircraft development roadmap, the heavy oil and rotor engine as the focus of research and development before 2010. Israel's HERMES-450 long-endurance tactical UAV uses AR801 rotor engines, and SCHIEBEL's newly developed VTOLUAV adopts rotor engines to reduce the weight and vibration level of the platform and improve high-speed performance. The Canadian CL-327 Guardian and the United States Fire Scout both use small and medium-sized turboshaft engines. It can therefore be said that, in many respects, the newest unmanned maritime helicopters developed by developed countries use more new technologies and materials than some manned helicopters.

(7) Ship-aircraft adaptability technology

Adaptability of aircraft and ship is a special requirement of maritime UAV, and also a very important tactical technology index. For shore-based UAVs, the UAV can occupy a large space, and the volume of the shipboard control station cannot be strictly required or restricted. However, due to the narrow space and harsh environment of ships, the applicability of maritime UAS must put forward high requirements, otherwise the maritime UAS will not be able to be used normally or the combat capability will be significantly reduced.

The general requirements for the adaptability of the machine and ship are: the maritime UAV adopts modular design, easy to install, dismantle and store, and small parking space. Shipboard control stations should be miniaturised, integrated and even portable.

The adaptability of machine and ship is also manifested in the following aspects: the wind resistance, anti-skidding and impact resistance of the maritime UAV should be strong; the anti-corrosion, anti-salt spray and anti-mould ability of the whole machine should be strong; the fuselage should be designed with a deck tethering ring, and the engine should be easy to flush; the wings of the fixed wing UAV with a large chord ratio should be foldable, and the rotor blades of the maritime unmanned helicopters with more than three blades should be foldable, and so on.

The adaptability of the aircraft and ship also requires that the maritime UAV can be recovered normally and conveniently on a small ship in a harsh environment with changing weather and sea conditions. This is particularly important and difficult for maritime fixed-wing UAVs.

All in all, good adaptability of aircraft and ships requires that maritime UAVs have compact characteristics, i.e., they would rather be smaller in size and sacrifice a little aerodynamic efficiency, but require a larger power reserve and stronger wind resistance, so as to ultimately realise a wider scope of application and stronger combat capability. In the author's view, the scope of use of maritime drones should not be lower than that of manned maritime aerial vehicles with the same functions. Only in this way can the maritime UAV be truly practical and useful.

3.2.2 Key Technologies for Unmanned Boat Platforms

Judging from the development of unmanned boats at home and abroad, unmanned boats, as a complex engineering system, cover a wide range of disciplines, such as cognitive psychology, material science, structural science, acoustics, optoelectronics, radar, and communications, etc. Combined with the application prospects of unmanned boats, the related theories and technologies are mainly pattern recognition, image processing, and control theory. Combined with the application prospect of unmanned boat, the related theories and technologies involved in unmanned boat mainly include pattern recognition, image processing, control theory and so on. In summary, they include the following

Key technologies

(1) Design and optimisation of unmanned crafts

The main research contents are: boat design suitable for the characteristics of high-speed unmanned boats; performance optimisation of high-speed unmanned boats and wave resistance research. The key technologies are: ① the research on the ship type suitable for the characteristics of high-speed navigation of unmanned boats, so that it has better wave resistance, anti-submergence and anti-subversion performance in complex sea conditions; ② the research on the optimisation of the performance of unmanned boats. In the case of not increasing the size of the propeller and the main engine horsepower, improve the load capacity and sailing speed.

3.2 Key Technologies of Marine Unmanned Platform

(2) Autonomous integrated driving control technology research

It is mainly related to the automatic control of the direction and speed of the unmanned boat under the condition of wave and current interference. The focus is on the water jet propeller, diesel engine and gearbox speed adjustment, water jet propeller water jet steering adjustment.

(3) Autonomous navigation technology of unmanned boat

Positioning technology represented by GPS/BeiDou navigation system, heading detection and control based on digital compass and combined navigation technology based on JieLian inertial navigation. In the case of unmanned operation, it provides navigation support for the unmanned boat to realise autonomous navigation.

(4) Target recognition and detection technology

Identification of stationary and moving targets on the water surface, motion trajectory prediction, tracking technology; multi-sensor target tracking technology.

(5) Intelligent path planning and decision-making technology

Global path planning based on electronic charts; local route planning based on automatic radar plotter and other sensors; emergency collision avoidance technology based on dynamic targets.

(6) Wireless data communication technology

Wireless data network link technology; using UHF spread spectrum technology and satellite broadband technology, real-time transmission between unmanned boat and shore-based monitoring system of information on the environment of the waters around the unmanned boat, still pictures, control instructions, etc.; encryption of signals in the process of data transmission, anti-attenuation, anti-jamming technology.

(7) Data fusion and system integration technology

High-speed water surface intelligent unmanned boat information integrated processing technology; to meet the rapid modular function replacement and rapid debugging technology; a variety of sensors to coordinate the work and information screening, fusion technology.

3.3 Typical Applications of Marine Unmanned Platform

3.3.1 Typical Applications of Marine Drones

UAVs can be applied to maritime patrol and law enforcement, investigation, evidence collection and emergency response, maritime search and rescue, maritime oil spill, sewage monitoring and emergency response, beacon inspection, waterway survey and other maritime regulatory business areas.

(1) Maritime patrol law enforcement, investigation and evidence collection and emergency response

At present, the trend of large-scale maritime vessels, fast has been very obvious, high-speed ships and large container ships speed has exceeded 28 kn, but most of the existing maritime system patrol boats can not reach this speed. At the same time, by the objective conditions of the patrol ship, the use of patrol ships to carry out cruising there are short sight range, slow response, difficult to grasp the overall situation, the illegal ship can not be sustained and effective tracking, some violations can not continue to collect evidence and deal with other problems. The high-speed and high-efficiency advantages of UAVs can effectively make up for the lack of speed of law enforcement vessels. Especially in the investigation and evidence collection and emergency response, through the use of drones, can ensure the rapidity of response and timely investigation, to prevent the escape of the ship, the use of airborne camera, photographic equipment can also be recorded and saved data, to facilitate the investigation and processing.

(2) Maritime search and rescue

General rescue at sea often use aircraft or drones to quickly arrive at the scene, and in the target area over the low-speed flight search. On-board photoelectric cooling infrared pods can be used to detect living targets, avoiding the uncertainty of manual search and rescue caused by the omission. The refrigeration infrared sensor of the optoelectronic pod can distinguish the colour of living targets and objects without temperature within the field of view, and the ground station staff can indicate the target for rescue helicopters and ships through the identification, and command the rescue helicopters, rescue ships and passing ships to carry out rescue in cooperation. Moreover, the unmanned aircraft is able to resist grade 8 winds, and can reach the dangerous areas where many personnel and ships cannot reach, and can transmit high-definition videos and pictures to the monitoring centre in real time, providing information guarantee for the relevant departments to deal with the situation quickly, and the use of unmanned helicopters can greatly improve the success rate of the rescue.

(3) Monitoring and emergency response to oil spill and sewage discharge from marine vessels

Marine environmental protection is one of the most concerned themes of today's marine countries. As the volume of oil transported by sea increases year by year, the oil tankers tend to be larger, and the risk of oil spill from marine vessels also increases. Statistics show that oil is the biggest source of marine pollution, and 42% of the oil discharged into the sea each year is caused by the oil transport process. For this reason, various marine countries have formulated oil spill contingency plans, and the International Maritime Organisation (IMO) has also adopted corresponding resolutions. With the increase of China's oil imports and the increasing efforts of marine oil exploitation, the real-time monitoring of marine oil spills is particularly important. The first few hours after the occurrence of marine oil spill is the best time to prevent the spread of pollution and its hazards, the use of drones on important routes, oil mining key areas for real-time monitoring, once the occurrence of crude oil leakage, with the help of airborne multi-spectral imaging radar on the sea surface inspection, which special multi-spectral imaging radar can be more in the night for oil spill monitoring. At the same time, for the gradually hidden night-time sewage operations, the unmanned helicopter-mounted multi-spectral imaging radar can determine the sewage acts through the temperature and colour value of the objects discharged by the illegal sewage vessels.

(4) Beacon Inspection

Beacon is the main means of navigation security, there are many important lighthouses and lightposts located on isolated islands along the coast of China, with many points, long lines, scattered, inconvenient traffic, and very difficult to supply and maintain. The use of UAV task equipment to achieve rapid inspection of navigation markers, timely and effective report on the working status of the beacon, to avoid purposeless inspection, can effectively improve the normal rate of the level of navigation markers.

(5) Fairway measurement

The method of using aerial photography to shoot the ground and water surface, obtaining image information, processing, handling and analysing to extract the spatial position of the object to be measured and related information has been widely used, especially the application of all-digital photogrammetric methods, so that unmanned aircraft aerial photography is also able to meet the requirements of aerial surveying.

3.3.2 Typical Applications of Marine Unmanned Boats

The United States, Israel and some European countries have paid considerable attention to the surface unmanned boat, and developed the 'Protector', 'Starfish', 'Spartan Scout', 'Stingray' and many other models, which can fulfil a variety of tasks. 'Stingray' and many other types of surface unmanned boats, which can fulfil a variety of tasks. According to the statistics of RAND Corporation in 2013, there are

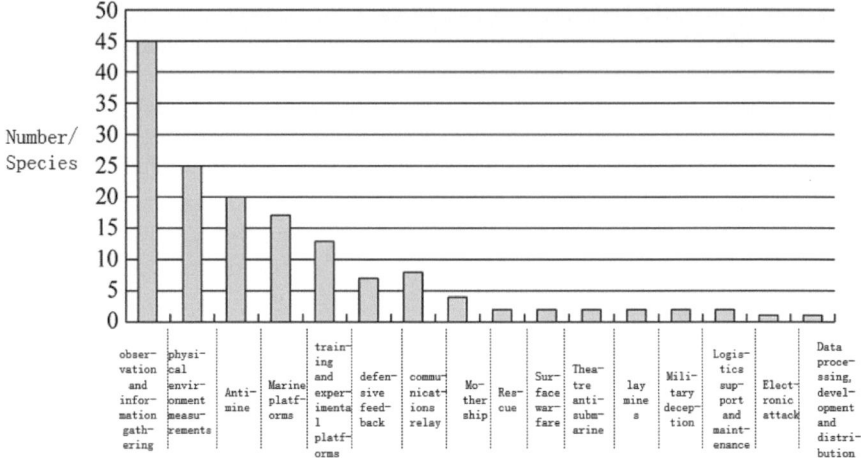

Fig. 3.7 Surface Unmanned Vehicle (USV) Application Direction Statistics

63 types of technologically mature surface unmanned vessels in the world, capable of carrying out 16 types of missions, as shown in Fig. 3.7.

1. Military field

Among the surface unmanned boats currently in service in various countries of the world and those whose technological maturity has reached the level of service, those for military use account for about 70%.

- Anti-mine warfare

Anti-mine warfare is one of the main combat missions of surface unmanned aerial vehicles. Through the mine hunting system, it can carry out the tasks of mine detection, channel clearing, and providing undersea images, and provide security for multiple ships to sweep for mines at the same time. Characterised by its fast, efficient and low-cost cluster approach to complete a wide range of mine detection and positioning tasks.

- Anti-submarine warfare

Facing the growing threat from underwater submarines, the surface unmanned boat provides a new means of anti-submarine warfare, which can carry anti-submarine equipment and form a mobile anti-submarine guard network around the main fleet, and can operate for a long time without stopping to improve the anti-submarine capability of the fleet. The long-endurance surface unmanned craft equipped with submarine-hunting detection sonar can carry out the tasks of detecting enemy submarines, providing target data to the mother ship or launching attacks on the target directly.

- Maritime security

The Intelligence, Surveillance and Reconnaissance (ISR) module/force protection module on board the UAV is well placed to provide this capability. In addition, it is capable of detecting and preventing frogmen from attacking when operating in shallow waters, as well as conducting in-theatre nuclear, biological and chemical reconnaissance.

- Surface warfare

Surface unmanned boats can be equipped with machine guns, small guns and small shipborne anti-aircraft and anti-ship missiles according to operational needs, providing fire support for the surface. Its characteristics are flexible and compact, good concealment, etc., under special circumstances can even be installed on the surface of the unmanned boat with high explosives or remote-controlled weapons, made of unmanned attack boats at sea, choose a favourable time to destroy the enemy's maritime combat platform.

- Supporting special forces operation

Surface unmanned boats support special forces operations, and are used to carry out anti-piracy and terrorist missions.

- Electronic warfare

Surface unmanned boat has the characteristics of modular design, compactness, manoeuvrability and good concealment, etc. It can be equipped with different mission modules according to different tasks, and with the help of waves, islands and reefs, etc. to make it difficult for the enemy to find out the enemy's shore-based radar station and the shipborne detection system, which is conducive to the realisation of the information warfare environment of the surface electronic warfare.

- Supporting maritime interception operations

The use of unmanned surface craft equipped with radio and optoelectronic systems to intercept unidentified ships, radio monitoring and long-distance observation through optoelectronic systems can prevent naval officers and soldiers from facing the threat of suicide attacks.

2. Civilian field

At present, in order to meet the requirements for performing various tasks related to maritime safety, civil surface unmanned craft should have the following performances.

(1) Sufficient speed, manoeuvrability and endurance.
(2) Survivability and stability under adverse marine conditions.
(3) Sufficient payload capacity to carry a variety of equipment and instruments.
(4) Ease of replacement of equipment and instruments to suit different missions.
(5) Reliable communication with base stations on the mother ship or on shore, and sufficient communication range.

(6) With the data interface to connect different equipment and instruments.

- Course data measurement

With the high manoeuvrability of the unmanned boat, fixed-speed cruise function, you can use the unmanned boat in the waterway to carry out a lot of data measurement.

- Maritime cruising

With the high manoeuvrability and efficiency of unmanned craft, it is possible to arrange for unmanned craft to carry out patrols and safety inspections in important waters, assist maritime authorities in maritime management, investigate and collect evidence from illegal vessels with the help of the equipment on board, and use non-lethal equipment such as loudspeakers and bright lights to warn and dissuade fishing vessels operating in violation of the law. Unmanned vessels can also be used to assist in the inspection and management of navigation markers.

- Harbour monitoring

Using the high mobility and efficiency of unmanned boats, it is possible to set up guards to monitor key positions in the harbour, providing an important means of reconnaissance to understand the dynamics of the harbour at the first time. Configure the unmanned boat with photoelectric reconnaissance equipment, so that it has the ability to obtain high-resolution images and videos, and then monitor illegal behaviours, the specific applications are as follows: ① to illegal behaviours such as piracy of sea sand, dumping, sewage, etc. to take pictures and videos for evidence; ② to carry out 24-h dynamic supervision of the sea area.

- Hydrographic survey

UAVs do not adversely affect the surrounding waters and can be positioned and tracked with high accuracy in shallow waters. In shallow waters, rivers and inaccessible waters, conventional survey vessels are not suitable. Using the stability and precise track control of the unmanned vessel, it is possible to measure the depth of water, the current and its flow profile at harbours and river entrances.

- Water quality sampling

With the stability and precise track control of the UAV, it is possible to measure the water depth, study the sediment and extract biological samples from water currents in harbours and river entrances.

- Maritime search and rescue

Most of the maritime disasters occur in bad sea conditions, and it takes a long time for search and rescue personnel to arrive at the rescue site. By using search and rescue helicopters to send unmanned boats to the scene of the disaster, on the one hand, the situation at the scene of the disaster can be transmitted back in real time, which is convenient for the search and rescue personnel to formulate a plan; on the other hand, using unmanned boats to search for the people who have fallen into the water can greatly improve the efficiency of the search and rescue. At the same time, the unmanned boat carries basic life-saving equipment, which can help the overboard personnel to wait for the arrival of rescue personnel.

Marine unmanned platform technology is a new research direction in the development of information technology, although the development of key technologies has made some progress, but there are still many problems in the positioning of strategic objectives, prototype development methods, results of the comprehensive use of the stakeholders in the division of responsibilities and other aspects need to be further resolved and improved. From the perspective of technology foundation and practical application, this book makes reference to the latest research technologies and achievements at home and abroad.

From the perspective of technology foundation and practical application, this book refers to the latest research technologies and achievements at home and abroad, and summarises the authors' research results and experiences in acoustic, optical, radar, communication and integration technologies of unmanned platforms at sea over the years.

From the second part of this book, the basic concepts, basic principles, key technologies and application scenarios of acoustic, optoelectronic, radar and communication technologies of unmanned maritime platforms will be discussed, and the book strives to be simple and easy to understand, hoping to be helpful to researchers and enthusiasts in the field of unmanned maritime platforms.

Bibliography

1. Yingqi, Xu. 2019. Practical Analysis of the Application of Drones in Maritime Management. *Decision-Making Exploration* 11: 94.
2. Yangsheng, Pan. 2018. Application of Unmanned Aircraft Systems in Maritime Law Enforcement. *World Maritime Transport* 41(10):5-9.
3. Zikun, Cheng. 2017. Analysis of the Application of Drones in the Yangtze River Maritime Supervision. In *Inland Waterway Maritime Professional Committee of China Nautical Society*, 8–13.
4. Cheng, X., and Y. T. Wang. 2017. Research on the Structure and Development Planning of Maritime Security System Based on Unmanned Aircraft System. *China Maritime Affairs* 7: 49–51.
5. Dahe, Lu, and Fan Wei. 2017. The Application of Drones in Maritime Supervision. *China Water Transport* 4:24-25.
6. Guijun, Duan. 2015. Discussion on the Application of Drones in Maritime Management. *World Maritime Transport* 38(2): 38-40.
7. Nan, Li. 2014. Wings of the Blue Sea—Challenges and Solutions of Drones in Maritime Supervision in the New Period. *China Water Transport* 7: 22-23.
8. https://www.163.com/dy/article/E7OG4PS20511DV4H.html

Part II
Acoustics

Chapter 4
Overview of Acoustic Technology

4.1 History of Acoustic Technology

4.1.1 Overview of the Development of Acoustics

Acoustics is derived from the Greek word ακούειν, meaning "to hear" [1]. Acoustics is a branch of physics, but closely related to engineering and technology. It is a discipline that studies the generation, propagation, reception and effects of sound and other issues of science. The propagation of sound is essentially the propagation process of sound waves. Sound waves act on the human ear caused by the feeling called sound. Visible sound waves are the essence of sound propagation. In a narrow sense, the sound is what the human ear can hear. From a broader sense, the sound is the mechanical disturbance propagating through an elastic medium, such as pressure, density, mass displacement, mass velocity and other changes. Before the seventeenth century, the study of acoustics was mainly concerned with music and musical instruments. In the seventeenth–nineteenth centuries, classical acoustics has been systematically studied and summarised, including the principles of vibration of objects and generation of sound waves, propagation and radiation of sound waves, standing waves and reflections, and diffraction. In classical acoustics, sound, tone and music all refer to phenomena that can be heard. Sound refers to sound waves, but also to the sensations caused by sound waves acting on the human ear. Sound waves are therefore understood to be synonymous with "audible sound". Since the 1920s, with the emergence of electronics and the progress of electronic products, acoustics has also been rapid development. Some acoustic equipment produce sound source frequency that are "inaudible sound" to the human ear. Sound waves can be classified into infrasound, audible sound and ultrasound according to their frequency range. The frequency range of audible sound is 20 Hz–20 kHz, the frequency of infrasound extends downwards from 20 to 10^{-4} Hz, sound waves extend upwards from 20 kHz to 5×10^8 Hz for ultrasound, and upwards to 10^{13} Hz for hyper-ultrasound. The division of sound waves by frequency is based on the frequency range of human acceptance as a reference, and people as a source of sound can be vocalised in the frequency range of 85–5000 Hz. People and animals are both vocalisers and receivers of sound, but the frequency range of their vocalisations and receipts are not the same. With the development of electronic technology, some sound-generating and receiving devices are widely used in physics, engineering, medicine, art and other human production practices. In essence acoustics is also the science of sound; that is to say, everything related to sound is within the scope of acoustic research. Sound emitted from various sources, propagated through different media, and received and perceived by organs that can hear sound (e.g., the ear, measurement transducers), is related to acoustics in every aspect of the series of processes. In terms of human hearing, the sound emitted by a sound source (vibration of an object) is divided into two aspects: firstly, how the sound is transmitted from human hearing to the human ear, thus making the tympanic membrane of the human ear vibrate; secondly, how the vibration of the tympanic membrane is made to be subjectively perceived as sound. In terms of sound measurement, the sound emitted by the sound source is also divided into two aspects: first, how the sound is transmitted from the source to the measurement point (measurement point). The second is how the sound is measured by the transducer. Whether from human hearing or sound measurement, the first part of the content is the same, that is, the transmission of sound in the medium is the same, but the second part of the obvious difference, that is, the human ear and microphone have a fundamental difference [2].

Modern acoustics was developed by applying electronics to acoustic research, with the experimental basis being electroacoustic measurement techniques. Electroacoustics has played a decisive role in the development of modern acoustics. At present, due to the development of digital technology and large-scale integrated circuits, the use of microprocessor measurement technology to make acoustic measurements to improve the speed and accuracy of acoustic measurements, and the realization of the past cannot be used in many new measurement methods, such as real-time analysis of the frequency spectrum, sound intensity measurements coherent measurements, sound source identification, signal processing technology, etc.. As modern acoustics involves a wide frequency range (10^{-4}–10^{12} Hz) and different media (gas, solid and liquid), generally speaking, in different frequency ranges and different media, although the physical principle is basically the same, but the technology and equipment varies a lot, and therefore formed a sub-discipline applied to various scopes, almost involving every aspect of human activities.

4.1.2 Relationship Between Acoustics and Other Disciplines

Acoustics is a discipline with a wide range of applications, involving all aspects of human production, life and social activities; at the same time, acoustics is a discipline with strong cross-penetration, interacting with a variety of new disciplines, new technologies, and promoting each other, and constantly absorbing, applying and developing new ideas, which enhances the vitality, competitiveness, and academic and artistic charm of acoustics. Medicine, psychology, physiology, biology, language and music, communication and broadcasting, computational science, mechanical engineering, oceanography, electro-technology, etc. are all linked to acoustics to varying degrees. Modern acoustics not only penetrates into various branches of physics, but also widely penetrates into other fields of science and technology. The cross-disciplinary nature of modern acoustics is very obvious, and the names of some sub-disciplines reflect the combination of two different disciplines to form a new sub-discipline, for example, acoustics and electronics to form electroacoustics, acoustics and architecture to form architectural acoustics, and others, such as music acoustics, psychoacoustics, bio-acoustics and so on, are cross-branching disciplines. In the development of acoustics, many new discoveries have changed the direction of research, resulting in new research topics. However, with the deepening of the research work some of the new research areas have gradually been separated from acoustics to become independent sub-disciplines or part of other disciplines. Therefore, the development of modern acoustics, on the one hand, includes the work of many other disciplines, and on the other hand, the research content is constantly transferred to other scientific and technological fields.

The famous "Lindsay's Wheel of Acoustics" was introduced by R. Bruce Lindsey in 1964, see Fig. 4.1. The "Wheel of Acoustics" is a clear representation of the relationship between the various branches of modern acoustics and their underlying theories, as well as other scientific techniques. The centre circle is the foundation of modern acoustics, outside the circle there are two concentric rings and divided into several sectors. Each sector of the inner ring represents each branch of the discipline, and the sectors of the outer ring correspond to the scope of application of each branch. The outer ring describes the four main areas of acoustic research as earth sciences, engineering, life sciences, and the arts. In fact, Lindsey, a physicist, did not specifically list physics as a separate field of study in the outer ring, probably because a background in physics provides the necessary foundational knowledge for almost all areas of acoustic research.

As a result, the science of acoustics is widely distributed in all aspects of human activity and human society, such as music, medicine, architecture, industry, the environment and even war. The Acoustical Society of America (ASA) has 13 technical committees in accordance with acoustical research areas, which correspond to the following research areas: aeroacoustics, audio signal processing, architectural acoustics, bioacoustics, electroacoustics, environmental acoustics, music acoustics, noise control, psychoacoustics, speech, ultrasound, hydroacoustics, vibration and dynamics. In addition to the ASA's classification of the above acoustics by the Acoustical Society of America, there are also medical acoustics, materials acoustics, and virtual acoustics.

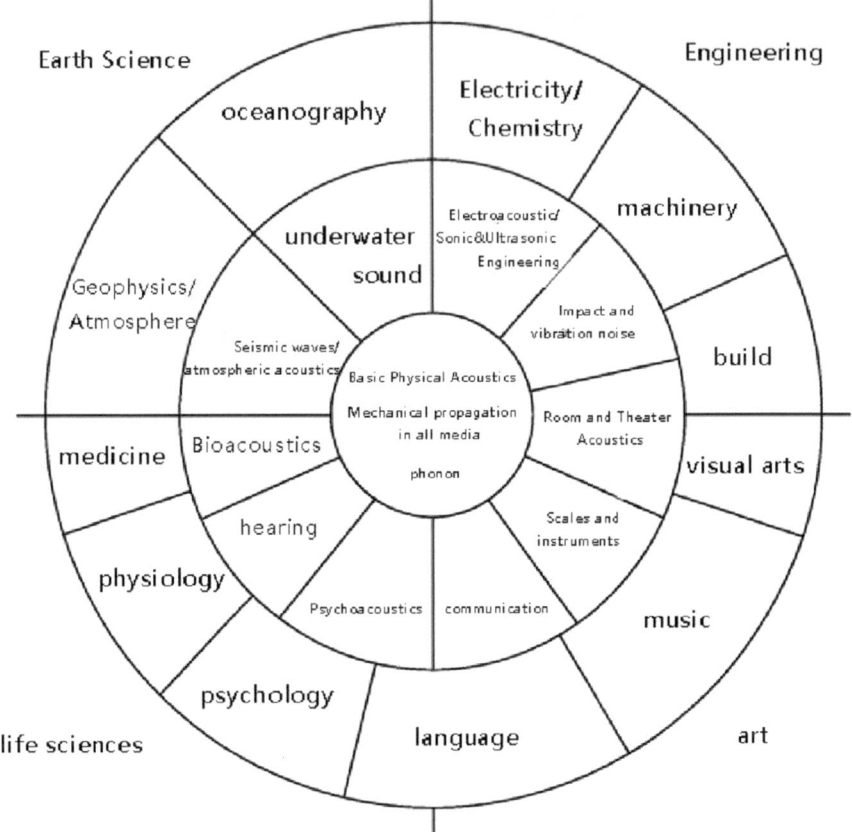

Fig. 4.1 Lindsay's wheel of acoustics

4.2 Main Content and Development of Modern Acoustics

Modern acoustics is developed on the basis of classical acoustics, which has a very rich content and a wide range of applications. The main features of modern acoustics are as follows:

(1) There are relatively few fundamental theoretical problems in modern acoustics, mainly because most of the the theories are already mature and have been fully developed in classical acoustics;
(2) The application fields of acoustics technology are very wide, and some applied basic theories have more research results in different ranges;
(3) The propagation medium of acoustics has become more and more extensive, including all gases, liquids and solids, while the environment in which the medium is located also extends to high or low temperature, high pressure or low pressure and other extreme conditions;

(4) Modern acoustics has been widely penetrated into other branches of physics and other scientific fields as well as cultural fields, forming a number of marginal subdisciplines, each of which is independent of the other, but there is cross-penetration between the disciplines;
(5) Modern acoustics research also involves the phonon motion and the interaction between phonons and matter, which can be used to study the internal structure of matter. Therefore, modern acoustics has both the nature of classical physics and quantum nature;
(6) The experimental conditions of modern acoustics are based on electro-acoustic measurement technology, and with the application of modern digital technology, computers and the development of information science, the research content of modern acoustics is constantly expanding and the discipline has new developments.

Modern acoustics is rich in content and has many subdisciplines, with obvious characteristics of crossover and penetration between various disciplines and other scientific fields. Based on the relationship between acoustic applications and the corresponding disciplines in the "Wheel of Acoustics", this chapter focuses on the contents and development of electroacoustics, ultrasonics and hydroacoustics related to the application of marine platforms.

4.2.1 Electroacoustics

Electroacoustics is the science of the principles and techniques of acoustic-electric interconversion and the storage, processing, transmission, measurement and utilisation of acoustic signals. It covers a wide range of frequencies, extending from very low-frequency infrasound all the way to multi-billion hertz ultrasound. The history of electro-acoustic technology can be traced back to the earliest nineteenth century, starting with Edison's invention of the phonograph and Bell's invention of the carbon microphone. In 1881, an experiment on stereo transmission with two-way channels was conducted using two carbon microphones connected to several pairs of headphones. Around 1919, the first experiment in sound amplification was carried out using an electronic tube amplifier and electromagnetic loudspeakers. After the First World War, scientists applied the research results in electromechanical to the field of electroacoustics, laying the theoretical foundation for electroacoustics. With the development of electro-acoustic transducer theory, various relatively sophisticated electro-acoustic equipment and electro-acoustic measuring instruments emerged successively. Especially in the 1970s, the applications of electronic computers and laser technology in the field of electro-acoustic, greatly promoting the development of electro-acoustics.

Important branches of electroacoustics research include electroacoustic transducers, electroacoustic technology, recording and playback technology, and digital audio technology. Electroacoustic transducers are devices that convert acoustic energy into electrical energy or electrical energy into acoustic energy, including infrasound, audible acoustic and ultrasonic transducers. Electro-acoustic converter according to the energy transfer method can be divided into electric, electrostatic, piezoelectric, electromagnetic, carbon particle type, ionic and modulated airflow type. The latter three are irreversible, carbon particles can only turn acoustic energy into electrical energy, ionic and modulated airflow can only produce acoustic energy, while the other types of transducers are reversible, can be used as acoustic receivers, but also as acoustic transmitters. In recent years, electro-acoustic transducers have made new progress in terms of new materials, new technology and new structure, and their research has developed towards the direction of wide bandwidth, high efficiency, high sensitivity and high power. Electroacoustic technology mainly includes sound recording and amplification technology, sound reinforcement technology and its related electroacoustic instruments and electroacoustic test technology. Electroacoustic technology is a fast developing branch of electroacoustic field, which has a wide range of applications in various fields such as politics, military, culture and so on. For example, it is used in wired or wireless communication systems, wired or wireless broadcasting systems, and sound reinforcement in meeting halls and theatres; recording studios, high-fidelity recording and playback systems, etc.; it is also used in the development of voice-controlled speech control technology; and new technologies such as language recognition and sound measurement. Recording and playback technology refers to the natural sound through a series of technical equipment (such as microphones, tape recorders, pickups, etc.) for the reception, amplification, transmission, storage, recording and reproduction of the processing, and then replayed for listening to the technology. It studies the main problem is how to maintain the excellent sound quality of natural sound, that is, in each ring band and the entire system, have the ability to realistically maintain the original appearance of the sound signal, including the necessary beautification and processing of the sound signal. Digital audio technology refers to the technology related to the digital processing of audio signals, including analogue-to-digital and digital-to-module conversion, digital data transmission, recording, storage, mixing and other processing techniques. With the development of digital technology and large-scale integrated circuits, digital audio technology has also developed rapidly, and its typical applications include digital recording, programme transmission, artificial reverberation and mixing.

With the application of electroacoustics and the needs of society and production, people put forward a large number of practical and theoretical problems of electroacoustics, promoting the continuous development of electroacoustics. Electroacoustics future general development trend is: electro-acoustic devices and electro-acoustic equipment towards high fidelity, stereo, high noise immunity, high efficiency, high call capacity of the direction of development; sound quality evaluation of the research to improve the recording and playback technology as well as sound processing technology; the new transducer mechanism research as well as the development of new materials; to improve the detection of acoustic signals is still the main direction of sound measurement technology research.

4.2.2 Ultrasonics

Ultrasonics is the study of ultrasound generation, reception and propagation in the medium of the law, ultrasound effects, as well as ultrasound in the basic research and national economy in various sectors of the application of the important branch of acoustics. Ultrasonics is the main research content is the design of ultrasonic transducer and its application, the early ultrasonic transmitter is the use of whistle and rotary flute ultrasonic transmitter, the emergence of electromechanical ultrasonic transmitter revolutionised the ultrasonic technology. Electromechanical ultrasonic transmitters mainly utilise three physical phenomena: magnetostriction, piezoelectric effect and electrostriction.

In 1883 for the first time made ultrasonic gas whistle, since then there are various forms of gas whistle, whistle and liquid whistle and other mechanical ultrasonic generator (also known as transducer). Due to the low cost of this type of transducer, so after continuous improvement, is still widely used in the ultrasonic treatment of fluid media technology. In the early twentieth century, the development of electronics has enabled people to utilize the piezoelectric effect and magnetostrictive effect of certain materials to make a variety of electro-mechanical transducer. In 1917, the French physicist Paul Langevin used the natural piezoelectric quartz to create the sandwich ultrasonic transducer, which is used to detect the submarines underwater. With the continuous development of military and national economic sectors in the application of ultrasound, and the emergence of greater ultrasound power magnetostrictive transducer, as well as a variety of different uses of electric, electromagnetic force, electrostatic transducer and other ultrasound transducer. The development of materials science has led to the evolution of piezoelectric transducers, which were initially made from natural piezoelectric crystals, to those utilizing piezoelectric ceramics, artificial piezoelectric single crystals, piezoelectric semiconductors, and piezoelectric plastic films that offer high electromechanical coupling coefficients, low costs, and excellent performance. The frequency of generating and detecting ultrasonic waves have also been increased from tens of kilohertz to thousands of megahertz. The types of waves generated and received have expanded from simple longitudinal waves to transverse waves, torsional waves, bending waves, surface waves and so on. For instance, the frequency of tens of megahertz to thousands of megahertz of micro-surface waves have been successfully used in radar, electronic communications and imaging technology.

In recent years, the generation and reception of ultrasound has been moving towards higher frequencies (above 10^{12} Hz) in the context of fundamental research on the structure of matter. For example, in the end face of the medium directly evaporated or sputtered on the piezoelectric film or magnetostrictive ferromagnetic film, ultrasound with frequencies ranging from hundreds of megahertz up to tens of thousands of megahertz of can be obtained; by utilizing concave microwave resonance cavities, ultrasound with frequencies of thousands of megahertz can be obtained in the quartz rod tens. In addition, higher-frequency ultrasound is generated or received by thermal pulses, semiconductor avalanches, superconducting junctions, and photon-phonon interactions. With the increasing frequency of generated and received ultrasound waves, which are now approaching the frequency of the thermal vibration of the dot matrix, the quantised acoustic energy of these very high frequency ultrasounds is being used to study inter-atomic interactions, energy transfer and other issues. Through the determination of very high frequency ultrasound sound speed and attenuation, can study the interrelationship between sound waves and dot matrix vibration and the coupling between the modes of dot matrix vibration, in addition to be used to study the metal and semiconductor phonons and electrons, phonons and superconducting junctions, phonon and photon interactions, etc.. Therefore, ultrasound, together with electromagnetic radiation and particle bombardment, are listed as three most important means of studying the microstructure of matter and microscopic processes.

Research on the propagation of ultrasonic waves in high-speed flowing fluid media, propagation in special liquids such as liquid crystals, and the nonlinear problems associated large-amplitude acoustic waves in fluid media continues to evolve and develop. Ultrasonics continues to enrich itself by borrowing from other disciplines such as electronics, materials science, optics and solid state physics. At the same time, the development of ultrasonics and the development of these disciplines provide some important equipment devices and effective research methods. Such as ultrasonic flaw detection and ultrasonic imaging technology is to draw on the principles and techniques of radar and the development of ultrasound, and the development of ultrasound and electronics, optoelectronics, radar technology development provides ultrasonic delay lines, filters, convolutions, acousto-optic modulators and other important body wave and surface wave devices.

Because ultrasound is easy to obtain a directional beam with excellent directionality, a high spatial resolution can be achieved by using ultrasound narrow pulses, together with the fact that ultrasound can be propagated in opaque materials, ultrasound can be used in an extremely wide range of applications including ultrasonic testing, ultrasonic flaw detection, power ultrasound, ultrasonic processing, ultrasonic diagnostics, ultrasonic therapy, ultrasound imaging and so on. Ultrasonication is a technology that changes or accelerates the change of some physical, chemical or biological properties or states of a substance through the action of ultrasound on the substance. Due to the use of appropriate transducers can produce high-power ultrasound, and by focusing, increase the rod and other methods, can also obtain high sound intensity of ultrasound, coupled with the cavitation phenomenon in the liquid, so that the use of ultrasound for processing, cleaning, welding, emulsification, pulverisation, degassing, promotion of chemical reactions, medical treatment, as well as seed treatment and so on has been widely used in industry, agriculture, medicine and health and other sectors, and continues to develop. By exploiting the link between the non-acoustic properties of the medium (e.g. viscosity, flow rate, concentration, etc.) and the acoustic quantities (speed of sound, attenuation, and acoustic impedance rate), detection and control of non-acoustic quantities can also be achieved through the detection of acoustic quantities.

At present, new research and new applications in this area of ultrasonic detection are still emerging, such as acoustic emission technology and ultrasonic holography and so on. The use of digital signal processing technology to solve the problem of ultrasonic detection has not yet been solved or not yet satisfactorily resolved the research work, is also a hot research issue in recent years.

4.2.3 Hydroacoustics

Hydroacoustics is the study of the process of generation, propagation and reception of sound waves under water, which is used to solve acoustic problems related to the process of underwater target detection and information transmission. Sound waves are the only known fluctuations capable of travelling long distances in water, and in this respect are far superior to electromagnetic waves (such as radio waves, light waves, etc.). Hydroacoustics has developed with the development and use of the oceans and has been widely used. Hydroacoustics can be used in the military to detect submarines, and in the civilian world it is possible to use hydroacoustic technology to develop and make use of marine resources and to carry out the detection and testing of underwater structures. Such as the use of hydroacoustic technology for underwater target detection, tracking and identification, the realisation of ultra-long-distance propagation of underwater information, investigation of marine and seabed resources, exploration of seabed oil and mineral deposits; underwater structures on marine platforms in the construction, operation cycle of detection and monitoring, such as the quality of the foundation construction, foundation around the scouring, the pipeline of the overhang as well as the change of position, such as the detection and monitoring.

Around 1827, Swiss and French scientists measured the speed of sound in water for the first time quite accurately. In 1912 "Titanic" passenger ship collided with an iceberg and sank, prompting some scientists to study the iceberg echolocation, which marked the birth of hydroacoustics. Fessenden in the United States designed and manufactured an electric hydroacoustic transducer, which can detect the icebergs as far as two nautical miles away in 1914. In 1918, Langevin developed a piezoelectric transducer that generated ultrasonic waves, and the applied the newly vacuum tube amplification technology at that time to detect long-distance targets in the water. He received the submarine's echoes for the first time, initiating the modern hydroacoustic, and consequently inventing sonar.

In 1919, Marty made the echo sounder recorded on the recording paper with a pen, and in 1932, a magnetostrictive echo sounder was successfully tested, which was composed of a central device, an oscillating transmitter and an oscillating receiver, and the central device was equipped with the key equipments such as transmitting pulses, reflecting signals processing, and an automatic depth recorder, etc. After 1930, the piezo-electric oscillating quartz crystal was used to put into the mass production. Mass production and widely used in the world. Subsequently, various kinds of echo sounder were introduced one after another, which led to the remarkable development of ocean acoustics. The international joint survey of the Indian Ocean, which is carried out in 1960, also began to use the precision echo sounder. The emergence of echo sounder can be said to be a leap forward in ocean bathymetry technology, with the advantages of rapidity and the ability to obtain continuous records. With the development of the oceans, the application of hydroacoustics in the investigation and development of marine resources, the monitoring of ocean dynamical processes and the environment, and the enhancement of human understanding of the marine environment are also constantly expanding and advancing.

With the innovation of hydroacoustic transducers, the achievements of hydroacoustic research on the mechanism of temperature gradients affecting the sound propagation path, and the variation of the sound absorption coefficient with frequency have enabled sonar to be continually improved and played an important role in the Battle of the Atlantic against German submarines during the Second World War.

After the Second World War, in order to improve the ability to detect long-range targets (such as submarines), the focus of hydroacoustic research shifted to low-frequency, high-power, deep-sea and signal processing. At the same time, the field of hydroacoustic applications is also more and more extensive, the emergence of many new devices, such as hydroacoustic guided torpedoes, acoustic mines, passive scanning sonar, hydroacoustic communication instrument, acoustic buoys, acoustic speedometer, echo sounder, fish detector, acoustic navigation beacons, geomorphological instrument, undersea stratigraphic profiler, hydroacoustic releaser, as well as hydroacoustic telemetry, controllers and so on.

The research contents of modern hydroacoustics mainly include: new type of hydroacoustic transducer; non-linear acoustics in water; spatial and temporal structure of hydroacoustic field; hydroacoustic signal processing technology; noise and reverberation in the ocean, scattering and undulation, target reflection and ship radiated noise; and acoustic properties of marine media. In particular, hydroacoustics is interpenetrating with marine, geological and hydrobiological disciplines to form research fields such as ocean acoustics.

The research of hydroacoustic transducers is an important part of hydroacoustic research, so the research of new materials, structures and mechanisms of transducers is also the focus of its research. Since the 1960s, in order to achieve the remote detection of sonar, the development of a number of new transducer materials, structural vibration mode and transducer mechanism; the development of the work in the low-frequency, broadband, high-power and deep-water transmitter, with high sensitivity, broadband, low noise and other properties of the hydroacoustic transducer; the emergence of new types of hydroacoustic transducers, such as composite piezoelectric ceramic hydrophone, concave bending tension transducers, the use of the Helmholtz resonator principle of the low-frequency hydrophone made by the application of the jet switching technique of the modulating fluid-type transducers, audio-optic transducers and so on. With the development of technology, new materials have been gradually used in hydroacoustic transducers, such as hydroacoustic transducers of ferroalloys of the super magnetostrictive materials terbium (Tb) and dysprosium (Dy), which can solve the problems of the magnetic circuit and the structure, and fibre-optic hydrophones, which have been developed rapidly due to their high sensitivity.

4.2 Main Content and Development of Modern Acoustics

The study of hydroacoustic channels is to reveal the influence of marine acoustic environmental factors on the acoustic field by studying the propagation laws of acoustic waves in the ocean. The research on acoustic field in the sea includes: the spatial structure of acoustic field and the attenuation law of acoustic wave, the distortion of waveform in the transmission process, and the technique of extracting useful signals from environmental noise. International research on deep-sea acoustic fields has been conducted for many years and is relatively well resolved. In recent years, due to the subject of marine oil and gas exploration, western scholars turn their attention to the more complex shallow sea sound field research. China possessing the most important continental shelf in the world, has dedicated itself to study the shallow sea acoustic field since the 1960s, and achieved considerable results, attracting the attention of Western counterparts.

Detection and identification of hydroacoustic information is also the key technology of hydroacoustic signal processing, using high-speed digital computing chip to process the received signal in the spatial and temporal domains, enhance the signal, filter out the interference, and finally realize identification and valuation. The spatial processing techniques include beam forming, phase control, digital multibeam, split-beam inter-correlation, etc., which help form the optimal receiving directivity. Based on the predicted acoustic field, a simple positive wave filter array can be formed, utilizing conditions such as sound channel and convergence zone. Currently, techniques such as artificial intelligence, neural networks and pattern recognition have also been widely used in the recognition of hydroacoustic information.

Using hydroacoustics to observe the marine environment is also an important application of hydroacoustics in ocean exploration instruments based on hydroacoustic principles are used for statistical sampling of marine environmental parameters. For example, Doppler current meter (ADCP) measure the Doppler frequency shift generated by the current, allowing for the remote measurement of the current profile at various depths from ships or seabed. Acoustic correlation current Profilers (ACCP) utlize the correlation of signals received at location on a ship to determine the sea currents. The backscatter of the acoustic pulse from the seawater suspended in the sediment, organisms, pollutants in seawater can be used to remotely measure the concentration of suspended profiles. Low-frequency acoustic waves can be used in a similar method to medical tomography to invert the vortices and temperature changes in the ocean.

Marine mapping and resource exploration is an important application of hydroacoustics in marine exploration, and its applied research includes electronic nautical charting, mapping of seafloor geomorphology, stratigraphic profiles and seawater depths, as well as exploration of seafloor deposits.

Hydroacoustic positioning technology is a maritime positioning technology and method based on ultrasonic propagation technology. Positioning at sea is carried out by determining the propagation time or phase difference of acoustic signals. Hydroacoustic positioning includes long baseline positioning, short baseline positioning and ultra-short baseline positioning. Currently, dynamic systems that can achieve very small deviations from automatic control systems have been used for well re-entry of drilling and offshore positioning of tension-legged oil and gas development platforms. Modern marine engineering has widely used manned or unmanned submersibles to carry out observation, measurement, inspection, operation and other tasks through hydroacoustic positioning technology.

4.3 Development Status of Foreign Acoustic Technology

An offshore platform is a structure that provides production and living facilities for activities such as drilling, oil extraction, gathering, observation, navigation and construction at sea. According to its structural characteristics and working status, the marine platform is divided into three categories: fixed, movable and semi-fixed. According to the Mechanic characteristics, structural form, working condition and special working environment of the marine platform, the materials used in the marine platform are mostly steel or fibre composite materials for marine engineering, and sometimes reinforced concrete or steel-reinforced concrete composite structure is also used for the fixed marine platform. For the monitoring and detection of structural material damage or defects on offshore platforms, the main types of structural damage and defects include structural cracks, weld cracking, internal material defects, structural corrosion and fatigue, and fracture damage. Due to the randomness of the marine environment and the complexity of marine platforms, incidents such as damage or even capsizing of marine platforms and their accessories occur from time to time. The main reason for this is that the commonly used non-destructive testing is unable to detect potential structural hazards of offshore platforms in a timely manner. Acoustic emission detection technology can implement real-time monitoring of the marine platform due to its own characteristics. For the deformation of the underwater part of the structure of the marine platform, the erosion of the seabed foundation, the erosion, and the measurement of the underwater part of the underwater part of the underwater platform, due to the technological advantage of the acoustic wave detection in the water, the sonar detection has become the most important detection method.

According to the characteristics of the structural detection and monitoring of marine platforms, acoustic emission technology and sonar technology are the main methods for marine platforms. In recent years these two detection technologies have been widely used in the detection of marine platforms, and this chapter will introduce the domestic and international development of these two technologies.

4.3.1 Acoustic Emission Techniques

Modern acoustic emission technology originated from the research work conducted by Kaiser in Germany in the early 1950s. He observed that the deformation of metallic materials occurs simultaneously with the generation of acoustic emission signals.

4.3 Development Status of Foreign Acoustic Technology

Additionally, Kaiser also discovered a universal rule that the acoustic emission phenomenon was irreversible, which is known as the Kaiser effect.

In the late 50's, American scholar Tatro studied the acoustic emission mechanism of metals and found that in the plastic deformation of metals, the cause of the acoustic emission phenomenon is mainly the dislocation motion of the metal crystals, and therefore concluded that the acoustic emission phenomenon occurs inside the material rather than on the surface of the material [3, 4]. Tatro's and Schofield's study on the physical mechanism of acoustic emission was pioneering in the field of acoustic emission and they predicted the application of acoustic emission technology, which is considered to have a promising future in solving engineering problems [5]. In 1959, Rusch first investigated the characteristics of acoustic emission in concrete and found that the Kaiser effect similar to that of metals also exists in concrete.

Into the 1960s, Americans Dunegan played a significant role in promoting the improvement of acoustic emission technology. He was the first to apply acoustic emission technology to the inspection of pressure vessel. In the early 1970s, Dunegan and others carried out the development of modern acoustic emission instrumentation, they increased the test frequency to 100 kHz–1 MHz. This was a major progress in acoustic emission test technology, enabling acoustic emission technology to transition from the laboratory to field inspections.

With the emergence of modern acoustic emission instruments, in the 1970s and early 1980s, extensive and in-depth systematic research was carried out on acoustic emission source mechanism, wave propagation and acoustic emission signal analysis. In the early 1980s, the U.S. PAC company introduced modern microprocessor technology into the acoustic emission detection system, and designed the second-generation source positioning acoustic emission detection instruments with smaller volume and weight, and developed a series of multifunctional advanced detection and data analysis software. Through the microprocessor control, the components can be detected for real-time acoustic emission source positioning monitoring and data analysis. Due to the second generation of acoustic emission instruments are small, lightweight and easy to carry, which greatly promoted the wide application of acoustic emission technology in the field detection [6]. Leveraging the powerful capabilities of modern computers and fully utilizing their hardware and software platforms, the digital signals of acoustic emissions were recorded onto computer hard drives at high frequencies and high sampling rates. The acoustic emission signals are collected, recorded and analyzed with high speed, all-digital and all-waveform.

Regarding the application of acoustic emission technology in offshore platforms, foreign scholars have conducted extensive applied research on the defects such as cracks, fatigue, corrosion in steel structures. In 1976, Exxon Nuclear Energy Company in the United States was the first to use acoustic emission detection systems for offshore underwater inspection activities. In the1980s, the Norwegian Nork Hydro Research Centre applied AE technology for many years for the monitoring of offshore structures, and research on the drilling platform's lateral stubs and risers crack test verification, as well as the monitoring of underwater pipe nodes. Rogers proposed a method for remote monitoring of cracks based on the use of acoustic emission technology to monitor fatigue-prone points and defective welds on offshore platforms.

Rogers et al. [7] verified that in fatigue damage, the amplitude of acoustic emission in the crack growth phase is higher than that in the crack closure phase. Roberts and Talebzadeh [8] carried out acoustic emission monitoring during fatigue crack expansion of a material, explored the relationship between crack extension characteristics and acoustic emission signals during fatigue loading.

In recent years, with the progress of signal acquisition and analysis technology, as well as the introduction of neural networks, wavelet analysis, pattern recognition and other technologies, have further propelled acoustic emission technology towards broader and deeper development.

4.3.2 Sonar Technology

1. History of sonar development abroad

In 1827, the Swiss physicists Daniel C and Charles S collaborated to accurately measure the speed of sound underwater [9], which allows for precise calculation the distance to underwater target. In the mid-nineteenth century, the carbon microphone, one of the earliest hydrophones, was invented [10]. The precise measurement of the speed of sound underwater and the invention of the hydrophone laid the foundation for the development of hydroacoustics.

Sonar was invented in 1906 by the British Navy's Lewis Nixon, who invented the first sonar instrument, a passive listening device used primarily to detect icebergs. This technology was used on the battlefield in the First World War to detect submarines hidden underwater. These sonobuoys can only passively listen to the sound, and are passive sonobuoys, or "hydrophones". After the Titanic collided with an iceberg and sank in 1912, some scientists began to research the location of iceberg echoes. In 1914, Fessenden in the United States designed and manufactured an electro-dynamic underwater acoustic transducer, which detected an iceberg two nautical miles away, marking the birth of hydroacoustics. The outbreak of First World War In 1914 greatly promoted the research and development of both civilian and military sonar [9]. The first anti-submarine sonar was introduced in the first world war, but at that time, due to theoretical and technical imperfections, the performance of this hydroacoustic echolocation system was very unreliable. In 1916, Langevin proposed using piezoelectric quartz to generate ultrasonic waves, and achieve the "echo-location" through the vacuum tube amplification technology. In 1918, he used this technology for the first time to detect remote targets in the water, thus inventing the world's first practical active sonar.

Subsequently, echo sounding equipment was used to make marine echosounders, which increased people's confidence in the application of sonar technology for military and civilian purposes. Around 1925, the German "Signal" company named its production of sonar equipment as "sounder", and in the United States and Britain have commercial sales. At the same time, the leaders of the U.S. Naval Laboratory and its members were actively improving the method of echolocation of submarines,

4.3 Development Status of Foreign Acoustic Technology

and they found a suitable transmitting transducer for echolocation by using a magnetostrictive transducer. At the same time, the development of electronics has made it possible to amplify sonar information for simple processing and display. In 1935, Germany, Britain and the United States developed several more practical sonobuoys, and the United States began mass production of sonobuoy equipment in 1938. By the Second World War, almost all military ships were equipped with sonar systems and played a very important role in naval warfare. After the Second World War, military sonar technology continued to develop, but each country classified the latest technology in this area as strictly confidential.

After the 1970s and 1980s, with the rapid development of marine development, sonar technology has been transformed to civilian use at an alarming speed, and various modern sonars for various purposes have appeared, such as navigation sonar, communication sonar, side-scanning sonar, long-range warning sonar, hydroacoustic countermeasures sonar, towed array sonar, torpedo self-guided sonar, mine self-guided sonar and so on, and the sonar technology has become more and more mature and perfect.

2. Areas of application of sonar technology

Sonar is an electronic device that uses sound waves in the water to detect, locate and communicate with underwater targets, and it is the most widely used and important device in hydroacoustics. Acoustic waves are known to mankind so far can be in seawater remote propagation form of energy, sonar marine detection has an extremely wide range of applications. The application fields of sonar mainly include: military field, ocean mapping field, current velocity measurement, marine fishery, underwater acoustic positioning and hydroacoustic communication and other fields.

(1) Application in Military Field

Sonar is the main technology used by navies for underwater surveillance, which is used to detect, classify, locate and track underwater targets, carry out underwater communication and navigation, and safeguard the tactical manoeuvres of ships, anti-submarine planes and anti-submarine helicopters, as well as the use of underwater weapons. With the development and progress of modern sonar technology, the new generation of sonar has more advanced detection performance and longer detection distance, and some high-tech sonars also have quite high resolution, capable of identifying frogmen and suspicious underwater bodies.

(2) Ocean mapping

With the intervention of marine high-tech and the continuous upgrading of equipment, the underwater topographic acoustic detection technology has gained rapid development, and has now become one of the important research fields of marine mapping in the marine countries of the world. The equipment for marine mapping using sonar technology includes: single-beam echo sounder, side-scan sonar, multibeam bathymetry, and shallow stratigraphic profiler.

(3) Measurement of Current Velocity

Modern sonar technology can use the Doppler effect for flow velocity measurement, this sonar system uses a pair of directional transducers mounted on the bottom of the ship tilted downward, by the Doppler shift in the seafloor echo can be obtained from the ship relative to the seafloor speed. On the other hand, if the sonar is fixed in a flowing sea, it can automatically detect and record the speed and direction of seawater flow.

(4) Marine fisheries

Fish finder is a sonar system that can be used to discove the movement of fish, the location and range of fish, using it can greatly improve the production and efficiency of fishing; fish-assisting sonar equipment can be used for counting, baiting, catching, or tracking and trailing a particular fish. Acoustic barriers have been used in mariculture farms to prevent shark invasions, as well as to discourage the escape of lobster fish.

(5) Underwater acoustic positioning

The underwater environment in the ocean is complex and harsh, so accurate positioning for underwater operating equipment is very important for mastering the working condition of the equipment and recovering marine monitoring data and equipment. Underwater acoustic positioning technology appeared early, the development speed is fast, now has been widely used in all aspects of marine engineering, applicable to submarine markers, seabed base and underwater deep submersible and other underwater operational equipment. As the acoustic release is integrated in the submarine marker system and seabed base system, it can complete the ranging work in the positioning operation without the need for additional installation of special hydroacoustic communication devices, so this technology has an inherent hardware advantage in the positioning application of submarine markers and seabed bases.

When detecting, searching and locating underwater target sound sources, setting up acoustic base arrays is one of the most widely used underwater positioning techniques. According to the length of the positioning baseline of the operating system, it can be divided into long baseline array (LBL), short baseline array (SBL) and ultra-short baseline array (SSBL/USBL). It mainly consists of an acoustic communication device and its deck unit, a shipboard depth sounder, GPS and underwater target positioning measurement software based on the VB platform.

(6) Hydroacoustic communication field

Hydroacoustic communication is an important means of mutual communication between surface ships, submarines, the use of sonar systems underwater can replace the wire connection, the use of acoustic beams to transfer information to achieve communication and exchange between ships. The working principle of the hydroacoustic communication system begins with converting information such as text,

4.3 Development Status of Foreign Acoustic Technology

voice, images and other information into electrical signals, which are then digitally processed by an encoder. Subsequently, an energy transducer converts electrical signals into acoustic signals. Acoustic signals propagate through the water as a medium, transmitting the information to the receiving transducer, where the acoustic signals are converted back into electrical signals. The electrical receivers then further convert these signals into sound, text and images. Sound is produced by vibration, and to transmit communication information over long distances in the ocean, air is replaced by seawater. Acoustic signals of different frequencies and intensities can be emitted and received in air, water, and solids.

When the submarine is submerged, radio and other means of communication are disabled and the only possible means of communication is hydroacoustic communication. Hydroacoustic communication is also used for the command and data transmission of underwater deep submarine, including the state control of the underwater robot and the state answer of the underwater robot, the data return of the underwater acquisition system or the acquisition of the image of the deep-sea target and so on. And in the waters near China's continental shelf and far sea area, establishing a reliable and wide-ranging underwater acoustic communication network will undoubtedly play a crucial role in safeguarding China's territorial sea defense and supporting future long-range naval operations.

3. Development trend of sonar technology

In the context of the high development of information technology, the future development of sonar will also be towards the direction of high precision, intelligence, integration, multi-data fusion, multifunctionality and so on. It is mainly reflected in the following aspects:

(1) Fully adaptive intelligent cognition

The traditional active sonar system does not consider the influence of the environmental information perceived by the sonar receiver and the a priori knowledge of the target characteristics on the transmitter when dealing with the target reflected echo, and the parameters of the transmitted signal are fixed. Therefore, it is difficult to obtain ideal detection results in complex underwater environments such as transmission attenuation, noise, reverberation, multipath, time-varying and large Doppler. Intelligent cognitive sonar based on knowledge theory can improve the detection and recognition of underwater target signals by joint adaptive control of transmitter and receiver based on the a priori knowledge of environmental changes and target characteristics.

Inspired by the rapid development of cognitive radio and cognitive radar in recent years, cognitive sonar is proposed by introducing a priori knowledge and continuous learning into the traditional sonar system, and establishing adaptive feedback control on the transmitter, the composition of which is shown in Fig. 4.2. A dynamic closed-loop system is formed between the transmitter and receiver, the environment and the target, which can dynamically adjust the transmit beam, power, frequency, re-frequency, intra-pulse modulation, and receive beam, detection threshold, and

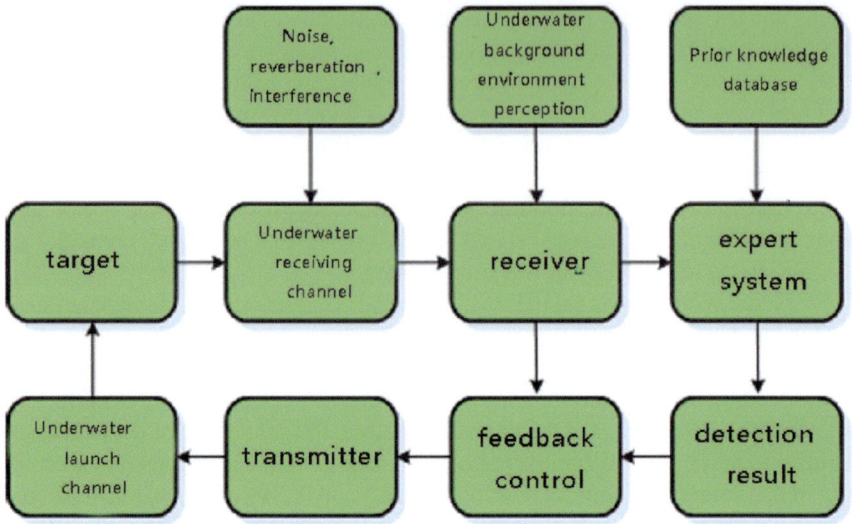

Fig. 4.2 Schematic of cognitive sonar

operating mode according to the environmental changes, performance requirements and a priori knowledge.

Cognitive sonar matches transmitter, receiver and environment adaptive matching according to the learning of the working environment and target information, constantly updating the receiver and adaptively adjusting the transmitter. The transmitter intelligently adjusts the parameters of the transmitter waveform according to the target distance, size for illumination. The entire cognitive sonar system constitutes a closed feedback loop of the transmitter, receiver and the environment. It leverages prior information about the environment and target to improve the performance of the sonar system.

(2)   Co-address and distributed MIMO sonar

MIMO technology is firstly applied in the field of communication and radar, and is divided into co-address MIMO and distributed MIMO. Co-address MIMO uses the diversity characteristics of the transmitted signal to expand the virtual aperture of the transceiver array and improve the target detection capability. Distributed MIMO arrays are arranged separately, transmitting orthogonal signals, irradiating targets from different angles, reducing undulation fading and improving detection stability. Underwater, especially offshore vessels, the number of large, noisy, complex acoustic field, multipath and Doppler effect is serious, the detection of mines, frogmen, silent submarines and other weak targets is difficult, the traditional active and passive radar are difficult to achieve the desired results, MIMO sonar provides a new way to solve this problem.

In 2006, Bekkerman and Tabrikian [11] proposed a processing framework for MIMO sonar target detection and localization, demonstrating that the introduction of virtual arrays by transmitting orthogonal waveforms can improve the target detection capability and deduce the generalized likelihood ratio detector and lateral CRBB performance limits. Li et al. [12] proposed the MIMO sonar processing model in 2008 and compared the performance with single-input–single-output respectively, single-input-multiple-output, and multiple-input–single-output processing, respectively. In 2009, Vossen et al. [13] improved the target detection capability by introducing virtual source information. In 2010, Song et al. [14] reduced the inter-correlation requirement between orthogonal transmit signals for MIMO processing by using space–time coding technique.

(3) Wide-area heterogeneous multi-sensor joint perception

The efficiency of single sensor detection is low, it is difficult to meet the wide range, long time underwater information acquisition needs, through the network technology will be vigilant monitoring of the sea area of a number of different locations of sonar, radar, laser, infrared and other sensors deployed for interconnection, to achieve the exchange of data, distribution and convergence, centralized or distributed data processing, you can form a distributed networked underwater vigilance detection system, to achieve the coverage of target detection, location, tracking and classification identification function. Distributed networked underwater warning and detection system has the advantages of mobility and flexibility, low cost and high cost-effectiveness ratio, and can effectively enhance the information perception capability of underwater battlefield.

In recent years, to counter potential submarine threats and mine threats in shallow and coastal waters, the United States has further developed the underwater detection system represented by the Seaeb, combined with the Distributed Agile Submarine Hunting DASH, Deployable Autonomous Distributed System DADS, and helicopter anti-submarine systems. This combination aims to achieve large-area underwater perception, and progress towards the Cross-Domain Maritime Awareness and Targeting System (CDMaST). The goal is to establish a cross-domain, distributed system for detection, identification, location, strike, and assessment in high-confrontation environments, utilizing radar, electro-optical, and sonar detection equipment from manned and unmanned systems underwater, at sea, and in the air, thereby enhancing operational effectiveness, see Fig. 4.3.

4.4 Current Status of Development Status of Domestic Acoustic Technology

Acoustic emission technology was introduced into China in the early 1970s, and it was hoped that acoustic emission would be used to forecast and measure crack opening points. Some research institutes and universities, such as the Shenyang Institute of Metals of the Chinese Academy of Sciences, the Beijing Institute of Aeronautical Materials, the Hefei Institute of General Mechanics of the Ministry of

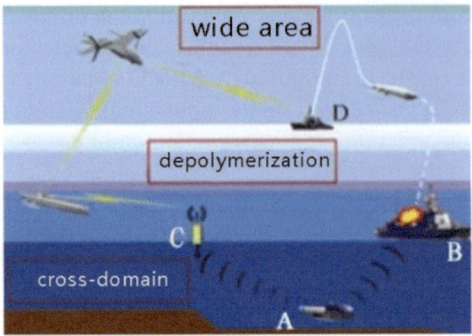

a) Schematic diagram of the cross-domain collaborative warfare system

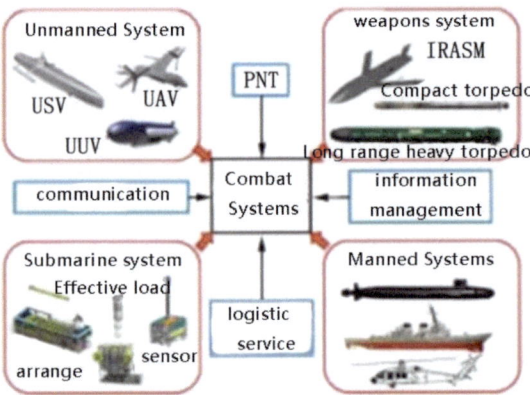

b) Components of a cross-domain collaborative warfare system

Fig. 4.3 U.S. army CDMaST platform, sensor, weapon, communication, and navigation collaborative warfare system architecture

Mechanics and Wuhan University, have carried out research on the acoustic emission properties of metals and composites.

In the early 1980s, China began to experiment with acoustic emission technology in engineering fields such as pressure vessel inspection. However, due to the limitations of the acoustic emission instruments and signal processing at that time, as well as the lack of understanding of transmission characteristics of the acoustic emission source and the process of acoustic emission waves reaching sensors, there were many issues with the repeatability and reliability of experimental results. Consequently, acoustic emission technology fell into a lull for a period.

In the 1980s, the Boiler and Pressure Vessel Inspection and Research Centre of the Ministry of Labour of the People's Republic of China took the lead in introducing the world's most advanced SPARTAN source-locating acoustic emission detection and

4.4 Current Status of Development Status of Domestic Acoustic Technology

signal processing analysis system from the PAC Corporation in the United States. They successfully conducted inspections on spherical tanks, horizontal tanks, and other pressure vessels in petrochemical and gas companies. Subsequently, many institutions such as the Wuhan Safety and Environmental Protection Research Institute of China Steel Corporation, Daqing Petroleum Institute, the Fourth Research Institute of the Aerospace Ministry, the 44th Research Institute in Xi'an, and China University of Petroleum also imported advanced acoustic emission instruments, including SPARTAN and LOCAN models, from PAC. These institutions carried out detection applications in pressure vessels, airplanes, metallic materials, composite materials, and rocks.

The fourth national acoustic emission conference in 1989 pointed out that from the 1990s onwards, the research and application of acoustic emission technology in China has entered a stage of rapid development. In the early 1990s, Yanshan Petrochemical, Tianjin Petrochemical, Daqing Oilfield, Shengli Oilfield, Liaohe Oilfield and Shenzhen Boiler and Pressure Vessel Inspection Institute of petroleum and petrochemical enterprises, such as the inspection unit and the professional inspection office imported a large-scale acoustic emission instruments and widely carried out pressure vessel inspection. In the mid-1990s, Beijing Aeronautical Engineering Technology Research Centre and Beijing Institute of Materials and Technology introduced the third generation of Mistras 2000 multi-channel acoustic emission instrument, which can store acoustic emission signal waveforms, from PAC Company of the United States. Thereby they initiating acoustic emission detection and signal processing of aerospace equipments based on waveform analysis.

In 2002, the Boiler and Pressure Vessel Inspection and Research Center of the General Administration of Quality Supervision, Inspection and Quarantine of China introduced the latest model of ASM5 36-channel acoustic emission instrument from VALIEN Company in Germany. This instrument can not only perform waveform-based pattern recognition analysis on acoustic emission signals but also detect leaks at the bottom of large atmospheric oil tanks. Currently, acoustic emission technology has been widely researched and applied in various fields in China, including petroleum, petrochemicals, electric power, aviation, aerospace, metallurgy, railways, transportation, coal, construction, machinery manufacturing and processing.

In the development and production of acoustic emission instruments, China's start was not too late. The Shenyang Institute of Electronics developed a single-channel acoustic emission instrument in the late 1970s. The Changchun Institute of Testing Machinery also developed a 32-channel acoustic emission positioning and analysis system using microprocessor control in the mid-1980s. The Boiler and Pressure Vessel Inspection and Research Centre of the Ministry of Labour successfully developed the world's first multi-channel (2–64) acoustic emission detection and analysis system using PC-AT bus for hardware and WNDOWS interface for software in 1995. In 2000, Guangzhou ShengHua Company developed a full-waveform, fully digital multi-channel acoustic emission detection and analysis system based on the full-scale programmable integrated circuit (FPGA) technology. In 2002, the Boiler and Pressure Vessel Inspection and Research Center of the General Administration of Quality Supervision, Inspection and Quarantine of China developed a fully digital multi-channel acoustic emission detection and analysis system based on signal processing integrated circuit technology.

China's acoustic emission technology has been successfully applied to pressure vessels, pipelines, metallic materials, non-metallic materials, aircraft, and marine platforms made of reinforced concrete materials. Geng Rongsheng of Aerospace Power Research Institute used acoustic emission technology to track and detect the formation and expansion of fatigue cracks in aircraft fatigue tests, and timely forecast the expansion of fatigue cracks in the aircraft spacer frames and bolt holes of the main beams. Guodong and Zhiming [15, 16] of Beijing Jiaotong University analyzed the acoustic emission characteristic parameters of fatigue test of 16MnR steel material and established an acoustic emission assessment model for the damage degree of 16MnR steel material.

Domestic scholars' research on acoustic emission detection of large bridge structures mainly focuses on the study of key metal components of bridges, such as steel strands, diagonal cables and booms. Li Dongsheng and Jinping [17] carried out monitoring tests on the stretching of steel strands in bridge tension cables and prestressed concrete structural systems through acoustic emission technology, and used the monitoring data to judge the moment and number of roots at which the breakage occurrs. The study showed that the characteristic parameter of strand damage can be expressed by AE cumulative energy, and established the Weibull cumulative distribution function equation that can be used to express the tensile damage of strand by defining the strand damage factor with the characteristic parameter of AE. Dongsheng et al. [18] also used AE technology to comprehensively monitor the suspension of Sichuan Ebian Dadu River Arch Bridge, and determined its damage using AE signal parameter analysis. In addition, they also carried out acoustic emission detection of corrosion fatigue of cable-stayed cables of a large-scale cable-stayed bridge of multiple ages in China, and determined three stages of cable-stayed fatigue damage through the cumulative energy of AE energy diagrams.

In the research of acoustic emission from marine platforms, some domestic scholars have also carried out a lot of research. For example, Guang et al. [19] carried out a study on the propagation characteristics of acoustic emission by seawater salinity. Wensheng et al. [20] obtained the characteristic information of different stages of material fatigue damage by constructing a bending fatigue test platform for ocean platform materials and using acoustic emission characteristic parameters and wavelet analysis technology. Hua et al. [21] applied the acoustic emission technology to real-time monitoring of the structure of the conduit rack ocean platform and proposed the structure based on the test data health assessment method. Li and Deyou [22] used recognition algorithm to locate the acoustic emission signals, and established the acoustic emission signal recognition system of conduit ocean platform based on the local wave method. Hongtao et al. [23] fused the acoustic emission signals with multiple signal processing methods, and established a system for rapid recognition of the structural health condition of ocean platform.

4.4 Current Status of Development Status of Domestic Acoustic Technology

4.4.1 Sonar Technology

The research on underwater acoustics in China started relatively late, and relevant units for underwater acoustics research, design, and production were established in the Chinese Academy of Sciences and other relevant ministries and academies only in 1958. The achievements of China's hydroacoustic research are firstly manifested in the shallow sea, where the seabed reflection loss model has been theoretically established by three methods. The representative one is established by Zhang Renhe according to the ray-simple positive wave theory in 1965. He established the universal formula linking the sound line span, group velocity and simple positive wave undersea reflection attenuation. In the acoustic field research, the inversion point dispersion problem adapted to both deep and shallow sea has been theoretically solved, which is a breakthrough in the acoustic field computation leading foreign research. Theoretical descriptions similar to these were not formally published abroad until 1974. China conducted a series of studies in the shallow sea field from the 1970s to the 1980s, such as the multi-way structure or multi-way drag dispersion and correlation loss problem of the shallow sea signal, and the problem of the extraction method of the simple positive wave at sea, the reverberation strength in the shallow sea, and the relationship between the simple positive wave and the ray expression of the acoustic field, and so on.

Deep-sea research in China began in the late 1970s. The theoretical representation of the gain, width and position of the deep-sea zone was given on the basis of the results of the divergence problem at the inversion point. China has made a number of achievements in deep-sea research theory, such as deep-sea acoustic field prediction research, acoustic field matching processing and other achievements were once in the international leading level. The resonance bucket method was used to study the problem of seawater absorption, and solved the difficult problem of accurate measurement of acoustic absorption of the low kilohertz band. In the field of nonlinear underwater acoustics, solved the issue of parametric array sound beams not following Snell's law after penetrating the water–sediment interface. In terms of target scattering, research was conducted on resonant scattering from finite-length elastic cylinders.

China has made a number of achievements in hydroacoustic theory research, especially in shallow sea theory, and has also made many important achievements in sonar technology research. There is still a big gap between China and the world's advanced level in terms of sonar technology and equipment, which is mainly manifested as follows:

(1) The research and development of core key technology is backward, the reliability is lower than that of imported products, and thus can only dominate the low-end market;
(2) The main research institutions of sonar and other technical equipment are universities and research institutes, which are not demand-driven, the industrialization level is low, and the technology research and development and market mechanism are not effectively combined.

The world's first side-scan sonar system was successfully developed by the British Institute of Oceanography in the 1960s. China's side-scan sonar research began in the 1970s to 1980s, the representative products are SGP type side-scan sonar of South China University of Technology and CS-1 type side-scan sonar of Institute of Acoustics, Chinese Academy of Sciences. After decades of development, the side-scan sonar technology has been more mature, but there are still some shortcomings compared with the international advanced equipment: it can only obtain the data of the relative undulation of the seabed, but can not obtain the accurate bathymetric data; the lateral resolution depends on the size of the sonar base array.

The research of multibeam bathymetric technology originated from a military research project of the US Navy in the 1960s. The world's first multibeam bathymetric sonar was developed in the 1970s, based on echo sounder. In China, multibeam bathymetric sonar research began in the mid-1980s, and the first experimental prototype was jointly developed by the Institute of Acoustics of the Chinese Academy of Sciences and the Tianjin Surveying and Mapping Institute at the end of the 1980s. The first sonar products was jointly developed by the Harbin Engineering University and the Tianjin Marine Surveying and Mapping Institute in 1998. Since the twenty-first century, with the support of national projects such as the "863" Program, Harbin Engineering University, the Institute of Acoustics of the Chinese Academy of Sciences, the 715th Research Institute of China Shipbuilding Industry Corporation, Zhejiang University, and other institutions have researched and designed multiple prototypes and products. At present, China's shallow-water multibeam bathymetric sonar has completed the development of a variety of products, while the deep-water multibeam bathymetric sonar is still in the stage of experimental prototypes, and has not been commercialized.

Before the 1990s, international research on synthetic aperture sonar was mainly in the theoretical research and experimental stage, and only a few organizations carried out experimental research. China began researching synthetic aperture sonar (SAS) technology in the 1990s. With the support of the national "863" Program, the Institute of Acoustics of the Chinese Academy of Sciences and the 715th Research Institute of China Shipbuilding Industry Corporation jointly developed a lake test prototype of SAS in 1997. In 2005, China successfully conducted sea trials of its first SAS with independent intellectual property rights. Currently, the leading domestic product is the SAS series produced by Suzhou Sangtai Marine Instrumentation Company, and the related technology has reached the international advanced level.

References

1. Turner, J.D., A.J. Pretlove. (1991). *Acoustics for Engineers*. Macmillan Education.
2. Xiangjun, Tan. (2018). *Learn NVH from Here: An Introduction and Advancement Guide of Noise, Vibration, and Modal Analysis*. Beijing: Machinery Industry Press.
3. Lemaster, R.A., K.E. Graff. (1978). Influence of Ceramic Location on Highpower Transduoers Performance. In *IEEE Ultrasonics Symposium Proceedings*, 296–299.

References

4. Tatro, C. A. 1957. Sonic Technique in the Detection of Crystal Slip in Metal. *Engineering Research* 1: 23–28.
5. Tatro, C.A., R.G. Liptai. (1962). In *Proceedings Symposium on Physics and Non-destructive Testing*. San Antonio: Southwest Research institute.
6. Gongtian, Shen, Dai Guang, and Liu Shifeng. (2003). Advances in Acoustic Emission Detection Technology in China—25th Anniversary of the Establishment of the Society. *Nondestructive Testing* 06:302–307.
7. Rogers, L. M., J. P. Hansen, and C. Webborn. 1980. Application of Acoustic Emission Analysis to the Integerity Monitoring of Offshore Steel Production Platforms. *Materials Evaluation* 38 (8): 39–49.
8. Roberts, T. M., and M. Talebzadeh. 2003. Fatigue Life Prediction Based on Crack Propagation and Acoustic Emission Count Rates. *Journal of Constructional Steel Research* 59 (6): 679–694.
9. Binghe, Wang, and Li Hongchang. 2001. Application and Latest Progress of Sonar Technology. *Physics* 30(8): 492–493.
10. Robert, J.U. 1990. *Principles of Underwater Sound*. Hong Shenze, Translation. Harbin: Harbin Engineering University Press.
11. Bekkrman, I., and J. Tabrikian. 2006. Target Detection and Localization Using MIMO Radars and Sonars. *Signal Processing IEEE Transactionson* 54 (10): 3873–3883.
12. Li, W.H., G. Chen, and E. Blash. 2009. *Cognitive MIMO Sonar Based Robust Target Detection for Harbor and Maritime Surveillance Applications*. Aerospace Conference IEEE, 7–14.
13. Vossen, R.V., L.T. Raa, and G. Blacquiere. 2009. Acquisition Concepts for MIMO Sonar. *Underwater Acoustic Measurments Proceedings*.
14. Song, X. F., S. L. Zhou, and P. Willett. 2010. Reducing the Waveform Cross Correlation of MIMO Radar with Space-Time Coding. *IEEE Transactions On Signal Processing* 58 (8): 4213–4224.
15. Guodong, Qin, and Liu Zhiming. 2004. The Development of Acoustic Emission Testing System. *Journal of Test and Measurement Technology* 18(3): 274–279.
16. Guodong, Qin, and Liu Zhiming. 2005. Development of LOCAN320 Data Format Recognition and Conversion Processing System. *Non-destructive Testing* 27(1): 12–14.
17. Dongsheng, Li, and Ou Jinping. Acoustic Emission Characteristics and Damage Evolution Model During the Tensile Process of Steel Strand. *Science and Technology of Highway and Transport* 24(9): 57–60.
18. Dongsheng, Li, Yang Wei, and Yu Yan. 2017. *Acoustic Emission Monitoring and Evaluation of Structural Damage in Civil Engineering: Theory, Methods and Applications*. Beijing: Science Press.
19. Guang, Jia, Yang Guoan, Shen Jiang, et al. 2013. Study on the Influence of Seawater on the Propagation Characteristics of Acoustic Emission from Offshore Platforms. *The Ocean Engineering* 31(3): 84–88.
20. Wensheng, Qu, Wang Shoujun, Mu Weilei, et al. 2016. Fatigue Damage Detection of Offshore Platform Materials Based on Acoustic Emission Technology. *Nondestructive Testing* 38(10): 10–13.
21. Hua, Zhang, Lv Tao, Xu Changhang, et al. 2016. Real-Time Structural Health Monitoring of Jacket Offshore Platforms Based on Acoustic Emission. *China Offshore Platform* 31(1): 86–90.
22. Li, Lin, and Zhao Deyou. Acoustic Emission Signal Recognition System for Jacket Offshore Platforms. *Non-destructive Testing* 31(1): 42–45.
23. Hongtao, Li, Liu Yue, Xu Changhang, et al. 2013. Experimental Study on Damage Localization Method for Offshore Platforms Based on the Fusion of Vibration and Acoustic Emission Information. *Natural Gas Industry* 33(4): 120–124.
24. Schofield, B.H. (1963). *Acoustic Emission Under Applied Stress*. Lessells and Associates Incwaltham ma.
25. Rusch, H. (1959). Physical Problems in Testing of Concrete. *Zement-Kalk-Gips(Wies)* 12(1).
26. Mengyuan, Li. (2010). *Acoustic Emission Testing and Signal Processing*. Beijing: Science Press.

Chapter 5
Principles of Acoustic Technology

5.1 Acoustic Emission Technology Principle

Acoustic Emission (AE) refers to a common physical phenomenon in which a solid material, under the action of external forces or environmental factors, produces elastic–plastic deformation, cracking, phase transformation and magnetic effects, accompanied by a rapid release of energy in the form of stress waves. Therefore it is also known as Stersswave emission [1–3]. The generation of stress waves leads to changes in the stress field, and these changes are also recorded in a structural evaluation and diagnosis monitor, which are also used for structural evaluation and diagnosis. Therefore, the acoustic emission meter is also known as the 'stethoscope' of the structure.

Acoustic emission is a common phenomenon in nature. For example, acoustic emission occurs when cracks are formed or cracks are extended in the steel structure of an offshore platform; acoustic emission occurs when cracks are formed in the reinforced concrete components of engineering structures, when they break and when they are fatigued; acoustic emission occurs when geological movement of the earth's crust occurs (e.g., earthquakes); and acoustic emission occurs when a tree is broken. However, different acoustic emission phenomena and processes produce acoustic emission wave frequency and amplitude varies greatly, its frequency range from infrasound, audible sound to about 50 MHz ultrasonic, amplitude can be from a few microvolts to hundreds of volts. Acoustic emission in general engineering differs from audible sound in that it refers to 'stress wave emission'. When materials and structures are subjected to stress, elastic deformation begins to occur in the form of elastic strain energy stored in the material, resulting in microstructural changes within the material, leading to a concentration of localised stress, resulting in an unstable stress distribution. When this unstable stress distribution in the structure accumulates to a certain extent, the unstable high-energy state must be transitioned to a stable low-energy state, the material appears to be a rapid phase transition, cracking and other phenomena, and in the process of the release of strain energy, which is the reason for the acoustic emission phenomenon [4].

In 1950, the German scholar Josef Kaiser carried out an exhaustive study on the acoustic emission phenomenon of a variety of metal materials, and discovered the irreversible effect of acoustic emission in the process of material deformation, which is the Kaiser effect, laying the foundation for the acoustic emission research. In 1959, Rusch carried out a study on the acoustic emission signals of the concrete subjected to force, and confirmed that in concrete materials, the Kaiser effect exists only in the range of below 70–85% of the ultimate stress [5]. In the 1960s, acoustic emission technology has been further developed in the United States, Germany. Dunegan discovered the significant advantages of acoustic emission technology in the detection of pressure vessels, raising the acoustic frequency to the ultrasonic range ultrasonic range (100 kHz–1 MHz) for the first time, which greatly reduced the background noise and ushered acoustic emission technology into a practical stage. In the 1970s, acoustic emission detection instrumentation system achieved computerized automatic monitoring and data processing, greatly expanding the field of application of acoustic emission technology. Acoustic emission technology was advanced rom the laboratory materials research stage to the engineering field of large-scale component integrity monitoring. At the same time, commercial acoustic emission detection instruments began to appear, further promoting the research on acoustic emission propagation theory, acoustic emission sensor calibration theory and experimental techniques, and so on. From the late 1980s to the present, people focus more on the application of technology research and the development of instrumentation, basic research involving acoustic emission mechanism is relatively less. However, with the development of computer technology and modern signal analysis technology, as well as artificial intelligence and pattern recognition technology, acoustic emission technology has been widely used in machinery, aerospace, civil engineering, water conservancy dams, offshore platforms, geotechnical, petroleum and other engineering fields of detection and monitoring.

This chapter mainly introduces the basic concepts of acoustic emission technology, the principle of acoustic emission and the main methods of signal data processing as well as typical acoustic emission products.

5.1.1 Basic Concepts

1. Basic Acoustic Emission Terminology

(1) Acoustic emission

Acoustic emission refers to the local area of stress concentration in the material, the rapid release of energy and produce transient elastic wave phenomenon, it is known as stress wave emission.

5.1 Acoustic Emission Technology Principle

(2) Acoustic emission source

The localized change in a material that causes acoustic emission is known as an acoustic emission event, while the acoustic emission source refers to the physical origin or the mechanism source of the acoustic emission wave that occurs during the event [6].

(3) Acoustic emission detection technology

Acoustic emission detection technique is a non-destructive testing (NDT) technique. The technique of detecting, recording and analyzing acoustic emission signals with an acoustic emission detector and inferring the source of acoustic emission using acoustic emission signals is called acoustic emission detection technique. We can receive and analyze these acoustic emission signals for the purpose of detection and diagnosis. Acoustic emission detection is a dynamic non-destructive testing method, and its signals come from the defects themselves, so the activity and severity of the defects can be judged by acoustic emission detection method.

(4) Acoustic emission detection system

Acoustic emission detection system usually consists of sensors, preamplifiers, data acquisition and processing system and record analysis and display system of four parts (see Fig. 5.1). In acoustic emission equipment, sensors receive and collect acoustic wave signals from the acoustic emission source, known as acoustic emission signals. These signals are amplified by a preamplifier and processed by a signal acquisition and processing system before being recorded, analyzed, and displayed by a recording and display system, achieving the purpose of detecting the acoustic emission source. Sometimes, certain components of the acoustic emission detection system are combined together, such as acoustic emission sensors with built-in amplifiers, or handheld acoustic emission instruments that integrate amplifiers, data acquisition and processing, recording, analysis, and display into a single unit.

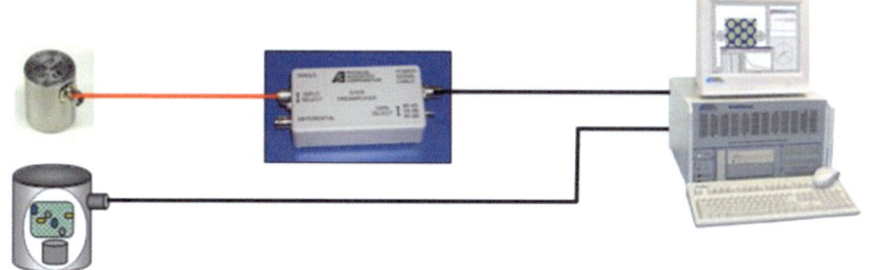

Fig. 5.1 Acoustic emission detection system

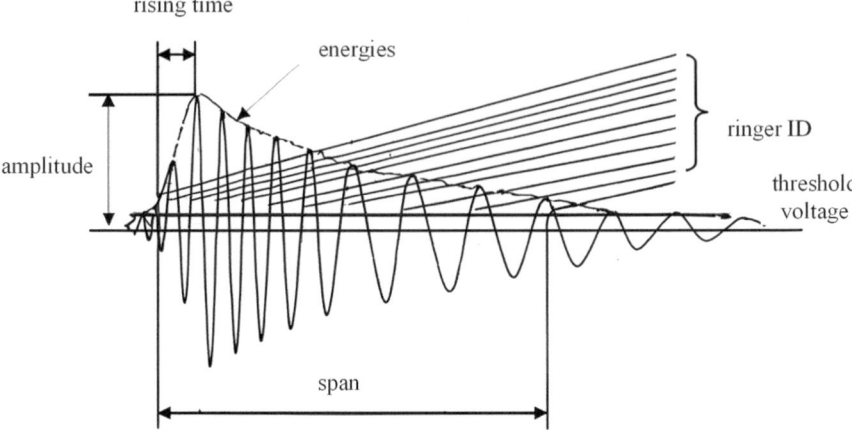

Fig. 5.2 Definition of burst-type signal characteristic parameters

2. Acoustic emission signal related terms

Acoustic emission detection usually involves analyzing the characteristic parameters of the collected signal, which include amplitude, energy, ringing count, duration, rise time, intensity, etc. There are two main types of acoustic emission signals: burst and continuous. The characteristics of a typical burst type model are shown in Fig. 5.2.

(1) Burst type signal

When an acoustic emission signal occurs at a low frequency, the waveform of each signal can be separated individually in the time domain, and this type of signal is a burst-type signal.

(2) Continuous signal

When the frequency of acoustic emission signals is so high that the signals cannot be separated from each other, forming a continuous e signal.

(3) Hits

Any signal that exceeds the threshold and makes a certain channel to acquire data is called a hit, which can be measured by the total number of counts and the count rate. It reflects the total number and frequency of acoustic emission activities, and is often used as an indicator of acoustic emission activity evaluation.

(4) Event

A local change in material that triggers acoustic emission activity is called an acoustic emission event, which can be measured by the total number of counts and count rate. It reflects the total number and frequency of acoustic emission events, and is

5.1 Acoustic Emission Technology Principle

often used as an indicator evaluating the activity and localization concentration of acoustic emission sources, which is easily affected by the sample geometry, sensor characteristics and coupling conditions.

(5) Ringing Counts (Counts)

The number of oscillations of an acoustic emission signal that exceeds the threshold value can be measured by total counts and count rate. It Can roughly characterize the signal intensity and frequency of the signal, is easy to handle, and is widely used as an indicator for evaluating acoustic emission activity. However, it is affected by the threshold value. A signal hit may produce a small number of counts or a large number of counts, depending on the size and shape of the signal.

(6) Amplitude

The maximum peak voltage that obtained from the signal waveform of an acoustic emission event is the absolute value of the peak voltage. Amplitude is an important indicator measuring the size of the signal energy, and it is usually measured in decibels (decibel, dB). Decibel is a relative quantity, and voltage can be converted to decibels using the following equation:

$$A = 20 \log 10 \left(\frac{V}{V_{ref}} \right) \quad (5.1)$$

where A is the amplitude (dB), V is the peak offset voltage, V_{ref} is the reference voltage.

The peak value of the acoustic emission signal can indicate the intensity of the wave source. Since the response of a transducer may be affected by a number of factors, the peak amplitude of a single transducer may not provide meaningful information about the wave source, whereas useful information may be obtained when amplitude-related data are evaluated by statistical methods.

(7) Energy

The area under the envelope of the acoustic emission signal waveform that reflects the intensity of the signal. The energy parameter is often used as an evaluation criterion for acoustic emission testing of materials or structures. However, energy is not sensitive to threshold, frequency and propagation characteristics, and can be used as a substitute for ringing counts and to identify the type of wave source.

(8) Rise time

The rise time of an acoustic emission signal refers to the duration from the signal's initiation to its peak value. Rise time provides stress wave characteristic information similar to duration. However, due to its significant influence by propagation, rise time does not have a clear physical meaning and is sometimes used to distinguish electromechanical noise.

(9) Duration

The duration of an acoustic emission signal is the time interval from when the signal first crosses the threshold to when it finally falls below the threshold. The duration of the signal is affected by the selected threshold voltage. In experiments, other parameters may also affect the duration of the signal, and different acoustic wave sources may result in different signal durations. Mechanical noise sources usually have long duration signals, while electronic pulse signals usually have durations of less than 10 μs. The relationship between duration and amplitude is a key feature of the acoustic emission signal waveforms, and is often used to identify specific types of wave sources and noise.

(10) Arrival time

The time it takes for a signal to travel from its source to a sensor, measured in microseconds, is known as the arrival time. The arrival time can be used to determine the location of the wave source, based on parameters such as the distance between sensors and the wave propagation velocity.

(11) RMS

The root mean square value of the voltage signal within a sampling time interval is denoted as V_{rms}. The effective voltage is related to the size of the acoustic emission, is easy to measure, snd is not affected by threshold value. It is primarily used as an evaluation metric for continuous acoustic emission activity.

(12) Frequency

The number of oscillation cycles per second under fluctuation is referred to as the frequency. Acoustic emission signal waveforms usually contain multiple frequency components, which can be used for identification, separation and noise removal of the acoustic source signal.

(13) Average Signal Level

The average value of the signal level within a sampling time interval, measured in dB, is known as the average signal level. Its role is similar to that of the RMS (root mean square) voltage. It has significant advantages for signals with high amplitude dynamic ranges and low time resolution requirements in continuous signals. It can also be used to measure background noise levels.

(14) Kaiser effect

The Kaiser effect [7] was discovered by the German scholar Kaiser when he studied the acoustic emission properties of metals. When a material is reloaded and no acoustic emission signal is generated until the stress value reaches the previous maximum level. In most metallic materials and rocks, a pronounced Kaiser effect can be obvserved. The Kaiser effect can be used to infer the maximum stress that a material or structure has previously subjected, as well as to monitor the initiation and propagation of fatigue crack.

5.1 Acoustic Emission Technology Principle

(15) Felicity effect

The Felicity effect [8] refers to the phenomenon where acoustic emission signals are detected before the stress level reaches the previous maximum value under already loaded conditions. The Felicity effect provides a more detailed description of the Kaiser effect. The Felicity ratio can be used as an indicator of damage, with a smaller Felicity ratio indicating a higher level of damage growth.

The general definition of the Felicity ratio is:

$$\text{The Felicity ratio} = \frac{\text{Acoustic emission threshold load during reloading}}{\text{Maximum load during previous loading}} \quad (5.2)$$

The Felicity ratio, as a quantitative parameter in acoustic emission monitoring, can effectively reflect the degree damage or structural defects in materials and serves as an important criterion for defect assessment.

3. Concepts related to acoustic emission detection

(1) Broken lead test

When using acoustic emission (AE) equipment for testing, it is typically necessary to determine threshold values, wave velocity measurements, and other parameter settings. Most of the AE acquisition parameter settings and parameters are determined through the lead break test. The lead break test utilizes the breaking of a pencil lead as a simulated pulse sound source to mimic the signals generated by deformation and fracture of structures or materials, and is used for pre-experimental testing and calibration of acquisition parameters in AE testing.

In simulating AE sources, the primary requirements for the simulated source are that the signal should be stable and have a broad frequency spectrum. Research has shown that for burst-type pulses, wave source simulations can be generated by electrical sparks, rupturing of glass capillaries, pencil lead breaking, falling balls, and laser pulses, among others. Among them, the lead break simulation of AE sources has the advantages of simplicity, economy, and good repeatability, and is a commonly used method in current AE testing.

(2) Peak discrimination time (PDT)

Peak discrimination time refers to the maximum peak waiting time interval set to correctly determine the rise time of the impact signal. The general principle of its selection is as short as possible, but setting it too short may mistake high-speed, low-amplitude precursor waves as the main wave. Peak discrimination time is a predetermined time parameter for identifying the true peak point of an acoustic emission waveform, and its main function is to avoid mistaking high-speed, low-amplitude leading waves as acoustic emission waves.

(3) Hit discrimination time (HDT)

Hit discrimination time refers to the hit signal waiting time interval set to correctly determine the endpoint of an impact signal. If it is selected too short, a single impact may be mistakenly identified as multiple impacts; if it is set too long, multiple impacts may be incorrectly considered as a single impact. The function of the hit discrimination time is to enable the system to determine the end of the impact, stop the measurement process and store the characteristics of the test data. The hit discrimination time circuit can be triggered once by an acoustic emission signal exceeding the threshold. In most detection systems, the hit discrimination time must be set to at least twice the peak discrimination time to more accurately identify and describe the acoustic emission signal. On the one hand, the hit discrimination time must be as long as possible to exceed the time interval when the signal is below the threshold; on the other hand, it must also be set as short as feasible to ensure signal throughput and reduce the risk of mistakenly identifying two separate signals as one.

(4) Hit lock time (HLT)

Hit lockout time refers to the time interval during which the measurement circuit is closed to avoid capturing reflected or delayed waves in the impact signal. It is a time parameter set to suppress the reflected waves and delayed acoustic emission signals. The hit lockout time circuit is activated after the hit discrimination time ends and remains untriggered by signals within the set duration. HLT (Hit Lock Time) is a crucial parameter for eliminating the influence of echoes and other noise. Its significance lies in that after the hit discrimination time, the system locks out a period during which no impact signals are processed to prevent noise interference. The hit lockout time must be sufficiently long to eliminate noise interference, but too long a duration may filter out genuine signals as noise interference.

4. Related concepts of sound waves

(1) Longitudinal wave

A longitudinal wave is a wave in which the direction of vibration of a mass is the same as the direction of wave propagation. In a longitudinal wave, the wavelength refers to the distance between two adjacent dense or sparse parts. Longitudinal waves can propagate in solids, liquids and gases.

(2) Transverse wave

A transverse wave is a wave in which the direction of vibration of a mass is perpendicular to the direction of wave propagation. In a transverse wave, the protruding parts are called crests, and the concave parts are called troughs. The wavelength usually refers to the distance between two adjacent crests or troughs. Since solids have of shear elasticity, transverse waves can propagate in solids, but cannot propagate in liquids and gases.

5.1 Acoustic Emission Technology Principle

(3) Body wave

Body waves refers to the longitudinal and transverse waves that propagate in a uniform medium, which can be divided into longitudinal and transverse waves.

(4) Guided waves

Guided waves are formed due to multiple reflections between discontinuous interfaces in a medium, which further produce complex interference and geometric dispersion. The essence of guided waves is a type of elastic wave that propagates at ultrasonic or acoustic frequencies in a waveguide (such as pipes, plates, rods, ropes, etc.) parallel to the boundary.

For elastic media with a certain thickness "layer," such as thin plates, long slender cylindrical rods, and hollow cylinders, their common feature is the introduction of one or more geometric characteristic dimensions (such as plate thickness, diameter, wall thickness, etc.). When ultrasonic waves excited by a signal propagate in such media, the longitudinal waves and transverse waves excited by the source propagate at their respective characteristic velocities. When they pass through the interfaces between "layers," the transverse waves and longitudinal waves will undergo reflection phenomena and simultaneously undergo waveform mode conversion. The waveforms couple with each other to form complex interference. We refer to such elastic bodies with a "layered" structure as waveguides, and the waves propagating in these waveguides are called ultrasonic guided waves. For example, Lamb waves in plates, circumferential and longitudinal guided waves in hollow cylinders, and longitudinal, torsional, and bending cylindrical guided waves in long slender cylindrical rods [4].

(5) Group velocity

Group velocity, also known as the 'speed of the wave packet', refers to the propagation speed of a point with a certain characteristic (such as the amplitude of the largest) on the envelope of an elastic wave. Alternatively, it can be described as the propagation speed of a family of waves with similar frequencies.

(6) Phase velocity

Phase velocity and group velocity are two fundamental but distinctly different concepts in guided waves. Phase velocity refers to the propagation speed of a point with a fixed phase on the wave along the direction of propagation, which is also the propagation speed of in-phase points of a certain frequency that maintain a constant phase.

(7) Lamb wave

Lamb wave refers to a kind of wave propagating in a thin layer, the signal is reflected and refracted by the upper and lower interfaces to propagate in the form of an ultrasonic guided wave. Lamb wave is a combination of longitudinal and shear waves.

(8) Rayleigh wave (Rayleigh wave)

When the transmission medium is a semi-infinite solid, if an acoustic emission source is generated at a certain point and propagates to a point on the surface, the mutual modal conversion between longitudinal waves and transverse waves will form surface waves (Rayleigh waves) in the theory of ultrasonic guided waves on the solid surface. Rayleigh waves exist at the interface between a semi-infinite solid medium and a gaseous medium, propagating along the solid surface with a depth of 1–2 wavelengths. The wave energy rapidly decreases as the depth increases.

(9) Stonely wave

When the transmission medium is the interface between two solids, what forms is the Stoneley wave, which propagates at the solid–solid interface in the theory of guided waves.

(10) Scholte wave

When the transmission medium is the interface between a solid and a liquid, what forms is the Scholte wave, which propagates at the solid–liquid interface in the theory of guided waves.

(11) Spreading effect

Due to the influence of waveguide geometry, the speed of ultrasonic waves propagating in the waveguide will depend on its frequency, resulting in the geometric dispersion of ultrasonic waves. This means that the phase velocity of the guided waves changes with different frequencies, a phenomenon known as dispersion effect. Here, the dispersion effect generally refers to geometric dispersion caused by geometric characteristics (for some polymeric non-metallic materials, the nonlinear physical properties of the material itself will lead to physical dispersion of the waves).

(12) Waveform effect

The stress waves released from the acoustic emission source are often broadband sharp pulses, not of a single frequency, but a group of waves (i.e., wave packet or wave train) composed of waves of different frequencies (and different wave velocities). Before this group of waves is received by the sensor, due to the coupling effect of the transmission medium, the distortion during their propagation is very complex. Due to the propagation characteristics of the medium and the geometric shape of the transmission medium, the signal waveform will inevitably undergo continuous reflection, refraction, and waveform conversion.

5.1.2 Basic Principles of Acoustic Emission

1. Mechanism of Acoustic Emission Wave Generation

The local changes in the material that causes acoustic emission are called acoustic emission events, and the acoustic emission source refers to the physical source of the acoustic emission event or the mechanism of acoustic emission wave generation. Acoustic emission source has a variety of mechanisms, including the formation and expansion of cracks in the solid (such as crack expansion during loading, constant load crack expansion, fatigue crack expansion, stress corrosion crack expansion and hydrogen embrittlement crack expansion, etc.), plastic deformation (dislocation movement, slip, twinning deformation and boundary movement, etc.), phase transition (martensitic phase transition), pressure leakage, friction and wear, crack surface closure and friction, impact, magnetic domain wall motion, combustion, boiling, solidification and melting, oxide film, rust skin and slag cracking. Although different source mechanisms produce different acoustic emission signals, they have in common the process that the material becomes unstable locally or partially due to changes in external conditions and releases energy to reach a new stable equilibrium [9]. Due to the existence of various defects and inhomogeneities within the material, stress concentration will occur and be locally unevenly distributed when the material is subjected to external forces. When the uneven stability of the accumulated strain reaches a certain level, and it will inevitably cause stress redistribution, often accompanied by dislocation movement in the process, slippage, crack generation and development, microscopic cracking, and ultimately make the material to achieve a new equilibrium state. This process is actually the release of strain energy to achieve a new equilibrium.

The mechanism of acoustic emission source may be different for different materials, but the essence of acoustic emission phenomenon is the release of transient elastic waves from the structure or material. For example, the mechanism of acoustic emission after loading of concrete is mainly due to the dislocation motion of crystals, slip between crystals, elastic and plastic deformation, crack generation and expansion and friction etc. And the dislocation motion, crack expansion and phase transformation is the acoustic emission source mechanism of metal materials. The cracking of rocks and crushing of soil particles under pressure in seismic prediction and monitoring of mine structures is the acoustic emission source mechanism. The mechanisms of acoustic emission sources for marine platform steel structural materials include the initiation and propagation of cracks, structural corrosion, and fatigue. For composite materials, the mechanisms of acoustic emission sources involve cracking of the matrix material and fiber fracture.

Based on the above mechanisms, acoustic emission signals and acoustic emission technology can be used to monitor the structure or material under load conditions of micro-deformation and cracking and crack generation and expansion to obtain their dynamic information. Various catastrophic damages, such as pitting corrosion caused by the destruction of the passive film on stay cables leading to eventual perforation, and the expansion of microcracks untill the breakage of cables, all generate acoustic emission signals. Acoustic emission monitoring of structures or materials is a passive, dynamic monitoring, which is the advantage of acoustic emission and other nondestructive testing methods. There are often acoustic emission phenomena before the destruction of the material, so if these early acoustic emission phenomena can be obtained and analyzed, not only can judge the current state of the acoustic emission source, and even its causes and future development trend prediction and analysis, so as to carry out structural condition monitoring and fault diagnosis.

2. Basic principle

Acoustic emission technology involves acoustic emission source, wave propagation, acoustic-electrical conversion, signal processing, data display and recording, interpretation and evaluation, etc. The basic principle is that the acoustic emission source generates elastic waves propagating in the material, causing the vibration of the surface of the specimen to be monitored, these vibrations are coupled in the specimen on the sensor to sense the piezoelectric effect generated by the elastic waves caused by the vibration of the surface converted into a voltage signal, and then amplified by the instrument. Processed in the form of parameters or waveforms collected, and then its signal processing [10]. The basic principle of acoustic emission technology is shown in Fig. 5.3.

3. Characteristics of acoustic emission technology

Acoustic emission technology is a unique, non-invasive and highly sensitive nondestructive testing technology, which can accurately reveal the state of change within

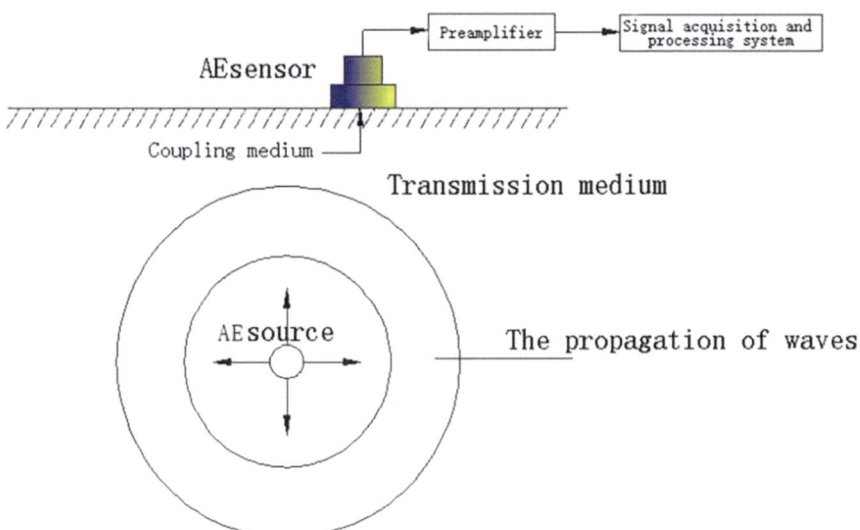

Fig. 5.3 Basic principles of acoustic emission technology

5.1 Acoustic Emission Technology Principle

the material and structure, compared with other conventional dynamic nondestructive testing has obvious advantages [11], and it is determined by its inherent physical properties. The main characteristics are as follows:

(1) Memory characteristics

The memory characteristics of acoustic emission technology are mainly manifested in the Kaiser effect and the Felicity effect. The memory characteristics of the Kaiser effect are manifested in the following: when the material is loaded again to the maximum load of the last loading, the obvious acoustic emission phenomenon will not occur, which is irreversible in nature. The Felicity effect is sometimes called the anti-Kaiser effect, that is to say, the obvious acoustic emission phenomena will occur before the material is repeatedly loaded to the original maximum load.

(2) Excitation characteristics

Acoustic emission dynamic monitoring is based on the material or structure due to deformation, cracking and other defects in the acoustic emission signal analysis and processing, to understand its health status, and according to the propagation of the AE signal in the medium to reasonably arrange the sensor for the location of the source of the damage and a wide range of monitoring, to improve the monitoring efficiency and reliability.

(3) Dynamic characteristics

Acoustic emission technology is a passive dynamic non-destructive testing methods, the detected energy from the object to be detected itself, without the need for external excitation to generate signals, and can capture the defects with the load and other factors change the real-time information. Acoustic emission technology is capable of non-destructive testing of internal structural defects or potential defects in components or materials. In addition, the acoustic waves be emitted from defects within the object under the action of external forces can be used to locate and diagnose the defects. According to the characteristics of the emitted sound waves and the external conditions that induce the acoustic emission, can we not only understand the current state of the defect, but also understand the history of the formation of the defect, as well as the defect in the future under the conditions of the actual use of the expansion of the tendency that will occur, which is not possible with other traditional non-destructive testing methods.

(4) Spectral characteristics

The signal distribution frequency of various types of acoustic emission is wide. There are a few hertz sub-acoustic frequency, tens of hertz to tens of thousands of hertz of acoustic frequency, and there are several megahertz of ultrasonic frequency. Furthermore, the signal intensity of most materials is relatively weak and inaudible to the human ear, requiring the use of acoustic emission detectors for detection, recording, and analysis.

Due to different materials acoustic emission phenomenon produced by different mechanisms, so different materials defects in the formation and development process to produce acoustic emission spectral characteristics are not the same. Therefore, we can use the spectral characteristics to locate and identify the defects.

(5) Attenuation characteristics

The essence of the acoustic emission signal is a transient elastic stress waves, which release the energy generated by of material or structure under the action of external conditions deformation, cracking, so the acoustic emission signal is essentially a mechanical wave, with the basic characteristics of wave volatility and attenuation [12], which has a certain influence on the spectrum analysis.

5.1.3 Acoustic Emission Data Processing

1. Acoustic emission characteristic parameter analysis method

Due to the limited functionality of early acoustic emission instruments, which could only acquire a small number of parameters such as counts, amplitudes, and energies, single-parameter analysis methods, such as count method, energy analysis method, and amplitude analysis method, were predominantly used. With the technological advancements of acoustic emission instruments, multi-channel acoustic emission instruments with powerful functions have been widely applied, leading to the evolution of various analysis methods including parameter list analysis, history plot analysis, distribution analysis, and correlation analysis [13, 14].

(1) Parameter list analysis method

The analysis method that arranges various acoustic emission characteristic parameters in chronological order is to directly display each acoustic emission signal's characteristic parameters in a list according to their time sequence, including the arrival time of the signals, parameters of each acoustic emission signal, and external variables.

(2) History plot analysis method

The acoustic emission history plot analysis method refers to the analysis conducted by establishing the variation of various parameters over time or with external variables. The most intuitive and common approach is to create graphical representations. The frequently used history plots and cumulative history plots include the changes in counts, amplitudes, energies, rise times, durations, etc., over time or with external variables.

5.1 Acoustic Emission Technology Principle

(3) Distribution analysis method

Acoustic emission distribution analysis refers to a method of statistical impact or event count distribution analysis based on the parameter values of the signal. The horizontal axis of the distribution graph represents the parameter, which parameter is chosen as the distribution graph of the parameter, the vertical axis is the impact or event counts, common distribution graphs are time distribution graph, energy distribution graph, rise time distribution graph, amplitude distribution graph and so on.

(4) Correlation analysis method

Acoustic emission correlation analysis method refers to the two arbitrary characteristics of the parameter to do correlation graph analysis method. The two coordinate axes of the correlation plot each represent a parameter, and each point in the plot corresponds to a hit or event count of an acoustic emission signal. Through the correlation diagram between different parameters, different acoustic emission source characteristics can be analyzed, so as to achieve the purpose of identifying the acoustic emission source.

2. Waveform analysis method

The sensors of early acoustic emission instruments were mostly resonant, high sensitivity type, which can be approximated as narrow-band filters. These sensors tend to mask or filter out the essential information from the acoustic emission source, capturing mostly attenuated sine waves. This inevitably leads to information loss, which is also the greatest limitation of parameter analysis methods. Based on the shortcomings of the parametric analysis method, people have long realized that the waveform contains all the information of the sound source and has important research value [15]. Common waveform analysis methods include modal acoustic emission (MAE), Fourier transform, wavelet analysis, neural network, and full waveform analysis.

(1) Modal Acoustic Emission

In 1991, American scholar Gorman [16] published a study of the plate wave acoustic emission (PWAE), deepening the researchers' understanding of Lamb waves and applying this theory more extensively to acoustic emission monitoring. PWAE is also known as Modal Acoustic Emission (MAE), MAE theory combines the physical mechanism of the acoustic emission source and the plate wave theory. This method is suitable for thin plate metal materials, The method is suitable for monitoring the acoustic emission signal from corrosion of thin plate metal materials and thin-walled long pipes. This is because the signal has the typical characteristics of extended wave and bending wave, which differ significantly from noise in waveform features, making it easier to identify the waveform of corrosion signals [17].

(2) Fourier transform

Fourier transform was first proposed in 1807 by the French mathematical and physical scientist Fourier (Jean Baptistle Joseph Fourier), and it was fully developed until 1966. It represents a milestone in the history of human mathematics, has always been regarded as the most basic, classical signal processing methods. The spectral information obtained through it has significant physical meaning and is widely applied in various fields. It is the promotion of Fourier series, which transforms the time domain signal into frequency domain for analysis, and makes a qualitative change in signal processing, which is very suitable for the analysis of periodic signals. However, because it is the average analysis of data segments, for non-smooth, non-linear signals lack of local information in the time domain, the processing results are unsatisfactory.

(3) Wavelet analysis

Wavelet analysis is a kind of analysis method that evolved, improved and developed from Fourier analysis, representing a form of double integral transformation. This method has adaptive function for the signal, that is, it can ensure that the window area (size) unchanged, by changing the window shape, time window and frequency window, to achieve the signal in different frequency bands, different moments of proper separation, and decompose the signal layer by layer into low frequency and high frequency part [18]. The frequency of the low-frequency part has higher frequency resolution and lower time resolution, while the high-frequency part has higher time resolution but lower frequency resolution. Therefore, it is named as the 'mathematical microscope', which provides a powerful and efficient tool for the extraction and analysis of non-stationary and weak signals.

Noise separation and extraction of useful weak signals are important aspects of wavelet analysis applied to signal processing. By decomposing the signal into different frequency bands, it is easy to separate the noise. At the same time, the time–frequency analysis capability of wavelet analysis has a great advantage when dealing with acoustic emission signals, which have non-stationary characteristics.

According to the characteristics of the acoustic emission signal, when using wavelet analysis, the selection principles of the wavelet base are as follows.

① Select discrete wavelet transform as much as possible

Compared with the discrete wavelet transform, the continuous wavelet transform can freely choose the scale factor, and the time–frequency space division of the signal is finer than the binary discrete wavelet, but the calculation volume is larger. The data volume of the acoustic emission signal is huge, so the discrete wavelet transform is more suitable for the acoustic emission signal from the perspective of processing speed. Since the purpose of analyzing the acoustic emission signal is to obtain the relevant information of the acoustic emission source, the wavelet analysis of the acoustic emission signal can realize the reconstruction of the characteristic signal of the acoustic emission source, which is conducive to obtaining the information of the acoustic emission source.

5.1 Acoustic Emission Technology Principle

② Preferentially consider selecting wavelet bases with compact support in the time domain

Acoustic emission signals have sudden transient nature, the ability to accurately pick up sudden acoustic emission signals is a prerequisite for obtaining the correct acoustic emission source information, so priority should be given to the selection of wavelet bases in the time domain with a tight support, and tightly supported wavelet bases can avoid computational errors [19]. To ensure the local analysis capability of the wavelet basis in the frequency domain, it is required that the frequency band of the wavelet basis exhibit rapid attenuation.

Comprehensive analysis of the above, in the time domain with tight support, in the frequency domain with fast decay is another rule of acoustic emission signal wavelet base selection.

③ Wavelet basis in the time domain have similar characteristics to the acoustic emission signal

Acoustic emission signals in the time domain is usually manifested as a class of waveform signals with certain impact characteristics and approximate exponential decay properties, and has a certain duration. Therefore, the wavelet basis with similar properties can provide a good analysis effect for the characteristics of the acoustic emission signal.

④ Select wavelet bases with certain order vanishing moments

The wavelet basis with a certain order vanishing moment can effectively highlight the singular characteristics of the signal. Acoustic emission signals have the characteristics similar to impact signals, so choosing the wavelet basis with a certain order vanishing moment can highlight the characteristics of the acoustic emission signals.

⑤ Select symmetrical wavelet bases as much as possible

The wavelet transform analysis of the acoustic emission signal should try to choose symmetric wavelet bases, and if it is difficult to obtain symmetric wavelet bases, it should try to choose approximate symmetric wavelet basis to reduce the distortion of the signal.

(4) Neural network analysis method

Neural network is an emerging discipline with the development of computers, with various functions such as self-organization, self-adaptation, self-learning functions, as well as strong robustness, and thus has a strong adaptability in the processing of data [20]. Each information processing unit (neuron) in an artificial neural network (ANN) communicates by sending excitation or inhibition signals to other neighbouring units to complete the information processing of the whole network system, which has a high degree of robustness and the ability to process information in parallel and distributed, and at the same time has the ability to express knowledge in a distributed manner. The system has the advantages of high robustness and parallel distributed processing information ability, at the same time, it also has the advantages of distributed expression of knowledge, automatic acquisition, automatic processing of adaptive and better fault tolerance and learning ability, etc. And it is widely used in speech recognition, image recognition, image classification and other fields [21].

(5) Full waveform analysis method

With the continuous development of the acoustic emission instrument, the mainstream of the third generation of digital acoustic emission monitors are multi-channel, and equipped with broadband sensors. It can perform real-time and comprehensive acquisition of acoustic emission signals, and achieve good results in the analysis of acoustic emission signals and signal-to-noise separation by adopting a method that combines the analysis of the time-domain waveform and frequency-domain analysis of the signals.

3. Acoustic emission for structural damage assessment

(1) b-value analysis

The b-value analysis was originally used to characterize the relationship between the magnitude and frequency of earthquakes, and the similarity between acoustic emission transient stress waves and seismic waves has led to the wide application of the b-value method in acoustic emission damage assessment [22]. In seismology, the probability of occurrence of small magnitude earthquakes is greater than the probability of occurrence of large magnitude earthquakes. Guenberg and Richter proposed the well-known G-R criterion to describe this magnitude-frequency relationship.

$$N = aM_L^{-b} \text{ or } \log_{10} N = a - bM_L \qquad (5.3)$$

where M_L denotes the Richter magnitude, N represents the number of earthquakes with magnitude higher than M_L, and a and b are constant coefficients, which can be obtained by line fitting.

Due to the similarity between acoustic stress waves and seismic waves, b-value analysis is widely used in the analysis of acoustic emission signals. The G-R criterion applied in the field of acoustic emission monitoring of materials can be interpreted as follows: the number of acoustic emission impacts released by slight cracking of the material is more than that released by more severe cracking, and the b-value can be defined according to Eq. (5.4):

$$\log_{10} N = a - b\left(\frac{A_{dB}}{20}\right) \qquad (5.4)$$

where, A_{dB} is the acoustic emission impact amplitude, N is the number of acoustic emission impacts whose amplitude exceeds A_{dB}, a and b are the linear fitting parameters, 20 is a constant used to keep consistent with the value of b in the G-R criterion, and the parameter b represents the slope of the curve. In finding the b-value, the data

5.1 Acoustic Emission Technology Principle

obtained from the acoustic emission experiments were first grouped, and then the corresponding b-value of each group was found separately to analyze the change rule of b-value with the loading process of specimen [23].

The physical significance of b-value is: when the crack expands with a large amplitude, the proportion of signal components with larger amplitude is larger, and the b-value is smaller. When the crack expands with a smaller amplitude, the proportion of signal components with smaller amplitude is larger, and the b-value is larger at this time. Therefore, b-value reflects the strength of acoustic emission signals and the combination of acoustic emission signals of different intensities, and the b-value is independent of the propagation distance [24]. The pattern of b-value changes shows that: the larger b-value corresponds to the development of structural micro-cracks, while the smaller b-value represents the generation of macro-cracks in the structure, so it is possible to achieve the monitoring of the whole process of structural damage by analyzing the changes of b-value. Schumacher et al. [25] believed that if a large number of lower amplitude acoustic emission impacts appear and b > 1.0, then there are micro-cracks in the internal material, and If there are a large number of high amplitude acoustic emission impacts with b < 1.0, then there are macro cracks in the material and there is a large amount of damages.

(2) RA-AF correlation analysis

RA-AF correlation analysis of acoustic emission parameter has been widely used in failure mode recognition of brittle materials such as concrete. The parameter AF is defined as the ratio of the acoustic emission impact ring counts to the duration (kHz), and the parameter RA is defined as the ratio of the rise time of the acoustic emission event to the amplitude (μs/V).

It is shown that for different failure modes (shear crack and tensile crack), the acoustic emission parameters AF and RA show obvious differences: tensile crack usually releases a large amount of energy in an instant, the signal rise time is short and the amplitude is large, the ringing counts (the oscillation frequency) are more, and the AF is higher; on the contrary, shear crack shows a long rise time and duration and fewer ringing counts, and the RA is higher; therefore, the AF and RA parameters AF are defined by the ratio of the rise time to the amplitude of the acoustic emission event (μs/V), RA is defined as the ratio of the rise time to the amplitude of the acoustic emission event (μs/V). Therefore, by correlating the parameters AF and RA, the damage failure mode of the structure can be determined qualitatively [4], see Fig. 5.4.

(3) Acoustic emission signal strength analysis

The analysis of acoustic emission signal strength requires the calculation of two metrics, historicindex (HI) and severity value (Sr) [26, 27]. These two indicators can be used to assess the loading capacity and the degree of degradation of structural elements. In order to analyze the strength of the acoustic emission sources, it is necessary to calculate HI and Sr from the recorded acoustic emission signals, to analyze the variation of their values and to represent them graphically.

Fig. 5.4 Acoustic emission crack classification

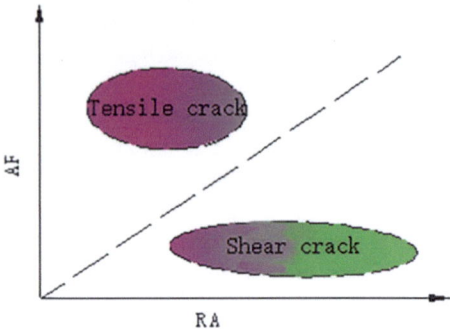

HI is an index used to assess the change in signal strength of a specimen during loading and is calculated using the following formula:

$$\mathrm{HI} = \frac{N}{N-K} \frac{\sum_{i=K+1}^{N} S_{oi}}{\sum_{i=1}^{N} S_{oi}} \qquad (5.5)$$

where N is the number of acoustic emission events from the beginning of loading to the moment t, S_{oi} is the signal intensity of the ith acoustic emission event, K is an empirical factor related to N. For concrete structural members, the value of K can be selected according to the following criteria: K = 0, N ≤ 50; K = N -30, 51 ≤ N ≤ 200; K = 0.85N, 201 ≤ N ≤ 500; K = N - 75, N ≥ 500.

For the severity value Sr, which is the average of the 50 loudest transmitter signal strength values up to the moment t, it can be calculated by the formula (5.6).

$$S_r = \frac{1}{J} M = \left(\sum_{m=1}^{J} S_{om} \right) \qquad (5.6)$$

where S_r is the severity value of the acoustic emission; S_{om} is the mth largest acoustic emission signal strength value in descending order, and J is an empirical constant related to the member material. For concrete structures, it can be selected as follows: J = 0, N < 50; J = 50, N ≥ 50.

(4) NDIS-2421 damage assessment guidelines

Acoustic emission monitoring is usually performed to obtain qualitative or quantitative results, it can assess the degree of damage to a structure by observing the trend of the characteristic parameters of the acoustic emission signals collected in real time. The NDIS-2421 quantitative rating standard recommended by the Japan Society for Non-Destructive Testing (JSNDI) can be used for structural damage assessment. This standard utilizes the 'Kaiser effect' of concrete materials during repeated loading, and the NDIS-2421 damage classification criterion proposes two indicators, CR (calm ratio) and LR (load ratio):

5.1 Acoustic Emission Technology Principle

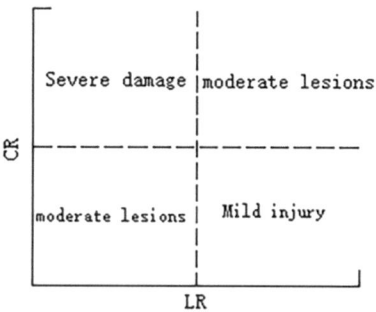

Fig. 5.5 Damage classification based on two NDIS-2421 ratios

CR indicator = cumulative number of AE activities during the unloading phase of this load cycle/total AE activities of the previous load cycle;

LR indicator = load value at the start of AE activity in this load cycle / load value in the previous load cycle;

The damage classification boundaries for the NDIS-2421 quantitative rating criteria are shown as dashed lines in Fig. 5.5. The degree of damage to the structure is determined based on the crack opening width displacement (CMOD), with the cut-off value between Intermediate Damage (Intermediate Damage) and Minor Damage (Minor Damage) being a crack opening width displacement (CMOD) of 0.1 mm, and Severe Damage (Heavy Damage) occurring with a crack opening width displacement greater than 0.5 mm. Ohtsu [28] suggested cut-offs of 0.05 and 0.9 for CR and LR respectively. Therefore, a structure is considered to have severe damage when the CR of its acoustic emission response characteristics under a particular load is greater than 0.05 and LR is less than 0.9.

5.1.4 Typical Acoustic Emission Technology Products

The research on acoustic emission technology began in the 1950s, with the German scholar JosefKaiser found the Kaiser effect as an important symbol. In the1960s, Poland, Sweden, Canada and other countries have successfully developed a single-channel, multi-channel bunker acoustic emission detector, and applied to the mine large-area ground pressure activities and local rock fall prediction, forecasting. Until the 1970s, with the development of the electronics industry and computer hardware and software technology maturity and progress, commercial acoustic emission detection instruments began to appear, such as the U.S. Physical Acoustics Corporation (PAC) created by Duengan and Bell Laboratory. In the early 1980s, PAC introduced modern microprocessor computer technology into the acoustic emission detection system, designed the second generation of source positioning acoustic emission detection instruments with smaller volume and weight, and developed a series of multifunctional detection and data analysis software. Through the microprocessor computer control, real-time acoustic emission source localization monitoring and data analysis and display can be conducted on the detected components. In 1985 PAC company merged with Dunegan, the most famous acoustic emission technology company at that time, making PAC company become the largest acoustic emission technology research and development company in the world.

Companies such as American PAC, DW, German VallenSysteme, and China's Guangzhou Shenghua successively developed third-generation digital multi-channel acoustic emission detection and analysis systems that were more computerized and had smaller volume and weight. Besides real-time measurement of acoustic emission parameters and localization of acoustic emission sources, these systems could also directly observe, display, record, and perform spectral analysis of acoustic emission waveforms.

At present, the main acoustic emission companies at home and abroad have developed a new generation of all-digital multi-channel acoustic emission measurement system, such as PAC's SAMOS system, Germany Vallen's AMSY-6 acoustic emission system and Beijing ShengHua company's SAEU3H detector.

(1) SAMOS system

SAMOS system is the third generation of fully digital system developed by PAC, the core of which is a parallel processing PCI bus acoustic emission function card—PCI-8 board, on a board with 8 channels of real-time acoustic emission feature extraction, waveform acquisition and processing capabilities, which is the PAC's more integrated and lower-priced system is more suitable for pressure vessel inspection and other engineering applications.

PAC acoustic emission system products include: acoustic emission sensor series, general acoustic emission system, special acoustic emission system, acoustic emission expert system, acoustic emission software and accessories.

(2) AMSY-6 system

AMSY-6 is a fully digital multi-channel acoustic emission (AE) measurement system. It consists of parallel measurement channels and system front-end software running on an external PC. The measurement channels consist of an acoustic emission transducer, a preamplifier, and a 1-channel ASIP-2 (dual-channel acoustic signal processor). Each channel consists of an analogue measurement section and a digital signal processing unit. It is possible to extract characteristic parameters of the acoustic emission signal with the ASIP-2, such as first threshold crossing time (arrival time), rise time, duration, peak amplitude, energy and count. While extracting the features, the complete waveform can be recorded by an optional transient recording module.

(3) SAEU3H Detectors

The SAEU3H series of benchtop detectors includes the SAEU3H centralized acoustic emission detector, the 4-channel SAEU3H acoustic emission detector, the 20-channel SAEU3H acoustic emission detector, the 48-channel SAEU3H detector, and the corresponding acoustic emission software system. The centralized SAEU3H multi-channel acoustic emission detector can be customized according to the number of channels required. The acoustic emission parameters that can be analyzed by the system include threshold arrival time, peak arrival time, amplitude, ring count, duration, relative energy, absolute energy, signal strength, rise count, rise time, RMS, average ASL, start phase, 12 external parameters, centre-of-mass frequency, peak frequency, 5 localized power spectra, original frequency, reverberation frequency, average frequency, and so on.

5.2 Principle of Sonar Technology

Sonar is an electronic device that uses sound waves in the water to detect, locate and communicate with underwater targets, and is the most widely used and important device in hydroacoustics. The term "sonar" is an acronym formed from the initials of the three English words "sound," "navigation," and "ranging," representing sound, navigation, and ranging. Sonar utilizes the propagation characteristics of sound waves underwater to detect, locate, and communicate with underwater targets or structures through electro-acoustic conversion and information processing. By employing sonar underwater imaging technology, it analyzes and judges the position, shape, and presence of defects in underwater targets or structures, thereby enabling the inspection and assessment of underwater structures.

Sonar system has many similarities with radar and electro-optical systems, sonar operation is based on the propagation of acoustic waves between the target sonar and the receiving sensor. However, it differs from radar and electro-optical systems because the energy observed by the sonar is propagated through the mechanical vibration of liquids, solids, gases, or plasma, rather than electromagnetic waves. That is to say, sonar technology is based on sound waves for propagation, and the wave speed of its sound waves is much smaller than the wave speed of electromagnetic waves. The propagation of sonar in water is only longitudinal wave propagation without transverse wave, which is due to the fact that transverse wave has no shear strength in water or other liquids.

Sonar system is mainly divided into two categories: passive sonar and active sonar, in addition to daylight/environmental sonar system. Passive sonar systems is one where the energy generated by the target that is transmitted to the receiver, similar to passive infrared detection. Active sonar systems are sound waves that are transmitted from a transmitter to a target and back to a receiver, similar to pulse-reflection radar.

In military applications, sonar systems are used not only for communication, navigation and identification of obstacles or hazards (e.g. polar ice), but also for detecting, classifying, localization and tracking submarines and mines. In commercial applications, sonar can be used in fish detectors, medical imaging, marine geomorphological mapping, channel bathymetry, seismic detection, marine platforms and submarine pipeline detection.

5.2.1 Basic Concepts

1. Basic concepts of sonar

(1) Sonar

Sonar refers to the technology that utilizes the propagation and reflection characteristics of sound wave in the water, through electro-acoustic conversion and information processing for navigation and ranging, also refers to the electronic equipment that employs this technology for underwater target detection (presence, location, nature, direction of movement, etc.) and communication.

(2) Active sonar

Sonar emits a certain detection signal, which, upon encountering obstacles or targets on its propagation path in water, reflects back to the emission point and is received. Since the target information is preserved in the echoes reflected by the target, the parameters of the target can be determined based on the received echo signals. The process of active sonar acoustic wave emission, acoustic signal propagation, and reception processing is illustrated in Fig. 5.6.

(3) Passive sonar

Passive sonar, also known as noise sonar, is a collective term for various types of sonar that acquire target parameters by receiving and processing the radiated noise or sonar signals emitted by underwater targets. The information flow of passive sonar is illustrated in Fig. 5.7.

2. Sonar parameters

In the sonar equation, sonar parameters refer to various physical quantities used to reflect the propagation law of sound waves in the medium, the nature of the target, the interference background and the basic performance of sonar equipment. The parameters related to active sonar include source level (SL), propagation loss (TL), target strength (TS), directivity index (DIR), noise level (NL), equivalent plane wave reverberation level (RL), and detection threshold (DT); while the parameters related

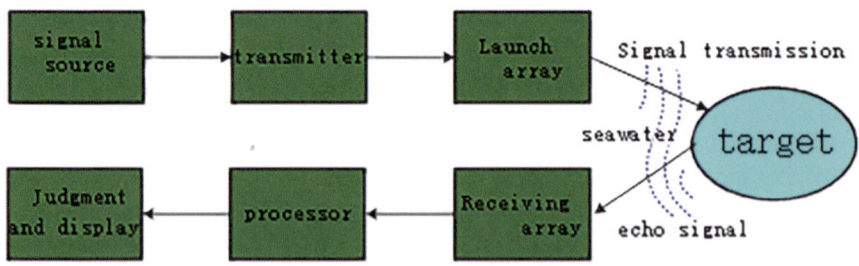

Fig. 5.6 Active sonar information flow

5.2 Principle of Sonar Technology

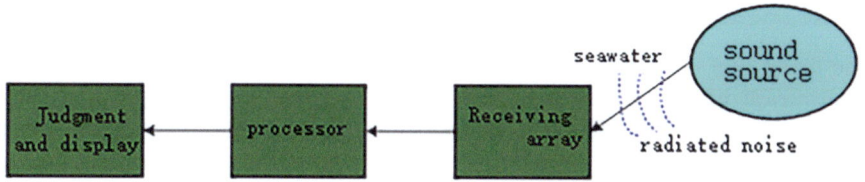

Fig. 5.7 Passive sonar information flow

to passive sonar include source level (SL), propagation loss (TL), directivity index (DIR), noise level (NL), and detection threshold (DT).

(1) Active sonar source level (SL)

Active sonar source level characterizes the strength of the acoustic signal emitted by the active sonar, and its value is the decibel number of the sound intensity generated at 1 m from the sound source on the acoustic axis in relation to the reference sound intensity, which is expressed by SL as follows.

$$SL = \frac{10 lg I}{I_0} \qquad (5.7)$$

where I is the sound intensity at 1 m from the source in the direction of the sound axis of the transmitter, and $I0$ is the reference sound intensity, i.e., the sound intensity of a plane wave with a RMS sound pressure of 1 μPa.

For a point source at a distance of 1 m from the source centre, the relationship between the sound intensity I (in W/m²) and the radiated power P is shown in Eq. 5.8.

$$I = \frac{P}{4\pi} \qquad (5.8)$$

The relationship between the radiated sound power of a non-directional source and the source level is as follows.

$$SL = 10 lg P + 170.8 \qquad (5.9)$$

The relationship between the radiated sound power of a directional sound source and the source level is as follows:

$$SL = 10 lg P + 170.8 + DI_T \qquad (5.10)$$

where, P refers to the sound power, unit is W; SL refers to the sound source level at 1 m from the centre of the sound source when the reference sound pressure is 1 μPa, unit is dB.

(2) Passive sonar source level (SL)

The target ship radiation noise of Passive sonar is broadband, with varying noise intensities at different frequencies, reflecting the frequency dependence of the sound source radiation noise intensity. The decibel value of the ratio between the target radiated noise intensity IN measured at a point 1 m away from the acoustic center of the target, along the acoustic axis of the receiving hydrophone, and a reference sound intensity is denoted as SL.

$$\text{SL} = \frac{10 lg I_N}{I_0} \tag{5.11}$$

where, I_N refers to the noise intensity within the operating bandwidth of the receiving equipment.

(3) Propagation Loss (TL)

The propagation loss quantitatively describes the attenuation change of sound intensity after a certain distance of sound wave propagation. It is defined as the loss of intensity between a point and the reference distance (usually 1 m), the unit is dB. When the acoustic signal propagation in the ocean, it will attenuate and deform due to a variety of mechanisms. The factors that affect transmission loss include: diffusion, absorption, scattering (in water bodies, sea surfaces, or seabeds), multipath effects, and wave influences such as waveguide leakage. If I_0 is the sound intensity at the reference distance, then the formula for calculating the propagation loss (TL) at distance R and depth D is as follows.

$$\text{TL} = -10 lg \left(\frac{I(R, D)}{I_0} \right) \tag{5.12}$$

where I(R,D) refers to the sound intensity at distance R and depth D; I_0 refers to the sound intensity at the reference distance.

(4) Target strength (TS)

In active sonar, target strength is the ability of a target to reflect echoes. In the sonar equation, the target strength is a specific distance from the centre of the target echo (usually 1 m) at the reflected intensity Ir and the ratio of the incident intensity, take the logarithm of the base 10 and multiply it by 10. The target strength is calculated in Eq. (5.13).

$$\text{TS} = \frac{10 lg I_r}{I_i} \tag{5.13}$$

where, Ir refers to the reflected sound intensity at the reference distance; I_i refers to the incident intensity of the sound wave.

5.2 Principle of Sonar Technology

(5) Marine environmental noise level (NL)

In the ocean, ambient noise is related to a specific environment. Ambient noise is defined as the noise remaining after removing all the individual discernible sound sources, which is composed of sound waves from a large number of various noise sources in the ocean, and is a kind of background interference of sonar equipment. Possible sources of environmental noise include turbulence, shipping, wave motion, thermal disturbances, earthquakes, rainfall, marine organisms, ice break-up, and so on. Environmental noise is a measure of the strength of the environmental noise, calculated in Eq. (5.14).

$$\text{NL} = \frac{10 lg I_N}{I_0} \quad (5.14)$$

where I_N is the noise intensity within the operating bandwidth of the receiving equipment.

(6) Equivalent plane wave reverberation level (RL)

When an active sonar transmits a signal, many sources other than the target of interest reflect the acoustic signal and transmit it to the receiver. Signals that are not from the target of interest but are reflected back from discrete target-like objects are collectively referred to as reverberation. Reverberation sources in the ocean include the surface, the seafloor, and the water column. Volume reverberation sources include marine organisms, air bubbles, and the inhomogeneous structure of the sea water itself.

It is known that the intensity of I plane wave axially incident on the hydrophone, the hydrophone output a certain voltage value. When the hydrophone be placed in the reverberation field, the acoustic axis pointing to the target, the hydrophone output a certain voltage value. If the two voltage values are exactly equal, the plane wave sound level is the reverberation level, the calculation of the formula (5.15).

$$\text{RL} = \frac{10 lg I_N}{I_0} \quad (5.15)$$

(7) Receiving Directivity Index (D_{IR})

Directivity is the gain obtained by the acoustic signal of a hydrophone in an isotropic constant noise field. For the hydrophone directivity index is the ratio of the noise power generated by the non-directional hydrophone to the noise power generated by the directional hydrophone, and then take the logarithmic value with a base of 10 after the value in dB.

(8) Detection threshold (DT)

Sonar equipment receiver receives sonar signals and background noise, the ratio of the two parts (signal-to-noise ratio SNR), that is the receiving bandwidth of the signal power and the working bandwidth (or 1 Hz bandwidth) of the noise power ratio, which affects the quality of the work of the equipment, the higher the ratio, the equipment can work properly, 'judgement' is more credible.

Detection threshold (DT) is defined as the signal-to-noise ratio (SNR) at the input of the processor required for the device to function properly. Therefore, for the same function of the sonar equipment, detection threshold is lower equipment, its processing power, performance is also good.

3. Sonar combination parameter

Sonar combination parameter is a parameter which is combined by basic sonar parameter and has clear physical meaning. Common sonar combination parameters are quality factor, echo signal level, noise masking level, reverberation masking level, echo margin, quality factor.

(1) Quality factor

Quality factor refers to the difference between the sound source level measured by the sonar receiving transducer and the noise background interference level, which is expressed $SL - (NL - DI)$.

(2) Echo signal level

Echo signal level refers to the sound level of the echo signal added to the active sonar receiver transducer, and its value is $SL - 2TL + TS$.

(3) Noise masking level

Noise masking level refers to the lowest signal level required for normal operation of sonar equipment working in noise interference, and its value is $NL - DI + DT$.

(4) Reverberation masking level

Reverberation masking level refers to the lowest signal level required for normal operation of sonar equipment working in reverberation interference, and its value is $RL + DT$.

(5) Echo margin

Echo margin refers to the number of active sonar echo level exceeding the noise masking level, and its value is $SL - 2TL + TS - (NLDI + DT)$.

(6) Quality factor

For passive sonar, the quality factor refers to the maximum allowable one-way propagation loss. For active sonar, when $TS = 0$, the quality factor specifies the maximum allowable two-way propagation loss. The value of the quality factor is $SL - (NL - DI + DT)$.

5.2 Principle of Sonar Technology

4. Sonar equation

The sonar equation refers to the necessity in engineering to design a sonar system reasonably or to accurately predict its performance during use by comprehensively considering the equipment performance, channel effects, target characteristics, and other factors of the sonar system as a whole. Based on certain criteria, a fundamental equation is utilized to quantitatively reflect the quantitative relationships among these three components, and this fundamental equation is known as the "sonar equation." Essentially, the sonar equation connects the seawater medium, sonar targets, and sonar equipment, and relates signals to noise. It is a relational expression that comprehensively considers the various unique phenomena and effects of underwater acoustics on the design and application of sonar equipment.

The sonar equation is divided into active sonar equation and passive sonar equation According to the classification of sonar.

(1) Active sonar equation

When calculating the active sonar equation, the signal is emitted from the transmitter, propagated to the target and reflected, and then transmitted back to the receiver, as shown in Fig. 5.8. The emitter and receiver of an active sonar are located in the same position, which is known as a colocated active sonar. The sonar equation for this type of sonar is expressed as follows, with the meanings of the symbols provided in the sonar parameter section.

When noise is considered as background noise, the active sonar equation is as follows:

$$(SL - 2TL + TS) - (NL - DI) = DT \qquad (5.16)$$

When the noise is considered as reverberant noise, the active sonar equation is as follows:

$$(SL - 2TL + TS) - RL = DT \qquad (5.17)$$

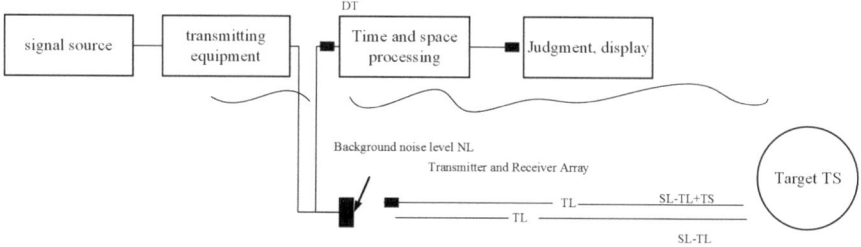

Fig. 5.8 Active sonar equation signal relationships

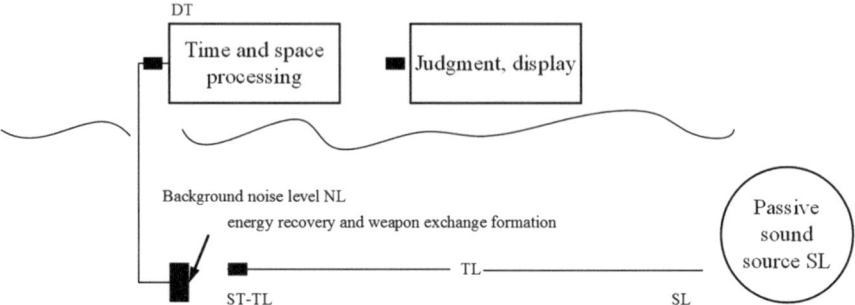

Fig. 5.9 The signal relationships of passive sonar equation

(2) Passive sonar equation

The sonar equation is the same in terms of signal-to-noise ratio calculation for active sonar and passive sonar. The difference between passive sonar and active sonar is that: the noise from the noise source propagates directly from the noise source to the receiver transducer, and the noise from the noise source is not reflected by the target, i.e., there is no TS. The background interference is the ambient noise, and there is no reverberation interference. The signal relationship of passive sonar equation is shown in Fig. 5.9, sonar equation is as follows

$$(SL - TL) - (NL - DI) = DT \tag{5.18}$$

5.2.2 Principle of Sonar Technology

With significant advancements in modern underwater acoustic signal processing technology and underwater transducers, fine detection and imaging sonar technology for underwater targets has become a research hotspot both domestically and internationally, playing an irreplaceable role in both civilian and military fields compared to other sonar technologies. In the civilian sector, imaging sonar technology can be utilized in marine engineering fields such as marine resource development, seabed geological exploration, seabed topographic and geomorphic surveying and mapping, and underwater object detection. In the military fields, there is an urgent need for high-resolution fine detection and imaging sonar technology for underwater targets in areas such as the detection and identification of highly concealed underwater military small targets (such as military unmanned underwater vehicles, torpedoes, mines, and frogmen), security and prevention for harbor anchorages and ships, and terrain-matched navigation. Currently, there are various advanced imaging sonar technologies available both domestically and internationally, with mainstream sonar technologies mainly including interferometric side-scan sonar, multi-beam echosounder, and synthetic aperture sonar. Their basic principles are introduced as follows.

5.2 Principle of Sonar Technology

1. Multi-beam sonar technology

The working principle of the multi-beam sonar system is to utilize an array of transmitting transducers to emit sound waves covering a wide fan-shaped area towards the seabed. By leveraging the orthogonality of the transmitting and receiving fan-shaped areas, it forms illuminated footprints on the seabed terrain. These footprints are appropriately processed, allowing a single detection to provide depth values for hundreds or even more seabed measurement points within a vertical plane perpendicular to the course. This enables the system to accurately and rapidly measure the size, shape, and elevation changes of underwater targets within a certain width along the voyage route, reliably depicting the three-dimensional characteristics of the seabed terrain. The basic principle of the multi-beam sonar system is illustrated in Fig. 5.10.

The beam forming principle of the multi-beam sonar system can be divided into two basic principles: beam control method and coherence method. The beam control method is to measure the round-trip time of the reflected signal at a specific angle, while the coherence method is to measure the angle of the reflected echo signal at a specific time. Therefore, there are two main variables to be measured in the multi-beam sonar system: one is the oblique distance or the distance from the acoustic transducer to each point on the seafloor, and the other is the angle from the transducer to each point on the bottom of the water. All multi-beam sonar systems use either or both beam control and coherence methods to determine these variables.

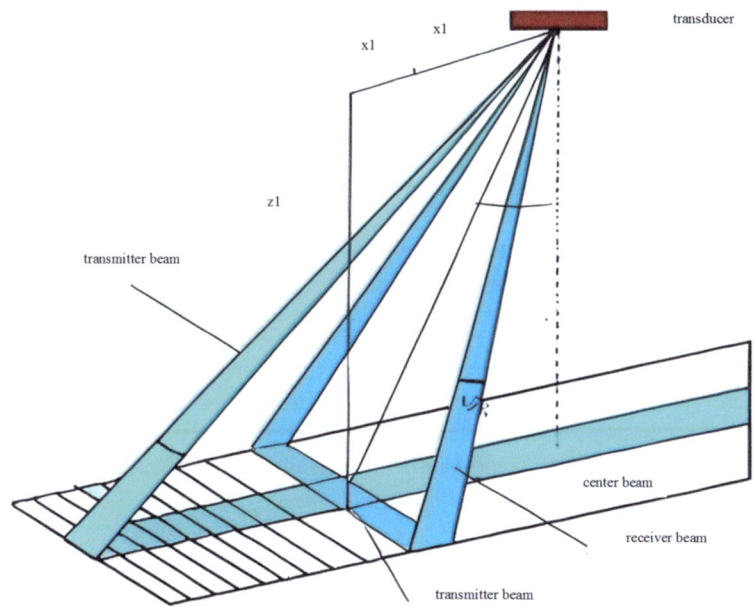

Fig. 5.10 Principle of operation of the multi-beam sonar system

Multi-beam echo sounders can accurately measure the depth of the seabed and obtain hydrographic imaging, enabling the generation of intuitive and precisely positioned full-coverage three-dimensional seabed topographic maps. However, the beam footprint of multi-beam echo sounders expands as the depth increases, resulting in lower target detection resolution at greater distances and making it more challenging to detect small targets.

Modern multi-beam bathymetric systems mainly include multi-beam acoustic systems, software systems and peripheral auxiliary sensor systems. The multi-beam acoustic system includes sonar signal processing system, transmitting/receiving transducer, display and control system, etc. The software system includes navigation acquisition software, data post-processing software, etc. The main equipment of the peripheral auxiliary sensor system includes compass, attitude sensor, sound velocity meter, tide gauge, GPS/underwater navigation system, etc.

The applications of multi-beam sonar systems in ocean depth detection and seabed topographic mapping are relatively mature. The factors that affect the accuracy and application of multi-beam sonar detection are mainly manifested in the following aspects:

(1) Detection resolution

Multi-beam bathymetric resolution refers to the minimum separation between two adjacent target points that a multi-beam bathymetric system can distinguish in the three-dimensional direction of the seabed space. It determines the fine detection capability for underwater small targets and complex topographies. The main factors affecting multi-beam bathymetric resolution include pulse width, ping sampling rate, beam width, and vessel speed.

Due to the limitations of the detection mechanism, multi-beam sonar systems exhibit lower target detection resolution at greater depths when exploring deep oceans, making it more challenging to detect small targets.

With technological advancements, researchers have improved the resolution of multi-beam sonar by enhancing the precision of the transmit/receive transducer array and refining algorithms. Additionally, multi-beam sonar systems improve their resolution through synthetic aperture techniques. For instance, the HydroSweep DS can be equipped with a transmit/receive transducer array with a beam width of $0.5° \times 1°$, capturing 320 feedback beams (hard beams) per transmission, which are then decomposed into 960 water depth points through high-order beamforming technology. Its dual-frequency transmission doubles the number of received beams to 640, resulting in 1920 water depth points, significantly enhancing bathymetric resolution.

(2) Bathymetric accuracy

The most important application of multi-beam bathymetric systems is for marine seabed topographic surveying and mapping, and bathymetric accuracy is undoubtedly a core indicator for measuring their performance. The International Hydrographic Organization has specific regulations regarding the accuracy of bathymetric measurements, requiring compensation for sound velocity refraction, motion attitude, and tidal levels for the measured water depth data. Therefore, the accuracy of multi-beam sonar can be improved by enhancing the precision of algorithms such as sound velocity compensation, motion attitude compensation, and tidal level compensation.

Surface sound velocity can be obtained in real-time using a surface sound velocity profiler, and combined with periodic measurements from expendable bathythermographs to obtain full-depth sound velocity profiles, providing accuracy in sound velocity refraction compensation. Motion sensors can provide three-dimensional offset parameters for the transmit and receive arrays, where the transmit beams undergo pitch, roll, and yaw corrections, while the receive beams undergo roll corrections. In situations where deep and distant seas lack tidal station support, GPS carrier phase measurement technology is utilized to determine instantaneous tidal variations, enabling tidal level compensation for bathymetric data.

(3) Measuring range

The coverage range of multi-beam bathymetry directly determines the surveying and mapping efficiency of multi-beam bathymetric systems. It is generally expressed in terms of several times (5 to 6 times) the water depth coverage or the swath fan angle (140°). However, in deep-sea environments, due to the impact of two-way propagation attenuation, the signal-to-noise ratio (SNR) of outer signals (with small grazing angles) is low, and waveform broadening is severe. Therefore, the coverage range is also constrained by the maximum coverage width (30 km). By optimizing the design of the transmit and receive arrays and using broadband signals, the SNR of outer signals can be improved. Using broadband signals can also suppress waveform broadening issues. Adopting new target azimuth estimation methods (such as multi-subarray detection methods) can improve the azimuth estimation accuracy of outer signals.

(4) Water body detection

When gases or fluids from beneath the seabed enter the vicinity of the seabed in the form of vents or leaks, they form localized anomalous seawater of various shapes such as plumes, columns, and whip-like forms, which differ in physical properties from the surrounding seawater. Plume flows serve as important indicators for the detection of seabed hydrothermal vents, cold seeps, and natural gas hydrates. Only when a deepwater multi-beam bathymetric system possesses powerful water body detection capabilities can it detect anomalous features within the water body, achieve seamless large-area seabed plume flow detection, and conduct full-coverage three-dimensional detection of the seawater within a certain depth range above the seabed.

(5) Detection results

Shipborne multi-beam bathymetric systems generally operate continuously throughout the entire voyage, generating a vast amount of survey data for each voyage/leg, amounting to hundreds of GB or even more. In the process of mapping and analysis using software for multi-beam bathymetric systems, a significant amount of manual intervention is still required, and the processing results may vary among data processors with different levels of experience. Therefore, for massive survey data, a unified standard for multi-beam bathymetric data mapping should be established to minimize the influence of human factors and enhance the intelligence, reliability, and convenience of mapping software.

2. Side-scan sonar technology

(1) Side-scan sonar system

Side-scan sonar is a kind of active sonar that emits acoustic waves laterally to detect the acoustic structure and medium properties of water bodies, sea surfaces, and seabeds (including upper strata). It uses the seabed backscatter to realize the acquisition of topographic and geomorphological information of the seabed, and constructs the image information of seabed topography and geomorphology, which is the basis of seabed imaging. A side-scan sonar system mainly includes five parts: transmitter array, receiver array, transmitter, receiver and signal processor. During operation, the sonar sends a pulse drive signal through the signal processor to drive the transmitter to generate a high-power transmit pulse. This pulse signal is characterized by being narrow in the horizontal direction and wide in the vertical direction. During the reception of acoustic signals, the antennas on each receive array receive the echo signals. After preliminary processing by the receiver, strong echo signals are obtained and ultimately sent to the computing and processing unit to obtain image-related information.

(2) Working principle

The side-scan sonar system conducts underwater target detection based on the principle of echo detection. It employs a transducer matrix to emit directional pulsed ultrasonic waves with a wide vertical beam angle and a narrow horizontal beam angle towards the seabed at a certain oblique angle and emission frequency, illuminating a narrow trapezoidal area of the seabed on both the left and right sides of the transducer, as shown in Fig. 5.11. After the acoustic pulse is emitted, the sound waves propagate in the form of spherical waves and, upon encountering the seabed, the reflected or backscattered waves return to the transducer along the original path. Echoes from nearby objects arrive at the transducer first, while echoes from distant objects arrive later. The equipment operates in a transmit/receive cycle at regular intervals, displaying each received line of data to produce a two-dimensional acoustic image of the seabed topography and landforms. Further data processing of this image through a computer results in a grayscale image of the marine landforms, which allows for the identification and interpretation of marine landform information.

Due to the unevenness of the seabed, some areas of the seabed or seabed targets are illuminated by sound waves, while others are not. This results in some areas appearing black and others appearing white (target shadows) on the recording paper, similar to the negative of a photograph taken by a camera, thereby reflecting the seabed's topographical conditions. The side-scan sonar converts the strength, frequency, and

5.2 Principle of Sonar Technology

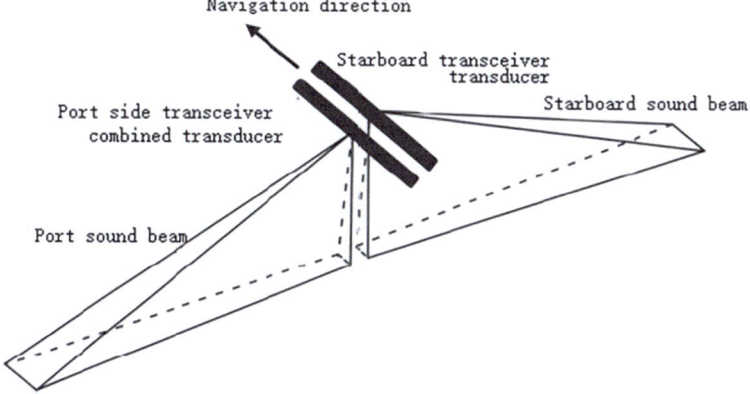

Fig. 5.11 Side-scan sonar operating principle

time intervals of echoes reflected from seabed targets into the length and blackness variations of lines drawn by the instrument's recording needle on the recording paper. Multiple transmissions and receptions form a side-scan sonar image, which reflects the seabed's topographical conditions (shapes, heights, and relative positions of seabed objects). Many current devices have incorporated matched filter digital signal technology (Chirp technology) and image processing techniques to enhance signal anti-interference capabilities, filtering, and visual effects, facilitating the identification and determination of seabed targets. Figure 5.12 shows a real-time side-scan sonar image of detected exposed seabed rocks.

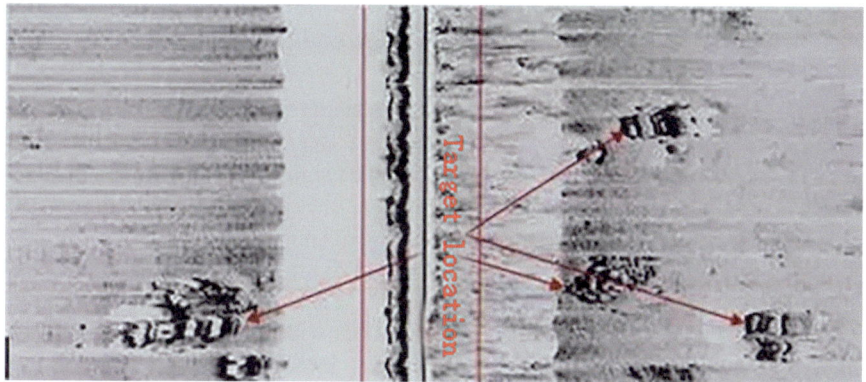

Fig. 5.12 Side-scan sonar imaging for detecting outcrops on the seabed

(3) Side-scan sonar technology characteristics

The side-scan sonar generally needs to be mounted on an underwater towed vehicle for operation. It features simple equipment installation and high cross-track resolution, enabling the identification and judgment of targets with the help of shadows. However, due to the limitations of its detection mechanism, it is not easy to obtain accurate seabed depths, and there are gaps in the measurement of the near-bottom area, which require separate sonar equipment or methods to fill. The side-scan sonar system does not provide direct height or depth information; instead, height and depth must be inferred from the images. Based on the ratio of the object's height to the towfish's height, these can be estimated through the ratio of the object's shadow length to the distance between the towfish and the end of the shadow.

According to the detection mechanism of side-scan sonar, the most important parameters in its technical specifications are range and resolution. The resolution of side-scan sonar determines the image quality of its imaging. Its resolution mainly includes along-track resolution, cross-track resolution, and backscatter resolution.

① Range of side scan sonar

The range of side-scan sonar determines its efficiency in operation, referring to the width of the seabed that can be scanned by a single line of the sonar. The range of side-scan sonar is actually a function of its transmission frequency: the higher the frequency, the smaller the range; the lower the frequency, the larger the range. There is a certain relationship between the range and resolution of side-scan sonar: for low-frequency sonar with a large range, its resolution is relatively low; whereas for high-frequency sonar with a small range, its resolution tends to be higher. Therefore, many modern side-scan sonars are equipped with dual-frequency or even triple-frequency operation capabilities, allowing for the simultaneous acquisition of large-range (low resolution) and small-range (high resolution) data without switching systems or requiring additional survey lines.

② Vertical Trajectory Resolution

Vertical track resolution or range resolution determines the minimum distance between two objects that can be seen in the beam direction of the side-scan sonar. In the early days, the transmission signal of side-scan sonar used CW mode, but now the transmission of side-scan sonar is FM or Chirp signal, the main advantage of Chirp is to take into account the larger range and better vertical track resolution.

The vertical track resolution of a side-scan sonar with a CW-type signal is determined by the pulse length of the signal, whereas the vertical track resolution of a Chirp-type side-scan sonar is determined by the bandwidth of the signal, which transmits longer signal pulses and thus allows for increased transmit energy. A high-frequency, low-range (1600 kHz, 35 m) Chirp-type side-scan sonar with sub-centimeter range resolution can obtain very good image detail in the vertical track direction.

5.2 Principle of Sonar Technology

③ Along-track direction resolution

Traditionally, the along-track resolution of a side-scan sonar is determined by the horizontal beam angle of the side-scan sonar (typically in the range of 0.2°–1.5°), the effective range and the towing speed. A small beam angle is capable of detecting small objects next to the track at short ranges. Given that 100% coverage of the seafloor is often required when searching for objects, the NOAA specification states that at least three scans are required for a 1 m^2 object, which limits the maximum speed limit of the side-scan sonar. At this speed, the SSS can detect objects at considerable distances with continuous beam/ping while towed, which is in the range of 4–5 kn for conventional side-scan sonar systems.

④ Multi-beam and multi-ping

If the side-scan sonar is made to receive more echoes from an object, the towing speed at which the sonar operates can be increased. Currently, side-scan sonar employs two different solutions to increase towing speed. The first is the use of multi-beam technology, where, for example, utilizing five beams on one side of the SSS can improve the along-track resolution by five times, thereby allowing for higher towing speeds, as illustrated in Fig. 5.13. The second solution is the adoption of multi-ping technology, which can be implemented in Chirp-type side-scan sonar. If there are two pings for a certain range in the water, the effective towing speed can also be doubled, as shown in Fig. 5.14.

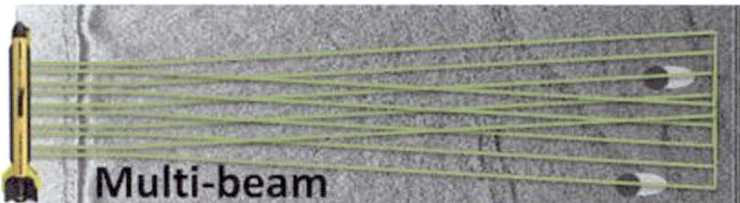

Fig. 5.13 Multi-beam schematic of side-scan sonar

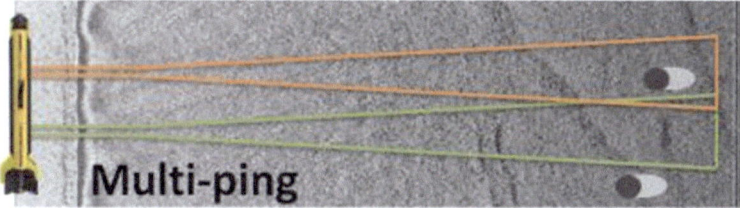

Fig. 5.14 Multi-ping schematic for side-scan sonar

⑤ Backscatter resolution

The last parameter in a digital system that determines the image quality is the level of digitization of the received acoustic signal into a usable digital signal. Currently, side-scan sonar can transmit the towed fish data to the sonar processor through the network for digital processing. In order to digitize the signal, there must be an A/D converter to convert the analogue signal into a digital signal, and the number of digital bits in the A/D converter determines the level of detail recognition. For side-scan sonar, increase the signal A/D converter digital bits can improve the resolution of the details of sonar scanning image.

3. Synthetic aperture sonar

(1) Basic principle

Synthetic aperture sonar (synthetic aperture sonar, SAS) is a new type of high-resolution underwater imaging sonar, the principle is to use the movement of the small aperture base array to obtain a larger synthetic aperture in the direction of the azimuthal high resolution. Interferometric synthetic aperture sonar (InSAS), based on the synthetic aperture sonar, adds one (or multiple) receiving arrays in the cross-track direction. By using phase comparison for height measurement, it obtains height information of the scene and processes it to produce a three-dimensional image of the scene. The advantage of synthetic aperture sonar is that its cross-track resolution is independent of operating frequency and distance, and is one to two orders of magnitude higher than that of side-scan sonar.

(2) Sonar interferometry principle

Interferometric Synthetic Aperture Sonar (InSAS) technology extracts three-dimensional spatial information of seabed targets based on the phase of sonar complex images. The basic idea is to equip a platform with a sound-emitting device and multiple receiving arrays in the cross-track direction to obtain pairs of sonar complex images of the same area. Due to the unequal distances between the receiving arrays and the seabed targets, a phase difference arises between corresponding image points in the pairs of sonar complex images, forming an interferometric fringe pattern. The phase values in the interferometric fringe pattern represent the measured phase differences between the two imaging sessions. Based on the geometric relationship between the imaging phase difference and the three-dimensional spatial position of the seabed targets, as well as the positional parameters of the platform, the three-dimensional coordinates of the seabed targets can be determined.

5.2 Principle of Sonar Technology

(3) Multi-beam Synthetic Aperture Sonar (MSSAS)

① Multi-beam synthetic aperture sonar development

Compared with side-scan synthetic aperture sonar, research on multi-beam synthetic aperture sonar started relatively late. The earliest appearance in literature was in 2001, when Japanese researchers applied synthetic aperture algorithms to the SeaBeam2000 multi-beam echo sounder and achieved excellent detection results. In 2002, American researchers applied for an invention patent for multi-beam synthetic aperture sonar with the United States Patent and Trademark Office, proposing the preliminary concept of multi-beam synthetic aperture sonar internationally for the first time. However, subsequently, there were no publicly published articles by researchers from this institution or internationally that conducted in-depth research on synthetic aperture algorithms using multi-beam echo sounders.

In 2015, Kongsberg Maritime, for the first time, applied synthetic aperture processing algorithms to data from its EM2040C shallow-water multi-beam echosounder system and compared the results with those obtained from the multi-beam echosounder. The comparison revealed that the synthetic aperture processing algorithm produced a more detailed underwater topographic image. The company named this system HISAS2040, which represents the latest example seen internationally to date of applying synthetic aperture processing algorithms to multi-beam sonar data.

Harbin Engineering University, through research on side-scan synthetic aperture sonar, was the first in China to propose the concept of multi-beam synthetic aperture sonar. The university independently conducted experiments using existing domestic multi-beam echosounder systems based on single-line arrays, demonstrating the feasibility of multi-beam synthetic aperture sonar. Compared with traditional multi-beam echosounder systems, it exhibits significantly improved resolution and is capable of achieving full-coverage mapping in a single survey. It provides good imaging of target depth information and cross-track coordinate information, and can effectively enhance the range resolution of synthetic aperture sonar while ensuring the same cross-track resolution as side-scan synthetic aperture sonar, completing seamless mapping directly below the survey area.

② Model of Multi-beam Synthetic Aperture Sonar

A multi-beam synthetic aperture sonar measurement model is proposed on the basic model of synthetic aperture sonar and multi-beam bathymetric sonar. It can complete the full coverage mapping of the mapping area at one time, without additional gap filling, while the multi-beam synthetic aperture sonar can get the target echo direction through the distance to the beam formation, so as to solve the depth of the target, forming a three-dimensional imaging sonar, the basic model is shown in Fig. 5.15. The key difference between multi-beam synthetic aperture sonar (MBSAS)

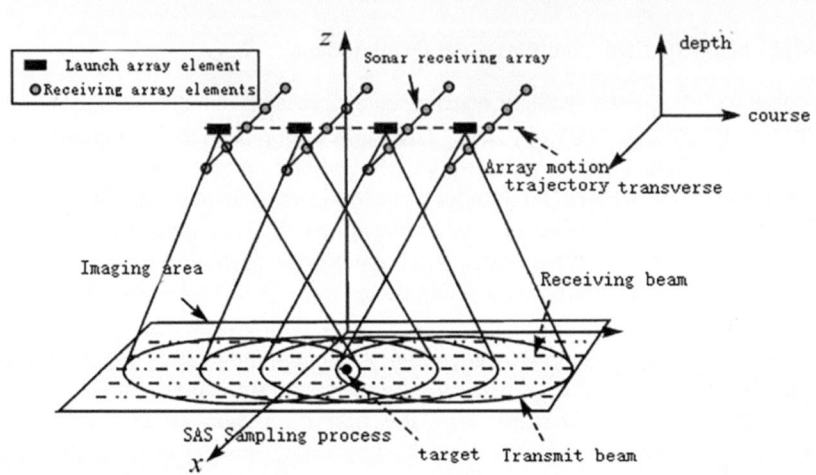

Fig. 5.15 Multi-beam synthetic aperture sonar model

and multi-beam echosounder is that the transmitted beam of the former has a large opening angle along the track direction. This allows the beam to illuminate the target multiple times at different positions along the track, enabling an improvement in cross-track resolution through synthetic aperture processing.

5.2.3 Data Processing

Sonar is the primary and effective method for marine underwater detection equipment, and its application results in various fields, whether it be multi-beam detection, side-scan sonar, or synthetic aperture sonar, are mainly presented in the form of sonar imaging. It not only provides detailed location information of underwater objects but also offers specific characteristics such as size and shape of the targets being explored, thereby enabling more intuitive and comprehensive research and analysis of underwater targets. Therefore, the research focus of sonar data processing lies in underwater imaging technology, including data denoising (image preprocessing), image segmentation, feature value matching, and data fusion techniques. Underwater imaging can be divided into array-based imaging and synthetic aperture imaging, among others. However, the underwater acoustic channel is an extremely complex random transmission channel in both time and space, and environments such as oceans and lakes are complex dynamic systems. Regardless of the form of underwater imaging, its imaging quality and resolution are far inferior to optical imaging. How to improve imaging quality during the post-processing of sonar imaging has always been the goal of sonar data processing and analysis research.

5.2 Principle of Sonar Technology

1. Noise reduction processing of sonar image

The marine environment is complex, the density of seawater in different layers in different regions is different, and a large number of particles are suspended in the ocean, coupled with the irregular surging of seawater, the echo received by the sonar inevitably contains noise. Moreover, the sonar itself is subject to interference from the carrier, such as the vibration of the ship's hull and engine operation, and electromagnetic interference from other equipment, etc. In addition, the sonar equipment itself is subject to interference from the carrier. In addition, the sonar equipment itself may also be disturbed by electronic equipment such as thermal noise. These noises can be classified into environmental noise, self-noise and radiated noise according to their sources.

Due to the influence of various factors in the process of collecting information by sonar, the image often has serious noise, multiple mismatches and other problems. Therefore, before performing sonar image post-processing, it is necessary to preprocess the images to be fused, including image denoising and image registration.

(1) The method of spatial domain filtering

The principle of spatial domain filtering is to decompose an image into two distinct vector spaces, namely signal and noise. Spatial domain denoising methods include mean filtering or weighted mean filtering, median filtering, Wiener filtering, bilateral filtering and so on.

① Mean value filter (weighted mean value filter)

Mean filtering, also known as linear spatial filtering, employs the neighborhood averaging method, where the grayscale value of each pixel is replaced by the average grayscale value of several neighboring pixels. It effectively suppresses additive noise but tends to cause image blurring, especially around the edges of the image.

② Median Filtering

Median filtering is a nonlinear signal processing technique based on the statistical theory of ordering, which can effectively suppress noise. The characteristic of median filtering is that it first determines a neighborhood centered on a specific pixel, then sorts the grayscale values of the pixels in the neighborhood, and takes the median value as the new grayscale value for the central pixel.

③ Wiener filter

Wiener filtering is an optimal estimator of a smooth process based on the minimum mean square error criterion. The mean square error between the output of this filter and the desired output is minimized, so it is an optimal filtering system. It can be used to extract signals contaminated by smooth noise.

4 Bilateral filtering

Bilateral filtering also employs a weighted averaging method, where the intensity of a certain pixel is represented by the weighted average of the brightness values of surrounding pixels, with the weights based on a Gaussian distribution. Most importantly, the weights in bilateral filtering not only consider the Euclidean distance between pixels (like in ordinary Gaussian low-pass filtering, which only considers the influence of position on the central pixel) but also take into account the radiometric differences in the pixel range domain (such as the similarity between pixels in the convolution kernel and the central pixel, color intensity, depth distance, etc.). Both weights are considered when calculating the central pixel. The kernel function of bilateral filtering is a combination of the spatial domain kernel and the pixel range domain kernel: in flat regions of the image, where pixel values change little, the corresponding pixel range domain weights are close to 1, and the spatial domain weights play a major role, equivalent to applying Gaussian blur; in edge regions of the image, where pixel values change significantly, the pixel range domain weights increase, thereby preserving edge information.

Mean filtering is the simplest filtering method and works well for zero-mean noise in homogeneous regions. However, since its operation essentially acts as a low-pass filter, it will inevitably result in some loss of detail and edge distortion. Wiener filtering is an older filtering method that effectively suppresses Gaussian noise and multiplicative noise. Median filtering and bilateral filtering are nonlinear filtering techniques that can compensate for the drawback of linear filters in blurring image edges. To improve filtering effectiveness, Buades proposed the method of nonlocal means denoising in 2007 [29].

(2) Frequency domain filtering method

Since the edge details of the image and the noise belong to the same high-frequency components, high-frequency filtering in the frequency domain will inevitably cause the loss of image edge details. Some scholars have studied the data characteristics of sonar images through visual data analysis language, and the image is transformed by 2D-FFT to make it filtered in the frequency domain. Shi Hong et al. [30] addressing the issue of unsatisfactory noise reduction and edge preservation in existing sonar images, proposed an image filtering method based on multi-resolution analysis using wavelet transformation in frequency domain filtering, which is a typical representative of this approach.

(3) Wavelet domain sonar image noise reduction

Due to its excellent time–frequency characteristics, wavelet transform can decompose different frequency components of a signal and is widely used in signal denoising. From a signal processing perspective, wavelet denoising is a signal filtering problem. Although it can largely be regarded as low-pass filtering, wavelet denoising has an advantage over traditional low-pass filters in that it successfully preserves image characteristics after denoising. Currently, the commonly used wavelet threshold denoising algorithms can be mainly classified into three types: hard thresholding function, soft thresholding function, and semi-soft/semi-hard thresholding function.

Anitha proposed a sonar image denoising algorithm based on the wavelet domain and compared it with spatial domain filtering algorithms such as median filtering, mean filtering, and Gaussian filtering. Experiments demonstrated that it has significant advantages in terms of denoising effect and edge preservation in the wavelet domain [31]. Nafornita conducted research on wavelet bases to find the most suitable one for sonar image decomposition, aiming to improve the denoising effect of sonar images [32].

(4) Methods of radar image filtering

Some scholars have proposed radar image filtering methods based on statistical models according to the types of noise, such as Lee filter, Kuan filter, and Frost filter. These filters, mostly belonging to adaptive filters, are designed based on the characteristics of image distortion in radar images acquired at oblique angles. Adaptive filters are designed to compress speckle noise while minimally reducing image resolution. They calculate a new pixel value using the standard deviation of each pixel value. These filters differ from traditional low-pass smoothing filters in that they suppress noise while preserving high-frequency information and details of the image.

The Lee filter is used to smooth noise data closely related to image brightness and additional or multiplicative types of noise. The enhanced Lee filter can reduce speckle noise while preserving the texture information of radar images. The Kuan filter is used to reduce speckle noise while preserving edges in radar images. The Frost filter can reduce speckle noise while preserving edges. The enhanced Frost filter can reduce speckle noise while preserving the texture information of radar images. The Gamma filter can be used to reduce speckle noise while preserving edge information in radar images.

(5) Other filtering methods

In addition to the above methods, there are other methods such as morphological filtering, fuzzy filtering, fusion of Hidden Markov Tree and morphology, denoising of sonar image by combining Bayesian maximum a posteriori probability estimation in Curvelet domain, and so on.

2. Sonar image alignment

Image alignment is a key step before image fusion. Image fusion refers to the best matching of two or more images, so that the images can be completely aligned in the geometric position. There exists a scaling, translation and rotation relationship between the image to be aligned and the reference image, and the goal of image alignment is to determine the correspondence of the three transformations and calculate the transformation matrix.

At present, according to the reference image information, image alignment can be divided into grey-scale information-based alignment algorithms and feature-based alignment algorithms.

(1) Alignment algorithm based on grey scale information

Alignment algorithm based on grey scale information is to use the grey scale information of the image to calculate some statistical information of the image, and take this information as the discriminative index of the image similarity, and then use a certain search algorithm to make the reference image and to be aligned to the maximum similarity of the image, and complete the alignment on this basis. Alignment methods based on grey scale information mainly include mutual information method, sequence similarity detection method and mutual correlation method. Mutual information method is to compare the statistical dependence of the images, and take the mutual information similarity as the criterion, when the mutual information reaches the maximum, the image to be aligned and the reference image are matched. Sequence similarity detection method is to calculate the residual sum of each point, and the point with the slowest growth of residual sum is the matching point. The correlation method calculates the value of the correlation between the search window and the reference image, and uses the degree of correlation to determine the degree of matching. Alignment algorithm based on grey scale information is simple to implement, but the use of a narrow range, large arithmetic, and also not applicable to the direct correction of non-linear changes in the image.

(2) Feature-based alignment algorithm

Feature-based alignment algorithm work by extracting some features of the image, such as feature points, straight line segments, edges, contours, etc., and then using these features of the image for feature matching. The matching relationship of the features reflects the mapping relationship between the reference image and the image to be aligned. This type of method is computationally small, fast, and robust to grey-scale changes in the image, so feature-based image alignment algorithms are currently more widely used methods. Commonly used feature matching methods include SIFT-based matching algorithm, Hough transform-based matching method, LSD method using line segment detection, SURF-based matching algorithm and so on. To speed up the algorithm, some researchers proposed ORB based matching algorithm. Later for fuzzy images and images with noise interference some, scholars proposed Kaze algorithm and so on. Due to the low contrast of the sonar image, if the grey scale information of the image is used for alignment, there will be many similarities and a high rate of mis-matching, thus affecting the alignment effect. Moreover, the noise of sonar images is complex, and the alignment method based on grey scale information is more sensitive to noise, so the feature-based alignment method is more suitable for sonar images.

5.2 Principle of Sonar Technology

① SIFT method

The scale invariant feature transform (SIFT) algorithm is suitable for sonar point feature extraction. The SIFT algorithm is used to accurately locate the position of the candidate feature points by establishing the scale space of the image, using the gradient to find out the magnitude and direction of the feature points, and finally constructing a 128-dimensional feature descriptor to complete the SIFT algorithm.

The SIFT algorithm introduces appropriate scale parameters in the image processing process, and by constantly changing the value of scale parameters and down sampling to get different blurring degree and size of the image, constituting a pyramidal scale space. The Gaussian convolution kernel is the only transformation kernel to realize the scale space. In order to ensure the robustness of the detected feature points, the image is differenced along the scale axis, and after the completion of the difference pyramid, the detection of feature points requires judging whether the candidate points in the difference pyramid meet the requirements of belonging to the extremely large or extremely small value points in the spatial neighbourhood, and if they do so, they will be marked as the preliminary feature points, and if they do not meet the requirements, they will be discarded. In order to improve the stability of the key points, it is necessary to calculate the curve interpolation of the DOG function of the scale space to further screen the feature points. The feature points are often used in the matching between images, SIFT algorithm in order to avoid the impact of lighting and visual changes on the matching, the use of feature vectors to construct a descriptor of the feature points, so as to improve the correct matching rate.

② Hough transform

The Hough transform is an effective method for straight line detection of binary images in the field of pattern recognition, and it is a global detection and extraction method for straight line feature extraction by analyzing the covariance of edge points. The Hough transform of straight line features mainly relies on the transformation between two coordinate spaces, mapping a straight line in one (image) space to a collection of points in another (parameter) coordinate space through the corresponding mathematical computation to form a peak, thus transforming the problem of detecting straight line features into a statistical peak problem.

Using the Hough transform for the extraction of linear features of sonar images, firstly, the sonar map needs to be denoised, and then the image is processed by edge processing, and on the basis of which the extraction of linear features is carried out. The methods for binary edge processing include Laplacian of Gaussian (LOG) operator, Robert operator and Canny operator. The edge points are mapped to the Hough space to get a series of curves combined by sine and cosine. The polar coordinate system is divided into grid cells according to the accuracy of radial distance and polar angle, with each cell corresponding to a coordinate of radial distance and polar angle. When points are mapped into curves and pass through each cell, the corresponding coordinate value is incremented by one, and this process is accumulated sequentially until all points have been mapped to their polar coordinates. The accumulation count

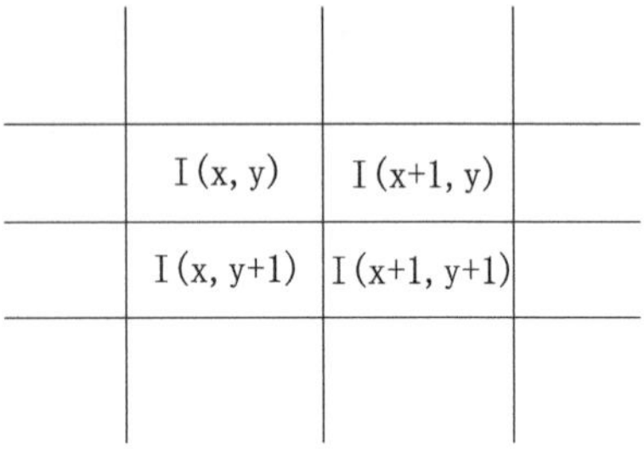

Fig. 5.16 Schematic of Pixel Point Location

	I (x, y)	I (x+1, y)
	I (x, y+1)	I (x+1, y+1)

for each cell is then tallied, and when it exceeds a set threshold, the cell is retained and inversely mapped back into the Cartesian coordinate system to obtain the final line features.

③ LSD method

The image line feature extraction in the Line Segment Detection (LSD) method is based on the gradient and magnitude of each pixel. The LSD algorithm uses a 2 × 2 template to calculate the gradient magnitude and direction of pixels, as shown in Fig. 5.16. Pixels with larger gradient magnitudes are selected as starting points, and adjacent points within a certain threshold of the gradient direction are grouped into the same region. This process is repeated for all points in the region as initial points until all pixels that meet the conditions are merged. To extract the line features of the region, a rectangular area is used to approximate the obtained region. To obtain the length, width, and center of the estimated rectangle, the center coordinate is located by calculating the pixel grayscale values. After determining the rectangular structure through calculations, the central axis of the rectangle is taken as the line feature to be extracted, serving as the result of line feature extraction for sonar images.

④ Kaze algorithm and its improvement

In order to ensure that the noise is removed without losing the detail information of the image, we make full use of the relationship between the conductivity coefficient and the local structural properties of the image. In the regions where the gradient of the image is very small, the conductivity coefficient will adaptively become larger so that the undulations caused by the presence of the noise are smoothed out in these regions, while in the regions where the gradient of the image is relatively large such as the edges of the image, the conductivity coefficient will adaptively become smaller, and the degree of smoothing will become smaller in these regions, thus achieving the purpose of protecting the edges. Perona and Malik defined a conduction function c based on the gradient magnitude of the image, which is defined as follows:

$$c(x, y, t) = g(|\nabla \mu(x, y, t)|) = \frac{1}{1} + \left(\frac{|\nabla \mu|}{K}\right)^2 \in [0, 1] \quad (5.19)$$

where the luminance function $\nabla \mu$ is the gradient of μ of the original image.

5.2 Principle of Sonar Technology

The PM equation is defined as

$$\begin{cases} \frac{\partial \mu}{\partial t} = div(c(\nabla \mu)\nabla \mu), & (x, y, t) \in R^m \\ \mu(x, y, t) = \mu(x, y), & (x, y) \in R^m \end{cases} \quad (5.20)$$

The equation of the regularized PM is:

$$\frac{\partial \mu}{\partial t} = div(g|\nabla \mu_\sigma|\nabla \mu_\sigma) \quad (5.21)$$

$$\mu_\sigma = G_\sigma * \mu \quad (5.22)$$

where G_σ denotes the Gaussian smoothing kernel with variance σ, which means that the gradient of the Gaussian smoothed image is used to replace the original image gradient in the original PM equation.

The conduction function c in Eq. (5.19) is a bounded function that depends on the monotonically decreasing magnitude of the image gradient, which is also called the diffusion function or edge stopping function. The parameter K is a constant term used to calculate the value of the diffusion function, which quantifies the magnitude of the image gradient and can usually be set manually. The transfer function can be changed adaptively according to the gradient information of the image region. When dealing with flat regions, the image gradient magnitude is much smaller than K, c(|u|) ≈ 1, and the image is smoothed; when dealing with regions containing detailed information such as the edges of the image, the image gradient magnitude is much larger than K, c(|u|) ≈ 0, and the degree of image smoothing will be very weak. When c(|u|) = 0, the smoothing operation stops and the image is not smoothed. Repeatedly iterating the above process, the noise in the image will be filtered out, and the detail information such as edges is not damaged.

The Kaze algorithm is improved by introducing the bootstrap filtering, through which the edge information of the image can be well protected while smoothing the noise. The bootstrap filtering algorithm performs different levels of filtering operations on the input image according to the gradient information of the bootstrap image during the filtering process. In the region where the gradient of the bootstrap image is large, the corresponding region of the input image is smoothed to a small extent, while in the region where the gradient of the bootstrap image is small, the corresponding region of the input image is smoothed to a large extent.

The bootstrap filter is a window-based adaptive filtering operation. If a pixel is located in a window with a large variance and a large gradient in the bootstrap image, the corresponding pixel in the input image is directly retained in the output image, while if a pixel is located in a window with a small variance and a small change in the pixel value in the bootstrap image, the pixel in the corresponding position in the output image is the weighted average of the pixels in the corresponding window in the input image. The bootstrap filter can be interpreted as two parallel box filters approximated as Gaussian filters when dealing with smooth regions, thus maximizing the retention of regions with large variations in the edges, and at the same time making the transition from smooth regions to edges smoother. This improved Kaze algorithm can better protect the edges of the image when filtering compared with the original PM algorithm, which is more suitable for the alignment of sonar images with weak features.

⑤ Other algorithms

Feature extraction of sonar image and image standard matching have been the focus of underwater image processing research at home and abroad, and there have been many research results in recent years. For example, scholars have proposed a feature extraction method to extract and classify features from the shadow of sonar image; Reed et al. use Markov random field (MRF) model combined with cooperating statistical snake (CSS) model to extract features from sonar image. Currently, an improved Hough transform method is proposed to address the issue of unsatisfactory results in extracting linear features using the traditional Hough transform. Additionally, a broken line feature fitting method is designed to address the problem of line features detected by the LSD algorithm being broken due to intersections.

(3) Sonar image fusion technology

The fusion method based on multi-resolution decomposition has been a research hotspot in the field of fusion, and this kind of method is widely used and applicable, and has achieved good fusion effect in remote sensing image, medical image and optical multi-sensor image fusion.

Sonar image is similar to optical image in nature, which is the distribution of energy in the spatial domain, but due to the special underwater environment and the different imaging mechanism of sonar, sonar image is different from the traditional optical imaging [33]. The main difference between sonar image and optical image manifest as follows: Sonar images underwater have low imaging resolution, are dominated by low frequencies, and have blurred details, making precise target identification difficult. The contours of underwater targets are incomplete and damaged, such as the edges of shipwrecks and aircraft wreckage being blurred due to corrosion from seawater and underwater organisms, making it hard to form detailed and precise edge features. Interference in the underwater acoustic channel, such as environmental noise, reverberation, and self-noise of the sonar, results in significant noise after sonar imaging. These noise sources, mainly speckle noise, have rich gray levels, while the gray levels of sonar targets account for a relatively small proportion, leading to poor sonar imaging quality and low contrast. Therefore, for sonar image processing, traditional image processing techniques may not yield good results, and instead, multi-resolution fusion techniques should be adopted for processing sonar images.

The fusion method based on multi-resolution decomposition is mainly composed of three parts: multi-resolution decomposition of images, fusion of images in different frequency sub-bands, and image reconstruction. The specific framework is shown in Fig. 5.17.

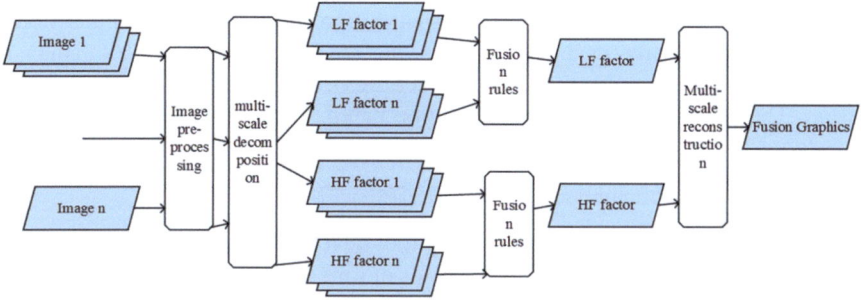

Fig. 5.17 Framework for sonar image fusion processing based on multi-resolution

Multi-resolution analysis is a flexible and efficient algorithm to deal with high-dimensional signals, often used to analyze non-stationary signals. The multi-resolution analysis of images is analogous to the principle of a camera, where distant objects appear small and vague within a wide field of view, while nearby objects can be observed in detail but within a narrower range. As the scale changes, object information is presented from coarse to fine, aligning with the human visual perception system. Applying this multi-resolution analysis method in the field of image fusion involves decomposing the images to be fused into multiple scales and selectively applying fusion rules to different frequency sub-band images to achieve optimal fusion results. Among the multi-resolution decomposition methods, the wavelet transform is typical and widely used, effectively representing one-dimensional signals but less adept at representing two-dimensional or higher-dimensional signals. This is because the wavelet transform can only express information in horizontal, vertical, and diagonal directions, and its two-dimensional transform basis has a square support interval, which does not well approximate the inherent singular curves in images. The proposal of multiscale geometric analysis overcomes the shortcomings of the wavelet transform. Typical multiscale geometric analysis methods include ridgelet transform, curvelet transform, contourlet transform, and nonsubsampled contourlet transform (NSCT). The ridgelet transform can well approximate multidimensional functions with straight-line singularities but has an approximation capability for multidimensional functions with curves equivalent to the wavelet transform. The curvelet transform can efficiently approximate linear singularities such as edges in images and is the sparsest representation of continuous curves in the C^2 space. However, its discrete implementation is difficult, limiting its application in multi-resolution decomposition. The contourlet transform uses rectangular structures to approximate images and can well describe geometric structural information in images. However, due to its low geometric regularity of basis functions and insufficient locality in both spatial and frequency domains, there is frequency aliasing in the frequency domain, and the implementation of this transform in the multi-directional decomposition stage is relatively cumbersome. The NSCT is an improvement of the contourlet transform, ensuring translation invariance of the algorithm. In recent years, Easley [34] proposed a new multiscale geometric analysis method called shearlet transform, which provides near-optimal sparse representation of multidimensional functions. Additionally, methods based on dual-tree dual-density wavelet analysis and dual-tree high-density wavelet analysis have been applied in sonar image data fusion. The commonly used multi-resolution based sonar data fusion methods are introduced below.

① Wavelet transform analysis

Wavelet analysis is a new data processing tool developed in the twentieth century. The Wavelet transform is a localized analysis of time (space) and frequency. It gradually refines signals (functions) at multiple scales through expansion and translation operations, ultimately achieving time subdivision at high frequencies and frequency subdivision at low frequencies. This automatically adapts to the requirements of time–frequency signal analysis, enabling focus on any detail of the signal. The advantage of the Wavelet transform is its ability to extract "specified time" and "specified frequency" changes in signals, which is why it is praised as the "mathematical microscope".

Wavelet transform includes continuous wavelet transform, discrete wavelet transform (DWT) and two-dimensional wavelet transform. Since the expansion and translation parameters in continuous wavelet transform are continuous real numbers, they usually need to be integrated in applications, so discrete wavelet transform (DWT) is often used in signal processing. In image processing, two-dimensional wavelet transform is usually used, and two-dimensional discrete wavelet function needs to be constructed when using two-dimensional wavelet transform. When performing two-dimensional wavelet decomposition on an image, the process involves constructing two-dimensional orthogonal wavelet basis functions to carry out decomposition at different scales. The process of image decomposition is illustrated in Fig. 5.18.

f(x, y) represents the original image, A represents the approximate image (low-frequency image), B represents the detailed image (high-frequency image), the subscript represents the number of layers of decomposition, and the mathematical expression for the decomposition is:

$$S \approx A_1 + B_1 \approx A_2 + B_2 + B_1 \approx A_3 + B_3 + B_2 + B_1 \tag{5.23}$$

Since the process of decomposition is iterative, theoretically it can go on continuously without limitation, but in practical image analysis, decomposition can be carried out until the details contain only a single sample. The more layers of decomposition, the more we can make full use of the correlation between the coefficients in the sub-bands of the details of each layer with the same direction and location, which is more conducive to the hierarchical analysis of the image. But at the same time, it should be noted that the more layers of decomposition, the signal-to-noise ratio of the reconstructed signal decreases, so in practice, the number of layers of decomposition can be selected according to the characteristics of the image or the appropriate criteria.

The following defects exist in the processing of sonar images using wavelet transform: Due to the band-pass filtering characteristic of the Wavelet transform, wavelet

5.2 Principle of Sonar Technology

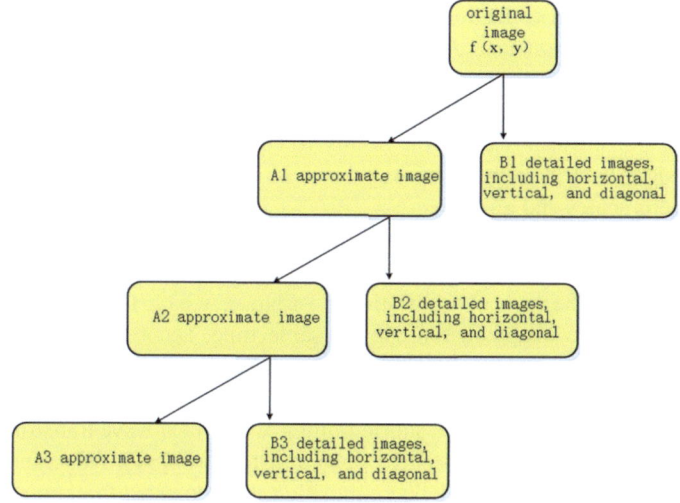

Fig. 5.18 Wavelet transform decomposition of a 2D image

coefficients exhibit a ringing-like fluctuation with positive and negative values at singular points, causing the wavelet coefficients at these points to be amplified. When the signal undergoes translation or even slight changes at singular points, the wavelet coefficients will fluctuate accordingly. The two-dimensional wavelet transform does not have good effect in the processing of the surface. The two-dimensional wavelet transform is not ideal for image processing of curved surfaces. The reason is that the separable wavelet transform only provides three wavelet directions: horizontal, vertical and oblique.

② Shear Wave Transform

Shear waveform transform refers to the expansion of one-dimensional wavelet transform to multi-dimensional form, to achieve anisotropic representation of image information in different scales, directions and positions. The shear wavelet transform is a new method of multiscale geometric analysis, which is close to the optimal sparse representation of multidimensional functions. It has the sparsest expression ability of high-dimensional information such as image edges similar to the curve wave transform, and also has a freer direction decomposition than the contour wave transform and non-downsampled contour wave transform. Moreover, since the forward transformation process of the shear wave transform only needs to simply sum the shear wave filters, it is simple to implement and has high computational efficiency. The mathematical structure of the shear wave transform is simple, and its basis function can be obtained by simple translation, expansion and rotation of a function. Moreover, the shear wave transform is layer-by-layer subdivided in the frequency domain space, and it is this layer-by-layer subdivided characteristic that makes it obtain better sparse expression of multi-dimensional functions.

The shear wave is a collection of functions with good localization in scale, direction and position parameters. As the scale parameter a decreases, the asymptotic decay property of the shearlet transform of an image not only describes the locations of edges in the image but also indicates the directions at the edges. Therefore, the shearlet transform possesses the ability to capture the directional information of image edges and can represent images rich in directional information optimally across various scales and directions.

Normally, the downsampling operation is performed in the two stages of multi-scale decomposition and localized direction decomposition of the shear wave transform, resulting in the lack of translation invariance, which is prone to cause the ringing effect in the image fusion. At the same time, according to the theory of multi-sampling rate, the operation of down-sampling the filtered image in rows and columns will easily lead to frequency aliasing.

In order to overcome the above shortcomings of the shear wave transform, Easley proposed the translation invariant shear wave transform (NSST). This transform also consists of two parts, multi-scale decomposition and localized direction decomposition. Among them, the multi-scale decomposition part adopts the non-subsampled Laplace pyramid filter (NSPF), in the decomposition process, the low-frequency sub-bands of the upper level should be filtered by the upper sampling filter to complete the filtering, and the sub-bands obtained by using the non-subsampled pyramid filter are the same in size as the input image.

The sonar image fusion algorithm based on the shearlet transform first applies the non-downsampled shearlet transform (NSST) to two original images separately to obtain multiband coefficients. Different fusion rules are applied to the low-frequency and high-frequency sub-bands. Specifically, a weighted averaging rule is used for the low-frequency sub-band, while a rule of selecting the larger absolute value is employed for the high-frequency sub-band. Finally, the fused image is obtained through the inverse non-subsampled shearlet transform. The flowchart of the image fusion algorithm based on the non-downsampled shearlet transform is shown in Fig. 5.19.

③ Improvement of Shear Wave Transform

When the noise of the sonar image is serious, the fusion algorithm based on the non-downsampled shear wave transform cannot get a good fusion image. The main reason is that in the multi-scale decomposition stage, the noise is decomposed into the high-frequency sub-band as the detail information, and the coefficients of the high-frequency sub-band corresponding to the noise are relatively large. If the absolute value of the high-frequency is taken as a rule, the noise in the original image will be fused into the final image while preserving the edges of the image.

The improved shearlet transform data fusion algorithm employs an enhanced PM diffusion equation to refine the multiscale decomposition phase within the shearlet transform. This refinement involves transitioning from non-downbsampled Laplacian decomposition to nonlinear diffusion filtering for multiscale decomposition, ultimately achieving better preservation of sonar image edges during the decomposition phase.

5.2 Principle of Sonar Technology

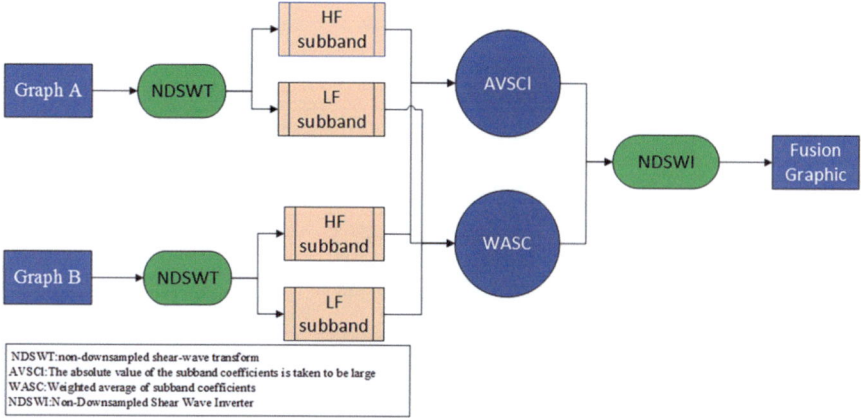

Fig. 5.19 The process of image fusion algorithm based on non-downsampled shearlet transform

The isotropic nonlinear diffusion filter has a smoothing effect, and the residual difference between the original image and the filtered image is a high-frequency signal, which provides a basis for the realization of the multiscale decomposition and reconstruction of the improved nonlinear diffusion filter. The specific steps of the decomposition and reconstruction of the improved nonlinear diffusion filter are as follows.

(a) Determine the number of decomposition scale layers n for the image to be processed, and adjust the number of diffusion filtering iterations N accordingly based on the number of decomposition scale layers to increase or decrease the iterations.
(b) Perform nN_t diffusion filtering on the image I_0 to be processed, and the obtained filtering result is denoted as $I_1^{nN_t}$, and the residual is denoted as $E_1^{nN_t}$;
(c) Perform $(n-1)N_t$ diffusion filtering on $E_1^{nN_t}$, and the obtained filtering result is recorded as $I_2^{(n-1)N_t}$, and the residual is recorded as $E_2^{(n-1)N_t}$t;
(d) Similarly, N diffusion filters are applied to $E_n^{N_t}$, and the obtained filtering result is denoted as $I_n^{N_t}$, and the residual is denoted as $E_n^{N_t}$.
(e) The reconstruction process is $I_0 = I_1^{nN_t} + I_2^{(n-1)N_t} + \cdots + I_n^{N_t} + E_n^{N_t}$

After the above multi-scale decomposition, $E_n^{N_t}$ is the information of the highest frequency, $I_1^{nN_t}$ is the information of the lowest frequency, and $I_1^{nN_t} \sim I_n^{nN_t}$ is the frequency increment of the signal.

The sonar image fusion algorithm based on improved shear wave transform firstly carries out the improved non-downsampled shear wave transform on the two original images, obtains the multi-frequency sub-band coefficients, gets the fusion coefficients according to a certain fusion rule, and finally obtains the fused image through the improved shear wave inverse transform. The flow of the image fusion algorithm based on the improved shear wave transform is shown in Fig. 5.20.

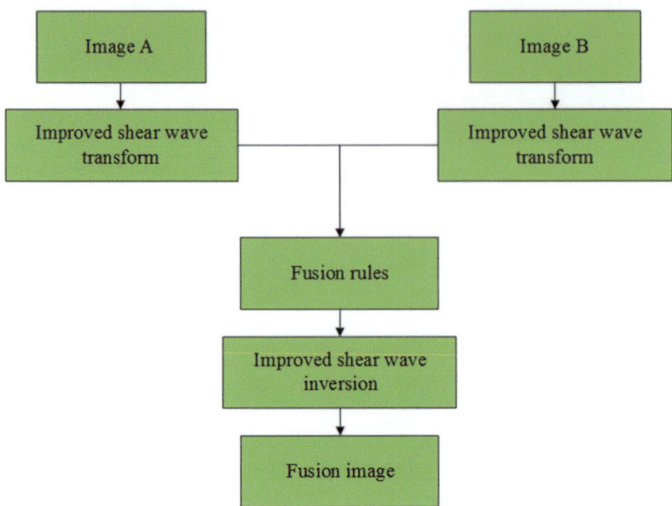

Fig. 5.20 Flow of image fusion algorithm based on improved shear wave transform

④ Curvelet transform (curvelet)

Curvelet transform is a new theory proposed on the basis of wavelet transform, with the characteristics of anisotropy, so that the geometric features of the image has a better ability to express, and can effectively extract the features of the image edge. The wavelet transform has gone through two versions of development. In 2000, Candes and Donoho [35] firstly proposed a generation of wavelet concept, which is mainly derived from the basic principles of Ridgelet analysis. This algorithm is a transform based on the combination of multiscale Ridgelet and bandpass filter banks. The first generation curve wave algorithm is based on Ridgelet analysis, and the Radon transform in the Ridgelet is inefficient to perform in either the time or frequency domain, and the algorithm is complicated to implement. Therefore, easy-to-implement second-generation curvilinear waveforms are proposed to discard the use of the Ridgelet transform as a preprocessing step for the Cuvelet transform and to reduce algorithmic redundancy to improve the speed of computation.

Construct a second-generation continuous curvilinear wave structure in a two-dimensional space R^2, set the space variable x and the frequency domain variable $w = (w_1, w_2)$, and further obtain the frequency domain polar coordinates $r = \sqrt{w_1^2 + w_2^2}$, $\theta = \arctan\left(\frac{w_1}{w_2}\right)$, Define a pair of window functions for the Curvelet transform: the radial window $\left\{w(r), r \in \left(\frac{1}{2,2}\right)\right\}$ and the angular window $\{V(t), t \in [-1, 1]\}$, which satisfy the tolerance conditions:

5.2 Principle of Sonar Technology

$$\sum_{j=-\infty}^{\infty} W^2(2^j r) = 1, r \in \left(\frac{\frac{3}{4,3}}{2}\right) \quad (5.24)$$

$$\sum_{l=-\infty}^{\infty} V^2(t-l) = 1, t \in \left(-\frac{\frac{1}{2,1}}{2}\right) \quad (5.25)$$

For each $j \geq j_0$, a radial window and an angular window in the frequency domain form the window function U_j in the frequency domain.

$$U_j(r, \theta) = 2^{-\frac{3}{4}} W(2^{-2j} r) V\left(\frac{2^{\left|\frac{j}{2}\right|} \theta}{2\pi}\right) \quad (5.26)$$

where $\left|\frac{j}{2}\right|$ is the integer part of j/2. The support interval of U_j is a wedge obtained by the restriction of the W and V support intervals, and the wedge conforms to the properties of anisotropic scales as shown in Fig. 5.21.

Let $\varphi_j(\omega) = U_j(\omega)$, noting the angle parameter as $\theta_l = 2\pi \times 2^{\left|\frac{j}{2}\right|} \times l, l = 0, 1, \ldots, 0 \leq \theta_l < 2\pi$, and the position parameter as $k = (k_1, k_2) \varepsilon Z^2$, the Curvelet function defined at scale 2^{-j}, direction θ_l, position at $k = (k_1, k_2)$ is:

$$\varphi_{j,l,k}(x) = \varphi_j\left(R_{\theta l}\left(x - x_k^{(j,i)}\right)\right) \quad (5.27)$$

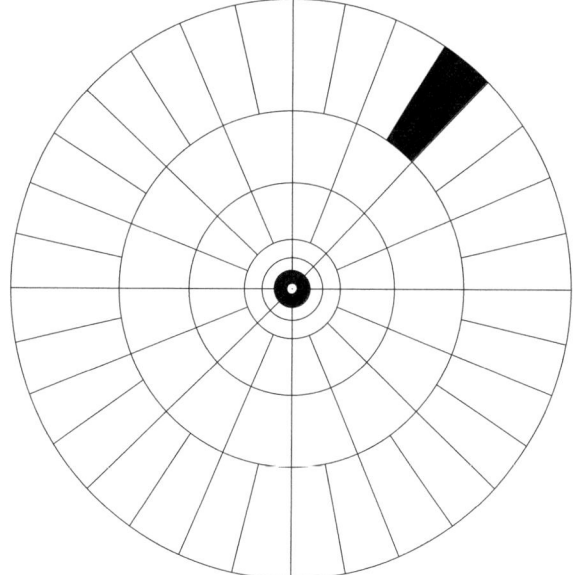

Fig. 5.21 Frequency schematic of curvelet function

where $x_k^{(j,l)} = R_{\theta l}^{-l}(k_1 \times 2^{-j}, k_2 \times 2^{-j})$, $R_{\theta l}$ denotes the rotation matrix.

Similar to the theories of wavelet transform and Ridgelet transform, the Curvelet transform is defined using the inner product form between the basis function and the signal (or function).

$$c(j, l, k) = <f, \varphi_{j,l,k} \geq \int_{R^2} f(x)\overline{\varphi_{j,l,k(x)}}dx \tag{5.28}$$

It can be derived from the above equation through Plachherel's theory:

$$c(j, l, k) = \frac{1}{2\pi^2} \int f(\omega)\varphi_{j,l,k(x)}d\omega = \frac{1}{2\pi^2} \int f(\omega)U_j(R_{\theta l}\omega)e^{j(x_\mu(j,i),\infty)}d\omega \tag{5.29}$$

For the algorithm implementation of the Curvelet transform on sonar images in computers, discretization and calculation of Curvelet coefficients are also required. There are mainly two algorithms to achieve the fast discrete Curvelet transform (FDCT): the two-dimensional FFT algorithm based on unequal spatial sampling (USFFT) and the Wrap algorithm. The main difference between these two implementations lies in the spatial grids selected by the FDCT at different scales and directional angles. The Wrap algorithm, based on USFFT, adds specified frequency sampling wrapping rules to achieve faster implementation of the algorithm.

5 Dual-tree dual-density wavelet transform

Due to the shortcomings of the traditional wavelet transform, such as translation invariance and poor directional selectivity, some scholars have constantly proposed to improve the problem of poor accuracy of traditional wavelet decomposition by increasing the redundancy of wavelets. To address the deficiencies of wavelets, some scholars have proposed a new wavelet framework and complex wavelet. The structure of complex wavelets extends from the real number field to the complex number field. The Daubechies wavelet is the earliest proposed complex wavelet, which not only inherits the multi-resolution analysis and time–frequency localization capabilities of traditional real wavelets for signals but also possesses excellent translation invariance and multi-directional selectivity. However, since this wavelet structure is in complex form, it is difficult to meet the perfect reconstruction condition during wavelet reconstruction, making it challenging to implement this algorithm in practice. Based on this, Kingsbury designed a new complex wavelet structure known as the double-tree discrete wavelet transform (Dt-Dwt) [36].

The Dt-Dwt is designed as a binary tree structure with two parallel Dwt paths, and its decomposition structure is shown in Fig. 5.22.

In the design of Dual-Tree Discrete Wavelet Transform (DT-DWT), one tree employs real wavelets, while the other uses imaginary wavelets. When performing the first-level decomposition, the designed filter delay in one tree is exactly one sampling interval different from that in the other tree, such that the samples taken by one tree correspond precisely to the values discarded by the other tree due to downsampling by two. For subsequent decomposition levels, it is only necessary to ensure that the

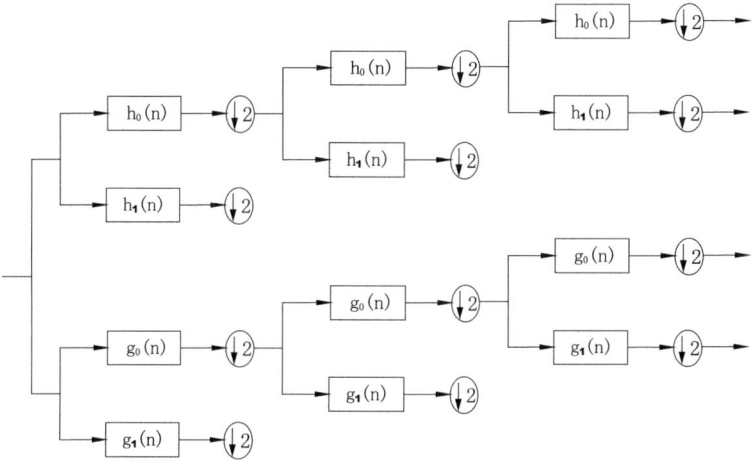

Fig. 5.22 Dt-Dwt decomposition structure diagram

sum of the delay differences at that level and all preceding levels, relative to the input, amounts to one sampling period. Based on this idea, it is relatively straightforward to design filter banks that satisfy perfect reconstruction. In the decomposition structure diagram, $h_0(n)$ and $h_1(n)$ represent the low-pass and band-pass filter pair of the tree A filter bank, respectively, while $g_0(n)$ and $g_1(n)$ represent the low-pass and band-pass filter pair of tree B filter bank. The dual-tree complex wavelet function is described as follows.

$$\phi_h(t) = \sqrt{2}\sum_n h_0(n)\phi_h(t) \tag{5.30}$$

$$\psi_h(t) = \sqrt{2}\sum_n h_1(n)\phi_h(t) \tag{5.31}$$

$$\phi_g(t) = \sqrt{2}\sum_n g_0(n)\phi_g(t) \tag{5.32}$$

$$\psi_g(t) = \sqrt{2}\sum_n g_1(n)\phi_g(t) \tag{5.33}$$

$$\psi(t) = \psi_h(t) + j\psi_g(t) \tag{5.34}$$

where $\psi_g(t)$ and $\psi_h(t)$ are approximate Hilbert transform relations: $\psi_g(t) = H\{\psi_h(t)\}$.

Double density Discrete wavelet transform (Dd-Dwt) is an improved discrete wavelet transform proposed by Selenick based on the traditional DWT. Compared to the traditional Discrete Wavelet Transform (DWT), the Dual-Density DWT (Dd-DWT) features an additional directional wavelet, encompassing one scaling function and two wavelet functions. One of the wavelet functions approximates a half-time-unit delay of the other, resulting in smaller frequency band intervals between adjacent wavelets within the same scale. This interleaved structure provides the wavelet with shift invariance and allows for a high degree of freedom in designing the filter for the two-channel wavelet functions [37]. During the decomposition and reconstruction process of Dd-DWT, the signal undergoes decomposition and reconstruction through a multi-channel filter bank composed of $h_0(n), h_1(n), h_2(n)$, which correspond to the low-pass, band-pass, and high-pass filters, respectively. The scaling function and wavelet functions are defined as follows.

$$\phi(t) = \sqrt{2} \sum_n h_0(n) \phi_h(2t - n) \tag{5.35}$$

$$\psi_i(t) = \sqrt{2} \sum_n h_i(n) \phi(2t - n), i = 1, 2 \tag{5.36}$$

$$\psi_1(t) = 2\psi_1(t - 0.5) \tag{5.37}$$

The dual-tree dual-density wavelet transform DtDd-Dwt is a tightly-supported structure with redundancy of 3. Its design combines the features of dual-tree wavelet approximate translation invariance and high design freedom of dual-density wavelet bandpass filter banks. The one-dimensional signal DtDd-Dwt two-layer decomposition structure has two parallel oversampling filter banks, where $h_i(n)$ and $g_i(n)$ (i = 0, 1, 2) are finite shock response filters; the wavelet basis functions of DtDd-Dwt consist of two scale functions and four wavelet functions. The two-dimensional DtDd-Dwt is an extension of the one-dimensional, and four dual-density oversampling filter banks are used to filter the image in rows and columns, respectively, and the image is decomposed by the DtDd-Dwt to obtain four approximate sub-bands and 32 detailed sub-band coefficients. The first layer of decomposition represents the real part, with six directions of for its real part has $\pm 15°, \pm 45°, \pm 75°$, the second layer of decomposition represents the imaginary part, also with 6 directions. Thrrefore, each layer of decomposition has 12 directions and each direction is represented by two wavelets, increasing the information redundancy; each layer of decomposition results in 32 detail sub-bands representing various directions, and each detail subband provides a fine-to-coarse description of the same edge or contour information in the image at different directions and resolutions.

⑥ Dual-Tree High-Density Wavelet Transform

Higher Density Discrete wavelet transform (Hd-Dwt) consists of a three-channel oversampling filter bank, whose decomposition structure is shown in Fig. 5.23, where the first two groups of filters need to be down-sampled while the third channel filter is non-extractive sampling. $h_i(k)(i = 0, 1, 2)$ is the low-pass, band-pass and high-pass filter bank, respectively, and c_i and d_i are the low-frequency coefficients and high-frequency coefficients of the filter bank after decomposition.

5.2 Principle of Sonar Technology

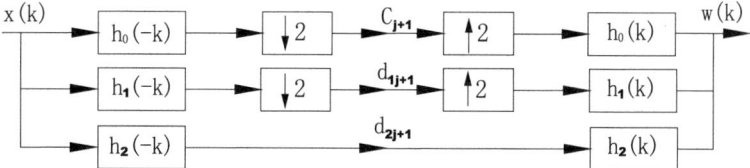

Fig. 5.23 High-density wavelet decomposition and reconstruction filter banks

The double-density wavelet transform only increases the sampling density in the time domain, while the high-density is sampled twice in the time domain and the scale at the same time, which is designed to make the extended wavelet have the middle scale characteristics, further increase the redundancy and improve the decomposition accuracy. The High-Density Discrete Wavelet Transform can achieve similar precision in signal transformation to the non-decimated wavelet transform, which has higher redundancy, while effectively reducing algorithmic complexity, thereby facilitating engineering implementation.

Considering the multi-directional selectivity on the basis of the high density wavelet transform, a new wavelet can also be obtained by combining the double tree complex wavelet with the high density wavelet: double tree high density wavelet transform. The framework structure of Double tree High density Discrete wavelet transform (DtHd- Dwt) consists of 2 scale functions and 4 wavelet functions. The one-dimensional filter bank structure of the double tree high density complex wavelet DtHd-Dwt is similar to the structure of the double tree dual density wavelet filter bank, with two parallel oversampling filter banks, where $\{h_i(k)\}$ and $\{g_i(k)\}$ (i = 0, 1, 2) are finite impact response filters (FIR), tree A outputs a sequence of real parts of the signal, tree B outputs a sequence of imaginary parts of the signal $h_i(k)$ and $g_i(k)$.

The basic principle of two-dimensional dual-tree high-density wavelet transform is as follows: firstly, the row sequence of the image is convolved with a one-dimensional DtHd-Dwt filter to perform row filtering, and then the column sequence of the image is convolved with a one-dimensional DtHd-Dwt filter (column filtering), and then the dual-tree wavelet tensor product is used to obtain the final two-dimensional DtHd-Dwt directional sub-band.

The Dual-Tree High-Density Discrete Wavelet Transform (DtHd-Dwt), due to its denser sampling on the time–frequency plane, results in smoother envelopes, which is more conducive to increasing the approximation of the original signal after decomposition and reducing the oscillation effect at signal singularities.

⑦ Other data fusion methods

Due to the differences in the characteristics of sonar images and optical images, the requirements of data fusion algorithms are relatively high. With the development of computer technology and artificial intelligence, some scholars also continue to research on algorithms for sonar image fusion technology. In addition to the algorithms introduced above, there are also studies on the combination of different algorithms for data fusion. For example, the algorithm of multi-scale GMRF level set segmentation combines the curvilinear waveform transform and Gauss-Markov random field; the sparse representation (SR) theory is used in the multi-focused image fusion; the image fusion algorithm based on the theory of higher-order singular value decomposition (HOSVD); the quad-tree-based method, genetic algorithm and difference algorithm, etc. In addition, the neural network is used in the image research for the integration of data. In addition, neural networks are used in image fusion, such as pulsed neural network (PCNN), and methods that combine PCNN with transform domains.

5.2.4 Typical Sonar Products

1. Multi-beam Sonar

The major international manufacturers of deepwater multi-beam echo sounder systems include companies such as L3 ELAC Nautik, Teledyne (ATLAS), and Kongsberg. Table 5.1 gives a comparison of typical deepwater multi-beam bathymetric systems.

(1) L3 ELAC Nautik company's SeaBeam3012

SeaBeam3012 is the latest generation of deepwater multi-beam bathymetric system of L3 ELAC Nautik Company. It operates at a frequency of 12 kHz, with a working depth range of 50–11,000 m, and features 301 beams with a minimum beam width of $1° \times 1°$. It offers both equal-angle and equal-distance beam spacing, a maximum coverage width of 5.5 times the water depth, and a maximum operating speed of up to 12 knots. The SeaBeam 3012 employs advanced beam-steering patent technology to fully compensate for yaw, pitch, and roll motions. It is the world's only full-ocean-depth multi-beam echo sounder system capable of real-time full-attitude motion compensation at all depths. The new beam-steering technology includes features such as wide coverage, shallow-water near-field focusing, multi-pulse, and chirp modulation, making its performance far superior to other conventional sector-scanning technologies.

The SeaBeam 3012 system is capable of collecting real-time bathymetric information, backscatter data, water column data, side-scan sonar images, etc., and presenting the measurement results to the operator in a good visual form. It has high application value in the fields of deep-sea topography mapping, seabed tectonic research, marine resources exploration, gas hydrate detection, geophysical exploration, and so on.

SeaBeam3012 multi-beam bathymetric system consists of underwater transmitting and receiving transducer arrays, surface receiving and transmitting units, data acquisition and data post-processing computer. Auxiliary equipment includes surface sound velocity meter, fibre optic compass motion sensor, uninterruptible power supply, acoustic synchronizer, graphic display and post-processing computer.

5.2 Principle of Sonar Technology

Table 5.1 Comparison of multi-beam sonar parameters

Parameter	Products		
	SeaBeam 3012	HydroSweep DS	EM122
Basic principle	Beam control method	Principle of coherent sonar	Beam control method
Transmission frequency/KHz	12	14–16	12
Measuring depth/m	50–11,000	10–11,000	20–11,000
Transmit beam width	1°, 2°	0.5°, 1°, 2°	0.5°, 1°, 2°
Receiving beamwidth	1°, 2°	1°, 2°	1°, 2°, 4°
Maximum strip width	5.5 times water depth	5.5 times water depth	6 times water depth
Maximum coverage width/km	31	28	30
Number of beams per strip	301	320	288
Maximum ping rate/Hz	3	10	5
Bathymetric resolution/cm	12	6	10–40
Bathymetric accuracy	0.2% water depth	0.2% water depth	0.2% water depth
Transmit waveform	CW	CW/LFM/Barker	CW/LFM
Beam spacing	Equal angle/equal distance	Equal angle/equal distance	Equal angle/spacing/encryption
Maximum operating speed/knot	12	10	16
Transmitter array length	Approx. 7.7 m (1°)	Approx. 5.6 m (1°)	Approx. 7.8 m (1°)
Receiving array length	Approx. 7.7 m (1°)	Approx. 5.6 m (1°)	Approx. 7.8 m (1°)

(2) Teledyne HydroSweep DS

HydroSweep DS belongs to the third generation of full-depth multi-beam bathymetric system, with an operating frequency of 14–16 kHz, a working water depth from 10 to 11,000 m, a number of 320 beams, a minimum beamwidth of 0.5° × 1°, two types of beam intervals of equiangularity and equidistant spacing, a maximum coverage of 5.5 times the width of the water depth, and a maximum working speed of up to 10 knots. It can continuously and uninterruptedly acquire the more demanding seabed data, which are characterized by water column, backscatter and sediment analysis. At the receiving end, each transmission is broken down into 320 received beams.

To overcome the limitations of conventional array apertures, HydroSweep DS uses a patented receiving beam formation technique known as Higher Order Beam Formation. As for the bathymetric data, it enables scanning at angles up to 140° (5.5

times the water depth), broken down into 960 narrow beams. The acoustic footprint is set to an isometric or iso-angular mode to suit specific survey needs. It possesses multi-frequency transmission capabilities, allowing for the simultaneous emission and reception of multiple frequencies, whereas traditional echo sounders typically operate on a single-frequency cycle. Multi-frequency significantly enhances survey efficiency, especially when beam accuracy is improved to 0.5° in the direction of the ship's track. To ensure 100% seamless seabed coverage, the vessel's speed would be limited to 4 n mile/h or lower in such cases. However, with multi-frequency transmission, in order to obtain higher beam resolution, the vessel can maintain high-speed navigation, or even increase its speed when very high accuracy is not required. Spatial resolution offers the greatest operational and scientific research value for a wide range of applications. Figure 5.24 presents the system block diagram of the HydroSweep DS.

(3) EM122 of Kongsberg Company

EM122 is a new-generation deep-sea multi-beam bathymetric system of Kongsberg Company, with an operating frequency of 12 kHz, a working water depth from 20 to11,000 m, 288 beams, a minimum beam width of 0.5° × 1°, three types of beam spacing: equirectangular, equidistant and encrypted, a maximum coverage width of 6 times the water depth, and a maximum working speed of up to 16 knots. The system adopts such technologies as high-efficiency noise-reduction front put, frequency coded grouping beam, active beam steering for longitudinal and transverse rocking and bow rocking, etc., to ensure to obtain the maximum seabed coverage

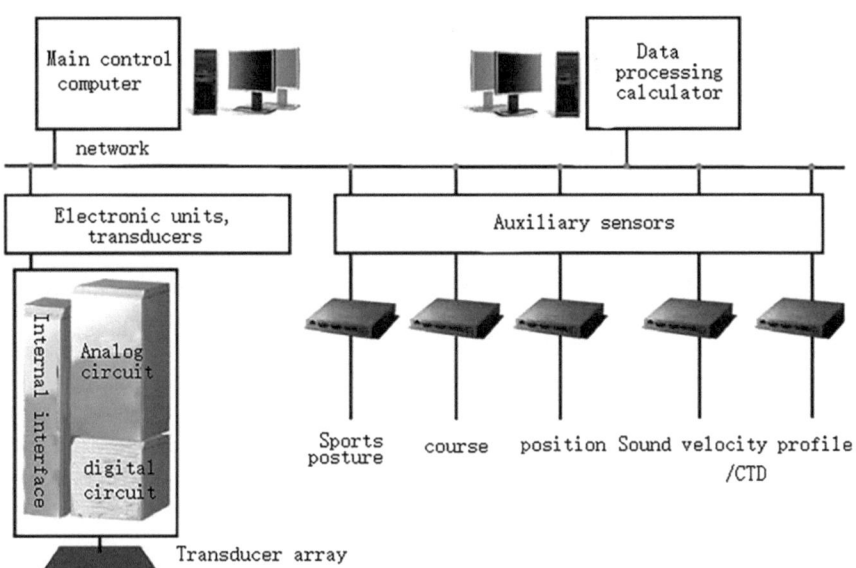

Fig. 5.24 Block diagram of teledyne (ATLAS) HydroSweep DS system components

width. To improve the measurement accuracy and resolution, the system adopts the combined phase and amplitude detection technology and the sound source level multi-directional separate automatic control technology. The system has many functions such as coverage sector and beam pointing angle can be changed automatically with the change of water, isometric beam, integrated seabed acoustic image and sound velocity profile interpolation, etc., which can obtain the maximum measurement benefit within one voyage.

(4) Domestic deep-water multi-beam

The Institute of Acoustics of the Chinese Academy of Sciences, in collaboration with several domestic institutions, jointly developed the first domestic deepwater multi-beam echo sounder system. This system obtained seabed topography and landform maps of a 6000 m-deep area in the Western Pacific Ocean and was practically applied in the major project of the National "863" Plan, "Experimental System for Seabed Observation Networks," completing the corresponding survey work. The system boasts 289 beams with a beam width of $1° \times 2°$ and a maximum coverage width of six times the water depth. It is equipped with functions such as three-dimensional attitude stabilization for transmission, roll stabilization for reception, and linear frequency modulation for edge beam transmission, satisfying the requirements for seabed topography and landform detection at full ocean depths ranging from 20 to 11,000 m.

2. Side-scan sonar

Typical products of side-scan sonar include the Edgetech4200 series from Laurel Technologies, the MA-XVIEW 600 and System 4000 from Klein Marine Systems, and the iSide series from Jiangsu Hi-Target Marine Information Technology Co., Ltd.

(1) Edgetech 4200 series

The Edgetech 4200 series, a new generation of side-scan sonobuoys from Laurel, include models such as Edgetech 4200-MP, Edgetech 4200-SP, Edgetech 4200-FS.

The 4200-MP side-scan sonar system applies EdgeTech's full-spectrum chirp technology to obtain broadband, high-energy transmit pulses and high-resolution, high signal-to-noise ratio echo data. The system employs a wide-band, low-noise preamplifier electronic circuit, which reduces phase error and drift introduced by the system to negligible levels. The 4200-MP towfish uses a splicable transmit/receive transducer array, and through the system's control software, two operating modes can be selected: High-Resolution Mode (HDM) or High-Speed Mode (HSM). The sonar data collected by the system is a highly correlated data set, which is suitable for post-processing by advanced users. The main working technical parameters in both modes are as follows:

① HDM mode allows simultaneous dual-frequency 100/400 kHz operation: at 100 kHz, the horizontal beam width is 0.64°, with a maximum single side range of 500 m; at 400 kHz, the horizontal beam width is 0.3°,a with a maximum single side range of 150 m.

② HSM mode operates with double pulse single frequency: 100 kHz or 400 kHz. At 100 kHz, the horizontal beam width is 1.26°, and at 400 kHz, the horizontal beam width is 0.40°. The range is the same as in HDM mode. The towfish has a rated operating depth of 2000 m and is equipped with standard heading, pitch, and roll sensors. The coaxial tow cable can be up to 6000 m long.

The 4200-SP side-scan sonar system is a streamlined version of EdgeTech's 4200-MP side-scan sonar system, with a high cost performance. Its pressure-resistant capacities are as follows: the stainless steel towfish is pressure-resistant to a depth of 2000 m, while the aluminum-housed towfish is pressure-resistant to a depth of 300 m.

The 4200-FS side-scan sonar system integrates full-spectrum and multi-pulse technologies into one unit, operating in dual modes: High-Definition Mode (HDM) and High-Speed Mode (HSM). Its main technical parameters and features include: an ultra-long array (90 cm) providing ultra-high resolution; dual-frequency operation at 120 kHz and 410 kHz; a rated operating depth of 1000 m; and a coaxial tow cable that can reach up to 6000 m in length.

(2) MA-X VIEW 600 and System 4000

In 2019, Klein Marine Systems has launched its new product MAX VIEW 600 product. It represents a revolutionary innovation in side-scan sonar products, being the first in the industry to integrate gap-filling sonar into a single-beam side-scan sonar. It boasts a scanning capability with an optimal range of 50 m on each side and a maximum range of 120 m, capable of generating high-definition images at 600 kHz.

KleinBLUE Technology has introduced innovative improvements in transducer design, signal conditioning, and post-processing, resulting in unparalleled image quality and range performance. The optimal design of KleinBLUE Technology has elevated acoustic performance to a new level. This technology is currently applied to two side-scan sonar models: KleinMA-XVIEW600 and System4000.

The MA-XVIEW600 side-scan sonar is the industry's first single-beam side-scan sonar integrated with gap-filling sonar. In addition to utilizing the latest BlueTechnology, it also employs the patent-pending MA-X technology, which uses a tail transducer to create echo imaging within the nadir region and display it on the MA-X window of SonarPro, saving approximately 40% of survey time. The main technical specifications of the MA-XVIEW600 side-scan sonar are shown in Table 5.2.

The S4000 side-scan sonar is another Blue Technology-enabled product that meets the need for long range, deep-water operations and optimized shallow-water performance. It encapsulates Klein's 50 years of design knowledge, combining Klein's signature imaging with unprecedented towing options and range performance, offering unparalleled search and survey efficiency. The main technical specifications are shown in Table 5.3.

5.2 Principle of Sonar Technology

Table 5.2 MA-XVIEW600 side scan sonar main technical parameters

Parameter	Numerical value
Frequency	600 kHz (side-scan sonar) 850 kHz (under-heaven MA-X)
Pulse type	FM CHIRP
Horizontal beamwidth	0.23°(side scan sonar)
Vertical beamwidth	40° (side scan sonar)
Vertical trace resolution	1.2 cm (side scan sonar)
Maximum range (each side)	120 m (side scan sonar)
Vertical beam centre	30°Downward from the horizontal (side scan sonar)
Output data format	SDF or XTF (simultaneous or optional)
Input voltage	12VDC or 110/220 VAC (50–60 Hz)
Power consumption	75 W

Table 5.3 S4000 side scan sonar main technical parameters

Parameter	Numerical value
Frequency	100 kHz/400 kHz (dual frequency simultaneous transmission)
Pulse type	FM CHIRP, selectable CW mode
Horizontal beamwidth	1°@100 kHz 0.3°@400 kHz
Vertical beamwidth	50°
Vertical Trajectory Resolution	9.6 cm@100 kHz 2.4 cm@400 kHz
Maximum operating range (unilateral)	600 m@100 kHz 200 m@400 kHz
Vertical beam centre	Tilt down 25° from horizontal
Output data format	
Frequency	Klein SDF (sonar data format) XTF (extended tricon format) Simultaneous or optional

3) iside series

In recent years, some domestic sonar manufacturers have also introduced new products. For instance, Jiangsu Zhonghaida Marine Information Technology Co., Ltd (hereinafter referred to as the marine company), a subsidiary of Zhonghaida, have launched four self-branded side-scanning sonar products iSide series at the same time, with the model number of iSide 1400/4900/4900L/400, respectively. The iSide series of high-resolution side-scan sonobuoys can work in dual-frequency at the same time, transmitting a variety of CW and Chirp signals, advanced digital circuit

Table 5.4 The technical parameters of the iSide1400 side scan sonar

Parameter	Numerical value
Operating frequency	100 and 400 kHz
Transmit pulse width	20–1000 μs (CW), 1–4 ms (LFM)
Signal type	CW/LFM
Horizontal beam angle	0.7°@100 kHz, 0.2°@400 kHz
Vertical beam angle	45°
Beam tilt	10°, 15°, 20°
Distance resolution	62.5px@100 kHz, 31.25px@KHz
Maximum range	600 m@100 kHz; 200 m@400 kHz
Working speed	2–6 kn
Working depth	1000 m
Dimension	105 mm × 1300 mm
Mass	30 kg (316 stainless steel)
Power consumption	≤ 40 W
Built-in transducer	Built-in attitude, bow, pressure and depth sensors
Tow cable	Kevlar reinforced cable, 50 m standard (250 m optional)

processing technology, and its technical performance has also reached the international technical indicators of the same type of sonobuoys. The technical parameters of the iSide 1400 are shown in Table 5.4.

3. Synthetic aperture sonar

In 2015, Kongsberg Company first utilized data from the company's EM2040C shallow-water multi-beam bathymetric system to perform synthetic aperture algorithm processing and compared the results with those obtained from multi-beam bathymetric sonar. The company refers to this system as HISAS2040, which represents the latest instance observed internationally to date of utilizing multi-beam sonar data for synthetic aperture processing.

The EM2040C is a shallow-water multi-beam sounder based on EM2040 technology, suitable for high-resolution mapping and seabed detection. The transducer incorporates both transmitting and receiving units and has the same dimensions as the EM3002. This system meets or exceeds the IHO-S44 Special Order standards and even the more stringent LINZ specifications. When the frequency range is between 200 and 300 kHz, a single transducer can achieve a coverage angle of up to 130°, with a sweep width that is 4.3 times the water depth. When using dual transducers, each inclined at 35° to 40°, the coverage angle can reach 200°, and the sweep width over a flat seabed can reach up to 10 times the water depth.

References

1. Rongbao, Xu. 1992. Principles and Applications of Acoustic Emission Dynamic Monitoring. *New Technology & New Process* 3: 20–21.
2. Niujun, Wang, and Chen Li. 2010. Principles and Applications of Acoustic Emission Testing Technology. *Journal of Shanxi Institute of Technology* 26 (2): 41–43.
3. Elbatanouny, M. K., A. Larosche, P. Mazzoleni, et al. 2014. Identification of Cracking Mechanisms in Scaled FRP Reinforced Concrete Beams Using Acoustic Emission. *Experimental Mechanics* 54 (1): 69–82.
4. Li, Dongsheng, Wei Yang, and Yan Yu. 2017. *Acoustic Emission Monitoring and Evaluation of Civil Engineering Structure Damage—Theory, Methods and Applications.* Beijing: Science Press.
5. Hongguang, Ji., Pei Guangwen, and Shan Xiaoyun. 2002. A Review of Acoustic Emission Technology in Concrete Materials. *Applied Acoustics* 21 (4): 1–5.
6. Mingwei, Yang. 2005. *Acoustic Emission Testing.* Beijing: Mechanical Industry Press.
7. Mengyuan, Li., Shang Zhendong, Cai Haichao, et al. 2010. *Acoustic Emission Testing and Signal Processing.* Beijing: Science Press.
8. Jun, Wei, and Zhao Jianhua. 1992. Felicity Effect of Acoustic Emission in Fiber Reinforced Composite Materials. *Journal of Composite Materials* 9 (1): 65–69.
9. Jingrong, Zhao. 2010. *Research on Acoustic Emission Signal Processing System and Source Identification Method.* Changchun: Jilin University.
10. Lijun, Ouyang. 2007. *Research on Bond Performance of Corroded Reinforced Concrete Members Based on Acoustic Emission Technology.* Nanning: Guangxi University.
11. Gongtian, Shen, Dai Guang, and Liu Shifeng. 2003. Progress of Acoustic Emission Testing Technology in China—In Celebration of the 25th Anniversary of the Society's Establishment. *Nondestructive Testing* 25 (6): 302–307.
12. Guangwei, Qing, Yue Lin, Feng Yuegui, et al. 2012. Application of Wavelet Transform in Feature Analysis of Acoustic Emission Signals. *Nondestructive Testing* 34 (11): 48–51.
13. Gongtian, Shen, Geng Rongsheng, and Liu Shifeng. 2002. Parameter Analysis Method for Acoustic Emission Signals. *Nondestructive Testing* 02: 72–77.
14. Congbing, Liu. 2010. Noise Reduction Processing Method for Acoustic Emission Signals Based on Wavelet Analysis. *Mechanical & Electrical Engineering Technology* 39 (7): 82–84.
15. Suikun, Ding. 2011. *Acoustic Emission Monitoring Technology for Corrosion Damage of Key Components of Cable-Stayed Bridges.* Dalian: Dalian University of Technology.
16. Gorman, M. R., and M. Ziolas. 1991. Plate Waves Produced by Transverse Matrix Cracking. *Ultrasonics* 29 (3): 245–251.
17. Fang, Jiangtao. 2007. *Research on Signal Recognition Technology of Metal Corrosion Acoustic Source.* Daqing: Daqing Petroleum Institute.
18. Shuang, Ju., Li. Xinci, Luo Tingfang, et al. 2018. Detection of Acoustic Emission Signal Characteristics on the Surface of Pinus massoniana Glued Wood Using Wavelet Analysis. *Journal of Northeast Forestry University* 46 (8): 84–90.
19. Zhikui, Chen. 1998. *Research on Wavelet Basis and Wavelet Transform Analyzer System in Engineering Signal Processing.* Chongqing: Chongqing University.
20. Oliveira, R. D., and A. T. Marques. 2008. Health Monitoring of FRP Using Acoustic Emission and Artificial Neural Networks. *Computers & Structures* 86 (3): 367–373.
21. Kwak, J. S., and M. K. Ha. 2004. Neural Network Approach for Diagnosis of Grinding Operation by Acoustic Emission and Power Signals. *Journal of Materials Processing Technology* 147 (1): 65–71.
22. Farhidrndch, A., E. Dehghan-Niri, S. Salamone, et al. 2013. Monitoring Crack Propagation in Reinforced Concrete Shear Walls by Acoustic Emission. *Journal of Structural Engineering* 139 (12): 113–134.
23. Xianzhen, Wu., Liu Xiangxin, and Liu Hongxing. 2011. Experimental Study on Acoustic Emission Characteristics and Dynamic b-Value Characteristics of Rockburst in Sandston. *Metal Mine* 40 (03): 13–18.

24. Yuli, Dong, Xie Heping, and Zhao Peng. 1996. Study on b-Value and Fractal Dimension of Acoustic Emission during the Whole Compression Process of Concrete. *Experimental Mechanics* 3: 272–276.
25. Schumacher, T., C. C. Higgins, and S. C. Lovejoy. 2011. Estimating Operating Load Conditions on Reinforced Concrete Highway Bridges with b-Value Analysis from Acoustic Emission Monitoring. *Structural Health Monitoring* 10 (1): 17–32.
26. Golaski, L., P. Gebski, and K. Ono. 2002. Diagnostics of Reinforced Concrete Bridges by Acoustic Emission. *Journal of Acoustic Emission* 20: 83–98.
27. Gostautas, R. S., A. M. Asce, and G. Ramirez. 2005. Acoustic Emission Monitoring and Analysis of Glassfiber Reinforced Composites Bridge Decks. *Journal of Bridge Engineering* 10 (6): 713–721.
28. Ohtsu, M., M. Uchida, T. Okamoto, et al. 2002. Damage Assessment of Reinforced Concrete Beams Qualified by Acoustic Emission. *Structural Journal* 99 (4): 48–57.
29. Zhang, Y., and G. Hong. 2005. An IHS and Wavelet Integrated Approach to Improve Pan-Sharpening Visual Quality of Natural Colour IKONOS and Quick Bird Images. *Information Fusion* 6 (3): 225–234.
30. Hong, Shi, Zhao Chunhui, and Shen Zhengyan. 2010. Sonar Image Denoising in Morphological Wavelet Domain Combined with Nonlinear Filters. *Journal of Harbin Engineering University* 31 (11): 1524–1529.
31. Anitha, U., and S. Malarkkan. 2015. A novel Approach for Despeckling of Sonar Image. *Indian Journal of Science & Technology* 8 (S9): 252–259.
32. Nafornita, C., D. Isar, and A. Isar. 2011. Searching the Most Appropriate Mother Wavelets for Bayesian Denoising of Sonar Images in the Hyperanalytic Wavelet Domain. In *Statistical Signal Processing Workshop*. 169–172. IEEE.
33. Qingwu, Li., Huo Guanying, and Zhou Yan. 2015. *Sonar Image Processing*. Beijing: Science Press.
34. Easley, G., D. Labate, and W. Q. Lim. 2008. Sparse Directional Image Representations Using the Discrete Shearlet Transform. *Applied and Computational Harmonic Analysis* 25 (1): 25–46.
35. Candes, E. J., L. Demanet, D. L. Donoho, et al. 2005. Fast Discrete Curvelet Transforms. *Applied and Computational Mathematics. California Institute of Technology* (6): 1–43.
36. Selesnick, I. W., R. G. Baraniuk, and N. C. Kingsbury. 2005. The Dual-Tree Complex Wavelet Transform. *IEEE Signal Processing Magazine* 22 (6): 123–151.
37. Selesnick, I. W. 2001. *The Double Density DWT. Wavelets in Signal and Image Analysis*, 39–66. Springer.

Chapter 6
Acoustic Key Technologies

6.1 Acoustic Emission Key Technologies

The purpose of acoustic emission detection is to find the acoustic emission source and to obtain as much information as possible about the acoustic emission source, and the results of acoustic emission signal processing and analysis can be used to locate, classify and evaluate structural defects. However, due to the influence of various factors such as the characteristics of the acoustic emission source, the propagation path from the source to the transducer, the characteristics of the transducer and the measurement system of the acoustic emission instrument, the acoustic emission waveform of the acoustic emission transducer is very complex, which is very different from the real acoustic emission source signal, and sometimes even completely different from the true acoustic emission source signal. Acoustic emission source signal refers to the original acoustic emission signal emitted by the acoustic emission source, while the acoustic emission signal usually commonly referred to is the signal received by the acoustic emission sensor after the acoustic emission source signal has passed through various propagation media. It can be seen that the received acoustic emission signal has a certain connection with the real acoustic emission signal, but in the propagation process due to the influence of various factors, the acoustic emission source signal undergoes distortion or deformation to a certain extent. Therefore, the key technology of acoustic emission technology in structural detection and monitoring is how to more accurately obtain the real information of the acoustic emission source based on the electrical signal output from the acoustic emission transducer, realize the localization of the acoustic emission source, make use of the extracted eigenvalues and adopt the pattern recognition method to classify the defects (e.g., cracks, corrosion, etc.) and the assessment of the degree of defects.

According to the mechanism of acoustic emission technology detection, the uncertainty in the propagation process of acoustic emission signals is the main reason for the difficulty of acoustic emission signal processing. The signal collected by acoustic

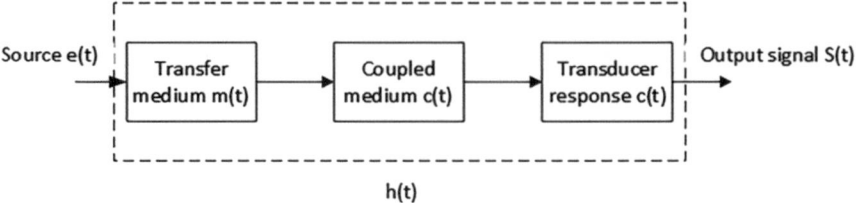

Fig. 6.1 Relationship between the output signal of the acoustic emission system and the source signal

emission instruments is jointly determined by the acoustic emission source, transmission medium, coupling medium and transducer and instrument characteristics, and in most cases, these response functions are unknown. Therefore, in general, the acoustic emission signal $s(t)$ will be quite different from the instantaneous elastic wave $e(t)$ generated by the acoustic emission source under the action of external force, as shown in Fig. 6.1.

The transmission process of the acoustic emission signal can be expressed as:

$$s(t) = e(t) * m(t) * c(t) * r(t) \tag{6.1}$$

If we combine a series of factors that distort the acoustic emission source signal, such as the propagation medium, coupling medium, transducer response characteristics into a transfer function $h(t)$, that is, $h(t)$ represents the transfer function for the entire process from the emission of the signal by the acoustic emission source to its reception by the instrument.

$$h(t) = m(t) * c(t) * r(t) \tag{6.2}$$

Then the propagation process of the acoustic emission signal (Eq. 6.1) can be expressed as:

$$s(t) = e(t) * h(t) \tag{6.3}$$

where $e(t)$ is the real acoustic emission signal generated by the acoustic emission source, and s(t) is the acoustic emission signal received by the transducer. Currently, the common practice in acoustic emission signal processing is to analyze and process the $s(t)$ signal directly. A valid acoustic emission signal $s(t)$ can be received only if the distortions and aberrations caused by the non-ideal characteristics of the propagation path and the transducer are not sufficient to completely overwhelm the information emitted by the acoustic emission source.

When acoustic emission technology is used for structural inspection of offshore platforms, in addition to mechanical and electromagnetic noise, the effects of environmental specific noise of the marine environment increase the difficulty of processing acoustic emission signals.

6.1.1 Signal Analysis and Processing Techniques

The previous section introduced the acoustic emission characteristic parameter method and waveform analysis method commonly using data processing methods. The key technologies of acoustic emission signal analysis and processing of include: acoustic emission signal noise reduction technology, signal recovery technology, spectral analysis of the signal waveform, characteristic parameter extraction and pattern recognition technology.

1. Acoustic emission signal noise reduction technology

1) Noise types of acoustic emission signals

According to the formation mechanism of acoustic emission and signal propagation process, the main difficulties encountered in the analysis of acoustic emission signals include the following two aspects: Firstly, the acoustic emission signal itself has a sudden, uncertain, weak and easy to distort. Secondly, the acoustic emission signal is mixed with a variety of types of noise signals, which has a certain impact on the accuracy of analysis results. Common types of acoustic emission noise mainly include mechanical noise and electromagnetic noise. Mechanical noise refers to the noise caused by the impact, friction and vibration between objects, while electromagnetic noise refers to the noise caused by electromagnetic induction and electrostatic induction.

Types of mechanical noise include: vibration due to device, man-made beat, wind, rain, snow, dust and other environmental factors triggered by mechanical impact generated by the collision noise; the equipment operation and connecting pipeline noise; the fluid noise generated by high-speed flow of fluids (including leakage, boiling, combustion, etc.) of pumps, valves, and containers; all the friction noise caused by the relative sliding of the machinery during the loading process; and the noise caused by the human and the surrounding animals; the unique noises in the marine environment include those from sea waves, tides, ships, fish schools, and other environmental factors.

Electromagnetic noise includes: inevitable white noise at the input of the preamplifier; ground loop noise caused by improper grounding of the instrument or structure; "pickup" noise generated within components and systems of the acoustic emission (AE) system; electromagnetic interference caused by wireless transmitters, power switches, motors, welding, sparks, lightning and other sources.

2) Noise reduction methods for different types of noise

The primary task of using acoustic emission (AE) technology for structural inspection and health monitoring is to "detect damage," with the ultimate goal of evaluating damage characteristics (including primarily the location, nature, and severity of the source) through "inverse source" analysis of received signals. In the process of discovering the damage, an important aspect of AE technology research is how to eliminate the interference caused by "false AE signals" due to noise. At present,

AE systems offer various hardware and software methods for noise reduction, which can be mainly categorized into two types: frontend noise reduction and terminal noise reduction. Frontend noise reduction utilizes instrument hardware or embedded software to reduce noise, allowing for interactive parameter settings but with limited openness. Terminal noise reduction, on the other hand, utilizes signal processing software on a computer terminal to reduce noise, enabling user customization based on the characteristics of the noise to be processed and offering high openness.

(1) Frontend noise reduction methods

Common frontend noise reduction strategies for acoustic emission signals include coincidence discrimination, primary-secondary discrimination, frequency discrimination, load-controlled gates, amplitude discrimination, time gates, leading-edge discrimination, and data filtering. Selecting the appropriate noise reduction method based on the characteristics of the noise is crucial for noise analysis. Additionally, attention should be paid to avoiding mutual interference between noise sources.

Coincidence discrimination and primary-secondary discrimination are suitable for noise reduction of mechanical noise outside specific areas. Coincidence discrimination is a zone-specific detection method that uses a time difference window gate circuit to only collect signals within a specific time difference range, allowing for effective discrimination. Primary-secondary discrimination mainly relies on the logical relationship of the order in which signals reach the primary and secondary sensors and their gate circuits to only collect signals originating near the primary sensor, achieving the purpose of filtering out noise. Frequency discrimination belongs to the frequency domain filtering method and is suitable for noise reduction of mechanical noise in any frequency band. Load-controlled gates are suitable for noise reduction of mechanical noise generated during fatigue experiments, as the load gate circuit only collects signals within a specific load range.

Amplitude discrimination is suitable for noise reduction of low-amplitude electromagnetic noise by adjusting fixed or floating detection threshold values to filter out noise. Time gates are applicable to noise reduction analysis of electrode and switching noise during spot welding, as their circuits only collect signals within a specific time range, which can be used to eliminate burst-type noise with long intervals.

Leading-edge discrimination and data filtering are both suitable for noise reduction analysis of both mechanical and electromagnetic noise. By setting a rise time filtering window for the signal waveform, distant mechanical noise or electrical pulse interference can be filtered out. Data filtering is mainly divided into two types: frequency domain noise reduction (including Gaussian filtering, band-pass filtering, and band-stop filtering) and time–space domain noise reduction (including Wiener filtering, Kalman filtering, and Extended Kalman filtering).

(2) Terminal noise reduction methods

The main terminal noise reduction methods include those methods based on wavelet analysis, empirical modal decomposition, independent component analysis (ICA) and neural network methods.

6.1 Acoustic Emission Key Technologies

Wavelet analysis is a commonly used method in noise reduction processing of acoustic emission signals, which belongs to the method of time–frequency analysis. This method is suitable for noise reduction mutant and non-smooth signals, and it mainly contains three types of wavelet transform mode maxima, wavelet coefficient scale correlation, and wavelet threshold noise reduction methods according to the different principles of the processing methods [1]. The choice of scale has a greater impact on the noise reduction effect of wavelet transform mode maxima, and the method has a unique advantage for the retention of signal singularities, and there is no need to know the variance of the noise, but the computational rate of the reconstruction is slower; wavelet coefficients scale correlation method for the analysis of the signal's edge characteristics has a significant advantage, but it need to predict the variance of the noise and the computational volume is large; The wavelet threshold denoising method is widely used because of its simplicity, small computational amount, wide range of application and many other advantages [2]. Soft thresholding method and hard thresholding method, there are differences between their processing of wavelet coefficients. The soft thresholding method utilize a slight reduction in all coefficients amplitude at the same time to reduce the noise, while the hard thresholding method thresholding method may damage some valid signals in the original waveform while eliminating noise. Wavelet denoising also has some limitations, and the effect of denoising is related to signal characteristics, wavelet basis function, signal-to-noise ratio and other factors. In the wavelet analysis of acoustic emission signals in the selection of wavelet basis function, you can refer to the acoustic emission data processing part of this book on the selection of wavelet basis function principles to select the appropriate wavelet basis for noise reduction.

Empirical Mode Decomposition (EMD) is an adaptive signal time–frequency processing method, which was proposed by Huang et al. [3] in 1998. Compared with wavelet, this method is not restricted by the choice of basis function and has higher resolution in time and frequency domains [4], and the decomposition process can be carried out based on the signal's own time scale characteristics. By integrating the advantages of wavelet thresholding and Empirical Modal Decompositionfor noise reduction, some scholars have proposed the wavelet thresholding-empirical modal integrated decomposition method for denoising crack acoustic emission signals, achieving remarkable results; Xu et al. [5] performed median filtering and SVD noise reduction on the original acoustic emission signals and EMD decomposition of the noise-reduced signals. The results of the processing of the simulated signals and the real acoustic emission signals showed that the noise is effectively filtered out, and the IMF has no frequency aliasing phenomenon, the number of decomposition layers is also reduced, and the accuracy and timeliness of EMD decomposition are also effectively improved.

Independent Component Analysis (ICA), developed gradually from Blind Source Separation (BSS), is an analysis method that can separate the mixed signals under the condition of unknown source and transmission mode. Because different source signals usually have statistical independence from each other, ICA makes use of this feature to estimate the source signals in the mixed signals [6]. When using Independent Component Analysis (ICA) for signal denoising, if the number of channels of the acquired signal is greater than or equal to the source signal, the signal can be separated directly, otherwise, denoising can be performed using virtual noise channels or sparse coding contraction (SCS) method [7]. The common criteria used in ICA include information maximiszation, mutual information minimization, non-Gaussianity metrics, and great likelihood estimation. The algorithms include adaptive algorithms, batch processing algorithms, projection pursuit algorithms (such as Fast ICA), and analysis algorithms based on neural networks. ICA is especially suitable for the extraction of weak signals under the strong background noise of the linear combination of the same source, but the method is difficult to apply to the acoustic emission signals with nonlinear aberration in transmission, which largely restricts the scope of its use.

The neural network method is widely used in acoustic emission signal processing due to its advantages of fully approximating arbitrary complex nonlinear relationships, high robustness and fault tolerance, fast computing ability, ability to learn uncertain systems, ability to acquire knowledge, and self-adaptive ability.

2. Signal recovery technology

From the point of view of the propagation process of the acoustic emission signal, the factors affecting the output signal are the transmission medium, the coupling medium and the response characteristics of the transducer. When the identification of acoustic emission signals becomes difficult due to the influence of the propagation process, in order to improve the signal resolution, reduce the impact of the inherent characteristics of sensors and detection instruments, and obtain more accurate information about the acoustic emission source, restoring the acoustic emission signals is an effective method.

Signal restoration does not mean recovering or reconstructing missing information, but rather reconstructing or restoring signals using known data and conditions that contain all the necessary information. According to the theory of signal restoration, certain conditions are required for reconstructing signals. If the practical application scenario meets these specified conditions, signal reconstruction can be achieved, which in this context refers to the restoration of acoustic emission signals. The restoration of acoustic emission signals is a prediction process for these signals, belonging to the category of predictive deconvolution in the field of signal processing.

To estimate the true signal of the acoustic emission source, the properties of the propagation medium and the transducer need to be known, i.e., the transfer function h(t) needs to be known. Since the received acoustic emission signal s(t) is a convolution of the transfer function h(t) and the original acoustic emission source signal e(t), the transfer function is unknown. Therefore, to predict the acoustic emission signal at the source, the response characteristics of the transfer function h(t) need to be estimated first.

According to the screening property of the discrete convolution:

$$x(t) * \delta(t) = x(t) \tag{6.4}$$

6.1 Acoustic Emission Key Technologies

If the source emits a unit pulse signal, the relationship between the signal received by the instrument and the source signal can be introduced:

$$s(t) = h(t) \tag{6.5}$$

Therefore, the signal received by acoustic emission instrument can be considered as the transfer function h(t).

The standard pencil-break signal is a unit pulse signal $\delta(t)$. Before conducting inspections using acoustic emission technology, a pencil-break test should first be performed to determine the parameters for signal acquisition and to measure the wave velocity. The pencil-break signal itself is an impulse signal with good repeatability. By using the standard pencil-break simulation signal as the unit impulse signal, the transfer function of the experimental system can be represented by the signal received by the standard pencil-break test instrument.

$$h(t) = s_{\text{pencil-break}}(t) \tag{6.6}$$

In other words, the standard pencil-break analogue signal received by the instrument can be considered as a transfer function. This can be used to recover the original acoustic emission signal from the acoustic emission source.

$$s(t) = e(t) * h(t) = e(t) * s_{\text{pencil-break}}(t) \tag{6.7}$$

3. Acoustic emission signal processing techniques based on wavelet analysis

Acoustic emission signal contains a series of frequency signal information, with randomness, uncertainty and non-stationarity. Wavelet analysis possesses the ability to characterize the local features of the signal in both time and frequency domains. It can effectively analyze the short-time high-frequency components of the signal and accurately estimate the low-frequency slow transformation components of the signal, which is most suitable for the analysis of acoustic emission signals containing transient phenomena and multi-modal characteristics of the spectrum.

At present, wavelet analysis has the following applications in the acoustic emission signal processing mainly.

(1) Signal source identification. The wavelet transform can be extracted in different frequency bands through wavelet analysis, so that the complex composition of the data waveforms can be separated into waves with single characteritics. Additionally, it enables simultaneous time–frequency analysis of acoustic emission data, obtaining more comprehensive characteristics of the signal source, and further identifying the signal source.
(2) Characteristic parameter detection. Wavelet analysis can effectively separate overlapping events and, combined with full-waveform data, minimize the loss of events. Meanwhile, using wavelet transform to detect acoustic emission event counts, independently of threshold levels, can significantly improve the accuracy of event counting.

(3) Noise rejection. The powerful decomposition and refinement capability of wavelet can be used to identify effective components from noisy signals. During decomposition and synthesis, undesirable channels can be removed to make the acoustic emission data more regular, thereby achieving the purpose of removing noise and extracting effective information.

(4) Source location. *Primarily, wavelet transform is utilized to extract single-frequency or narrow-band waveforms from acoustic emission data, and the peaks of these waveforms are selected to effectively compensate for the decaying signals, thereby enabling high-precision calculation of time differences. Since the time-difference localization method is widely used for acoustic emission source localization, the application of wavelet transform can significantly improve the accuracy of source localization. Some scholars abroad have proposed a new wavelet packet family that combines wavelet analysis with Fourier analysis, enabling the detection and localization of unknown waveform acoustic emission sources with the required accuracy.*

1) The basic theory of wavelet transform

(1) Definition of wavelet transform

The basic idea of wavelet transform is the consider with Fourier transform in that it also represents a signal's function using a set of functions, which are known as the wavelet function system. But the wavelet function system is different from the sine function used in Fourier transform, it is composed of a basic wavelet function of translation and expansion.

① Continuous wavelet transform

Suppose the function $\psi(t) \in L^1 \cap L^2$, the Fourier transform of $\psi(t)$ be denoted as $\hat{\psi}(t)$, if the following equation is satisfied.

$$\int_{-\infty}^{\infty} \frac{|\hat{\psi}(\omega)|^2}{|\omega|} d\omega < \infty \tag{6.8}$$

Then the $\psi(t)$ is named as a fundamental wavelet or wavelet mother function. Equation (6.8) is called the tolerability condition.

$$\psi_{a,b}(t) = |a|^{-\frac{1}{2}} \psi\left(\frac{t-b}{a}\right) \tag{6.9}$$

The above equation will be referred to as the fundamental wavelet or wavelet mother function. $\psi(t)$ depends on the continuous wavelet generated by a, b (a, b ∈ R, a ≠ 0), a is called the scale parameter and b is called the translation parameter. The scale parameter a changes the shape of the continuous wavelet and the translation parameter b changes the displacement of the continuous wavelet.

The wavelet transform of the function f(t) is given by

6.1 Acoustic Emission Key Technologies

$$W_f(a, b) = |a|^{-\frac{1}{2}} \int_R f(t)\overline{\psi}\left(\frac{t-b}{a}\right)dt \qquad (6.10)$$

where: $\overline{\psi}(t)$ is the complex conjugate of the function $\psi(t)$. From the admissibility condition

$$\int_{-\infty}^{\infty} \psi(t)dt = 0 \qquad (6.11)$$

The inverse transformation of $W_f(a, b)$ is

$$f(t) = \frac{1}{c_\psi} \int_R \int_R \frac{1}{a^2} W_f(a, b) \psi_{a,b}(t) dadb \qquad (6.12)$$

where:

$$C_\psi = \int_{-\infty}^{\infty} \frac{|\hat{\psi}(\omega)|^2}{|\omega|} d\omega < \infty \qquad (6.13)$$

The continuous wavelet $\psi_{a,b}(t)$ is localised in both the time domain space and the frequency domain space and its action similarly to the function $g(t-\tau)e^{-j\omega t}$ in the windowed Fourier transform. The essential difference between them is that: with the decreasing of $|a|$, the time-domain window of the $\psi_{ab}(t)$ becomes smaller and the spectrum is concentrated towards the high-frequency part, thus the time-domain resolution increases. That is to say, the wavelet transform for different frequencies in the time domain of the sampling step is adjustable. For low-frequency signals wavelet transform time resolution is lower, while the frequency resolution is higher. For high-frequency signals wavelet transform time resolution is higher, while the frequency resolution is lower, which is in line with the low-frequency signals change slowly, while the high-frequency signals change rapidly. Wavelet transform can decompose the signal into a variety of scale components intertwined together, and for the size of different scale components using the corresponding coarse and fine time domain or frequency domain sampling step, so that it can constantly focus on the object of any small details.

② Discrete Wavelet Transform

The discrete wavelet transform is often used for acoustic emission signal processing, defined as follows:

$$\psi_{m,n}(t) = a_0^{m/2} \psi(a_0^m t - nb_0), \quad m, n \in Z; a_0 > 1; b_0 > 1 \qquad (6.14)$$

Equation (6.14) is the discrete form of the continuous wavelet $\psi(t)$.

For $f \in L^2(R)$, the factors a and b in Eq. (6.14) are discretised, i.e., by taking $a = a_0^j (a_0 > 1)$, $b = ka_0^j b_0$ ($b_0 \in R; k \in Z$), the corresponding discrete wavelet transform is:

$$c_f(m, n) = \int_{-\infty}^{\infty} f(t)\overline{\psi}_{m,n}(t)dt \quad (6.15)$$

Its reconstruction formula is:

$$f(t) = C \sum_{m=-\infty}^{+\infty} \sum_{n=-\infty}^{+\infty} c_f(m, n)\psi_{m,n}(t)dt \quad (6.16)$$

C in Eq. (6.16) is a constant independent with signal.

The following is an analysis of whether the unique function $f(t)$ can be determined from $c_f(m, n)$ for the discrete wavelet transform.

Taking the discretisation parameters as $a_0 = 2$, $b_0 = 1$ (at this point also known as binary wavelet), Eq. (6.14) becomes:

$$\psi_{m,n}(t) = 2^{m/2}\psi(2^m t - n) \quad (6.17)$$

If $\{\psi_{m,n}(t)\}_{m,n \in Z}$ constitutes a set of canonical orthogonal bases in $L^2(R)$ space.

$$\int_{-\infty}^{\infty} \psi_{m,n}(t)\overline{\psi}_{m'n'}(t)dt = \begin{cases} 1, & m = m', n = n' \\ 0, & else \end{cases} \quad (6.18)$$

Then for any $f(t) \in L^2(R)$, there is the expansion

$$f(t) = \sum_{m,n \in Z} c_f(m, n)\psi_{m,n}(t) \quad (6.19)$$

It can be seen that as long as a set of canonical orthogonal bases (wavelet mother functions) in $L^2(R)$ space are constructed, the function $f(t)$ can also be uniquely determined for the discrete wavelet transform. Mallat unifies the construction of various orthogonal wavelet bases, and gives a decomposition algorithm and a reconstruction algorithm for the signals under the framework of the construction of the orthogonal wavelet bases – Mallat's algorithm.

(2) Mallat's algorithm

Mallat proposed the concepts of Multiscale Analysis and Multiresolution Approximation in 1989. He put the construction of orthogonal wavelet bases into a unified framework and gave a fast wavelet algorithm. The process of signal decomposition through the Mallat algorithm is the process of wavelet decomposition.

6.1 Acoustic Emission Key Technologies

① Multi-scale analysis

If $u(x) \in L^2(R)$, then a column of subspaces $\{V_j\}_{j \in Z}$ in the $L^2(R)$ space satisfying the following conditions is called a multiscale analysis of $L^2(R)$:

Monotonicity: $V_j \subset V_{j-1}, (j \in Z)$;
Approximation: $\cap V_j = \{0\}, \cup V_j = L^2(R)(j \in Z)$;
Stretch: $u(x) \in V_j \Leftrightarrow u(2x) \in V_{j-1}$;
Translational invariance: $u(x) \in V_j \Rightarrow u(x - 2^{-j}k) \in V_j (k \in Z)$;

Similarities: Let A_j be the operator that approximates the signal $u(x)$ with scale 2^{-j}. Among all approximation functions $g(x)$ with scale 2^{-j}, for any given $G_{(x)} = V_j$, the following equation holds:

$$\|g(x) - u(x)\| \geq \|A_j u(x) - u(x)\| \tag{6.20}$$

Riesz basis: There exists $g(x) \in V_j$ such that $\{g(x - 2^{-j}k) | k \in Z\}$ constitutes a Riesz basis of V_j, for any given $u(x) \in V_j$, there exists a unique sequence $\{a_k \in l^2\}$(square summable column),so that the following equation holds.

$$u(x) = \sum_{k \in Z} a_k g(x - k) \tag{6.21}$$

$$A\|u\|^2 \leq \sum_{k=-\infty}^{+\infty} |a_k|^2 \leq B\|u\|^2, A, B > 0 \tag{6.22}$$

The orthogonal basis of the vector space $\{V_j\}_{j \in Z}$ can be realized by stretching and translating a certain function $\varphi(x)$, and the function $\varphi(x)$ is unique.

If $\{V_j\}_{j \in Z}$ is a multiscale analysis of $L^2(R)$, then there exists a unique function $\varphi(x) \in L^2(R)$, such that its stretching and translating system

$$\{\varphi_{j,k}(x) = 2^{-j/2} \varphi(2^{-j}x - k) | k \in Z\} \tag{6.23}$$

constitutes a canonical orthonormal basis of the space V_j. The function $g(x)$ and the function $\varphi(x)$ is called the scale function of $\{V_j\}_{j \in Z}$.

For different multiscale analysis, the scale functions are different. Therefore, the key problem of multiscale analysis is how to construct its scale function.

② Multi-resolution analysis

The decomposition function of the signal is to decompose the original signal into low-frequency approximate components and high-frequency detail components by wavelet transform, and the multi-resolution analysis is only to further decompose the low-frequency part. Figure 6.2 shows the number of 3-layer multi-resolution analysis structure.

Fig. 6.2 Multi-resolution analysis structure tree

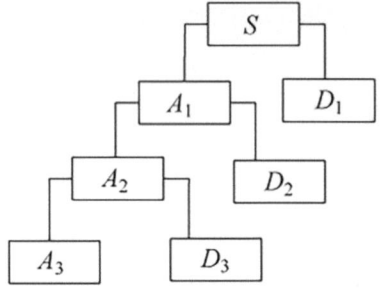

The original signal S is decomposed by 3-layer multiresolution analysis and its decomposition relation is:

$$S = A_3 + D_3 + D_2 + D_1 \qquad (6.24)$$

After the decomposition of the original signal S, its approximate component A is the low-frequency component of the signal on a large scale, and the detail component D is the high-frequency component on a small scale. In wavelet multiresolution decomposition, if the highest frequency component of the signal is regarded as 1, then each layer of the wavelet decomposition is a band-pass or low-pass filter, and the specific frequency bands occupied by each layer are.

$$A_1 : 0{-}0.5; \qquad D_1 : 0.5{-}1$$
$$A_2 : 0{-}0.25; \qquad D_2 : 0.25{-}1$$
$$A_3 : 0{-}0.125; \qquad D_3 : 0.125{-}1$$

③ Implementation of Mallat Algorithm

Inspired by Burt's Pyramidal Algorithm for image decomposition and reconstruction, Mallat put out the Mallat algorithm based on the multiresolution framework, which is a tower-like multiresolution decomposition and synthesis algorithm. The algorithm has a very important position in wavelet analysis.

The basic idea of Burt's tower algorithm is to decompose a discrete approximation $A_0 f$ at resolution 1 into an approximation $A_1 f$ at a coarser resolution of 2^{-J} and a detail-by-detail signal $D_j f$ $(0 < j \leq J)$. Assuming that a function $f(t) \in L^2(R)$ has been computed to discretely approximate $A_j f$ at a resolution of 2^{-j}, the discrete approximation $A_{j+1} f$ of $f(t)$ at a coarser resolution of 2^{-j+1} can be obtained by filtering the $A_j f$ with a discrete low-pass filter.

Let $\varphi(t)$ and $\psi(t)$ be the scale function and wavelet function of the function $f(t)$ under 2^{-j} resolution pass approximation respectively, then its discrete approximation $A_j f(t)$ and detail part $D_j f(t)$ can be expressed as:

6.1 Acoustic Emission Key Technologies

$$A_j f(t) = \sum_{k=-\infty}^{\infty} C_{j,k} \varphi_{j,k}(t) \tag{6.25}$$

$$D_j f(t) = \sum_{k=-\infty}^{\infty} D_{j,k} \psi_{j,k}(t) \tag{6.26}$$

where: $C_{j,k}$ and $D_{j,k}$, are roughness coefficients and detail coefficients at 2^{-j} resolution, respectively.

According to the idea of Mallat's algorithm, there are

$$A_j f(t) = A_{j+1} f(t) + D_{j+1} f(t) \tag{6.27}$$

where:

$$A_{j+1} f(t) = \sum_{k=-\infty}^{\infty} C_{j+1,m} \varphi_{j+1,m}(t) \tag{6.28}$$

$$D_{j+1} f(t) = \sum_{k=-\infty}^{\infty} C_{j+1,m} \psi_{j+1,m}(t) \tag{6.29}$$

Therefore

$$\sum_{k=-\infty}^{\infty} C_{j+1,m} \varphi_{j+1,m}(t) + \sum_{k=-\infty}^{\infty} C_{j+1,m} \psi_{j+1,m}(t) = \sum_{k=-\infty}^{\infty} C_{j,k} \varphi_{j,k}(t) \tag{6.30}$$

where: the scale function $\varphi(t)$ is the standard orthogonal basis, $\psi(t)$ is the standard orthogonal wavelet, there are

$$\begin{aligned} \varphi_{j+1,m}(t) &= 2^{-(j+1)/2} \varphi \big[2^{-(j+1)/2} t - m \big] \\ &= 2^{-(j+1)/2} \cdot \sqrt{2} \sum_{i=-\infty}^{\infty} h(i) \varphi(2^{-j} t - 2m - i) \end{aligned} \tag{6.31}$$

Multiply both sides of Eq. (6.31) by $\varphi_{j,k}^{*}(t)$, and do the integrate over t. Using the orthogonality of $\varphi_{j,k}(t)$, we have:

$$\langle \varphi_{j+1,m}, \varphi_{j,k} \rangle = h^{*}(k - 2m) \tag{6.32}$$

Similarly, there are

$$\langle \varphi_{j,k}, \psi_{j+1,m} \rangle = g^{*}(k - 2m) \tag{6.33}$$

Multiply both sides of Eq. (6.30) by $\varphi_{j+1,k}^*(t)$, and integrate over t, using Eq. (6.33), we have

$$C_{j+1,m} = \sum_{k=-\infty}^{\infty} h^*(k-2m)C_{j,k} \qquad (6.34)$$

Multiply both sides of Eq. (6.30) by $\psi_{j+1,k}^*(t)$, and integrate over t, using Eq. (6.33), we have

$$D_{j+1,m} = \sum_{k=-\infty}^{\infty} h^*(k-2m)C_{j,k} \qquad (6.35)$$

Multiply both sides of Eq. (6.30) by $\varphi_{j,k}^*(t)$, and integrate over t, using Eq. (6.32) and (6.33), we have

$$C_{j,k} = \sum_{k=-\infty}^{\infty} h(k-2m)C_{j+1,k} + \sum_{k=-\infty}^{\infty} g(k-2m)D_{j+1,k} \qquad (6.36)$$

Introducing the infinite matrices $H = [H_{m,k}]_{m,k=-\infty}^{\infty}$ and $G = [G_{m,k}]_{m,k=-\infty}^{\infty}$, where $H_{m,k} = h^*(k-2m)$ and $G_{m,k} = g^*(k-2m)$, the Eqs. (6.34)–(6.36) can be written as follows respectively:

$$\begin{cases} C_{j+1} = HC_j \\ D_{j+1} = GC_j \end{cases} j = 0, 1, 2, \ldots, J \qquad (6.37)$$

$$C_j = H^*C_{j+1} + G^*D_{j+1}, j = J, \ldots, 2, 1, 0, \qquad (6.38)$$

where: H^* and G^* are the dual operators (conjugate transpose matrices) of H and G respectively.

Equation (6.37) is the well-known one-dimensional Mallat tower decomposition algorithm, and Eq. (6.38) is the one-dimensional Mallat tower reconstruction algorithm. Thus, the implementation of the Mallat tower algorithm is converted into the design of filter banks G and H. The role of filter H is to realize the approximation of the function $f(t)$, while the role of filter G is to extract the details of the $f(t)$, so H can be regarded as a low-pass filter, and G can be regarded as a band-pass filter.

2) Acoustic emission signal processing based on wavelet decomposition

(1) Selection of wavelet base

According to the characteristics of commonly used wavelet bases, combined with the characteristics of the acoustic emission signal and the requirements of acoustic emission signal analysis in the project, the wavelet base used for acoustic emission signal analysis should meet the following conditions:

6.1 Acoustic Emission Key Technologies

① For large volume data signals can meet the requirements of rapid processing. One of the characteristics of the acoustic emission signal is sudden and multi-channel process monitoring, so in the practical application of the sampling time is often longer, the signal data volume is larger. Although the time–frequency spatial division of the signal by continuous wavelet transform is finer than binary discrete wavelet, the computational volume is larger than that of discrete wavelet transform. Therefore, from the perspective of processing speed, discrete wavelet transform is more suitable for acoustic emission signal processing than continuous wavelet transform, and the wavelet base that can be discrete wavelet transform should be selected.

② The wavelet basis should be sensitive to the defect signal and insensitive to the structural noise, i.e. the transformed scale should contain and characterize the defect information better. The better the correlation between the wavelet basis and the signal, the higher the feature extraction of the signal by wavelet transform, and the more accurate the features of the signal analyzed by wavelet basis. Acoustic emission signals in the time domain usually show shock oscillation attenuation, and has a certain duration. Therefore, the wavelet basis used for wavelet transform of acoustic emission signals should have similar properties.

③ The wavelet basis should have at least first order vanishing moments. In order to extract the acoustic emission signal of real interest from the received signal, the first step is to eliminate the various interference signals. Wavelet analysis is an effective means to reduce the influence of noise, and it can be known from wavelet theory that wavelet bases with certain order vanishing moments can effectively highlight the various singular characteristics of the signal, and acoustic emission signals have characteristics similar to impact signals. Therefore, choosing wavelet bases with certain order vanishing moments can highlight the characteristics of acoustic emission signals.

④ Effectively enhance the useful information and suppress the useless information. Selecting wavelet bases with linear phase to decompose and reconstruct the signal can avoid or reduce the distortion of the signal, and from the related theory, it is known that the symmetric or antisymmetric wavelet function has linear phase. Moreover, the decomposition with symmetric wavelet bases can make a sign-coherent and neatly-positioned representation of the mutation points of the signal. Therefore, when wavelet processing the acoustic emission signal, symmetric wavelet bases should be chosen as much as possible, and if it is difficult to get symmetric wavelet bases, approximate symmetric wavelet bases should be chosen as much as possible.

⑤ Good time–frequency analysis performance. The acoustic emission signal has the characteristics of transient and diversity, and the wavelet base with good local characteristics in time domain can effectively represent every sudden change of the acoustic emission signal, while the wavelet base with good local characteristics in frequency domain is good for analyzing multiple modes of acoustic emission signals in different frequency ranges, so as to extract the information related to the acoustic emission source. Previously mentioned the characteristics of the wavelet transform time–frequency analysis window, Heisenberg's principle of inaccuracy tells us that the width of the time window and the frequency window are mutual constraints, it is not possible to take a very small at the same time. When the width of the window decreases, the width of the frequency window will increase, and vice versa. For the study on the characteristics of acoustic emission signals, wavelet bases are required to have certain local analysis ability in both time and frequency domains. Therefore, wavelet bases with tight support in the time domain and fast decay in the frequency domain should be selected.

Considering the above requirements, wavelet basis commonly used in the current engineering can choose symmetric include double orthogonal wavelet (such as B-spline wavelet) and orthogonal wavelet with certain approximate symmetry, such as Coiflet wavelet, Symlets wavelet, Daubechies wavelet.

(2) Selection of maximum decomposition scale

The selection of wavelet decomposition scale can be can be analyzed and selected based on the actual signal analysis needs, combined with the frequency band decomposition characteristics of wavelet transform. Firstly, the frequency range of each decomposition scale in wavelet decomposition should be clear.

If the sampling frequency of the signal is fs, the measurable frequency range of the signal is $[0, f_s/2]$. Since the ranges of the detail and approximate signals are symmetrical in the frequency range of the signal, the frequency ranges of the approximate and detail signals are $[0, f_s/4]$ and $[f_s/4, f_s/2]$ respectively at scale 1. The next scale is a further decomposition of the approximate signal, i.e., $[0, f_s/4]$ is further decomposed into two symmetrical parts. By analogy the frequency ranges at all scales can be obtained, i.e., for a signal $f(n)$ with sampling frequency fs the signal can be decomposed into $j + 1$ frequency ranges after j wavelet decompositions, and the formula for each frequency range is.

$$\left[0, \frac{f_s}{2^{j+1}}\right]\left[\frac{f_s}{2^{j+1}}, \frac{f_s}{2^j}\right] \tag{6.39}$$

It can be seen that the larger the scale, the finer the frequency division, but its computation will also increase accordingly. Only by choosing a suitable decomposition scale, the superiority of wavelet transform for signal analysis can be reflected. The specific number of decomposition layers should be determined according to the frequency characteristics of the analyzed signal. For example, the frequency range of the acoustic emission of metal materials is generally from 100 to 500 kHz, so the frequency range of the detailed signal under the maximum scale of wavelet decomposition is about 50 kHz, which should be able to meet the requirements of acoustic emission analysis. For some acoustic emission activities with wider frequency range, such as acoustic emission signals whose activity frequency is concentrated in 10–550 kHz, the minimum decomposition frequency range should be no more than 10 kHz.

6.1 Acoustic Emission Key Technologies

(3) Wavelet analysis to achieve signal-to-noise separation of acoustic emission

① Wavelet threshold denoising

The general steps are as follows: Firstly, select an appropriate wavelet basis and the number of decomposition levels to perform multi-scale wavelet decomposition on a noisy acoustic emission signal. Secondly, apply threshold quantization to the noisy coefficients by selecting an appropriate threshold for multi-scale wavelet decomposition. This involves setting coefficients below the threshold to zero and adjusting coefficients above the threshold to the difference between their original values and the threshold. Finally, perform wavelet reconstruction to obtain the denoised acoustic emission signal.

Discrete binary wavelet transform of the noise signal, the larger the wavelet decomposition scale, the more conducive to the elimination of noise, but the scale is too large sometimes lose some important local singularities of the signal. Therefore, the choice of decomposition scale should be determined according to the actual requirements, and choosing the appropriate scale not only removes the white noise but also preserves the local singularity of the signal.

② Non-white noise removal

In the acoustic emission detection, some noise is actually generated by external disturbances in the acoustic emission signal (such as part of the mechanical noise), it is not the material itself defects issued by the signal, which belongs to a kind of interference signal, that is, also a kind of noise. In the signal processing process, should also try to remove such signals. For this type of noise separation can take the following two methods:

Firstly, in the case of understanding the frequency components of interest in the acoustic emission signal, through the wavelet decomposition, only to retain the frequency band of interest in the results of the wavelet transform, the results of the transform of the other frequency bands will be set to zero, and then re-synthesize the signal. Since the energy of the acoustic emission signal is mainly concentrated in the key frequency bands (the key frequency bands can be adjusted according to the specific object of study), we mainly focus on the key signal analysis. In the multi-scale analysis, the signals of each scale represent the signals within a certain frequency range, and we are only interested in the signals within the key frequency band, and set all the detailed signals of all scales outside the key frequency band to zero, and then the signals are reconstructed. This reconstructed signal filters out most of the non-white noise, and also ensures that most of the acoustic emission signal is not lost, highlighting the signal-to-noise ratio.

Secondly, in the case of understanding the frequency range of the noise component in the acoustic emission detection process, the noise can be removed by zeroing the wavelet transform coefficients of the frequency band in which the noise component is located, and then resynthesizing the signal. Using this method, the noise is first acquired and its characteristics analyzed. The analysis is also done by using wavelet multi-scale decomposition, supplemented by spectral analysis of the Fourier transform to find its main spectral range (e.g., noise such as mechanical vibration and friction collisions usually has a frequency of less than 20 kHz). This allows the wavelet decomposition of certain scales to be zeroed, and then reconstructing the signal can remove the interference. Of course, there is also a small amount of broadband noise, so it is necessary to analyze the spectral characteristics of various types of noise and understand their distribution patterns in order to eliminate them in a targeted manner.

6.1.2 Acoustic Emission Source Localization Technology

Localization of the acoustic emission source (often defects or potential defects) is one of the most important purposes of acoustic emission testing. According to the detection object, acoustic emission signal characteristics of the different positioning requirements, acoustic emission source positioning methods vary. Figure 6.3 shows the various types of acoustic emission signal source positioning methods in common use. Among the most commonly used source positioning techniques are two types: time difference positioning and regional positioning.

The time difference positioning method utilize the geometric relationship between the time difference of the acoustic emission signals arriving at different sensors and the sensor position to form a system of joint equations and solve them, and ultimately derive the relative position of the defect and the sensor, which is an accurate point positioning method. However, due to the acoustic emission wave in the propagation process attenuation, mode conversion, etc., coupled with different modes of the wave

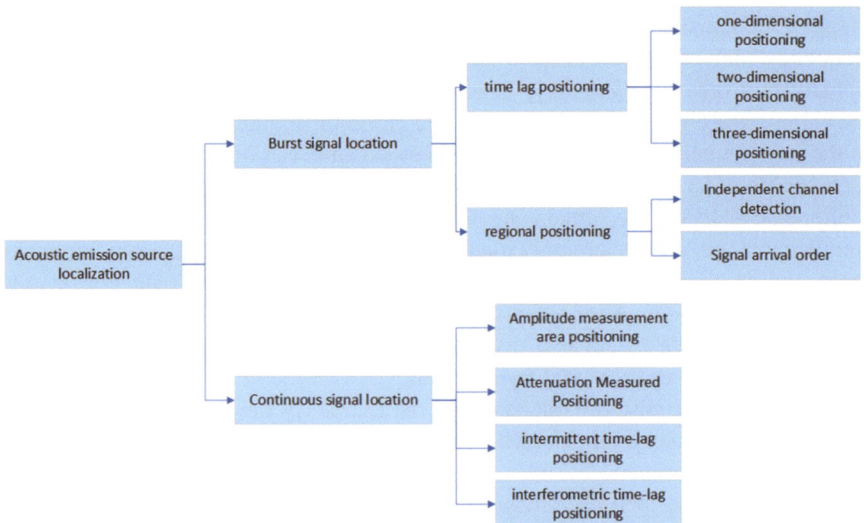

Fig. 6.3 Classification of acoustic emission source positioning methods

6.1 Acoustic Emission Key Technologies

in the same medium may have different speeds, and therefore its accuracy in practical applications are subject to many limitations.

Area localization is to determine roughly the area where the acoustic emission source is located according to the way different sensors detect different areas or according to the order in which the acoustic emission wave reaches each sensor. This is a fast, simple and rough positioning method, mainly used for composite materials and other acoustic emission frequency is too high due to the propagation of the attenuation is too fast or the detection of the number of channels is limited and difficult to use the time difference positioning occasions. With the development of acoustic emission technology and the improvement of the precision of instrumentation, the time difference method is now commonly used for positioning.

Acoustic emission source positioning method according to the test object or test positioning requirements can be divided into line positioning (one-dimensional positioning), surface positioning (two-dimensional positioning) and three-dimensional positioning.

1. Identification of signal arrival time

In the acoustic emission source positioning system, a very important quantity is the signal arrival time. Signal arrival time is the moment when the signal from the sound source is captured by the sensor, and the accuracy of this time has a great relationship with the accuracy of the acoustic emission source positioning results.

For the discrimination of arrival time of acoustic emission signals, the most accurate method is manual identification, which involves observing each waveform to identify the initial vibration time of the signal. However, for large amounts of data, manual identification is obviously unsuitable despite its accuracy guarantee due to its low efficiency. Existing acoustic emission detection systems generally use the threshold value to identify the arrival time of the signal. Once the voltage amplitude of the signal reaches a preset threshold, the moment of crossing the threshold is considered to be the time at which the signal arrives at the sensor. This way of identifying the arrival time can solve the problem of balancing efficiency and accuracy to a certain extent, but there are some shortcomings. The identification of the arrival time is affected by the setting of the threshold value, and from a rigorous point of view, the arrival time of the signal is not the time of crossing the threshold value, but the time when the waveform of the signal deviates from the equilibrium position and starts to vibrate.

Considering the identification of the arrival time of the signal time series, the more commonly used theory is the AIC criterion. Akaike Information Criterion (AIC), which is a theory proposed by the Japanese scholar, Hiroji Akaike, in 1973, and has been successfully applied in the ordering and selection of the AR model [8, 9].

(1) AIC criterion

The time series consists of a number of locally smooth time series, and each local time series is an autoregressive process that can be fitted with an AR model. For a time-domain signal of an acoustic emission waveform, the sampled time period consists of a time series after the onset of the signal as well as a period of time before the onset. Using the AIC criterion, the waveform of an acoustic emission signal can be divided into two smooth time sequences, and the waveform signals of the two time sequences represent the signal when the system is quiet (the noise signal) and the real acoustic source signal, respectively. The optimal interval point between the two time series is the arrival time of the acoustic emission signal, and the optimal interval point is determined by the minimum value of the AIC. In AR processing of time series, the order of AR processing needs to be determined experimentally and then the AR coefficients are determined by the Yule-Walker equation. In 1985, Maeda [10] proposed to directly calculate the AIC function using the time series of acoustic emission signals and determine the arrival time of the signals. The AIC function is as follows:

$$\text{AIC}(k) = k \cdot \log(\text{var}(R(1, k))) + (N - k - 1) \cdot \log(\text{var}(R(1 + k, N))) \quad (6.40)$$

where R denotes the time series under the time window, generally the absolute value of the voltage of the waveform sampling point as a time series; N is the number of points under the time window; var is the variance function of the sequence, var(R(1, k)) is the variance of the 1st to the kth parameter point in the time series. For the determination of the time window, the basic requirement is to place the starting position within the time window. Then according to the minimum value of the AIC function within the time window, determine the split point of the two smooth time series to get the starting moment of the waveform, which is the arrival time of the acoustic emission signal, as shown in Fig. 6.4, the minimum value of the AIC function value of the AIC function at the point 252 is the starting moment of the signal [11].

(2) Identification of Acoustic Emission Signal Arrival Time by AIC Criterion

A typical waveform from a uniaxial compression test was chosen as an example of the AIC criterion for identifying the arrival time. The parameters of the acoustic emission system were set as follows: sampling frequency of 5 MHz, pre-sampling length of 200 μs. The sampling frequency determines the time interval between every two sampling points, and the time interval between two points is 0.2 μs at the sampling frequency of 5 MHz.

Figures 6.5 and 6.6 give the waveforms of typical acoustic emission signals and the AIC function curves. Figure 6.5 is the complete waveform, and Fig. 6.6 is a local zoom in intercepting the sampling points 600–1400. The blue line in the figure is the signal waveform, the yellow line is the threshold of the voltage, and the purple curve is the AIC function curve. From Fig. 6.5, it can be visualized that there is a clear minimum in the AIC curve over the range of the waveform, and the minimum point divides the waveform into two parts: the background noise phase and the acoustic emission signal phase. To further analyze the local zoomed-in graph, the acoustic emission system identifies the signal arrival time using the moment over the threshold, and only the positive semi-axis of the waveform curve is considered, and the T1 moment in the graph is the signal arrival moment given by the system. From the waveform graph, there has been a waveform oscillation before the T1 moment, just because

6.1 Acoustic Emission Key Technologies

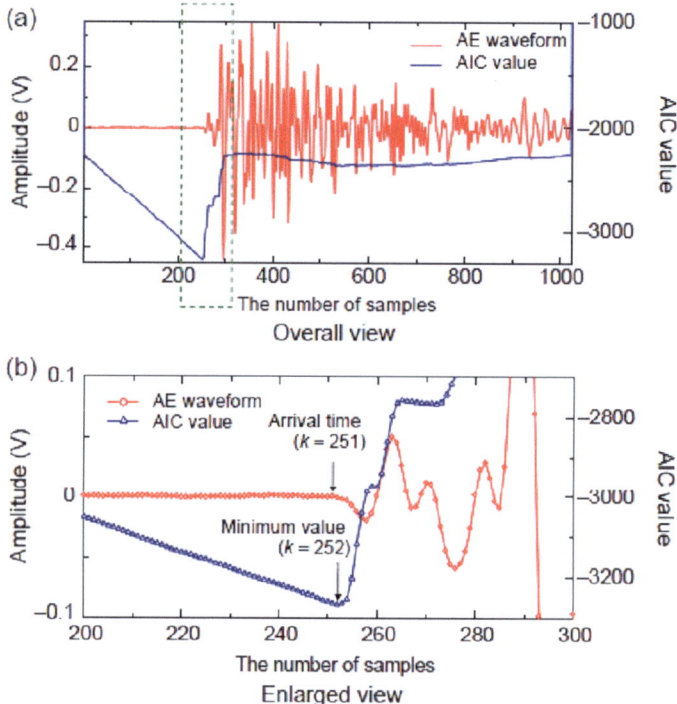

Fig. 6.4 Calculation of the AIC value of the acoustic emission signal

the voltage did not reach the positive threshold and did not trigger the identification mechanism of the signal arrival time. Therefore, the way of determining the signal arrival time through the threshold value has some errors.

After the calculation of the AIC function, the position of the minimum value of the AIC is marked in Fig. 6.6 (indicated by T2), and it can be seen from the red dotted line in the figure that this moment is very close to the starting moment of the waveform. The corresponding sampling point at the moment of T1 is 1002, and that at the moment of T2 is 952, with a difference of 50 sampling points, which is converted into a time scale of 10 us.

There are some differences between the AIC method and the gate threshold method in determining the arrival time of the acoustic emission signal, and the size of the difference depends on the threshold setting and the characteristics of the onset waveform. Through the typical waveform examples, it can be concluded that the AIC algorithm can more accurately identify the moment of oscillation of the waveform signal, which is comparable to the arrival time of the signal identified manually, and the interference of the threshold value can be excluded. Compared with the existing acoustic emission system using the positive threshold value of the signal arrival time identification method, the AIC method greatly improves the accuracy and precision of identification.

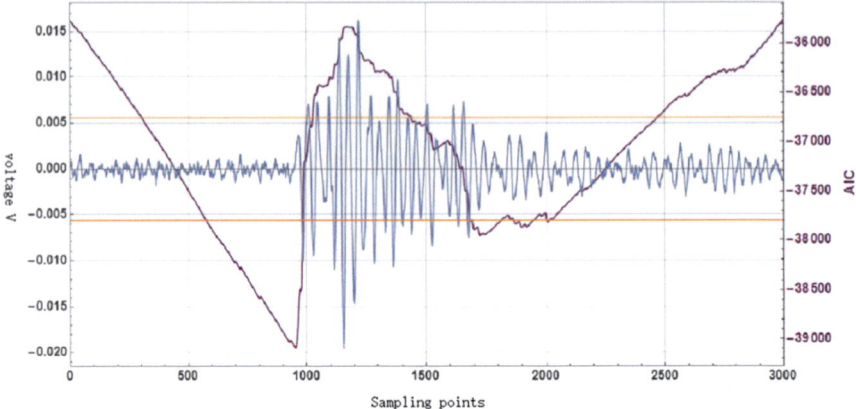

Fig. 6.5 Typical acoustic emission waveforms and AIC curves

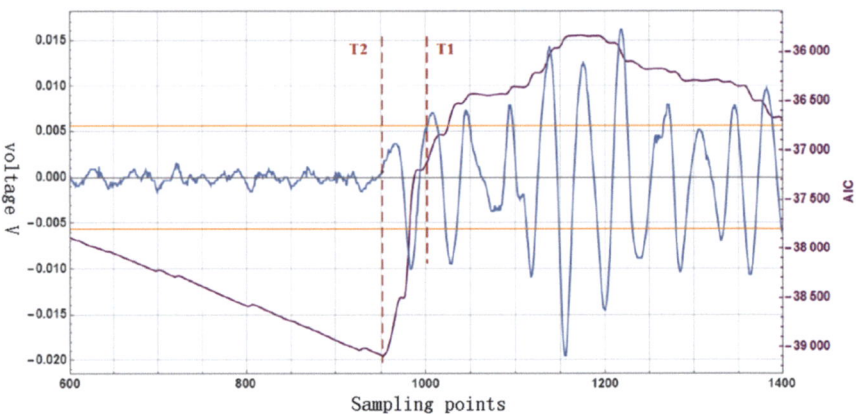

Fig. 6.6 Local amplification waveform and AIC curve

2. Acoustic emission source geometric positioning methods

Acoustic emission source location of geometric position positioning according to the detection of the object and test positioning requirements are divided into linear positioning, planar positioning and three-dimensional positioning, of which linear positioning and planar positioning are respectively called one-dimensional positioning and two-dimensional positioning. Acoustic emission signal geometric positioning is usually used time difference positioning method, usually assuming that the material sound propagation isotropic, the speed of sound is a constant, is currently the most common linear and planar positioning application of acoustic emission source positioning method. Based on the time differences in the arrival of acoustic emission signals from the same source to different sensors and the spatial positions of

6.1 Acoustic Emission Key Technologies

the sensors, it lists equations through their geometric relationships and solves them to obtain the precise location of the acoustic emission source. This section focuses on the commonly used linear positioning and planar positioning method. For three-dimensional localization, it can be performed in a Cartesian coordinate system. For pressure vessels commonly used in the petroleum industry, spherical coordinates can also be used for three-dimensional positioning.

(1) Linear positioning method

When the ratio of the length to the radius of the object being inspected is very large, the time difference line localization can be used for acoustic emission detection (such as pipelines, steel beams, etc.), and its positioning principle is shown in Fig. 6.7. If there is an acoustic emission source between Probe No. 1 and Probe No. 2 generating acoustic emission signals, and the time of arriving at Probe No. 1 is t_1, while the time of arriving at Probe No. 2 is t_2, then the time difference between the arrivals at the two probes is $\Delta t = t_2 - t_1$. Assuming the distance between the two probes is D, and the propagation speed of the sound wave in the specimen is V, the distance d from the acoustic emission source to Probe No. 1 can be calculated can be found by the following formula.

$$d = \frac{1}{2}(D - V\Delta t) \tag{6.41}$$

From Eq. (6.41), it can be seen that: when $\Delta t = D/V$, the acoustic emission source is located at Probe No.1; when $\Delta t = -D/V$, the acoustic emission source is located at Probe No. 2; when $-D/V < \Delta t < D/V$, the acoustic emission source is located in the area between the two probes. When the source is outside the probe array, regardless of how far the acoustic emission source is from the nearer probe, there is a time difference $\Delta t = -D/V$, and the acoustic emission source is located at the nearer probe.

(2) Planar positioning method

For planar localisation, the positioning can be carried out using the planar positioning calculation method of three or four probe arrays. When performing planar acoustic emission source positioning using an arbitrary triangular probe array, sometimes two

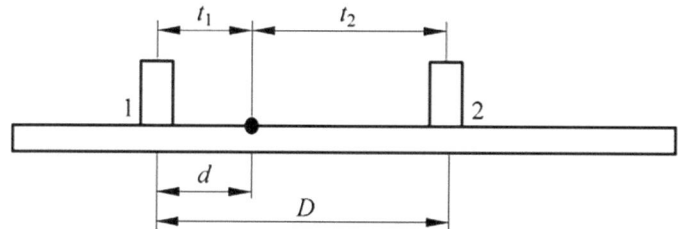

Fig. 6.7 Schematic diagram of the linear positioning method

sources may be obtained, namely one real AE source and one pseudo-AE source. However, if a rhombic array composed of four probes is used for planar positioning, it is equivalent to adding a constraint, and only one real AE source will be obtained, as shown in Fig. 6.8. If the hyperbola 1 is obtained from the time difference Δt_1 between probes S_1 and S_3; and hyperbola 2 is obtained from the time difference Δt_2 between probes S_2 and S_4. If the acoustic emission AE source is Q, the distance between probes S_1 and S_3 is a, and the distance between probes S_2 and S_4 is b, and the wave velocity is V, then the AE source is located at the intersection of the two hyperbolas, Q(x, y), whose coordinates can be expressed as:

$$x = \frac{\Delta t_1 V}{2a}\left[\Delta t_1 V + 2\sqrt{(X - a/2)^2 + y^2}\right] \quad (6.42)$$

$$x = \frac{\Delta t_2 V}{2b}\left[\Delta t_2 V + 2\sqrt{(y - b/2)^2 + x^2}\right] \quad (6.43)$$

3. Algorithms for acoustic source localization

When using the time difference positioning method for acoustic emission source positioning, the common solution methods are divided into the least squares method, simplex algorithm, Geiger algorithm and so on. The least squares method is a direct solution method, using the acoustic time difference, probe coordinates and other parameters to directly solve the equation to get the coordinates of the acoustic emission source. The simplex algorithm and Geiger algorithm are indirect solution methods, in which the calculated sound source positioning points are gradually converged to the real position of the sound source through iteration.

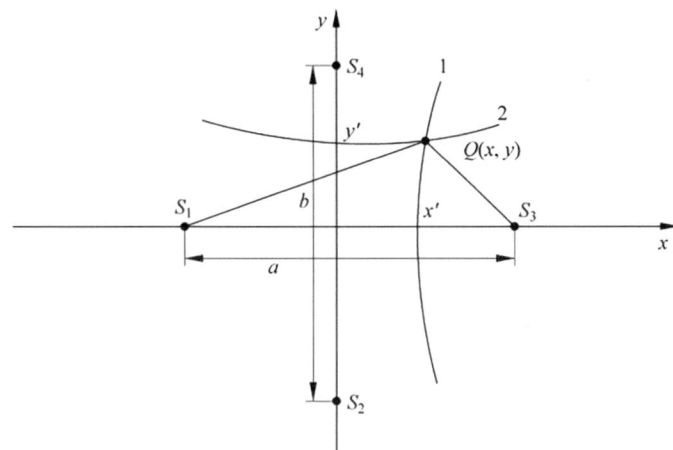

Fig. 6.8 Schematic diagram of plane positioning method

6.1 Acoustic Emission Key Technologies

The actual acoustic emission signal is non-smooth, multi-mode wave composition, and each mode of the wave is composed of a wide band of waves, each mode species of different frequencies of the wave speed is different, so the acoustic emission signal will appear dispersion phenomenon. The accuracy of acoustic emission positioning mainly depends on the accuracy of wave velocity and time difference. In order to improve the accuracy of acoustic emission localization, scholars at home and abroad have carried out a lot of research on acoustic emission localization algorithms: the wavelet transform is used to reduce the influence of the dispersion effect in order to improve the accuracy of acoustic emission localization; the wavelet transform is used in conjunction with the correlation relationship for the localization of sound sources; the spectral entropy energy product is introduced into the measurement of the arrival time of acoustic emission signals for the localization of sound sources; Some scholars have also proposed the exhaustive method, as well as based on artificial intelligence neural network and the use of time reversal focusing method for acoustic emission source localization.

This chapter focuses on the least squares, Geiger, and wavelet transform methods combined with correlation, which are commonly used at present.

(1) Least Squares

The least squares method is a direct solution method that uses the arrival time and sensor coordinates received by multiple sensors to establish a fixed set of equations, the essence of each equation is that the distance between the sound source and the sensor is equal to the wave velocity multiplied by the acoustic wave transmission time.

$$\sqrt{(x - x_i)^2 + (y - y_i)^2 + (z - z_i)^2} = v(t_i - t) \qquad (6.44)$$

where (x_i, y_i, z_i) denotes the coordinate value of the ith sensor, t_i denotes the moment when the signal arrives at the ith sensor; (x, y, z) denotes the coordinates of the acoustic emission source, t denotes the moment when the acoustic emission source generates the signal, and v is the acoustic wave velocity in the material to be localized, and it is assumed that the material is homogeneous and isotropic and that the wave velocity v is taken to be a constant value.

Equation (6.44) is the basic equation for the least squares time-difference localization. The unknowns in the equation are the acoustic emission source coordinates (x, y, z) and the initial time of acoustic wave generation t. The system of equations itself is a nonlinear system of equations, and the number of equations in the system of equations depends on the number of sensors, and for every sensor involved in the positioning detection an equation can be written. The method adopted to solve this nonlinear system of equations is to subtract one equation from each simultaneously, eliminating the second-order terms and thereby linearizing the system of equations. After linearization, the system of equations for N sensors becomes (N − 1) equations. For a three-dimensional positioning problem, since there are 4 unknowns, 4 linear equations are needed, which means at least 5 sensors are required to form the system of equations. For a two-dimensional positioning problem in a plane, the number of unknowns decreases by one. From Eq. (6.44), we can obtain:

$$(x - x_i)^2 + (y - y_i)^2 + (z - z)^2 = v^2(t_i - t)^2 \tag{6.45}$$

For three-dimensional positioning if 5 sensors are used for positioning, there are 5 systems of equations consisting of the above formulas. For the system of equations composed of 5 sensors, by subtracting the same equation from each equation and eliminating the quadratic term, 4 equations can be obtained to form a system of linear equations. For the obtained system of linear equations a matrix expression can be used:

$$AX = B \tag{6.46}$$

where A is the coefficient matrix and B is the constant term vector. The location of the sound source can be calculated by solving the system of four linear equations using the least squares method.

(2) Geiger localization algorithm

The acoustic wave emitted by the acoustic emission source is received by each sensor, and the sensor can record the time when the acoustic wave arrives at each sensor, and this time is the time that we measure, which is called the "measured arrival time". At the same time, by assuming the position of the acoustic emission source and defining the wave speed artificially, we calculate the arrival time of the signal emitted by the acoustic emission source to the sensor, which is called the "calculated arrival time". The difference between the measured arrival time and the calculated arrival time is due to the deviation between the assumed and the real position of the acoustic emission source. The difference between the two times is used to correct the assumed acoustic emission source position by the least squares method, and the real acoustic emission source position is obtained by iterative calculation, which is the basic idea of the Geiger iterative positioning algorithm. The calculated arrival time t_i^c is given as follows:

$$t_i^c = t(x_i, y_i, z_i, x_0, y_0, z_0) + t_0 \tag{6.47}$$

The calculated arrival time t_i^c consists of two parts: the first part is the acoustic transmission time t, which is a function of the known coordinates of the transducer (x_i, y_i, z_i) and the assumed coordinates of the acoustic emission source (x_0, y_0, z_0), i.e., the time of acoustic propagation from the acoustic source position (x_0, y_0, z_0) to the transducer position (x_i, y_i, z_i). The second part t_0 is the original time at which the acoustic emission source produces the signal. For a three-dimensional problem, there are four unknowns in this equation and at least four sets of arrival times are needed to solve, and for a two-dimensional problem, at least three sets are needed. If the number of sensors is greater than the number of unknowns in the localization,

6.1 Acoustic Emission Key Technologies

then the whole system is called a hyperdetermined system, forming a contradictory set of equations. For hyperdetermined systems, where the number of unknowns is less than the number of equations, it is necessary to solve the system by minimizing the residual, which is the difference between the Measured Time of Arrival and the Calculated Time of Arrival.

$$r_i = t_i^0 - t_i^c = \text{Min} \tag{6.48}$$

Linearization is required in the iterative localization method. In the process of iterative solution, the acoustic emission source position is to be assumed as the initial step of iteration, and the selection of the first assumed acoustic emission source position (x_0, y_0, z_0) is more important. Generally speaking, the assumed source position should be as close as possible to the real acoustic emission source position, while the real acoustic emission source position cannot be determined in advance. For small specimens, it is reasonable to select the first assumed acoustic emission source position at the center of the specimen, and for larger specimens, this assumed position can be selected at the position of the first sensor that receives the signal.

In the acoustic emission localization analysis of concrete specimens, the material can be considered homogeneous and isotropic, with acoustic signals emitted by the acoustic emission source and the time received by the transducer represented by the following function:

$$t_i = \frac{\sqrt{(x - x_i)^2 + (y - y_i)^2 + (x - x_i)^2}}{v} + t_0 \tag{6.49}$$

where the coordinate (x, y, z) is the computational target point, and the formula is used to calculate the propagation time of the acoustic signal from the target point to each sensor (x_i, y_i, z_i). The v in equation denotes the wave velocity in the material, and t_0 is the original time at which the source position is assumed for the first time.

The initial acoustic emission source position is assumed at the beginning of the calculation, so there must be a difference between the calculated arrival time obtained by the calculation and the measured arrival time obtained by the real measurement of the sensor. Therefore the residual difference between the calculated and measured arrival times is minimized by introducing a correction $(\Delta x, \Delta y, \Delta z, \Delta t)$ to the initial source position and the initial time. If the corrections are quite small, then Eq. (6.49) can be linearized. By retaining the first order term in the Taylor expansion, Eq. (6.48) takes the following form:

$$r_i = \left(\frac{\partial t_i}{\partial x} * \Delta x\right) + \left(\frac{\partial t_i}{\partial y} * \Delta y\right) + \Delta t \tag{6.50}$$

Change the expression to a matrix as follows:

$$r = G * \Delta x \tag{6.51}$$

where G is the partial derivative matrix and Δx is the vector of correction quantities, and the last column of the partial derivative matrix G is constant equal to 1 considering the correction term for the original time of the sound source.

For the three-dimensional localization problem, Eq. (6.51) is a set of linear equations with four unknowns. The result of taking the partial derivatives for every term in the partial derivatives matrix G is as follows:

$$\frac{\partial t_i}{\partial x} = \frac{(x - x_i)}{v} * \frac{1}{\sqrt{(x - x_i)^2 + (y - y_i)^2 + (z - z_i)^2}} \quad (6.52)$$

To solve for the correction Δx, we can use the matrix generalised inverse to solve Eq. (6.51) with the following expression.

$$\Delta x = (G^T G)^{-1} G^T r \quad (6.53)$$

After calculating the correction amount Δx, the correction amount is accumulated to the initial value, and $(x + \Delta x)$ is used as the new calculation data for iterative solving. When the iteration accuracy reaches the accuracy set in advance to meet the requirements, the iterative calculation is stopped. The iteration accuracy can be set as follows: the residual difference is less than the set value; the correction value of the sound source position obtained in the last iteration step is less than a certain set value; the number of iteration steps reaches a certain order of magnitude.

(3) Positioning algorithm using signal correlation relationship

Signal correlation localization method is based on the correlation characteristics of signals from the same acoustic emission source at the same moment, and measures the time difference between two channels of acoustic emission signals through the correlation between the waveforms of the two channels of acoustic emission signals, which is suitable for both bursty and continuous acoustic emission signals.

It is a signal processing method that is more suitable for relatively narrow frequency bands after decomposition. The specific idea is as follows: For the same acoustic emission source $s(t)$, the signals received by different sensors i are denoted as $f_i(t)$, for any two-channel signals, the correlation function $r_{ij}(\tau)$ can be expressed as:

$$r_{ij}(\tau) = \frac{1}{N} \sum_{n=0}^{N-1} f_i(t) f_j(t + \tau) \quad (6.54)$$

The correlation coefficient between the two signals is:

$$\rho_{ij} = \frac{\sum_{n=0}^{\infty} f_i(t) f_j(t + \tau)}{\left[\sum_{n=0}^{\infty} f_i^2(t) \sum_{n=0}^{\infty} f_j^2(t + \tau)\right]^{1/2}} \quad (6.55)$$

6.1 Acoustic Emission Key Technologies

By Schwartz inequality, there are $0 \leq \rho \leq 1$. The correlation coefficient ρ_{ij} presents the similarity between the signal $f_i(t)$ and $f_j(t)$. The closer the value $|\rho_{ij}|$ is to 1, then the better similarity between the two waveforms. By finding the maximum value, the time difference between the arrival of the signal at the two sensors can be calculated, ultimately achieving the purpose of locating the acoustic emission source.

Due to the propagation of the acoustic emission signal arriving at different sensors contains a variety of frequency components, the initial phase of the waveforms of different frequencies are different, the propagation speed is not the same, there is a dispersion phenomenon, and there are a variety of interferences in the propagation process, resulting in the arrival of the sensors of the waveform to a different extent have undergone a greater degree of change. If the multi-frequency mixed waveforms received by the sensor are directly used to the operate, the time difference is not accurate, will naturally have an impact on the positioning accuracy. Therefore, a more effective method is to perform correlation operations on the waveforms of the corresponding frequency bands of the acoustic emission signals after wavelet packet decomposition. This approach not only mitigates the impact of waveform distortion during propagation but also reduces the influence of noise.

6.1.3 Pattern Recognition of Acoustic Emission Signals

In the process of using acoustic emission (AE) technology to test materials or components, the experience of the testing personnel has always played a very important role, and to some extent, it can compensate for the shortcomings of the performance of existing testing instruments and signal processing methods. However, human factors in the testing process can affect the objectivity and accuracy of the analysis results, leading to different conclusions being drawn by different testing personnel for the same AE phenomena. This issue has severely restricted the development and promotion of AE technology for a considerable period of time. In response, many scholars and instrument manufacturers have attempted to use pattern recognition technology for automated AE detection of material (and component) damage. This is not only a need for the development of AE technology but also an inevitable result of the development of related technologies, especially computer technology.

1. The concept of pattern recognition

Pattern recognition is an interdisciplinary field that rapidly developed in the early 1960s. Generally speaking, pattern recognition refers to the classification and description of a series of processes or events, which can be either physical objects or abstract concepts. In some application areas, some experts refer to pattern recognition as quantitative (or numerical) taxonomy. However, strictly speaking, pattern recognition is not simply taxonomy; its goals should include the description, understanding, and synthesis of systems.

A complete pattern recognition system consists of two parts: design and implementation. Design refers to the creation of a classifier using a certain number of samples (each recognition object is called a sample, equivalent to an individual or sample in mathematical statistics, and their collection is called a sample set). Implementation refers to the use of the designed classifier to classify and make decisions about the samples to be recognized. Taking statistical pattern recognition methods as an example, its recognition system mainly consists of four parts: data acquisition, preprocessing, feature extraction and selection, and classification decision-making.

During the design phase, a large amount of data is obtained by measuring the training samples, and these data undergo preprocessing such as dimensional standardization. A small number of features most suitable for classification are then extracted or selected. Based on the distribution patterns of these features, a reasonable classifier is designed to complete the training process. For the samples to be recognized, they undergo data acquisition, preprocessing, and feature extraction before entering the classifier, where they are classified and recognized according to the discrimination rules.

2. Classification of pattern recognition

According to the different mechanisms and rules used in the recognition process, pattern recognition can be divided into four categories: statistical decision-making method, syntactic structure method, fuzzy recognition method and artificial intelligence method.

(1) Statistical decision method

Pattern recognition using statistical decision methods involves representing patterns as feature vectors. Based on the statistical laws governing the spatial distribution of these feature vectors, classification and discrimination are carried out using distance similarity criteria or Bayes' rule. Essentially, statistical decision methods utilize the mathematical characteristics of patterns.

According to the sample a priori knowledge, statistical decision method address two kinds of classification problems. Firstly, When the number of categories is known and a set of classified training samples is available, a discriminant rule is sought based on these samples. This rule is then used to classify samples of unknown categories. This type of problem is called a classification and discrimination problem, and the corresponding method is known as supervised learning. Secondly, When the categories and the number of categories of the samples are unknown, direct classification or the design of classifiers using these samples is required. Such problems are called clustering or cluster analysis, and the methods used are known as unsupervised learning.

Template matching in pattern recognition using statistical decision methods is the most primitive and fundamental method. The quality of the match depends on the statistical similarity between the template and various parts of the sample.

(2) Grammatical structure method

For some recognition objects, such as pictures, language, scenery, etc., structural information is its main feature, and the pattern of such objects is more complex. To perform recognition, it needs to be divided into several simpler sub-patterns, which are further divided into primitives. By recognizing these primitives, one can then recognize the sub-patterns, and ultimately the complex pattern. Since the structure describing such patterns is similar to the syntax of a language, its pattern recognition method is called syntactic structure method. The synthetic operational relations between primitives are called syntactic rules. When each primitive of the pattern to be identified is identified, its syntax is analyzed and its pattern is finally determined according to the primitives and syntax rules.

(3) Fuzzy recognition method

The concept of fuzzy mathematics is introduced in the process of pattern recognition, which becomes fuzzy pattern recognition. For the case of uncertain extension, a fuzzy set can be used to describe it, and then classify according to the principle of proximity. This method is mostly used in the recognition process affiliation degree, closeness and fuzzy relationship as the means of identification. In practical application, fuzzy recognition method is more often used for cluster analysis.

(4) Artificial Intelligence

Pattern recognition technology based on artificial intelligence has been an important development in the field of artificial intelligence in recent years. Recently, there are two main types of artificial intelligence methods introduced into the field of pattern recognition: one is logical reasoning; the other is artificial neural network. Logical reasoning methods rely on the knowledge representation of objects and use reasoning rules to identify them. The knowledge involved is the symbolic representation of objects obtained through statistical (or structurally fuzzy) recognition techniques or artificial intelligence technologies, which serves as the basis for classification. Based on this, by referring to the way humans classify objects, corresponding reasoning rules are determined, and the reasoning mechanism is activated to determine the category of the object.

Artificial neural network (ANN) pattern recognition is a method that mimics the principle of brain neural systems in processing information, using ANNs to recognize patterns. It first establishes a neural network of a certain structure, then learns from typical samples to obtain the corresponding recognition algorithm, and finally classifies and recognizes the remaining samples.

The application research of neural networks in the pattern recognition of acoustic emission signals includes the identification of cracks, corrosion damage stages, and fatigue processes in materials such as metals and reinforced concrete. Currently, the neural networks commonly used in the field of acoustic emission include: BP neural networks, improved BP neural networks, self-organizing feature maps (SOFM), probabilistic neural networks, and others.

3. Structural damage pattern recognition based on acoustic emission.

Structural damage recognition based on acoustic emission signals [12] involves utilizing collected acoustic emission signals to recognize the different development stages of structural damage (such as cracks, different damages during the structural loading stage, etc.) through pattern recognition methods. The process of structural damage pattern recognition using acoustic emission signals includes: determination of characteristic parameters, acquisition of information, pre-processing of data, extraction and selection of eigenvalues, and damage pattern recognition.

(1) Determination of characteristic parameters

The first step in utilizing acoustic emission signals for structural damage pattern recognition is to select and determine the characteristic parameters of the acoustic emission signals. For example, when monitoring and analyzing the initiation and propagation of cracks in metallic materials using acoustic emission, each acoustic emission event can be represented by a set of numerical values for characteristic parameters (such as signal amplitude, duration, etc.). In other words, each acoustic emission event can be represented as a point in a multidimensional space formed by the characteristic parameters of the acoustic emission signals. For the process of crack initiation and propagation composed of a large number of acoustic emission events, the distribution of these events is generally described according to statistical laws. Therefore, recognition methods based on statistical laws are effective for identifying crack damage in metallic materials.

When the parametric analysis method is used for the acoustic emission detection of crack damage in metal materials (30CrMnSi high-strength alloy steel, etc.) specimens, the acoustic emission mechanism of crack damage in metal materials is not very clear, and the characteristic parameters of the acoustic emission signals are difficult to comprehensively reflect all the information of crack damage. Therefore, pattern recognition of crack damage in metal materials should be based on the large amount of measured data to extract and select features that can reflect the differences between the different expansion stages of cracks, and design appropriate classifiers based on their specific distributions to achieve the recognition of crack damage patterns in metal.

For different materials and structures, due to the different mechanisms of acoustic emission, the statistical laws of their characteristic parameters in different damage stages are different. Therefore, for pattern recognition, suitable acoustic emission characteristic parameters that can characterize the differences between different damage stages should be selected according to the damage formation mechanism of different materials or structures.

According to the research results of scholars at home and abroad, for 30CrMnSi high-strength alloy steel, seven acoustic emission characteristic parameters can be used for crack damage pattern recognition, namely amplitude, energy, pre-peak count, rise time, duration, ring-down count, and average frequency. The results of Li Dongsheng's study on acoustic emission signals during the damage process of FRP-concrete filled steel tubular columns indicate that characteristic parameters such as amplitude, energy, ring-down count, rise time, duration, and RMS can well characterize the damage process.

6.1 Acoustic Emission Key Technologies

Therefore, when using acoustic emission signals to identify the structural damage mode, suitable acoustic emission characteristic parameters should be selected and determined according to the structural material damage acoustic emission mechanism.

(2) Acquisition of information

In pattern recognition, each analyzed object is called a sample. For structural or material damage (such as cracks, corrosion, fatigue, etc.) pattern recognition, each acoustic emission impact signal is a sample. A description of the characteristic parameters for each sample, such as the amplitude, energy, rise time, duration, ringing counts of the acoustic emission impact, etc., constitutes the most primitive measurement data. Of course, additional parameters can be obtained by combinations of parameters such as amplitude/rise time, ringing count/duration, etc.

Taking the crack pattern recognition of 30CrMnSi high-strength alloy steel as an example to introduce information acquisition, seven parameters can be adopted, namely amplitude, energy, pre-peak count, rise time, duration, ring-down count, and average frequency of the acoustic emission hits. Then, each sample (acoustic emission hit) can be represented as a point in a 7-dimensional space, denoted as:

$$X = [x_1 x_2 \ldots x_7]^T \qquad (6.56)$$

If N samples are obtained during the experiment, they can be expressed in the following form:

$$X_n = [x_{1n} x_{2n} \ldots x_{7n}]^T, n = 1, 2, \ldots, N \qquad (6.57)$$

For 30CrMnSi specimens with prefabricated cracks, the acoustic emission shows different characteristics at different stages of crack expansion. In the crack initiation and early expansion stage, due to the slow crack expansion, the number of acoustic emission impacts is relatively small, signal amplitude is low, and the energy released by each acoustic emission impact is also low, and the duration is short; with the acceleration of crack expansion, the degree of acoustic emission activity is enhanced accordingly, and the number of impacts is greatly increased, and the signal amplitude is elevated, and the duration becomes longer; When the crack approaches the damage stage, the crack expansion is further accelerated, the degree of acoustic emission activity is also increased, and the signal amplitude of each acoustic impact is very high and the duration is very long. In related studies, the acoustic emission signals of crack damage in metallic materials show similar characteristics.

When loading tests were carried out on 30CrMnSi specimens with prefabricated cracks, the fracture process of the specimens under stress could obviously be divided into four stages: plastic deformation, crack formation, stable extension and unstable fracture, with strong acoustic emission phenomena at each stage. These four stages actually reflect the different degrees of crack damage in the specimen, so they can be regarded as four patterns of crack damage development in the specimen. For the acoustic emission signals of a certain stage of specimen fracture damage (sample to be identified) collected during the test, a specific pattern recognition method can be used to determine the stage of the crack corresponding to the emission signal, i.e., using acoustic emission test signals to identify the corresponding stage of the damage (pattern).

(3) Data pre-processing

Normally, data preprocessing should be carried out for the original acoustic emission data obtained in the experiment or engineering inspection before pattern recognition. Its content includes two aspects: one is the selection of typical data, and the other is the standardization of data processing.

In order to simplify the analysis, only the acoustic emission signal pattern recognition of 30CrMmSi crack damage under uniform loading conditions is discussed here. Since the data obtained from the crack damage acoustic emission detection is a reflection of the actual damage condition of the specimen, all of them can be used as typical data without considering the elimination of bad values.

The purpose of data standardization is to eliminate the effect of the difference in magnitude between different parameters, so that the comparison and selection of features can be carried out. Commonly used data standardization methods include polar deviation standardization, standard deviation standardization and so on.

(4) Feature parameter extraction and selection

For each sample some factors related to classification identification must be determined, which are called features, can be used as the basis for identification. Feature selection is a key issue in pattern recognition, which directly affects the design of the classifier and the classification effect. Feature selection can be carried out according to different rules, and there are more methods to choose, but the quality of sample selection should be judged according to the results of recognition.

Usually, in the information acquisition stage, we should try to list as many factors as possible that may be related to classification in order to make full use of all kinds of useful information, and the features at this time are called original features. However, the original features cannot be completely used for classification recognition, and need to be analyzed and selected through further analysis. The reasons for this are the following three aspects: Firstly, too many features can pose difficulties for computation; secondly, the features may contain many interrelated factors, resulting in duplication of information; thirdly, the number of features is related to the number of samples, and an excessive number of features can deteriorate the classification effect. Therefore, it is necessary to extract a small number of feature parameters from a larger number of original features that are more effective for classification and recognition. This is what is known as the extraction and selection of feature parameters.

There are two methods for feature extraction and selection: one is the selection of individual features, i.e., each feature parameter is evaluated separately to find out those feature parameters that play the most important role in recognition; the other

6.1 Acoustic Emission Key Technologies

is to construct a suitable amount of new feature parameters from a larger number of original features. These two methods can be used separately or simultaneously, which needs to be determined according to the effect of structural acoustic emission pattern recognition.

① Selection of individual characterization parameters

For 30CrMnSi high-strength alloy steel, there are seven parameters that can characterize the features of different stages of crack damage: amplitude, energy, pre-peak count, rise time, duration, ring-down count, and average frequency. However, these seven characterization parameters may not all have good effects on pattern recognition, and further feature parameter extraction and selection are required. Therefore, the cluster analysis method mentioned in statistical decision-making can be adopted, using intra-class and inter-class distances as separability criteria to select individual feature parameters. In the experiment, the crack fracture damage process of 30CrMnSi specimens was divided into four stages: plastic deformation, crack initiation, stable propagation, and unstable propagation. Therefore, the crack damage of the specimen was defined as four patterns. Let x_{jn}^i represent the jth characteristic parameter value of the nth sample of the ith (i = 1, 2, 3, 4) pattern, and the mean value of the jth characteristic multiparameter for the sample of the ith class of pattern is:

$$\widetilde{m}_j^i = \frac{1}{N_i} \sum_{k=1}^{N_i} x_{jn}^i \tag{6.58}$$

where: $i = 1, 2, 3, 4$, respectively, represent four different stages in the stress fracture damage process of the specimen: plastic deformation, crack formation, stable extension and unstable extension; $j = 1, 2, \ldots, 7$, respectively, represent the characteristic parameters of the acoustic emission signals selected for pattern recognition, which are, in order, the amplitude, energy, pre-peak counts, rise time, duration, ringer counts, and average frequency of the vibration signals; N_i is the number of samples in the ith damage mode.

The total average value of the jth characteristic multi-parameter of all samples in the above four damage modes is:

$$\widetilde{m}_j = \sum_{i=1}^{4} \frac{N_i}{N_1 + N_2 + N_3 + N_4} \widetilde{m}_j^i \tag{6.59}$$

For the four damage patterns: interclass variance s_{jb} and intraclass variance s_{jw}, respectively:

$$s_{jb} = \sum_{i=1}^{4} \frac{N_i}{N_1 + N_2 + N_3 + N_4} \left(\widetilde{m}_j^i - \widetilde{m}_j\right)^2 \tag{6.60}$$

$$S_{jw} = \frac{1}{N_1 + N_2 + N_3 + N_4} \sum_{i=1}^{4} \sum_{n=1}^{N_i} \left(x_{jn}^i - \tilde{m}_j^i \right)^2 \tag{6.61}$$

For the divisibility criterion of the acoustic emission signal the following equation can be used:

$$J_j = \frac{S_{jb}}{S_{jw}} \tag{6.62}$$

The effect of selecting feature parameters can be determined based on the divisibility criterion J_j value, so as to select a single feature parameter or a feature space vector composed of multiple feature parameters suitable for pattern recognition. The larger the J_j value, which indicates that the dispersion between the classes is larger, while the dispersion within the classes is smaller, the selected feature value is suitable for classification of pattern recognition, otherwise the opposite.

Yang et al. [12] verified the acoustic emission detection of 30CrMmSi alloy steel specimens through the experimental data of acoustic emission detection, and designed the number of samples of acoustic emission signals is 960. He produced 96 specimens of six different length crack specimens, of which each length of the test is four, and the crack damage pattern includes four stages of crack forms. The J_j value of the separability criterion of the samples was calculated by seven parameters of the acoustic emission signal, namely amplitude, energy, pre-peak counts, rise time, duration, ringing counts and average frequency, respectively, and is shown in Table 6.1.

As can be seen from Table 6.1, the values of the five feature parameters of amplitude, energy, ringer count, rise time and duration are significantly different in order of magnitude compared with the remaining two feature parameters, among them the energy, ringer count and amplitude are the most obvious three characteristic parameters. Therefore, the five characteristic parameters of energy, ringer count, amplitude, duration and rise time are selected as the characteristic parameters for pattern recognition. Whether the selection of feature parameters is reasonable or not can be verified by combining the two-dimensional map of the acoustic emission signal and the statistical distribution law of the feature parameters.

In addition, there are many clustering analysis methods for the selection of acoustic emission feature parameters, such as the simple Bayesian model (NBM) method, K-mean, fuzzy C-mean and other clustering methods.

Table 6.1 Calculated values of the separability criterion J_j for the characteristic parameters of the design sample

Characteristic parameters	Magnitude	Energy	Pre-peak count	Rise time	Duration	Ring count	Average frequency
J_j value	3.120	4.833	0.126	1.117	1.856	4.124	0.089

6.1 Acoustic Emission Key Technologies

② Determine the combination of feature parameters

When determining the feature parameters of pattern recognition, the combination of feature parameters obtained by selecting individual features of the sample may not be the best, and in some cases even fail to achieve the expected results. Therefore, it is necessary to evaluate the combinations of characteristic parameters (such as amplitude, energy, ring count, duration, and rise time) obtained previously.

Before carrying out the optimal combination of feature parameters, it is necessary to normalize the data samples of the selected feature parameters, and then judge them according to the selected effective judgement index of clustering, and then determine the optimal combination of feature parameters for pattern recognition based on the judgment results.

(a) Normalization of sample data

When determining the combination of feature parameters, the first step is to be normalize the sample data. The purpose of sample data normalization is to convert the data with different scales or orders of magnitude into data with the same scales and orders of magnitude, so that the data are comparable with each other. Data normalization methods mainly include linear function method, paradigm method and mean value method, etc. The linear function method includes the maximum and minimum value method, mean value method, and so on.

The maximum-minimum method, also known as the extreme difference standardization method, involves normalizing the sample vector data to a specific range using the max–min method, typically the range of [0, 1].

The mean method adjusts the sample vectors to an arbitrary range by dividing each sample data by the mean and then multiplying by an adjustment factor to obtain new vector data.

The norm method involves transforming the vector composed of sample data into a unit vector with the same direction but a length of 1 using the 2-norm method.

(b) Evaluation of feature parameter combination clustering

When performing clustering analysis on combinations of feature parameters, it is often necessary to predetermine the number of clusters. This involves analyzing the structural characteristics of the sample data through certain computational indicators, evaluating the clustering results with reference to relevant experience, and subsequently determining the optimal number of clusters. At present, there are many indicators to judge the effectiveness of clustering, such as DB indicator, XB indicator, PC indicator, CE indicator, DI indicator, CH indicator and so on. The following briefly introduces several commonly used feature parameter combination clustering effectiveness evaluation indicators.

DB indicator is proposed by Davies and Boudian, and its evaluation objective function is defined as follows:

$$DB(k) = \frac{1}{k}\sum_{i=1}^{k}\max\left\{\frac{S(i)+S(j)}{M(i,j)}\right\} = \frac{1}{k}\sum_{i=1}^{k}\max R(i,j)(i \neq j) \qquad (6.63)$$

where k is the number of clustering categories, S(i) and S(j) represent the degree of dispersion within the class (intra-class distance), the larger the value indicates that the data are more dispersed. M(i, j) represents the degree of dispersion between the classes (inter-class distance), the larger the value indicates that the distance between different classes is larger. R(i,j) is defined as the degree of similarity between the ith class and the jth class of the sample data, where a higher value indicates a higher degree of similarity. The smaller the DB index value, the more the corresponding number of clusters aligns with the characteristics of the samples. Generally, based on experience, the number of categories corresponding to the local minimum of the DB index is selected as the optimal number of clusters.

XB index is a kind of effective evaluation index of clustering based on fuzzy clustering, which is defined as:

$$XB(U, V, C) = \frac{\sum_{i=1}^{C}\sum_{j=1}^{n} u_{ij}^{m} v_i - x_j^2}{n * \min v_i - v_j^2} \qquad (6.64)$$

where C represents the clustering category; n is the number of samples in the ith category; U represents the optimal affiliation matrix when divided into C classes; V represents the corresponding clustering centers; u_{ij} stands for the membership value of the jth sample data belonging to the ith class; x_j represents the jth sample in the ith class, and v_i represents the clustering centre of the ith class. The XB metrics search for a balance point between the intraclass compactness and the interclass dispersion to make the XB metrics obtains the minimum value, at this time the corresponding cluster number C is the optimal number of clusters.

Dunn index (DI) is a hard cluster internal evaluation index proposed based on data compactness and separation of clusters. The evaluation result solely depends on the classification data itself.

$$DI(c) = \min_{i \in c}\left\{\min_{j \in c, i \neq j}\left\{\frac{\min_{x \in c_i, y \in c_j} d(x, y)}{\max_{k \in c}\{\max_{x, y \in c} d'(x, y)\}}\right\}\right\} \qquad (6.65)$$

where $\min_{x \in c_i, y \in c_j} d(x, y)$ denotes the minimum value of the distance between elements in two different categories, which is used as a measure of inter-class dispersion, and $\max_{x, y \in c} d'(x, y)$ is defined as the maximum value of the distance between elements within the same clustering category.

There are many discriminative indexes that can be chosen in the feature parameter cluster analysis, and in this section, the separability criterion of the vector composed of the combination of acoustic emission feature parameters is used for analysis. The normalization of the data is processed using the maximum-minimum value method (polarity normalization method), and the vector divisibility criterion J is used for evaluation.

6.1 Acoustic Emission Key Technologies

(c) Evaluation of combinations of acoustic emission characteristic parameters

According to the results of calculating the separability criterion index of a single indicator, five of the seven characteristic parameters of the acoustic emission signal are arbitrarily selected for combination, and C = 21 feature combinations can be obtained. Among these feature combinations, if the combination consisting of the selected feature parameters (amplitude, energy, ringing count, duration and rise time) is optimal, then this choice is feasible. For a small number of combinations, sometimes the exhaustive method may be used for calculation, but the computational effort is relatively large.

Here, the separability criterion generated by the intra-class variance array S_w and the inter-class variance array S_b is used to evaluate the classification performance of the feature combinations.

The normalized eigenvalues $\dot{x}^i_{jn}$ for each eigenvalue are normalized using the maximum-minimum method, and the normalized eigenvalues obtained are as follows:

$$\dot{x}^i_{jn} = \frac{x^i_{jn} - \min x^i_{jn}}{\max x^i_{jn} - \min x^i_{jn}} (1 \leq n \leq N_i) \tag{6.66}$$

The vector formed by combining any of the above 5 feature parameters is denoted as:

$$z'_n = \left[\dot{x}^i_{1n} \dot{x}^i_{2n} \dot{x}^i_{3n} \dot{x}^i_{4n} \dot{x}^i_{5n}\right]^T, \quad i = 1, 2, 3, 4; n = 1, 2, \ldots, N_i \tag{6.67}$$

where, z'_n represents a vector of five acoustic emission parameters, i represents four damage modes in the crack development process of 30CrMmSi alloy steel specimen, and N_i represents the number of samples of the ith type of damage composed of the four damage modes.

If the mean value of the samples in the ith category is represented by m^i, as follows:

$$m^i = \frac{1}{N_i} \sum_{n=1}^{N_i} z^i_n \tag{6.68}$$

The total mean vector of all samples is denoted by m as follows:

$$m = \sum_{i=1}^{4} \frac{N_i}{N_1 + N_2 + N_3 + N_4} m^i \tag{6.69}$$

Therefore, the within-class variance matrix and the between-class variance matrix can be respectively expressed as:

$$S_w = \frac{1}{N_1 + N_2 + N_3 + N_4} \sum_{i=1}^{4} \sum_{n=1}^{N_i} (z_n^i - m^i)(z_n^i - m^i)^T \quad (6.70)$$

$$S_b = \sum_{i=1}^{4} \frac{1}{N_1 + N_2 + N_3 + N_4} (m^i - m)(m^i - m)^T \quad (6.71)$$

Therefore, the separability criterion J using a combination of characteristic parameters can be expressed as:

$$J = tr(S_w^{-1} S_b) \quad (6.72)$$

The value of J reflects the classification performance of the selected feature parameter combination, and the feature combination with the maximum value of J is the optimal combination. The samples are still designed using the previous design samples for selecting individual feature parameters, and the J-values of different feature combinations are calculated according to the above formula. Among all the 21 combinations of feature parameters: the combination of energy, ring counts, amplitude, duration and rise time has the largest value ($J = 5.881$), followed by the J-value of the combination of energy, ring counts, amplitude, duration and average frequency ($J = 5.127$), and then by the combination of energy, ring counts, amplitude, rise time and pre-peak counts ($J = 4.235$). This proves that the chosen combination of features (energy, ringer counts, phase value, duration and rise time) is optimal under the conditions of the combination of the five characteristic parameters and has the best classification performance.

(5) Acoustic emission pattern recognition of metal cracks

There are various methods of structural crack identification using acoustic emission technology, and the following describes the identification of metal cracks based on the nearest-neighbor recognition method, which is common in cluster analysis.

① Nearest Neighbor Recognition Method

The Nearest Neighbor Recognition Method includes the Nearest Neighbor method and the K-Nearest Neighbor method. If samples are regarded as points in a multidimensional space, a simple and intuitive classification method is to assign a sample to the category of the closest sample, which is known as the Nearest Neighbor method. When performing pattern recognition using the Nearest Neighbor method, one first selects the minimum distance between the sample to be recognized and each pattern sample, then selects the smallest of these minimum distances, i.e., finds the nearest neighbor sample to the sample to be recognized, and determines that the sample to be recognized belongs to the pattern represented by that sample.

Assuming that that the research objects (samples) have a total of a feature parameters, and all samples belong to c patterns $\omega_1, \omega_2, \ldots, \omega_c$ according to some rule; for $\omega_i (i = 1, 2, \ldots, c)$, there are known samples $N_i (i = 1, 2, \ldots, c)$. The samples belonging to the c patterns can be represented in the form of vectors as follows:

6.1 Acoustic Emission Key Technologies

$$X_n^i = [x_{1n}^i x_{2n}^i \ldots x_{an}^i]^T, \quad i = 1, 2, \ldots, c; n = 1, 2, \ldots, N_i \quad (6.73)$$

$$X = [x_1 x_2 \ldots x_a]^T \quad (6.74)$$

By applying the min–max normalization method to all samples, we can obtain the standardized and normalized values as $\dot{x}_{jn}^i$.

② Nearest Neighbor Method

As a measure of pattern distance, the Euclid distance between the sample X to be identified and the nth sample of the ith class pattern can be expressed as:

$$d_n^i = \sqrt{\sum_{j=1}^{a} \left(\dot{x}_j - \dot{x}_{jn}^i \right)^2} \quad (6.75)$$

According to the definition of the Nearest Neighbor method, its discriminant function is the minimum distance between the sample X to be recognized and each point in the pattern ω_i.

$$g_i(x) = \min d_n^i, \ 1 \leq n \leq N_i \quad (6.76)$$

$$g_{j*}(x) = \min g_i(x), \ 1 \leq i \leq c \quad (6.77)$$

③ K-Nearest Neighbor Method

K-nearest-neighbor method is a promotion of the nearest-neighbor method, the nearest-neighbor method is based on the distance from the special identification sample X nearest a sample of the class to determine the class of X; and K-nearest-neighbor method is based on the K-nearest-neighbors of X to determine the class of X.

Among the N known pattern samples, where $N_i (i = 1, 2, \ldots, c)$ samples respectively from the pattern c in the $\omega_i (i = 1, 2, \ldots, c)$ class, if the number of samples belonging to $\omega_i (i = 1, 2, \ldots, c)$ pattern among the K nearest neighbors of the sample X to be recognized are $(k_1, k_2, \ldots, k_c)$ respectively, then the decision rule according to the K-Nearest Neighbor method is:

$$g_{j*}(x) = \max k_i 1 \leq i \leq c, x \in \omega_j \quad (6.78)$$

④ Metal crack pattern recognition

Taking 960 acoustic emission signals of 30CrMmSi alloy steel specimens with prefabricated cracks as the samples for pattern recognition, and using half of them, namely 480 acoustic emission signals from 4 damage pattern samples, as

Table 6.2 Pattern recognition results of metal cracks using K-nearest neighbor method

Crack damage mode	Mode1	Mode2	Mode3	Mode4	Accuracy (%)
Plastic deformation	116	2	1	1	96.67
Crack formation	4	110	6	0	91.67
Stable expansion	1	1	118	0	98.33
Unstable fracture	1	0	1	118	98.33

the test samples, the K-Nearest Neighbor method was employed to conduct pattern recognition testing. The results are shown in Table 6.2.

As can be seen from Table 6.2, the K-Nearest Neighbor method achieves good recognition results. For all 480 test samples, the overall accuracy rate is (116 + 110 + 118 + 118)/480 = 96.25%. Among the four patterns of crack damage, the acoustic emission signals caused by the crack formation stage are more prone to misjudgment, often confused with Pattern 1 and Pattern 3. There may be two reasons for this: on the one hand, the distribution range of the acoustic emission signal characterization parameters during the crack initiation phase is relatively close to those of plastic deformation and stable crack propagation; on the other hand, the selected feature parameters are limited and do not encompass all the information of the signals, and they are also influenced to a certain extent by the chosen K value (the larger the K value, the more accurate the recognition results, but the computational load will also increase rapidly). However, overall, the K-Nearest Neighbor method can effectively recognize the acoustic emission characteristic parameters, thereby determining the stage of damage in the sample.

6.2 Key Technologies of Sonar

6.2.1 Beam Forming Technology

Beam forming technology is based on array signal processing technology. It is relatively mature and stable, with early development and extensive application, and is also a widely used acoustic imaging technology. Early imaging sonar systems were mainly two-dimensional, such as side-scan sonar systems and forward-looking sonar systems, which mainly employed technologies including fixed beam, single beam, and multi-beam. Among them, the side-scan sonar system uses a fixed beam to scan the seabed and collect echo signals as the system platform moves. Both single-beam and multi-beam systems can scan the seabed independently, but with different efficiencies. In recent years, three-dimensional imaging sonar technology has also developed rapidly. Currently, some advanced countries in the world already possess excellent three-dimensional sonar imaging technology, which has been applied in fields such as seabed survey and seabed imaging.

Beam forming is mainly divided into two methods: data-independent algorithm (conventional beam forming algorithm); data-dependent algorithm (adaptive or partially adaptive algorithm).

1. The principle of beam forming

Beamforming technology involves processing the outputs of multiple array elements arranged in a certain geometric shape (such as a straight line, cylinder, etc.) through operations like weighting, delaying, summing, etc.), to create spatial directivity. This spatial directivity aims to maximize the amplitude of the beam output in a specific direction in space while suppressing the beam output in other directions, thus directing the beam output towards a certain direction in space. Generally, a beam forming can be regarded as a spatial filter that filters out incoming signals from other directions and allows only signals from the specified direction to pass through [13]. To improve the imaging performance of imaging sonar, the receiving transducer and transmitting transducer in the acoustic imaging system typically adopt an array structure composed of multiple array elements. Beamforming technology is used to make the array form directivity in a predetermined direction, so that the transmitting transducer can concentrate the energy of the acoustic wave in a certain direction for emission, while the receiving transducer receives echoes from the spatially directed direction. Taking the simplest and most basic of a uniformly spaced linear array of point sources as an example, as shown in Fig. 6.9, for an N-element linear array with each element spaced d apart, assuming the same receiving sensitivity for each element and a plane wave incident direction of θ, the output signal of each element is as follows.

$$\begin{cases} F_0(t) = A \cos \omega t \\ \vdots \\ F_n(t) = A \cos(\omega t) + n\varphi = A\mathrm{Re}\left[e^{-j\omega t} \cdot e^{-jn\varphi}\right] \end{cases} \quad (6.79)$$

where A is the signal amplitude, ω is the signal angular frequency; τ is the time difference between the received signals of neighbouring arrays; φ is the phase difference

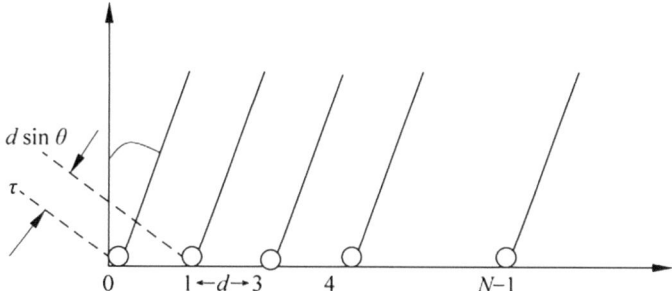

Fig. 6.9 Uniform linear array

between the received signals of neighbouring arrays, there are

$$\tau = \frac{d \sin \theta}{c} \tag{6.80}$$

$$\varphi = 2\pi f \tau = \frac{2\pi d}{\lambda} \sin \theta \tag{6.81}$$

Re(·) is the notation of the real part, $\lambda = c/f$ is the wavelength of the signal, and c is the speed of sound in seawater, the output of the array can be deduced as:

$$S(\theta, t) = \sum_{n=0}^{N-1} F_n(t) = A \cdot \text{Re}\left[e^{-jwt} \sum_{n=0}^{N-1} e^{-jn\varphi} \right] \tag{6.82}$$

due to

$$\sum_{n=0}^{N-1} e^{-jn\varphi} = \frac{1 - e^{-jN\varphi}}{1 - e^{-i\varphi}} = e^{-j\left[\frac{(N-1)\varphi}{2}\right]} \frac{\sin \frac{(N\varphi)}{2}}{\sin \frac{(\varphi)}{2}} \tag{6.83}$$

Thus

$$s(\theta, t) = A \frac{\sin(N\varphi)/2}{\sin(\varphi)/2} \cos[wt + (N-1)\varphi/2] \tag{6.84}$$

The amplitude normalization is achieved by dividing it by the maximum amplitude value of the array output, i.e. NA, to obtain an output amplitude:

$$R(\theta) = \frac{\sin\left(\frac{N\varphi}{2}\right)}{N \sin\left(\frac{\varphi}{2}\right)} = \frac{\sin\left(\frac{N\pi d}{\lambda} \sin \theta\right)}{N \sin\left(\frac{\pi d}{\lambda} \sin \theta\right)} \tag{6.85}$$

Equation (6.85) represents the normalized natural directivity function of an N-element equally spaced line array. It can be seen that when $\theta = 0$ and $R(\theta) = 1$, the output signals of all array elements are in-phase and added together, causing $R(\theta)$ to reach its maximum value [14]. The output amplitude of a multivariate array varies with the incident angle of the signal. If the linear array is to be oriented in the θ_0 direction, the signal delay of the ith array element relative to the reference array element becomes.

$$\tau_i(\theta_0) = (i - 1) \frac{d \sin \theta_0}{c} \tag{6.86}$$

6.2 Key Technologies of Sonar

At this point the directivity function of the line array changes accordingly:

$$R(\theta) = \frac{\sin[\frac{N\pi d}{\lambda}(\sin\theta - \sin\theta_0)]}{N \sin[\frac{\pi d}{\lambda}(\sin\theta - \sin\theta_0)]} \tag{6.87}$$

Line arrays can achieve spatial DOA estimation, and the more the number of array elements, the smaller the width of the main flap, and the higher the DOA estimation accuracy.

The above introduction is the principle of time-delayed beam formation, and the principle of phase-shifted beam formation is similar to time-delayed wave formation. When the signal is a single-frequency or narrow-band signal, due to the narrow frequency band and the absence of frequency changes, phase-shift beamforming can be approximated as time-delay beamforming.

In summary, due to the existence of time delay or phase difference between the received signals of each array element, by performing time delay or phase compensation on the output signals of each array element before superposition, the incident signals from the predetermined direction can be added in-phase, maximizing the beam output amplitude in that direction while suppressing the beam output in other directions, thereby achieving the function of spatial filtering.

2. Multi-beam formation technology

Despite the simplicity, ease of understanding, and straightforward implementation of single-beam sonar, it can only transmit one fan-shaped beam for scanning at a time, limiting the observed scanning space to the coverage of a single beam during one transmission and reception process. To observe a larger sector area with single-beam sonar, the beam is typically rotated using mechanical or electronic scanning methods, allowing the single beam to progressively search and cover the scanning area. However, this method can only observe targets in one azimuth direction, resulting in low efficiency. In fields such as seabed bathymetry, exploration, and marine surveying and mapping that demand high accuracy, single-beam sonar cannot meet the requirements. Multi-beam systems, however, can not only increase search speed but also achieve wide-area coverage scanning of the seabed. They divide the target area into multiple sections, each corresponding to a pixel, and each pixel has a beam associated with it. Therefore, multi-beam sonar has a wide range of applications in technical fields such as seabed bathymetry and seabed mapping. A multi-beam system sonar consists of multiple beam formers (see Fig. 6.10), and the beams formed by multiple beam formers are sequentially arranged and fill this scanning space, dividing the scanning space into several blocks.

When a phase shift β or a time delay $\tau\left(\tau = \frac{\beta}{2\pi f}\right)$ is inserted between neighbouring array elements, the summation output of the array in Eq. (6.82) becomes:

$$s(\theta, t) = \sum_{n=0}^{N-1} A \cos[wt + n(\varphi - \beta)] \tag{6.88}$$

Fig. 6.10 Principle of multi-beam formation

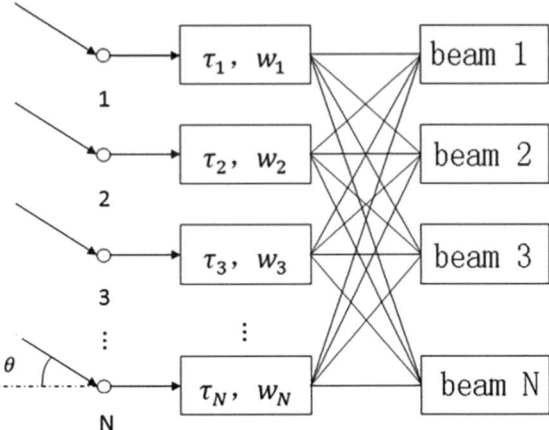

The normalized output amplitude in the form of Eq. (6.85) thus becomes

$$R(\theta) = \frac{\sin\left[\frac{N}{2}(\varphi - \beta)\right]}{N \sin\left[\frac{1}{2}(\varphi - \beta)\right]} \tag{6.89}$$

When the obtained main wave direction (main beam) satisfies $\varphi - \beta = 0 = 0$, i.e., $\frac{2\pi d}{\lambda} \sin \theta = 0$, the θ at this point is denoted as θ_0.

$$\sin \theta_0 = \frac{\beta \lambda}{2\pi d} \tag{6.90}$$

$$\theta_0 = \sin^{-1} \frac{\beta \lambda}{2\pi d} \tag{6.91}$$

Equations (6.90) and (6.91) show that the principal polar direction changes from $\sin \theta = 0$ to $\sin \theta_0 = \frac{\beta \lambda}{2\pi d}$, or there is $\theta_0 = \arcsin \frac{\beta \lambda}{2\pi d}$, due to $\beta = 2\pi f \tau$, so Eqs. (6.91) can be written as:

$$\theta_0 = \arcsin \frac{c\tau}{d} \tag{6.92}$$

From the above analysis, it can be seen that inserting different phase shifts β (corresponding to phase-shift beamforming) or different delays τ (corresponding to time-delay beamforming) between the array elements makes the main beam point in different directions. From Fig. 6.10, we know that the ith beamformer corresponds to a time delay of τ_i (corresponding to a phase shift of β_i), and the time delays (phase shifts) of each beamformer are different, then multiple beams pointing in different directions in space can be formed.

When spatial segmentation is performed using multiple beams, the number of parts to be segmented directly affects the pixels of the resulting image. In a two-dimensional imaging sonar system, the two-dimensional image is determined by pixels in two directions, perpendicular and parallel to the acoustic axis. The multiple beams formed are perpendicular to the direction of the acoustic axis, and the higher the requirement of pixels perpendicular to the direction of the acoustic axis, the finer the spatial segmentation, the higher the number of beams to be formed. The pixels parallel to the acoustic axis direction are determined by the time interval of the active sonar transmitting acoustic waves and the sampling frequency of the echo signal by the receiving array. With pixels at each point in these two directions, a complete two-dimensional image is determined. For three-dimensional imaging sonar, each pixel in space has coordinate information in three directions (x, y, z), the resolution along the acoustic axis direction is determined by the echo sampling frequency, and the pixels perpendicular to the acoustic axis direction of each cut are proportional to the number of beams formed in space. Generally the number of pixels of an imaging sonar will be larger than the number of array elements N of the base array, which will not affect the quality of the image too much. However, if the pixel count is less than N, the image quality will be compromised. The beam formation will also directly affect the quality of the image, usually using the pre-formed multi-beam method. When the number of array elements and the number of pre-formed beams are equal, the smaller the scanning angle range, the higher the corresponding angular resolution. When the signal comes from a certain scanning angle, the corresponding beam will output the maximum value, and the more beams are formed in a specified scanning space region, the higher the angular resolution corresponding to that region.

3. Several key issues in beam formation

(1) Interpolated beam formation [15].

According to the principle of beam formation, let the direction of the incoming wave be φ_b, $\delta_s = \frac{df_s}{c}$ is the maximum number of delayed sampling points of adjacent array elements, and let the delayed sampling point of the mth array element be q_{mb}. In order to form a beam pointing to φ_b, then

$$q_{mb} = m\delta \sin \varphi_b \qquad (6.93)$$

where, q_{mb} is the number of integer delay points, which is an integer multiple of the sampling interval t_s. Let the spatial beam index number be b (b is an integer, for example, $b = 0 \pm 1, \pm 2, \pm 3 \ldots$). And b beam corresponds to the incoming wave direction φ_b, there are

$$\delta_s \sin \varphi_b = b \qquad (6.94)$$

The beam pointing angle φ_b is obtained as follows:

$$\varphi_b = \arcsin\left(b \cdot \frac{1}{\delta_s}\right) \tag{6.95}$$

However, such beam directivity is far from meeting the realistic requirements. For example, the signal frequency f_0 is 100 kHz, the sampling frequency f_s is 500 kHz, the spacing between array elements d = 0.5 cm, so that $\delta_s = 1.67$, and the maximum delay between array elements is rounded to 1 sampling point. In this case, if we pre-form 3 beams between 0° and 180°, it is evident that the spatial resolution is very low, and the accuracy of DOA (direction of arrival) testing is far from sufficient. The above issues can be addressed through beam interpolation. Assuming that the beam interpolation factor is D, then

$$q_{mb} = mD \cdot \delta_s \sin \varphi_b \tag{6.96}$$

$$\varphi_b = \arcsin\left(b \cdot \frac{1}{D \cdot \delta_s}\right) \tag{6.97}$$

From Eq. (6.96), we know that the beam interpolation can be an integer multiple to increase the sampling frequency, the sampling frequency becomes $D \cdot f_s$, and the beam pointing angle is also changed to the original D times accordingly.

(2) Spatial resolution of the base array

The spatial resolution of an array can be represented by the main beam width Θ of the beam. The main beam width is defined as follows: starting from the main beam direction (where $R(\theta)$ reaches its maximum value), the angular interval at which $R(\theta)$ decreases to $\sqrt{R(\theta)_{max}/2}$ is called the half-power beam width $\Theta/2$, and Θ is referred to as the main beam width, also known as the half-power point or the 3 dB beam width. Spatial resolution is an important indicator in sonar imaging. The narrower the main beam width, the stronger the beam directivity, the higher the target spatial resolution, and the clearer the resulting image. When $\sin\theta$ is used as the coordinate axis, half of the main beam width is denoted as $\triangle \sin \theta = \frac{\lambda}{Nd}$, the wavelength of the emitted sound wave and the length of the base array $l = Nd$ determines the main flap width [14]. In practical engineering, the spatial resolution of the base array is often improved by increasing the frequency of the emitted sound wave or by increasing the length of the base array. This is the reason why acoustic signals of several hundred kilohertz to several megahertz are mostly used in sonar imaging systems.

(3) Improvement of the beam main lobe to side lobe ratio

Imaging sonar systems have high requirements for the directivity of the array, desiring a lower side lobe level to reduce the relative amplitude of the main lobe and side lobes, thereby avoiding interference in the main lobe direction and degrading the imaging quality of the sonar image. Amplitude weighting of the array element outputs can improve the directivity of the array and reduce the side lobe level. Common amplitude weighting criteria include: obtaining the narrowest main lobe given a required side lobe height; achieving the lowest side lobe given a required main lobe width; and satisfying a given main lobe-to-side lobe height ratio with a certain number of array elements. The most commonly used method is the Dolph-Chebyshev weighting method, which can produce equal side lobe levels. Dolph-Chebyshev weighting has the following two characteristics:

(1) Achievement of the narrowest main lobe width for a given arbitrary side lobe level;
(2) Making the side lobe level lowest for a given main lobe width.

Chebyshev weighting method is widely used in beam forming algorithms [16]. If the signal frequency f_0 is 300 kHz, the sampling frequency f_s is 3000 kHz, and the spacing between the array elements d is 0.25 cm. Compared with the equivalent weighting, after the Chebyshev weighting with the main-para-valve ratio of 35 dB, the side lob level decreases significantly, and the side lobe level decrease obtained by amplitude weighting is at the cost of increasing the main lobe width. In practice, the main lobe width and the main-sidelobe ratio should be considered as a compromise.

(3) Selection of signal frequency and array element spacing

According to the basic principles of sonar technology beam formation, the directivity of the base array exists in the first sub-maximum λ/d. When the spacing between the array elements is large, the main beam rotation, the first sub-maximum will fall into the scanning sector, resulting in the target orientation fuzzy or confusing. If the half-width of the scanning sector is θ_s, it is necessary to satisfy $2 \sin \theta_s \leq \frac{\lambda}{d}$, so that the scanning can be done in this sector, and the sub-maximum value will not appear in it. If the scanning range is $\pm 90°$, then the condition $\lambda/d \leq 1/2$ must be met. Sometimes limit the search range, according to the requirements of the search within a limited range, it should be appropriate to choose d/λ, so that the sub-extreme enough to stay away from the edge of the sector.

For example, the signal frequency f is 200 kHz, the selected beam scanning range of $\pm 30°$, we can calculate the signal wavelength λ using the formula, where $c = 1500$ m/s, $\lambda = 0.75$ cm, then get the array spacing $d \leq \frac{\lambda}{2 \sin \theta_s} = 0.75$ cm, and get the main lobe width of the beam by the array spacing and the number of array elements.

6.2.2 Beam-Forming Acoustic Imaging Technology

Beam-forming acoustic imaging technology is a widely used acoustic imaging technology, which is an important part of the beam-forming acoustic imaging system, the beam formation will directly affect the quality of the final image generated. The earlier development were mainly focused on two-dimensional imaging sonars, such as side-scan sonar and forward-looking sonar. The two-dimensional acoustic image contains only the distance and orientation information of the target, and does not contain three-dimensional information such as the vertical distance change of the scanned area. With the continuous progress and development of technology, three-dimensional sonar imaging technology has also advanced rapidly. Currently, some advanced countries in the world have already possessed excellent three-dimensional sonar imaging technology. Three-dimensional sonar imaging has been used in fields such as seabed survey and seabed imaging.

1. Two-dimensional acoustic imaging technology

In a two-dimensional acoustic imaging system that employs beamforming, the primary objective is to obtain distance and azimuth information of the observation area through beamforming. The distance information is derived by converting the backscatter echo time of the target into distance, while the azimuth information is realized based on the intensity of beam output at various scanning angles. Two-dimensional acoustic imaging mainly utilizes linear arrays, which have the advantages of simple circuit structure, fewer channels, and convenient hardware implementation, making them widely used in acoustic imaging systems. The following are two common imaging methods for linear arrays:

(1) single-beam scanning imaging

Single-beam imaging sonar can only emit one beam at a time and can only observe a beam scanning area. To obtain complete information about the observed area, a single-beam linear array can be mechanically rotated to obtain information about the specified sector or all-round area, and ultimately a two-dimensional acoustic image is generated by fitting a two-dimensional point map.

(2) Multi-beam pre-formed electronic scanning imaging

Using the multi-beam formation technique, multiple beams are pre-formed during a pulse emission to achieve simultaneous scanning of a wide area, obtaining two-dimensional information of the observation area and generating a two-dimensional image of the target. The advantage of pre-formed multiple beams is that a single scan can cover a wide area, because of the digital electronic scanning technology, its imaging speed is fast, and it can carry out large area imaging.

1) Composition of the imaging system

Two-dimensional sonar imaging often employs single-beam scanning sonar and pre-formed multi-beam sonar for scanning and imaging, and uses a linear array as its base array. For two-dimensional imaging sonar, its underwater target acoustic imaging system is mainly composed of four parts: target echo signal reception, pre-preprocessing, post-processing and terminal display. The functions of each module in the system are illustrated in Fig. 6.11:

6.2 Key Technologies of Sonar

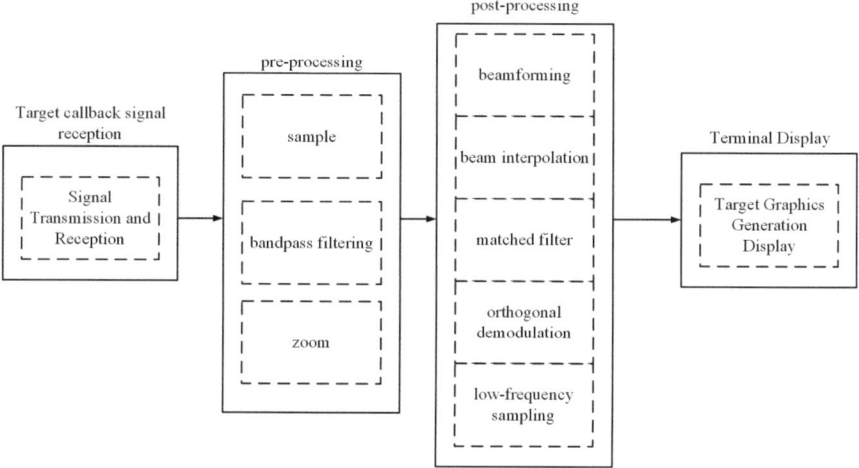

Fig. 6.11 2D sonar imaging system components

(1) Target echo signal reception: the transmitting transducer transmits a specific acoustic signal to the detection waters, after reflecting from the underwater environment reflector, the receiving transducer base array receives and scatters the echo, the acoustic signal is transmitted through the underwater channel and then reflected by the reflector to form the echo signal of the underwater target, which is superimposed by the target echo and reverberation.

(2) pre-processing: complete the pre-processing of the echo signal, to provide reliable and effective data for the subsequent beam formation and image generation, mainly including: sampling of the echo signal, band-pass filtering, amplification and other functions.

(3) Post-processing: On the basis of pre-processing, the echo signal is processed by beam formation, beam interpolation, matched filtering, orthogonal demodulation, low-frequency sampling and other data processing, and then the target's orientation, distance, and intensity and other information is estimated to ultimately generate the echo acoustic imaging (intensity-angle-intensity acoustic image) such as two-dimensional top-view, side-view, and three-dimensional map of the target, which is then displayed by the terminal.

(4) Terminal display: the results of sonar imaging are displayed through the terminal display.

Performing acoustic imaging on underwater targets can comprehensively obtain multi-dimensional spatial information about the targets and their underwater environment, thereby providing a more sufficient basis for target classification and identification. Underwater acoustic imaging relies on the processing of data and images from target echo information to form pseudo-images of the target composed mainly of bright spots, thereby enabling effective target identification.

2) The realization of beam formation

The two-dimensional imaging sonar system primarily employs a linear array and utilizes conventional beamforming algorithms for implementation. By measuring the round-trip time difference of the transmitted acoustic waves, distance information about the target is obtained, and the target's azimuth information is then acquired through beam space scanning. According to beamforming theory, the system's output can be expressed as:

$$y(t) = \sum \omega_i s_i(t - \tau_i) \qquad (6.98)$$

where ω_i is the weighting coefficient of the ith array element, s_i is the output signal of the ith array element, and τ_i is the time delay of the ith array element. The weighting coefficients are used in order to improve the side-lobe ratio or inhibit a specific interference. The commonly windowing functions include Hanning window, Hemming window, Chebyshev window and other weighting methods, and the weighting coefficients should be selected according to the actual situation in the actual project. Delay time compensation accuracy depends on the system signal sampling frequency, better system performance requires a higher sampling frequency, generally required to reach the Nyquist frequency of 5–10 times, and requires a higher data transmission rate. To obtain a higher spatial angular resolution, it can be achieved by increasing the sampling frequency f_s or by increasing the beam interpolation multiplier D.

3) Matching filter

In active sonar signal processing technology, methods such as beam forming and matched filtering are commonly used to improve the signal-to-noise ratio of the echo signal and improve the confidence in determining the presence or absence of a target. The echo signals received by active sonar first undergo beamforming to achieve spatial gain; further processing of the beamformed output is then conducted to obtain sufficient time gain.

Considering passing the echo signals through an optimal linear filter, the output with the maximum SNR can be obtained after passing through this filter. In waveform detection, matched filters are often used to construct the optimal detector, and matched filters occupy an important place in signal detection theory [13–15]. In communication systems, commonly used receivers are simplified to consist of two parts: a linear filter and a decision circuit.

Let the input to the receiver system be a definite, energy-limited signal $u_e(t)$ with complex form $u(t)$, the receiver system response function is $h(t)$, the transfer function is $H(f)$, and the output of the system at $t = t_0$ is:

$$y(t_0) = \int u(t) h(t_0 - t) dt \int U(f) H(f) e^{j2\pi f t_0} df \qquad (6.99)$$

6.2 Key Technologies of Sonar

The system that maximizes $y(t_0)$ is defined as a matched filter system that matches the signal $u(t)$ is referred to as a matched filter. It follows from Schwarz's inequality and the condition that the system outputs a maximum value:

$$y(t) = K \int u(t')u^*(t' - t + t_0)dt' = KR_u(t_0 - t) \qquad (6.100)$$

where K is the relative amplification of the filter, $R_u(t_0 - t)$ is the correlation function of the input signal $u(t)$, and at the moment $t = t_0$ is the maximum output moment of the matched filter. For general input signals, the matched filter is equivalent to an intercorrelator that can compute the cross-correlation function, and the matched filter is the best receiver for effective signal detection in white noise background [16].

2. Three-dimensional acoustic imaging technology

With the rapid development of sonar technology in recent years, people have been able to use sonar to complete a variety of tasks such as underwater navigation, positioning and ranging. In acoustic imaging sonar, the two-dimensional images generated by 2D imaging sonar only contain distance and azimuth information. However, the shape of any object is three-dimensional. In underwater acoustic imaging, the received sonar signals are processed to form spatial three-dimensional information about the target, which is then displayed as a 3D image at the terminal. This will facilitate a comprehensive understanding of the target's three-dimensional information and help people make more intuitive and accurate judgments. Therefore, underwater target three-dimensional acoustic imaging technology is receiving increasing attention, and its applications in fields such as seabed surveying and mapping, seabed imaging, and underwater target detection are becoming more widespread.

1) Three-dimensional imaging methods

Currently, there are three main methods used for underwater three-dimensional acoustic imaging: ① Utilizing a one-dimensional linear array to form multiple beams in the vertical (or parallel) direction along the track to obtain two-dimensional information of the observation area, and conducting mechanical scanning in the parallel (or perpendicular) direction to the track. The acquired two-dimensional information is then fitted by a computer to ultimately generate a three-dimensional image, such as the SeaBat8125 imaging sonar from RESON in the United States. ② Employing a planar array with two scanning angles to conduct multi-beam scanning in space, directly obtaining the resolution of the target in three directions: horizontal, vertical, and distance, thereby forming a three-dimensional image, such as the Echoscope series sonar from CodaOctopus. ③ Leveraging the interference principle in acoustic lens technology to obtain height information of the observation area and form a real-time three-dimensional image of the target, such as interferometric bathymeters.

The amount of data collected by three-dimensional imaging is substantial, and the hardware requirements for processing all the data at once to generate a three-dimensional spatial image are extremely high. Generally, a two-dimensional slice (or vertical section) within a vertical (or horizontal) cross-section is first formed, and multiple two-dimensional slices are then sequentially arranged to form the echo matrix required for the final three-dimensional image display. The display of three-dimensional images also comes in various forms, which can either directly show the three-dimensional target in a stereo space or be supplemented by top views, side views, etc., for more accurate judgment.

2) Three-dimensional acoustic image processing technology

The sonar system transmits acoustic signals to the underwater area of interest, and uses the acoustic imaging method to process the received backscattered echoes to obtain a series of 2-D slices of the sonar image in the area, and the synthesis of this series of slices can be obtained as a three-dimensional sonar image. The sonar image slices can have two types, one is the amplitude (intensity) image, and the other is the distance image, and the amplitude image and the distance image at the same moment are one-to-one correspondence in orientation. The three-dimensional data obtained from the distance image is called range data. For a three-dimensional acoustic imaging, these two types of images can be obtained at the same time, in which the amplitude map gives the target echo amplitude (intensity) information on the imaging orientation, and the distance map gives the distance information between the target and the reference point on the imaging orientation. By processing two types of sonar images simultaneously for target detection and recognition, the three-dimensional sonar image processing workflow includes: image filtering, image separation and reconstruction, eigenvalue extraction, image classification and identification, and terminal display.

(1) Image filtering

Image filtering aims to reduce the impact of non-object surface scattering points, eliminate side lobes and noise interference, the purpose is to reduce the colour point (speckle) and enhance the contrast of the image. Filtering of sonar images primarily leverages amplitude information, and there are various methods available, including commonly used ones such as FIR filtering, thresholding, and statistical methods.

FIR filtering method is mainly used to eliminate the colour point (speckle) in the image, by setting a continuous window function, using the same window function within the pixel weighted sum to realize the filtering. FIR filtering method will make the edge of the image fuzzy, so that the target recognition becomes difficult, and also can not effectively filter out the interference and signal-related noise. The threshold method is a simple and effective method in distinguishing backscattered echoes from interference signals, which is widely used in underwater 3D image processing. The threshold method is generally applied to the amplitude image, and by setting the threshold of the echo amplitude, the resolution units with amplitude higher than the threshold are left, and the resolution units with amplitude lower than the threshold are filtered out. The same process is applied to the distance map through the pairwise relationship between the amplitude image and the distance image. This process works directly on the beam signal and has a significant effect. Statistical theory is used to simulate real-time physical processes, and then "inversion" (reversal) techniques are used to attenuate the noise. Threshold methods and FIR filters are more effective in removing noise and improving image quality, while statistical methods are more advantageous for real-time image restoration.

(2) Image Segregation and Reconstruction

There are different segmentation methods for data from different sonar images. In general, segmentation is considered as the process of setting up pixel groups (regions). Different pixel points have different features, and by grouping image clusters with similar features together, and labelling different regions with different labels, it is possible to classify regions, or simply describe different regions. Image segmentation mainly uses statistical methods, especially using MRF (Markov Random Field) model. Another method is the geometric method, which deals mainly with distance images.

The Markov Random Field (MRF) is essentially a conditional probability model combined with a Bayesian criterion that reduces the problem to solve the maximum a posteriori probability estimation of the model, which is further transformed into a combinatorial optimization problem of solving for the minimum energy function, involving detection and classification. Although the MRF model is suitable for processing acoustic signals, its arithmetic is extremely large and is not conducive to the real-time display of 3D images. Geometric methods are mainly used to segment the distance image. Currently, the main methods for segmenting range images include edge-based segmentation, region-based segmentation, and hybrid region segmentation. Edge-based segmentation detects multiple continuous surfaces based on locally enclosed boundaries of image data points. Due to the low resolution and low signal-to-noise ratio of sonar images, edge-based segmentation is not suitable for processing sonar images. Region-based segmentation mainly involves dividing areas with the same geometric differential characteristics, typically following a segmentation sequence from local to global. In the first local stage, the image undergoes over-segmentation, where all points are labeled according to different differential characteristics. The subsequent global stage involves a series of merging processes where locally similar features are combined to achieve better segmentation. The global stage is actually an iterative process aimed at improving the results of the initial over-segmentation. During the computation, the first step is to determine which surface function to use for classifying (modeling) different locals. Generally, region-based segmentation methods can achieve good segmentation results, but their computational complexity is unpredictable. This method is widely used in 3D sonar image processing [17]. Hybrid region segmentation methods combine edge-based and region-based approaches. First, quadratic surfaces are fitted to the measured data points, and then the Gaussian curvature and mean curvature of the surfaces are calculated. These two parameters are used for initial region segmentation. Edge-based methods are then employed to extract boundaries from the initial regions, resulting in the final segmented regions.

The purpose of image 3D reconstruction is to construct a 3D geometric representation of the object to be measured and to recover the real surface condition of the object.

Image reconstruction mainly uses geometric methods and is often performed iteratively with image segmentation. Based on different data types, geometric methods include fusion of different intersection regions, surface fitting, and reconstruction.

(3) Image recognition and classification

Image recognition primarily focuses on the classification and description of images, with the objective of identifying the shape and texture characteristics of various parts of the image, which is referred to as feature extraction. The features of an image are the original characteristics or fundamental attributes that make an image different from other images. Some features correspond to the visual appearance of the image and have primitive nature, such as brightness, shape description, grey scale texture; some features lack natural correspondence, such as grey scale histogram, colour histogram and so on.

If the original target image is directly classified, it requires a large amount of computation, and the noise contained in the target image also has a great impact on the recognition. The basic task of feature extraction and selection is to find out those features that are the most effective for classification and recognition from a large number of features, so as to realize the compression of the feature space dimension. According to different mechanisms to extract multiple target features, complete the mapping of the target to different feature spaces, these features form a feature vector that constitutes the mapping of the target space to the multi-dimensional feature space, this multi-dimensional feature recognition technology can often achieve very good recognition results. The feature extraction methods mainly include colour-based feature extraction, texture-based feature extraction and shape-based feature extraction, and the feature extraction of colour is generally not involved in sonar images. After feature extraction, the conversion of target data from image domain to feature domain is completed, and the image data can be represented by features, and then the feature data will be sent to the designed classifier for target classification.

(4) Image Visualisation

The information obtained by the sonar is usually displayed centrally with some display devices, and the content of the display generally includes the position of the target and its movement, and various characteristic parameters of the target, etc. These display devices are a kind of sonar. These display devices are a type of terminal equipment for sonar, which can be called sonar terminal display. The sonar imaging processing software on the display terminal processes the image and displays the three-dimensional imaging result through the display terminal.

3) Focusing beam formation

Sonar systems make an assumption when imaging in the far field, namely, that the wavefront surface of the signal is a plane wave. However, when imaging underwater targets in the near field, the potential for error is larger. Assuming the scale of the base array is L, the target distance is r, and the signal wavelength is λ, when $r \leq 2E/n$ the target is located in the near field of the base array. At this point, the signal propagates as a spherical wave, and the sonar system needs to be focused. Generally, focusing and beamforming are combined for processing. There are typically two methods to address the focusing issue for near-field imaging: single-beam focusing directly ahead and focusing all beams.

1) Single Beam Focusing

Taking Line array as an example, let's introduce the principle of front single beam focusing. As shown in Fig. 6.12, $\vec{X_i}$ represents the position of the line array elements, F represents the focus position (along the acoustic axis from the centre of the array R), the spacing of the array elements is d. Assuming that the reference array element is $\vec{X_k}$, to achieve in-phase superposition of signals from all elements at F, the first step is to calculate the phase differences $\phi_1, \phi_2, \ldots, \phi_N$ between the signals received by each element and the reference element. Before beam forming, $\phi_1, \phi_2, \ldots, \phi_N$ is used to compensate for the phase of the beam in other directions in space, this focusing method is called forward single-beam focusing, and $\phi_1, \phi_2, \ldots, \phi_N$ is called the focusing correction factor. For planar arrays, the phase compensation factor $\phi_{n,m}$ for forward single-beam focusing is the acoustic range difference between an arbitrary array element and a reference point (along the acoustic axis direction away from the array center point R).

Because the beams in all directions in space are compensated by the focusing correction factor of the single beam directly in front, only the beams along the acoustic axis can achieve the in-phase summation of the true F. The larger the beams deviate from the acoustic axis, the larger the error caused by the focusing, and the beam width will become wider and the main-parallel flap ratio will be reduced. Therefore, this focusing method is only applicable when the scanning angle of the base array is small.

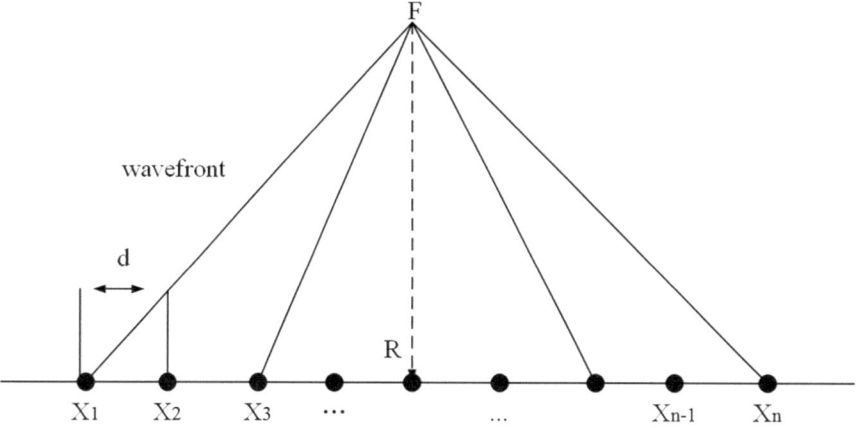

Fig. 6.12 Geometric relationship of beam formation for single-beam focusing in the forward direction

(2) Focusing method for all beams

The method of focusing all beams involves calculating the focusing correction factors for all beams within the scanning range and compensating for them using their respective focusing correction factors before beamforming. Taking a linear array as an example, for a sound source located at an arbitrary position F in the near field, the phase differences between any array element of the base array and a reference point are $\phi_1, \phi_2, \ldots, \phi_N$. Before beamforming, phase compensation is applied to the output of the base array so that the output signals of all array elements are in-phase and added constructively at the sound source direction. This process of calculating $\phi_1, \phi_2, \ldots, \phi_N$ and applying compensation is performed for all beam directions within the scanning space. An N-element uniformly spaced linear array has N independent beams in space, thus requiring phase compensation for N directions in space. A uniformly spaced planar array requires compensation for N × M directions, with its focal surface located on a tangent plane centered at the array's center and with a radius of R. This focusing method involves a significant amount of computation, which is unfavorable for real-time imaging. In practical engineering applications, a more feasible approach is to select only a few focal planes and compensate the phases for distances close to these focal planes R as if they were all at R.

6.2.3 Multi-beam Sonar Technology

1. Multi-beam detection technology

The multi-beam imaging algorithm is based on the algorithm of a multi-beam bathymetric sonar bathymetry. In practical engineering applications, the imaging algorithm is combined with the bathymetry algorithm, and the multi-beam bathymetry algorithm is used to estimate the TOA and DOA for each beam. And the TOA and DOA information obtained is applied in the imaging algorithm to calculate the echo strength information, horizontal displacement and angle of incidence of each beam, and other information, so that the information required to form the undersea sonar graph can be obtained. Among the multi-beam bathymetry algorithms, the more classic and widely used algorithm is the amplitude-based WMT algorithm.

(1) FFT beam formation

The number of array elements in the receiving array of a multi-beam sonar system is generally fixed. To obtain more information about beam control directions, beamforming calculations must be performed on the received signals before processing.

The essence of beamforming is spatial filtering, with the goal of responding to signals from specific angles while ignoring signals from other angles. According to the theoretical knowledge of linear systems, beamforming can also be essentially regarded as a type of convolution operation. Therefore, beamforming can be achieved through the product in the frequency domain, which is known as frequency-domain beamforming. When performing beamforming calculations on data from a multi-beam sonar system, the spatial Fourier transform can be directly applied to the signal at each sampling point as the output of beamforming. Due to the low computational complexity, high speed, and ease of implementation in hardware, most digital signal processing devices currently have mature FFT algorithms that are convenient to use. Therefore, FFT is widely utilized for beamforming in multi-beam sonar systems.

The result of multi-beam formation can be expressed in the form of a matrix:

$$\begin{bmatrix} B(\theta_1) \\ \vdots \\ B(\theta_M) \end{bmatrix} = \begin{bmatrix} D_{11} & \cdots & D_{1N} \\ \vdots & \ddots & \vdots \\ D_{M1} & \cdots & D_{MN} \end{bmatrix} \times \begin{bmatrix} S_1 \\ \vdots \\ S_N \end{bmatrix} \quad (6.101)$$

where B denotes the beam output vector, D matrix denotes the phase compensation matrix, and S vector denotes the weighting coefficients.

And the general expression of FFT operation is:

$$H_k = \sum_{i=0}^{N-1} h_i e^{j\frac{2\pi i k}{N}} \quad (6.102)$$

where the value of k ranges from 0 to N − 1, can be replaced as follows:

$$H_k = B(\theta_k)$$
$$\frac{2\pi}{N} ik = \frac{2\pi}{\lambda} id \sin\theta_k$$

Then it can be obtained:

$$\theta_k = \sin^{-1}\left(\frac{\lambda k}{dN}\right) \quad (6.103)$$

where θ_k is the kth control angle of the beam. Therefore, by utilizing FFT operations for beamforming, the control angle of each beam can be calculated, which is the Direction of Arrival (DOA) required for depth information calculation. Consequently, when a multi-beam sonar employs the Weighted Matrix Transform (WMT) method to measure depth and uses FFT for beamforming, the DOA can be rapidly computed, allowing the primary focus to be placed on the estimation of Time of Arrival (TOA).

(2) WMT bathymetry algorithm

The basic principle of multi-beam WMT bathymetry algorithm is to cross-sample the seabed by transmitting beams and receiving beams for Mills, form the beam output sequence distributed according to the angle of the beam, and then estimate the TOA of each beam to get the arrival time of the echo in the direction of the main axis of each beam, and then get the same depth information as the number of beams.

The multi-beam bathymetric sonar system receives echo signals from the seabed, and the data received in one measurement period (1 ping) are distributed in the following matrix:

$$D_1 = \begin{bmatrix} a_{11} & \cdots & a_{1T} \\ \vdots & \ddots & \vdots \\ a_{N1} & \cdots & a_{NT} \end{bmatrix} \quad (6.104)$$

where N represents the number of received array elements in the receiving array, and T represents the number of sampling points in each measurement cycle, which is the length of sampling time in each measurement cycle. Since the received signals are real signals, they are not utilized for beam forming calculations. Therefore, in order to facilitate the beamforming calculation, the time series of the received signal of each receiving array element is subjected to the Hilbert transform, which changes the real signal into a complex signal and obtains the resolved signal of the original signal, so that the following resolved signal data block arranged in a matrix manner can be obtained:

$$D_2 = \begin{bmatrix} S_{11} & \cdots & S_{1T} \\ \vdots & \ddots & \vdots \\ S_{N1} & \cdots & S_{NT} \end{bmatrix} \quad (6.105)$$

For each column in the data matrix obtained after the Hilbert transform (the signal obtained at each sampling moment), do the FFT operation at M (generally an integer power of 2) points.

Where, A_{ij} represents the amplitude value of the signal received by the array element at the beam steering angle in the direction θ_i and at the sampling time t. According to the data block, for each signal in the beam control angle direction, the dynamic threshold for each sampling moment is calculated by the following formula:

$$\text{DOOR}_j = \frac{1}{M} \sum_{i=1}^{M} A_{ij} \quad (6.106)$$

For each beam steering angle direction θ_i, the echo signal amplitudes A_{ij} at various sampling points are arranged in chronological order of sampling. Then, using a start gate and an end gate, data outside the dynamic threshold range are eliminated, leaving only data within the dynamic threshold range, as shown in Fig. 6.13. The time range represented by the start gate and the end gate corresponds to the Time of Arrival (TOA) range for that beam. The role of the dynamic threshold is to eliminate sampling points with excessively small amplitudes in the beam amplitude sequence. The dynamic threshold for the beam is determined using the central energy

6.2 Key Technologies of Sonar

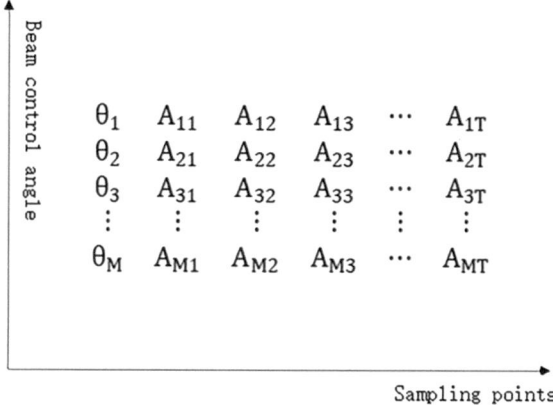

Fig. 6.13 Data obtained after FFT beam formation

convergence method. Subsequently, the data retained within the dynamic threshold is used to calculate the Time of Arrival (TOA) of the echo. The calculation method is amplitude-weighted averaging, as shown below:

$$\hat{t}_{TOA} = \frac{\sum_{i=1}^{M} A_i t_j}{\sum_{i=1}^{M} A_I} \qquad (6.107)$$

In each beam control direction, the corresponding time of arrival (TOA) of the echo can be calculated, and the angle value (DOA) of each beam control direction can be calculated when performing FFT beam formation. With the TOA and DOA information, the depth information and horizontal displacement corresponding to each beam can be calculated using the triangulation model, providing information for the calculation of the multi-beam imaging algorithm.

2. Multi-beam sonar imaging technology

(1) Multi-beam sonar imaging methods

In submarine imaging technology, multi-beam imaging technology is a hot research area both domestically and internationally due to its advantages of wide coverage and high resolution. Many imaging algorithms based on multi-beam sonar have been proposed, with typical examples including snapshot methods, pseudo-side scan methods, beam amplitude methods, and Snippet methods.

1) Snapshot method

In typical multi-beam seabed imaging algorithms, the snapshot method completes sampling of echo signals within each narrow depth-sounding beam, allowing for the acquisition of a complete time series reflecting echo intensity within each beam. During the imaging processing, all echo data are utilized to achieve complete echo intensity data sampling that includes both seabed and water column information.

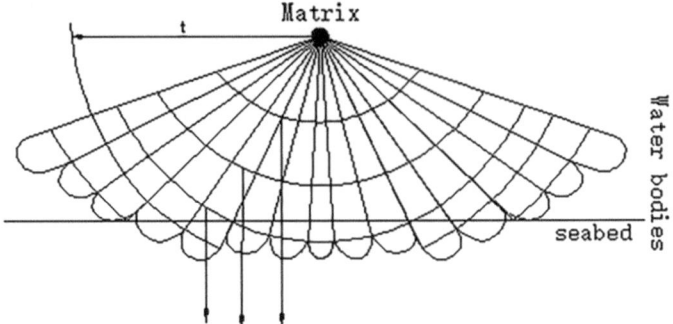

Fig. 6.14 Geometric schematic of the snapshot method

Figure 6.14 is a geometric schematic diagram of beam sampling using the snapshot method. Therefore, the snapshot method is relatively effective for water column imaging research. However, in seabed imaging, it is not necessary to use all echo intensity information, which means that the snapshot method generates a relatively large amount of data. Data transmission and processing can impact the performance of detection equipment, directly manifesting as a reduction in measurement frequency. Even with the increasing advancement of signal processing capabilities, it is still necessary to minimize data volume in practical engineering to meet the requirements of high efficiency and reliability.

(2) Pseudo side-scan method

Pseudo side-scan method use the two wide beams on both sides of the base array to obtain the seabed echo intensity data. The geometric schematic diagram of its operation is shown in Fig. 6.15. At a specific moment, the coverage areas of the two wide beams are sampled to obtain echo intensity data, and then each echo intensity value obtained is converted into a pixel value. This method of generating echo intensity time series is similar to side-scan sonar. Since the wide beams of the pseudo side-scan method are independent of the narrow depth-sounding beams and are very wide on the elevation plane, the horizontal displacement or incident angle corresponding to the echo intensity information must be indirectly calculated through depth profiles.

(3) Beam amplitude method

The beam amplitude method is similar to the snapshot method in that the echo intensity data are acquired by means of a bathymetric narrow beam, and the echo intensity time series generated for each beam corresponds to the inverse echo intensity value of that beam, which is then converted to a pixel value. The geometrical schematic diagram of the beam amplitude method is shown in Fig. 6.16. The pixel value corresponds to the echo intensity value in the direction of the main axis of the beam, and therefore the position information of each pixel value corresponds to the center

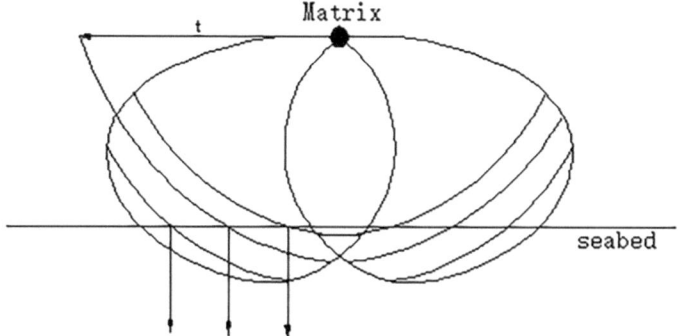

Fig. 6.15 Geometry of the pseudo side-scan method

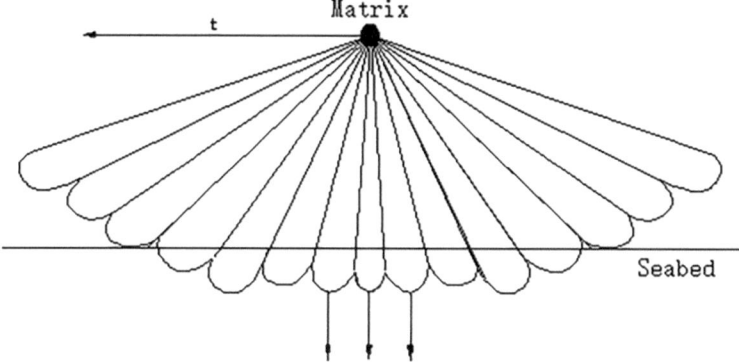

Fig. 6.16 Schematic diagram of beam amplitude method measurement

position of the beam footprint, and Fig. 6.17 represents a schematic diagram of the measurement of the echo intensity data with the position information.

(4) Snippet algorithm

Snippet algorithm is a relatively mature and advanced multi-beam imaging algorithm abroad. Using Snippet algorithm for multi-beam imaging has many advantages: ① During the measurement process, the same narrow beam is used for both echo intensity measurement and bathymetry, allowing the positional information of the echo intensity to be combined with the azimuth information of the received beam. This facilitates the acquisition of positional information for echo intensity samples in subsequent processing and the accurate fusion of backscatter data with bathymetric data. ② Each Snippet calculates a series of echo intensity values, generating seabed sonar images with a resolution comparable to that of side-scan sonar. However, multi-beam bathymetric sonar provides more accurate estimates of geographical locations, and multi-beamforming offers higher spatial resolution than side-scan sonar. ③ The Snippet algorithm can resolve backscatter intensity information within the beam

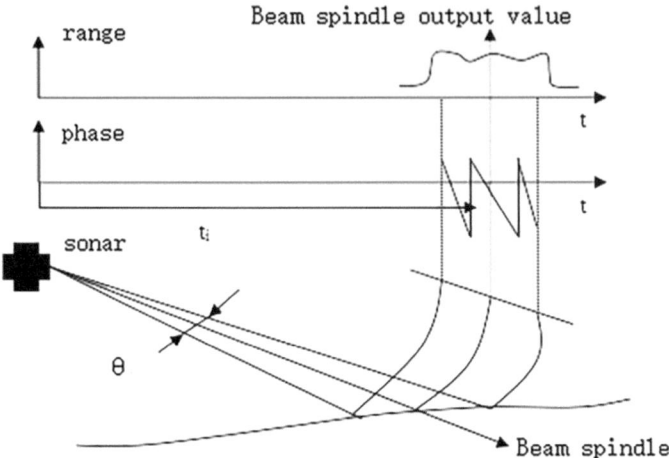

Fig. 6.17 Schematic diagram of measurement of data

footprint of each beam, thereby significantly enhancing the signal-to-noise ratio of the backscatter echo intensity data.

Each of these imaging algorithms has its own advantages and disadvantages, and the specific imaging algorithm needs to be selected according to the conditions of use.

(2) The basic principle of multi-beam Snippet imaging algorithm

Unlike several other multi-beam imaging methods, the Snippet method outputs a complete signal envelope for each bathymetric narrow beam, takes the seafloor detection point in the direction of the main axis of the respective beam as the sampling reference point, and performs time sampling within the footprint of the respective beam to generate the backscattered intensity time series, as shown in Fig. 6.18. Because the sampling is only carried out for each beam output sequence within the beam footprint on the seafbed, the echo intensity time series obtained by this method is only a slice intercepted from the complete signal envelope, which is called the Snippet slice. The Snippet method is a good way to overcome the low resolution, low information utilization and failure to merge with the topographic information existing in several other seafloor imaging methods. So it is the most important means and research hotspot for the acquisition of seabed echo intensity data by multi-beam bathymetric sonar.

(1) Segment Interception

Due to the differences in transducer structure, beam angle, and control among various multi-beam sonar systems, there will be certain variations in the methods used to intercept snippets within beams when utilizing the multi-beam Snippet imaging algorithm for imaging processing. However, there is a prerequisite: the echo intensity time series intercepted within each beam must be seamlessly connected, meaning there

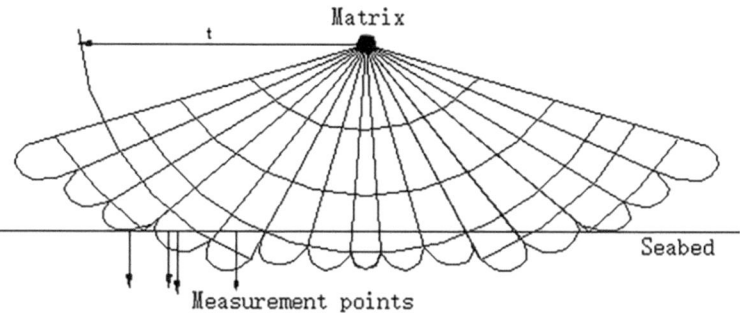

Fig. 6.18 Measurement Schematic of the Beam Amplitude Method

should be no gaps between the snippets intercepted by adjacent beams. If gaps occur, measurement blind spots will arise, making it impossible to connect the backscatter intensity information within adjacent beams.

In practical applications, the sonar system processing unit generally pre-sets the starting time and length of each snippet based on the estimated slant range value of each beam. When using the multi-beam Snippet imaging algorithm, time is generally represented by the number of sampling samples. Therefore, the starting time and length are also referred to as snippet offset and the number of sampling samples, respectively.

There are two modes of data acquisition for the multi-beam Snippet imaging algorithm: uniform distance mode and flat seafloor mode. When data are acquired in the uniform distance mode, the length of the time window is proportional to the beam slant length, and the position of the detection point corresponding to the main axis of the beam is the center of the sampling time window for that beam. If the depth corresponding to the direction of the main axis of a certain beam is h, the angle of the beam is θ, and the width of the beam is θ_R, then the proportionality between the length of the sampling time window and the slant distance corresponding to the center position of the beam footprint when collecting data in the uniform distance mode can be expressed by the following formula:

$$prop = \frac{h \cdot \tan\left(\theta + \frac{\theta_R}{2}\right) - h \cdot \tan\left(\theta - \frac{\theta_R}{2}\right)}{\frac{h}{\cos\theta}} = \frac{2 \sin\theta_R \cos\theta_R}{\sin 2\theta + \cos\theta_R} \quad (6.108)$$

From Eq. (6.108), it can be seen that the value of the proportionality is not directly related to the depth information, but mainly related to the beam width and beam angle, and the value of the proportionality will gradually become larger with the increase of the beam angle in the case of fixed width of each beam.

In practical engineering applications, if each beam is intercepted in pieces strictly according to the corresponding proportionality relationship, it will increase the workload of the system, so the same proportionality value is taken for each beam angle, which can simplify the calculation more. However, in order to achieve the purpose of making each sampling time window cover the whole beam footprint as much as possible, and considering the characteristic that the beam footprint changes with the beam angle, the value of the proportionality relationship is generally calculated by taking the leftmost or rightmost beam. After the proportionality relationship is determined, according to the sampling length interval of the sonar system d_s, the number of samples N_s on each side of each sampling time window can be calculated. And according to the number of samples N_s, the sample number corresponding to the sampling time window opening and closing moments can be calculated, and the length of the sampling time window can be determined.

Therefore, the sample numbers corresponding to the sampling time window opening moment and the window closing moment can be calculated, respectively:

$$N_{start} = N_{center} - N_s$$
$$N_{end} = N_{center} + N_s \qquad (6.109)$$

The length of the sampling time window can be calculated using the following equation:

$$W_{in} = 2N_s + 1 \qquad (6.110)$$

When the seafloor topography within the measurement range of the multi-beam sonar changes significantly, it becomes very difficult to calculate the beam footprint width accurately, in which case the uniform distance mode is generally chosen for data acquisition. However, other problems may occur when data acquisition is carried out in the uniform distance mode, which is directly manifested in the possibility of gaps between the coverage areas of neighbouring beam snippet segments. The segments of each beam cannot be seamlessly connected, resulting in measurement blind spots.

The premise of data acquisition in the flat seafloor mode is to assume that the seafloor within each beam footprint is flat, the topography within the beam footprint is considered to have no undulation, so that the corresponding depth information and beam angle of each beam can be used to calculate the sample number corresponding to the opening and closing of each beam snippet sampling time window, and the calculation formula is shown below:

$$\begin{cases} N_{start} = \left[2 \cdot \dfrac{h}{\cos\left(\theta - \frac{\theta_R}{2}\right)} \cdot \dfrac{1}{d_s} \right] \\ N_{end} = \left[2 \cdot \dfrac{h}{\cos\left(\theta + \frac{\theta_R}{2}\right)} \cdot \dfrac{1}{d_s} \right] \end{cases} \qquad (6.111)$$

The length of the sampling time window can be obtained:

$$W_{in} = N_{end} - N_{start} + 1 \qquad (6.112)$$

6.2 Key Technologies of Sonar

It should be pointed out that, according to the triangular model, when data acquisition is carried out in the flat seabed mode to obtain a slice of each beam, the sample number corresponding to the detection point on the seabed in the direction of the main axis of each beam does not correspond to the center of the sampling time window, except for the case in which the direction of the main axis of the beam is perpendicular to the seabed.

Both the flat seafloor mode and the uniform distance mode share a common characteristic: they are both affected by changes in seabed topography and variations in beam incidence angles, resulting in unequal numbers of echo intensity samples collected from the segments intercepted within each beam. The number of sampling samples increases with the increase in seabed depth and beam incidence angle. When seabed topographical changes are not significant, the flat seabed mode is generally adopted for data acquisition. It should be noted that there is no significant difference between the results obtained from the two modes.

(2) Fragment processing

After the snippet is intercepted, the echo intensity value corresponding to each beam can be obtained by the snippet of each beam. Arranging the processing results in accordance with the horizontal displacement, the echo intensity of one measurement cycle (1ping) can be obtained. There are two main processing methods for Snippets: the first method is to calculate one echo intensity per Snippet, and the second method is to calculate a series of echo intensities per Snippet.

Method one is essentially the same as the beam amplitude method, but the echo intensities of the Snippet method are calculated from the peak or average intensities of the slices, so the first processing method of the Snippet method can be categorized as one of the beam amplitude methods.

Because the information recorded in the sampled snippet is a time series of echo intensities within a beam, in order to improve the information utilization rate and also make the calculated results more accurate, the method generally adopted is to calculate the average intensity value, which is more reasonable. The method of calculating the average intensity value involves first integrating the square of the echo signal envelope and then normalizing it, with the normalization process using the pulse width of the transmitted signal.

The primary purpose of using the multi-beam Snippet imaging method is to improve the resolution of imaging, and using this processing method makes it possible to obtain only one echo intensity value within a beam, which does not do any good for improving the resolution of imaging, and also does not accurately reflect the characteristics of the change in backscattered intensity within the beam footprint. Therefore, it is not reasonable to use the first method to process the multi-beam slice data.

The second method involves utilizing the intercepted snippets to calculate a series of echo intensities within each beam, which not only enhances the information utilization rate of the sampling snippets but also improves the imaging resolution. During the imaging process, if all the echo intensity results calculated in each beam are used, the generated geomorphological image is similar to the geomorphological image generated by the side-scan sonar, but compared with the side-scan sonar has an obvious advantage, that is, the side of the flap and the noise in the water will be much smaller. In each beam segment, the echo intensity sampling interval is set according to the imaging needs, and then the main axis of is used as the reference point, and the echo time series segments are sampled forward and backward, respectively.

If the sampling interval is Δt, then the number of sampling points ΔN at the interval can be expressed as:

$$\Delta N = \left[\frac{\Delta t}{t_s}\right] \tag{6.113}$$

The sampling length L_{sampel} on the corresponding slant distance can then be expressed as:

$$L_{sanpel} = \frac{c \cdot \Delta t}{2} \tag{6.114}$$

Furthermore, the number of samples of the echo intensity within each slice can be calculated as:

$$N_{sampel} = \left[\frac{N_{center} - N_{start}}{\Delta N}\right] + \left[\frac{N_{end} - N_{center}}{\Delta N}\right] + 1 \tag{6.115}$$

The first two terms on the right side of Eq. (6.115) are the number of samples sampled before the beam main axis direction and after the beam main axis direction, respectively. If the sample number corresponding to the beam main axis direction in the echo intensity sequence is N^s_{center}, then the sample number corresponding to each echo intensity TOA can be calculated as:

$$N_{sampel} = N_{center} + \left(N^s_{sampel} - N^s_{center}\right) \cdot \Delta N \tag{6.116}$$

where, $N^s_{sompel} = 1, 2 \ldots, N^s_{compel}, \ldots N^s_{sompel}$. If the moment corresponding to this sample number is the opening moment of the echo intensity sampling time window, then the closing moment of each echo intensity sampling time window can be calculated using the following equation:

$$N_{over} = N_{sanpel} + \Delta N \tag{6.117}$$

In this way, the acoustic intensity information of the corresponding samples can be calculated using the average amplitude values of all sampling points within each echo intensity time window.

Since either the amplitude or phase detection methods can only be used to estimate geographic information in the beam principal axis direction, and cannot estimate the geographic location information for points outside the beam principal axis direction.

The multi-beam Snippet algorithm assumes a flat seabed when estimating the horizontal displacement of each echo intensity. The DOA for each echo intensity can be calculated using the following equation:

$$\theta_{sampel} = \arccos\left(\frac{h}{\frac{h}{\cos\theta} + L_{sampel} \cdot \left(N^s_{sampel} - N^s_{center}\right)}\right) \quad (6.118)$$

Then the horizontal displacement corresponding to each echo intensity is:

$$x_{sampel} = h \cdot \tan(\theta_{sampel}) \quad (6.119)$$

From the above equation, it can be seen that the sonar sampling in slant range is linearly spaced, but it appears nonlinear on the seabed. When the incident angle is smaller, this nonlinear phenomenon is more obvious, and as the incident angle becomes larger, the sampling interval gradually becomes smaller. when the angle becomes larger, the sampling interval tends to a fixed value, then the sampling can be regarded as linear.

3. Multi-beam sonar imaging influencing factors and corrections

Multi-beam sonar imaging algorithm can get higher spatial resolution than side-scan sonar, and the signal-to-noise ratio is higher than side-scan sonar. However, the echo intensity does not directly reflect the seafloor geomorphological features, and the parameter that reflects the seafloor geomorphological features is the backscatter intensity, so it is necessary to calculate the seafloor backscatter intensity according to the echo intensity. In the actual detection, sound wave propagation in seawater will be interfered by many factors, such as propagation loss, sound line bending and so on. Therefore, in the process of data processing, the sound intensity information should be compensated, so as to make the obtained seabed echo scattering intensity information more accurate.

1) Scattering intensity influencing factors

According to the active sonar equation, the key step in the calculation of scattering intensity in the direction of multi-wavelength sonar is to calculate the propagation loss of sound waves in seawater and the acoustic irradiation area on the seabed. When sound waves propagating in seawater, the propagation speed changes with the change of depth, and it is known by the refraction theorem of sound waves that sound waves will be bent when they propagate in seawater. When calculating the propagation loss, it is necessary to accurately calculate the actual distance travelled by the sound wave propagating in seawater, so it is very necessary to carry out acoustic line tracking for the acoustic line bending phenomenon. In addition, during the sound tracking process, the depth of the seabed and the horizontal offset of the sound waves can be accurately calculated, which is essential for the accurate integration of imaging and topographic information in the subsequent processing. Therefore, sound line bending is also an important influence on the backscatter intensity. In the following, the propagation loss, the sound line bending phenomenon and the acoustic irradiation area will be analyzed and the corresponding correction methods will be introduced.

(1) Propagation loss

As an acoustic signal travels through seawater, the farther it travels, the more energy is lost and the weaker the sound intensity becomes. There are two types of lost energy, one is expansion loss, also known as geometric attenuation, and the other is attenuation loss. Expansion loss arises from the attenuation caused by the gradual expansion of the wavefront as it moves away from the sound source and increases in distance, leading to a decrease in sound intensity. Attenuation loss includes absorption loss and scattering loss. Absorption loss refers to the attenuation caused by medium viscosity, thermal conduction and other relaxation processes when the sound wave propagates in the medium; scattering loss is the attenuation of sound intensity caused by the scattering of acoustic waves due to the presence of suspended ions and inhomogeneity in the seawater medium.

The expansion loss of sound waves is caused by the expansion of the wavefront in the process of propagation. Apart from the loss caused by wavefront expansion, when calculating the propagation loss, it is also necessary to consider the attenuation loss resulting from the significant absorption of acoustic wave energy as it propagates through seawater. This absorption is caused by the large number of free ions and suspended particles present in the seawater, as well as the viscosity and thermal conduction of the seawater itself.

2) Sound line bending

According to the acoustic dynamic equation, the calculation of acoustic wave propagation loss in seawater needs to know the actual propagation distance R of acoustic wave in seawater medium. Therefore, in the process of calculating the propagation loss, the accurate calculation of the actual propagation distance of sound waves in seawater medium is a very critical step. The propagation speed of sound waves in seawater is not constant. At different depths in the ocean, factors such as temperature, salinity, and seawater pressure vary, and these factors affect the propagation speed of sound waves in seawater. Therefore, the propagation speed of sound waves varies with the depth of the seawater.

Due to the difference in sound velocity of sound waves at different depths, there exists an interface between water layers with different sound velocities. When sound waves propagate to this interface, refraction occurs. This refraction phenomenon can be described by Snell's law for sound wave propagation in different media. According to Snell's law, during multi-beam imaging detection and calculation, the difference in sound velocity of sound waves in water layers at different depths cannot be ignored in terms of the accuracy of the results.

3) Effective irradiation area

The effective irradiation area A_E of the beam is another important influence factor on the calculation of the backscattered intensity B_S. In order to make the calculation of backscattering intensity more accurate, the accurate solution of the effective irradiation area of the beam is also a very important step. When a sound wave is emitted by a transducer and propagates to the bottom of the water, the beam expands and forms a beam footprint on the bottom of the water. The effective irradiation area of the beam is related to the beam footprint area A_{fpa} on the water bottom in addition to the instantaneous acoustic irradiation area $A_{intonif}$, which is jointly determined by these two factors. The overlap between the footprint area and the acoustic transient irradiation area is the effective irradiation area of the beam, and its area is the effective acoustic irradiation area of the beam.

Figure 6.19 shows a diagram of the effective acoustic exposure area. Wherein, the beam footprint area A_{fpa} is a physical quantity related to the beam angle width θ_R, the beam width φ received by the receiving array transducer, the actual propagation distance R of the acoustic wave on the seafloor, and the beam width ϕ of the signal emitted by the transmitter, with the expression:

$$A_{fpa} = \frac{R^2 \phi \sin(\varphi)}{\cos(\theta_R)} \tag{6.120}$$

The acoustic instantaneous irradiation area $A_{intonif}$ is not only related to the beam control angle θ_R, the actual propagation distance R of the acoustic wave on the seafloor and the beam width ϕ of the signal emitted by the transmitter, but also affected by the acoustic velocity C of the acoustic wave propagating on the seafloor and the pulse width τ of the emitted signal, which can be expressed as follows:

$$A_{insonif} = \phi R \frac{C\tau}{2\sin(\theta_R)} \tag{6.121}$$

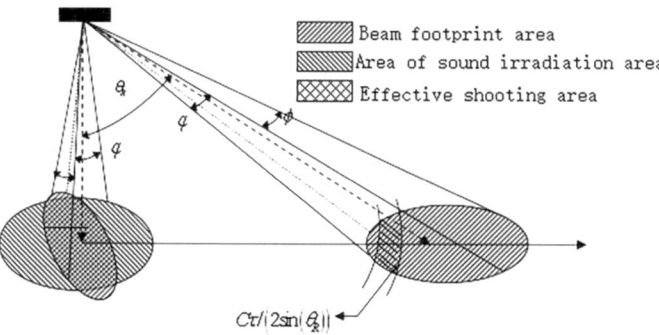

Fig. 6.19 Schematic of effective acoustic irradiation area

From Eqs. (6.120) and (6.121), it can be seen that both the beam footprint area A_{fpa}, or acoustic instantaneous irradiation area $A_{intonif}$, have a close relationship with the actual propagation distance of the sound wave on the seabed, and the latter is also related to the actual propagation speed of sound on the seabed. So in order to calculate the effective irradiation area of the beam, first of all, we need to accurately calculate the acoustic propagation of the sound wave on the seabed. So in order to calculate the effective irradiation area of the beam, the first step is to accurately calculate the actual distance of the sound wave propagation on the seabed and the speed of sound, which requires to use the acoustic tracking technology.

2) Scattering intensity correction

From the multi-beam sonar bathymetry and imaging, it is known that the factors affecting the scattering intensity are propagation loss, sound bending phenomenon and effective acoustic irradiation area, which should be corrected to ensure the accuracy of the results in the calculation.

(1) Snell's Law

Due to the variation in sound velocity at different depths, an interface exists between water layers with different sound velocities. When sound waves propagate to this interface, refraction occurs. This refraction phenomenon can be described by Snell's law of sound wave propagation in different media. Figure 6.20 illustrates Snell's law in the process of sound wave propagation.

When sound waves incident upon an inhomogeneous interface of a medium, refraction occurs, causing a deviation in the propagation path of the sound waves. In the scenario illustrated in Fig. 6.20, the sound velocity C_1 in the medium above the interface is greater than the sound velocity C_2 in the medium below the interface. Additionally, the incident angle θ_1 in the medium above the interface is larger than the refraction angle θ_2 in the medium below the interface. When sound waves undergo refraction during propagation, the sound rays bend towards the direction of slower sound velocity in the medium. This is the implication of Snell's law, which can also be quantitatively expressed using Eq. (6.122).

$$\frac{\sin \theta_1}{C_1} = \frac{\sin \theta_2}{C_2} = P \qquad (6.122)$$

Fig. 6.20 Schematic of Snell's Law

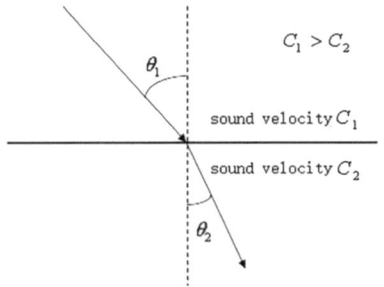

where P is Snell's constant.

From Snell's law, if the sound wave propagates in seawater without being incident at a vertical angle, then the propagation path of the sound wave in seawater is not a straight line, but propagates along a folded line at the interface of different sound velocities. Therefore, it is inaccurate to calculate the depth of the seabed and the horizontal displacement of the sound wave by using DOA and TOA according to the linear propagation model of the sound wave, so the calculation of the propagation distance of the sound wave in seawater in this way is obviously inaccurate as well, which leads to errors in the calculation of propagation loss, and thus the imaging results will be affected. Therefore, in the process of calculating the propagation loss, it is necessary to consider the acoustic bending of the sound wave in the process of seawater propagation. The acoustic line tracking method is used to correct the acoustic line bending during the propagation of sound waves in seawater.

The acoustic line tracking method is to divide a water layer between two neighboring acoustic velocity sampling points in the acoustic velocity profile, so that the seawater is divided into a number of thin water layers in the longitudinal direction. The acoustic velocity varies in a constant gradient when the sound wave propagates in such a model. In this way, in each water layer, the vertical displacement, horizontal displacement and propagation distance can be deduced by using the trigonometric method, and finally, by adding up the calculation results of each water layer, the more accurate water depth, horizontal displacement and actual propagation distance can be obtained.

Assuming that the sound wave propagates in seawater through the number of layers M with different sound velocity gradients, and the sound velocity gradient g of sound wave propagation in each layer is a constant, then

$$g_i = \frac{C_i - C_{i-1}}{h_i - h_{i-1}} \tag{6.123}$$

where the sound velocity at the upper interface of layer i is C_{i-1}, the sound velocity at the lower interface is C_i, the depth of the upper interface is h_{i-1}, and the depth of the lower interface is h_i. As shown in Fig. 6.21, according to Snell's law, it is obtained as follows.

$$P = \frac{\sin\theta_i}{C_i} \tag{6.124}$$

In summary, the depths of the upper and lower interfaces of layer are h_{i-1} and h_i, respectively, and the thickness of layer i can be calculated as Δh_i. When the sound wave propagates in this layer, the actual sound speed is also changing, so the actual propagation trajectory should be an arc with a certain curvature, and the radius of curvature R_i, can be expressed as:

$$R_i = \frac{1}{Pg_i} \tag{6.125}$$

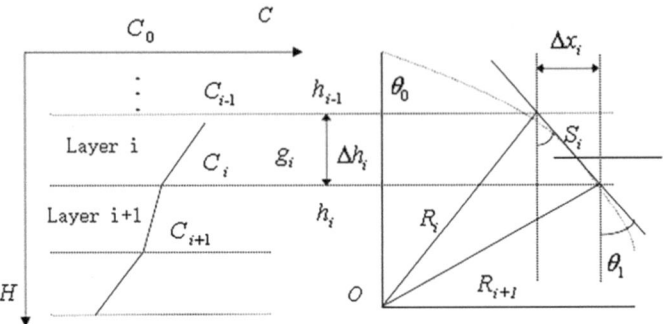

Fig. 6.21 Schematic of voice tracking

Then the horizontal displacement of the sound wave in layer i is:

$$x_i = R_i(\cos\theta_i - \cos\theta_{i-1}) = \frac{\cos\theta_{i-1} - \cos\theta_i}{Pg_i} \quad (6.126)$$

The angle of incidence of the acoustic wave on the layer is θ_{i-1}, which can be expressed as:

$$\cos\theta_{i-1} = \sqrt{1 - (pC_{i-1})^2} \quad (6.127)$$

The thickness of the layer $\Delta h_i = h_i - h_{i-1}$, then Eq. (6.124) can be expressed as:

$$x_i = \frac{\sqrt{1 - (pC_{i-1})^2} - \sqrt{1 - (pC_i)^2}}{pg_i} \quad (6.128)$$

In Fig. 6.21, S_i represents the actual distance travelled by sound waves within the layer, which can be expressed as:

$$S_i = R_i(\theta_{i-1} - \theta_i) \quad (6.129)$$

The time experienced by a sound wave travelling through the layer can be expressed by Eq. (6.130).

$$t_i = \frac{R_i(\theta_{i-1} - \theta_i)}{C_{Hi}} = \frac{\theta_{i-1} - \theta_i}{pg_i^2 \Delta h_i}\ln\left(\frac{C_i}{C_{i-1}}\right) = \frac{\arcsin(pC_i) - \arcsin(pC_{i-1})}{pg_i^2 \Delta h_i}\ln\left(\frac{C_i}{C_{i-1}}\right) \quad (6.130)$$

where C_{Hi} is the Harmonic mean speed of sound, which can be calculated by the following equation:

$$C_{Hi} = (h_i - h_0) \left(\sum_{n-1}^{i} \frac{1}{g_n} \ln\left(1 + \frac{g_n}{C_n} \Delta h_n \right) \right)^{-1} \quad (6.131)$$

In this way, after dividing the water layer, the vertical displacement, horizontal displacement, actual propagation distance and propagation time of the sound wave in each water layer can be obtained by calculating in each water layer according to the above method. By superimposing the results obtained in each water layer, the vertical displacement, horizontal displacement, actual propagation distance and propagation time of the sound wave incident on the seabed can be obtained, which is the basic principle of correcting the imaging by using the acoustic line tracking technique.

(2) Sound tracking method

Before using the multi-beam Snippet imaging algorithm for imaging processing, a series of seafloor echo DOA and TOA information can be obtained through multi-beam WMT bathymetry algorithms. And the data required for imaging can be calculated through the acoustic tracking processing method. In the actual measurement, using the sound velocity profiler, the sound velocity in the ocean and the depth of the corresponding function C (h) can be measured, the ocean is divided into n layers, then you can get n + 1 sound velocity values $C_{0,1,2,...,n}$ and n + 1 sound velocity interface.

Where C_0 represents the sound velocity at the surface of the transducer, and C_n represents the sound velocity at the seafloor.

The specific steps of the acoustic tracking process are described below.

In the first step, Taking θ_0 as the angle at the first layer's upper interface (the surface of the receiving transducer), calculate the horizontal displacement x_i, the actual propagation distance S_i, and the actual propagation time t_i of the sound wave in each layer, respectively. For each layer, after calculating its actual propagation time t_i, accumulate it with the actual propagation times calculated for all previous water layers to obtain a cumulative time value $\sum_{i=1}^{m} t_i$ (assuming accumulation up to the mth layer).

In the second step, the size of the time accumulation value $\sum_{i=1}^{m} t_i$ is compared with the value of TOA:

If the time-accumulated value of $\sum_{i=1}^{m} t_i$ is less than the value of TOA, then it is necessary to continue to accumulate the next layer. Then repeat the first step, and compare the time-accumulated value with the value of TOA.

If the cumulative value of time $\sum_{i=1}^{m} t_i$ is equal to the value of TOA, then do not need to continue to accumulate, at this time the cumulative value of time $\sum_{i=1}^{m} t_i$ is the occasion of the propagation time of the sound wave in the ocean. Therefore, by summing up the horizontal displacements x_i, actual propagation distances S_i, and layer thicknesses h_i calculated for each accumulated water layer, we can obtain the horizontal displacement, the actual propagation distance of the sound wave in seawater, and the depth of the seawater, respectively.

If the time accumulated value $\sum_{i=1}^{m} t_i$ is greater than the value of TOA, there is no need to continue to accumulate, but the following formulas should be used to calculate the excess part:

$$\Delta x' = \left(\sum_{i=1}^{m} t_i - t\right) C_m \sin\theta_m \tag{6.132}$$

$$\Delta S' = \left(\sum_{i=1}^{m} t_i - t\right) C_m \tag{6.133}$$

$$\Delta h' = \left(\sum_{i=1}^{m} t_i - t\right) C_m \cos\theta_m \tag{6.134}$$

Then the actual horizontal displacement, the actual propagation distance and the depth can be calculate, respectively:

$$x = \sum_{i=1}^{m} x_i - \Delta x \tag{6.135}$$

$$S = \sum_{i=1}^{m} S_i - \Delta S' \tag{6.136}$$

$$h = \sum_{i=1}^{m} h_i - \Delta h' \tag{6.137}$$

The above describes the case where the angle of incidence of the sound wave is not perpendicular to the surface of the water, i.e. $\theta_0 \neq 0$; if $\theta_0 = 0$, then the sound wave does not undergo the phenomenon of refraction as it propagates from the upper to the lower layer of water, and there is no bending of the sound line. The sound wave is incident vertically onto the seabed. In this case, the time taken for the sound wave to propagate through each water layer can be expressed as:

$$t_i = \frac{1}{g_i} \ln\left(\frac{C_i}{C_{i+1}}\right) \tag{6.138}$$

The actual distance traveled by the sound wave in each water layer is equal to the thickness of that layer, i.e., $S_i = h_i$, and the horizontal displacement remains zero throughout.

The specific calculation process is the same as the first and second steps above, and in the case that the time cumulative value $\sum_{i=1}^{m} t_i$ is greater than t of the value of TOA, $\Delta h' = \Delta S'$. Then the calculation can get the actual propagation distance, propagation time and horizontal displacement when the sound wave propagates in seawater.

6.2 Key Technologies of Sonar

The above description outlines the specific process of ray tracing. Whether it is for depth sounding using a multi-beam sonar system or for imaging processing, utilizing ray tracing can make the calculated results closer to actual values and improve the accuracy of measurements. Therefore, ray tracing technology has extensive applications in multi-beam sonar equipment. Figure 6.22 illustrates the entire process of line tracing processing.

It is important to note that the actual propagation distance calculated through ray tracing processing is the one-way propagation distance of the sound wave. When calculating the propagation loss, the two-way propagation loss, which is twice the one-way propagation loss, is used. By calculating, the propagation loss of the sound wave in seawater can be obtained. Combining the propagation distance of the sound wave in seawater and the sound speed when it reaches the seabed, the beam footprint area and the instantaneous acoustic illumination area can be derived. Further, the effective acoustic illumination area of each beam on the seabed can be determined, and subsequently, the backscatter intensity of the seabed can be calculated.

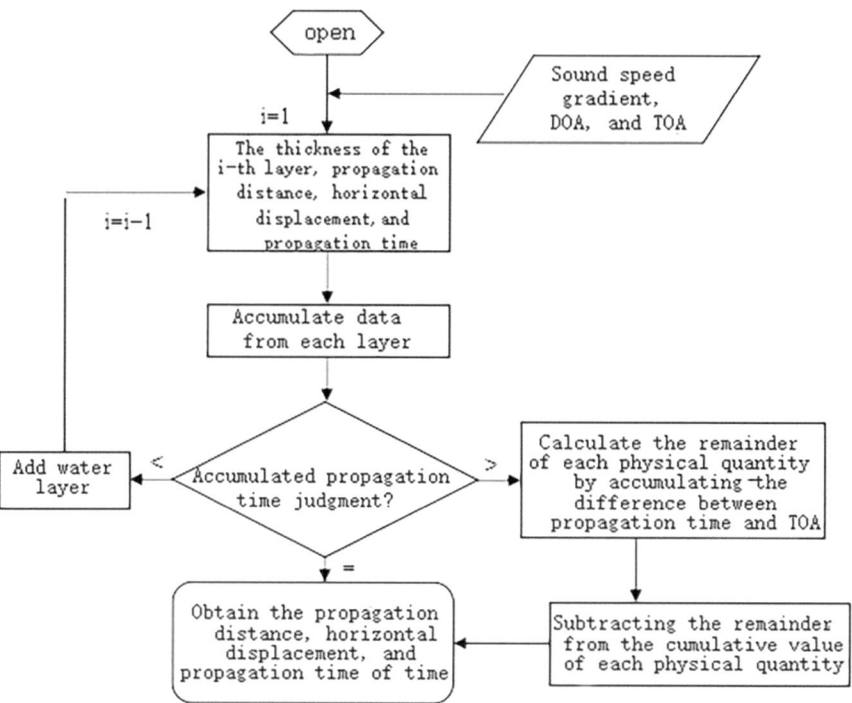

Fig. 6.22 Flowchart of sound line tracking calculation

6.2.4 Side-Scan Sonar Technology

Multi-beam sounder and side-scan sonar are both underwater acoustic devices capable of realizing full-coverage sweeping of the seabed, with a coverage several times that of the water depth. They have similar transmitting directional patterns, transmitting acoustic pulses to the seafloor at a certain angle of inclination, receiving seafloor backscattered echoes, and extracting the required seafloor geometric information from the seafloor backscattered echoes. Due to the different receiving beam patterns and different ways of processing the received echo signals, the multi-beam sounder is able to achieve spatial precision orientation through the receiving beam formation technology, and use certain characteristic parameters of the echo signal to detect the echo time delay to determine the echo round-trip time, so as to determine the slant distance in order to obtain the accurate bathymetry data, and to draw out the seafloor topographic map. The side-scan sonar only realizes the rough spatial beam orientation. It detects and records the amplitude energy of echo signals according to the natural sequence of the reverse scattering time of the echo signals on the seabed, displaying only the relative echo intensity information of seabed targets and obtaining seabed topographic sonograms. The differences in their working principles lead to different methods for detecting seabed reverse scattering echo signals.

In summary, multi-beam bathymetric sonar and side-scan sonar have similar testing principles, but side-scan sonar is designed to obtain the geomorphological acoustic map of the seafloor, and its transducer only completes the rough spatial orientation, and does not need to achieve precise orientation like multi-beam sonar. Therefore, the side-scan sonar measures the relative echo intensity information of the seafloor target, and does not need to detect the echo time delay, which is recorded on the acoustic map according to the natural order of time. The following focuses on the key technologies of side-scan sonar that different from that of multi-beam sonar, the principle of similar parts can refer to the key technology of multi-beam sonar.

1. Side-scan sonar echo detection

The key of signal detection in side-scan sonar is the amplitude processing of the echo signals. The core of amplitude processing for side-scan sonar echo signals is the normalization of echo signal amplitudes, which mainly includes three methods: time gain control (TDC), automatic gain control (AGC) and manual gain control (MGC).

(1) Time gain control TDC

In order to compensate for the loss of sound waves in the propagation process and avoid recording the uneven grey scale between the far and near ends of the sound map, the amplification level of the echo signal is made to change according to the exponential law of time. Consider the propagation loss and reverberation level related to distance time:

$$\text{EI}_r = 2\text{TL} - \text{RL} = 40\lg r_m + 2ar_m/10^3 - 10\lg \frac{CT}{2} r_m \emptyset + S_b \qquad (6.139)$$

6.2 Key Technologies of Sonar

Considering the relative amount of submarine echoes, the compensation is:

$$GL = 30\lg r_m + 2ar_m/10^3 \qquad (6.140)$$

According to Eq. (6.140), if the TGC starts to compensate at 10 m, and the maximum side-scan distance is 750 m, then for an echo signal with an operating frequency of 100 kHz, the echo signal needs to be compensated from 30 to 129 dB.

(2) Automatic Gain Control AGC

After the time gain control TGC, the signal amplitude still exhibits a large dynamic range, which can result in signal amplitude clipping or excessively small amplitudes. Automatic gain control is to compensate for the random fluctuations of the acoustic signal. Due to the side-scan sonar echo is a continuous seabed backscatter signal and is influenced by the transducer's directionality, the fluctuations in signal amplitude within a single line and between lines can be quite large. In order to obtain a uniform grey acoustic map, the echo amplitude signal needs to be controlled within a certain range. Automatic gain control AGC includes automatic gain control within a line and between lines.

Automatic gain control within the line: according to the change period of the envelope signal, a small time period is selected for integration to get the energy of the signal in this time period, and the amount of automatic gain control for the next moment is determined according to this value.

Automatic gain control between the lines: Since the seabed does not change significantly in a short period of time, the seabed echo signals have a certain degree of correlation. By smoothing and averaging the mean values of signals from adjacent lines, we can obtain an estimated mean value for the signal of the next line. Based on this estimated mean value, the automatic gain control amount for the next moment is determined.

(3) Manual gain control MGC

Manual intervention is performed to obtain the best display of the output acoustic map. Changes in the seabed substrate will cause a large change in the echo amplitude, which can reach 40 dB. In order to maintain a certain output amplitude, when the seabed substrate changes are large, you can use the manual gain control MGC.

Gain control in the circuit implementation primarily involves generating a voltage control curve that has an exponential relationship with time. This curve controls the gain of the amplifier. The control methods include: controlling the amplifier through a voltage charging and discharging curves; controlling the amplifier through a voltage charging and discharging curve generated by the microprocessor; and using a voltage-controlled gain amplifier with the dynamic range of up to 80 dB, broadband, linear continuity.

After gain control, you can get a record sonogram, which uniformly reflect the relative echo signal intensity of the seabed targets at both near and far distances in grayscale.

2. Ultra-short baseline positioning

When conducting seabed morphology surveys using side-scan sonar, to obtain high precision measurement data, it is necessary to accurately determine the underwater position of the side-scan sonar's tow fish. The positioning of the side-scan sonar's tow fish is typically achieved using an ultra-short baseline (USBL) positioning system. The ultrashort baseline positioning system is mainly composed of transmitting base array, transponder, receiving base array and data processing unit. In the side-scan sonar system, the transmitting base array and receiving base array are generally made into the same probe to form the transceiver base array, at least three transceiver base arrays are installed to different positions on the hull of the ship, and the transponder is installed to the tow fish itself. The system determines the relative azimuth angle from the receive arrays to the tow fish by measuring the phase differences of the signals received by each receive array. By measuring the time it takes for the sound waves to reach the receive arrays and then correcting the wave velocity using the sound velocity profile, the system ultimately determines the relative distance from the receive arrays to the tow fish, thereby determining the relative position of the tow fish.

The following is an example of the simplest ultrashort baseline system, which consists of three transceiver base arrays to introduce its working principle, the principle is shown in Fig. 6.23.

Using a right-angled triangular array with array element spacing of d on the legs of the right angle, and with the x-axis pointing in the direction of the bow, the coordinates of the three arrays are as shown in Fig. 6.23, namely (0, d, 0), (0, 0, 0), and (d, 0, 0). Taking the array located at the origin of the coordinates as the reference point, the formula for calculating the phase differences with the other two arrays can be obtained as follows, with the transponder on the tow fish having coordinates (x, y, z):

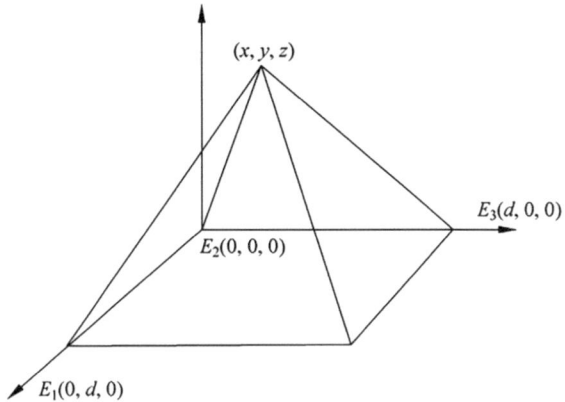

Fig. 6.23 The working principle of the ultrashort baseline system

6.2 Key Technologies of Sonar

$$\varphi_{12} = \frac{2\pi}{\lambda}\left[\sqrt{x^2+y^2+z^2} - \sqrt{x^2+(y-d)^2+z^2}\right] \quad (6.141)$$

$$\varphi_{23} = \frac{2\pi}{\lambda}\left[\sqrt{x^2+y^2+z^2} - \sqrt{(x-d)^2+y^2+z^2}\right] \quad (6.142)$$

where the diagonal distance r between the coordinate origin and the target on the tow fish is.

$$r = \sqrt{x^2+y^2+z^2} = v \times \Delta t \quad (6.143)$$

The slope distance can be obtained by taking the slope distance into the phase calculation formula:

$$\varphi_{12} = \frac{2\pi}{\lambda}\left[r - \sqrt{r^2+d^2-2dy}\right] \quad (6.144)$$

$$\varphi_{23} = \frac{2\pi}{\lambda}\left[r - \sqrt{r^2+d^2-2dx}\right] \quad (6.145)$$

Since the slant distance r of the side-scan sonar is much larger than the array element spacing d during detection, the following approximate formula can be derived:

$$\varphi_{12} = \frac{2\pi}{\lambda}\left[r - r\left(1 + \frac{d^2-2dy}{2r^2}\right)\right] \approx \frac{2\pi}{\lambda} \times \frac{y}{r} \quad (6.146)$$

$$\varphi_{23} = \frac{2\pi}{\lambda}\left[r - r\left(1 + \frac{d^2-2dx}{2r^2}\right)\right] \approx \frac{2\pi}{\lambda} \times \frac{x}{r} \quad (6.147)$$

With the above equations it is possible to derive an approximate formula for the tow fish coordinates x, y:

$$x \approx \frac{\lambda r \varphi_{23}}{2\pi} \quad (6.148)$$

$$y \approx \frac{\lambda r \varphi_{12}}{2\pi} \quad (6.149)$$

where λ is the wavelength of the received sound wave, v is the propagation speed of the received sound wave in the water, and Δt is the time interval between the transmission of the sound wave from the transponder on the tugboat and the reception of the sound wave by the receiving array on the ship.

Since the phase difference $\varphi 12$ and $\varphi 23$ can be measured by the instrument, the wave speed v and time interval Δt can also be measured, therefore the slant distance r can be calculated. Given that the sonar wavelength λ is a known quantity, and the coordinates (x, y, z) of the tow fish can be calculated using the formula of the ultrashort baseline positioning.

6.2.5 Synthetic Aperture Sonar (SAS) Key Technology

Synthetic aperture sonar (SAS) is a newly developed branch of underwater detection and imaging sonar, and it is one of the hot spots of international hydroacoustic high technology research. The basic principle of synthetic aperture sonar is to use the small size base array to virtually create a large aperture base array along the space of uniform linear motion, in the position of the motion track sequential emission and reception of echo signals, according to the spatial position and phase relationship between the echo signals of different positions for coherent superposition processing, so as to form the equivalent large aperture, to obtain the high resolution along the direction of motion (azimuthal direction). Synthetic aperture sonar is a high-resolution sonar, mainly demonstrated in two aspects: high range resolution and high azimuth resolution. Synthetic aperture sonar enhances the distance resolution through the pulse compression technology and improve the azimuthal resolution of the sonar through the synthetic aperture principle. The high range resolution of SAS is achieved by emitting large linear frequency-modulated signals with a large time-bandwidth product and applying pulse compression technology during reception. The high azimuth resolution is obtained through the synthetic aperture principle, where the azimuth echo signal of the sonar is approximated as a linearly frequency-modulated signal.

1. Interferometric synthetic aperture measurement (InSAS) of sonar

Sonar synthetic aperture interferometry (InSAS), based on synthetic aperture sonar, adds one (or multiple) receiving arrays in the cross-track direction. It obtains height information of the scene through phase comparison and height measurement, and processes this information to produce a three-dimensional image of the scene. Its principle is similar to Interferometric Synthetic Aperture Radar (InSAR). The advantage of InSAS is that its lateral resolution is independent of operating frequency and distance, and is one to two orders of magnitude higher than that of side-scan sonar.

The InSAS (Interferometric Synthetic Aperture Sonar) technology extracts three-dimensional spatial information of seabed targets based on the phase of sonar complex images. The basic idea is to mount a sound-emitting device and multiple receiving arrays in the cross-track direction on a platform to obtain pairs of radar complex images of the same area. Due to the unequal distances between the receiving arrays and the seabed targets, a phase difference arises between corresponding image points in the pairs of sonar complex images, forming an interferometric fringe pattern. The phase values in the interferometric fringe pattern represent the measured phase

6.2 Key Technologies of Sonar

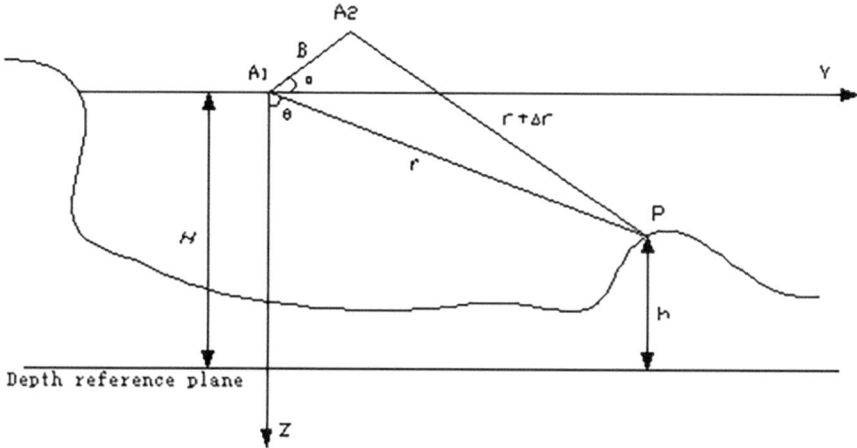

Fig. 6.24 The basic principle of InSAS technology

differences between the two imaging instances. Based on the geometric relationship between the imaging phase differences and the three-dimensional spatial positions of the seabed targets, as well as the positional parameters of the platform, the three-dimensional coordinates of the seabed targets can be determined.

As shown in Fig. 6.24, suppose the sonar receiving arrays image a specific seabed area P, with the spatial positions of the two arrays being A_1 and A_2. There exists a spatial interference baseline vector **B** with a length B, referred to as the baseline length. The angle between the baseline vector B and the horizontal direction is α, called the baseline inclination angle; θ is the incidence angle; the slant ranges from A_1 and A_2 to the ground point P are r and $r + \Delta r$, respectively; H is the height of A_1 above the reference plane; and B_x and B_y are the horizontal and vertical components of the baseline, respectively.

The acoustic signal with wavelength λ emitted from A_1 is reflected by the target point P and then received by A_1, resulting in the measured phase φ_1. The backward reflection from the target will cause a phase shift, which is denoted as φ_0, thus we have:

$$\varphi_1 = 4\pi \times r/\lambda + \varphi_0 \tag{6.150}$$

Similarly, for another array S_2, the measured phase φ_2 is obtained:

$$\varphi_2 = 4\pi \times (r + \Delta_r)/\lambda + \varphi_0 \tag{6.151}$$

Since r is much larger than Δr and B, Δr can be approximated using the cosine theorem:

$$\Delta_r = B_x \sin\theta - B_y \cos\theta \tag{6.152}$$

Thus the phase difference φ12 between A1 and A2 about the target P point is obtained.

$$\varphi_{12} = \varphi_2 - \varphi_1 = 4\pi \times \Delta_r/\lambda = 4\pi \times (B_x \sin\theta - B_y \cos\theta)/\lambda \quad (6.153)$$

φ_{12} is usually called the interferometric phase or absolute phase difference, which can be obtained by the interference of SAS images.

By performing SAS (Synthetic Aperture Sounding) image processing on the echo signals received by A_1 and A_2, two single-look complex image signals, C_1 and C_2, can be obtained. By registering these two signals and performing interference processing, an interference phase map can be formed. The interference processing involves multiplying C_1 and C_2 with their complex conjugates to create an interference image with alternating bright and dark stripes, from which φ_{12} can be derived, and subsequently, Δr can be calculated.

The following relationship can be obtained from the geometric relationship in Fig. 6.24:

$$\sin(a - \theta) = \left[(r + \Delta r)^2 - r^2 - B\right]/2rB \quad (6.154)$$

$$h = H - r \times \cos\theta \quad (6.155)$$

The unknown height difference h can be solved according to the above formula, which is the height of the seabed topographic point.

2. Imaging algorithm

(1) Range Doppler algorithm

Range Doppler (RD) algorithm is the most intuitive and basic classical method in SAS. Its concept originates from synthetic aperture radar, but the application in the field of sonar needs to consider the specific characteristics of underwater communication for parameter selection and implementation. The basic idea is to decompose the two-dimensional processing into two one-dimensional processing cascade form, which is characterized by the distance compressed data along the azimuthal direction for FFT, transformed to the distance Doppler domain. Then, range migration correction and azimuth compression are performed. The RD algorithm consists of three main steps: distance compression, distance migration correction, azimuth compression, to complete the focusing process.

The distance compression of RD algorithm is a matched filtering process, and its response function is:

$$f_r(t) = \exp\left\{j2\pi\left[f_c t - \frac{1}{2}\gamma t^2\right]\right\}\left(-\frac{T_p}{2} < t < \frac{T_p}{2}\right) \quad (6.156)$$

where, f_c is the carrier frequency, γ is the modulation frequency, T_p is the pulse width and distance compression results in sinc function.

6.2 Key Technologies of Sonar

The distance migration correction is divided into distance travelling correction and distance bending correction. The distance travelling correction is carried out in the time domain after the distance compression, and the expressions for the distance travelling quantity (set the phase center $t_0 = 0$) and the distance bending quantity are as follows.

$$R_{walk} = \frac{-\lambda f_{dc} t}{2} \quad (6.157)$$

$$R_{cur} = \frac{-\lambda f_{dr} t^2}{4} \quad (6.158)$$

where f_{dc} is defined as the Doppler center frequency and f_{dr} r is defined as the Doppler slope. Here, an interpolation method is used for the travelling correction and a straight line fitting method is used for the bending correction. On the one hand, the influence of range walk is significant, with a large number of range bins being shifted, necessitating precise correction results. In comparison, the influence of range curvature is relatively small. On the other hand, compared to time-domain shifting, spectral shifting is easier to implement.

After the distance migration correction, the trajectory of the signal along the azimuthal direction is changed from a curve to a straight line, and the azimuthal compression is changed to one-dimensional processing. The reference function is:

$$f_a(t) = \exp\left\{j2\pi\left[f_{dc} - \frac{1}{2}f_{dr}t^2\right]\right\}\left(-\frac{T_s}{2} < t < \frac{T_s}{2}\right) \quad (6.159)$$

where T_s is the synthetic aperture time. The amplitude of the output signal after azimuthal compression is a two-dimensional sinc function, the SAS echo signal is processed by the RD algorithm to obtain the complex image domain data, and the SAS image can be obtained according to the data amplitude.

(2) Chirp Scaling (CS) Algorithm

The RD algorithm uses interpolation to correct the distance migration, which not only reduces the computational efficiency of imaging, but also causes phase and amplitude errors in the SAS image. The CS algorithm uses a phase factor to change the spatial shift characteristics of the distance migration, so that the distance migration correction avoids interpolation. This not only eliminates the complex calculation, but also well maintains the phase accuracy of the image with good imaging effect.

Assuming a linearly frequency-modulated signal is denoted as f(t), with a frequency modulation slope of K and a phase center located at t_0, then it can be expressed as:

$$f(t) = \exp\{j\pi K(t - t_0)^2\} \quad (6.160)$$

Multiplying the signal by a linear frequency-modulated signal with a tuning frequency of KC_s (C_s is the bending factor) and a phase centre at t_0, changes the phase characteristics of the original signal, and the result of the multiplication is:

$$f_n(t) = \exp\{j\pi K_{new}(t - t_{new})^2\} \Delta \exp\{j\theta\} \quad (6.161)$$

where K_{new} is the new tuning frequency, t_{new} is the new phase center, and θ is the newly generated residual phase term. The above is the basic idea of the CS (Chirp Scaling) algorithm, which utilizes a phase factor to alter the phase characteristics of the LFM (Linear Frequency Modulation) signal, achieving a scale transformation of the range migration curve for point targets. This simplifies the range migration correction process.

3. Synthetic Aperture Sonar (SAS) phase correction

Usually synthetic aperture sonar (SAS) imaging algorithms are derived based on the Stop-and-Hop assumption, which assumes that the receiver is stationary between two adjacent pulses and that the transition from one pulse transmit/receive position to the next pulse transmit/receive position is instantaneous. This assumption is considered reasonable when the SAS system acts at a close distance. However, when the SAS imaging system acts at a longer distance, due to the slow propagation speed of sound in water, the aperture synthesis time becomes longer, allowing the receiver to move significantly within one pulse repetition interval. Therefore, the system phase error introduced by the Stop-and-Hop assumption cannot be ignored. Additionally, in the development of Vernier array technology, it is assumed that there exists an equivalent displaced phase center (DPC) right in the middle of each receiver/transmitter, where the sonar transmits and receives signals. This approximation also inevitably leads to system phase errors. Therefore, to improve the accuracy of SAS imaging algorithms, it is necessary to correct the system's phase errors.

1) Error caused by Stop-and-Hop assumption and correction

In the right-angle coordinate system (X, Y, Z), the point target P(x, y, 0) is fixed in the (X, Y, 0) plane, and the reflection coefficient is σ. The sonar is in a plane at $z = h$, moving uniformly and linearly along a direction parallel to the Y-axis with a velocity of v, and emits Linear Frequency Modulation (LFM) pulses p(t) with a pulse repetition period T towards an isotropic target area to be imaged. As shown in Fig. 6.25, the distance between the sonar transmitter and receiver is Δ_{ry} (for a colocated transmit/receive scenario, this distance is approximately zero, i.e., $\Delta_{ry} \approx 0$). Pulses are emitted at point A(0, u, h) and the corresponding target echoes are received starting at point C(0, u + Δry, h), where the variable u represents its azimuth (sonar movement direction) coordinate. Considering a Strip-map SAS imaging system, the path distance from the center of the mapping swath to the sonar platform is r_0. The distances from a point target to point A (the pulse emission location) and point C (the receiver's location when the sonar is in the state of emitting the signal) are R_1 and R_2, respectively.

6.2 Key Technologies of Sonar

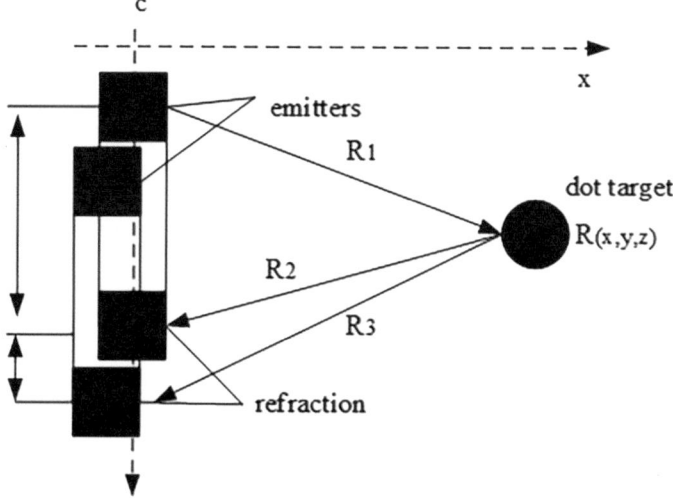

Fig. 6.25 Geometric Relationship in the Stop-and-Hop scenario

$$R_1 = \sqrt{x^2 + (y-u)^2 + h^2} \tag{6.162}$$

$$R_2 = \sqrt{x^2 + (y-u-\Delta_{ry})^2 + h^2} \tag{6.163}$$

In an aperture synthesis time, the target echo received by the sonar is the delay of the transmit pulse [18, 19].

$$s(t, u) = \sigma \cdot p[t - (R_1 + R_2)/c] \tag{6.164}$$

In the case of transmitter/receiver combining, the echo model can be expressed as:

$$s(t, u) = \sigma \cdot p\left[t - 2\sqrt{x^2 + (y-u)^2 + h^2}/c\right] \tag{6.165}$$

Equations (6.162), (6.163) are the mathematical model of the SAS system without phase correction [19] (without taking into account the transducer effect). In the derivation of this mathematical model, the so-called Stop-and-Hop approximation has been made. In fact, when the receiver receives the echo corresponding to the pulse at point A, it has already moved to point D, as shown in Fig. 6.25, and the distance between the receiver and the target at this time is R3:

$$R_3 = \sqrt{x^2 + (y-u-\Delta_{ry}-vt')^2 + h^2} \tag{6.166}$$

Its position at the time of receiving the echo has a displacement vt', relative to the position at the time of transmitting, and t' is the time of propagation of the sonar pulse:

$$t' = (R_1 + R_3)/c \tag{6.167}$$

The motion error due to this displacement is:

$$\varepsilon R = R_3 - R_2 \tag{6.168}$$

The resulting phase error is:

$$\Delta\varphi = 2\pi f_c \cdot \varepsilon R/c = \left[2\pi f_c/(c^2 - v^2)\right] \cdot \left[v^2 R_1 + R_2/c + 2v(\Delta_{ry} + u - y)\right] \tag{6.169}$$

where, f_c is the center frequency of the signal transmitted by the SAS system. Obviously, as the operating distance of the system increase (SAS systems generally operate in the near-field region of the target, and currently, the operating distance of most SAS imaging systems is around 100 m), the phase error caused by the Stop-and-Hop approximation will become increasingly severe, and must be corrected.

During sonar motion, there is generally more than one target in the area covered by a pulse beam, and the instantaneous distance between each target and the sonar will be different, so it is almost impossible to accurately estimate the motion error εR of all target echoes. However, when the system acts at a long distance, it can be assumed that the two-way delay induced by the pulse within the r_0 distance is:

$$t' = 2r_0/c \tag{6.170}$$

Since $\varepsilon R = R_3 - R_2$ is far less than the operating distance r_0, and considering that:

$$\sqrt{1+x} \approx 1 + x/2 - x^2/8 \tag{6.171}$$

Therefore, the kinematic error can be estimated by incorporating the two-way delay times into Equation (neglecting higher order quantities):

$$\varepsilon R = R_3 - R_2 \approx 2vr_0(y - u)/(cR_b) \tag{6.172}$$

where $R_b = \sqrt{x^2 + h^2}$. Thus, by multiplying a phase term on the echo signal Eq. (6.164), a modified echo model is obtained:

$$s(t, u) = \sigma \cdot p[t - (R_1 + R_2/c)] \exp\{-j2\pi f_c \cdot \varepsilon R/c\} \tag{6.173}$$

6.2 Key Technologies of Sonar

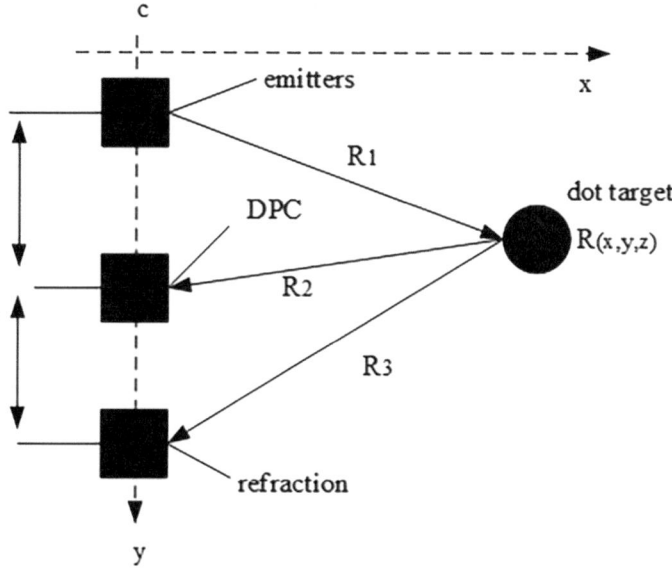

Fig. 6.26 Vernier array phase correction geometry relationship

(2) Errors caused by DPC approximation and corrections

In the development of the Vernier array technology, it had been assumed that there exists an equivalent phase center (DPC) between each transmitter/receiver, at which the sonar transmits pulses and receives echoes, as shown in Fig. 6.26. The distance from the target to the DPC is:

$$R' = \sqrt{x^2 + (y - u - \Delta_{ry}/2)^2 + h^2} \qquad (6.174)$$

In fact, the sonar transmits pulses and receives echoes at distances R_1 and R_2 from the target, respectively, and this approximation induces motion errors:

$$\varepsilon R = (R_1 + R_2) - 2R' \qquad (6.175)$$

The system phase error caused by this motion error must also be corrected.

For the DPC approximation, the kinematic error can be simplified by using the following equation:

$$\varepsilon R \approx \frac{\Delta_{ry}^2}{4R_b} + \frac{1}{R_b^3}\left\{ \frac{3y\Delta_{ry}^2}{8} + \frac{3yu\Delta_{ry}^2}{4} - \frac{7\Delta_{ry}^{24}}{64} - \frac{3y\Delta_{ry}^2}{8} - \frac{3u^2\Delta_{ry}^2}{8} - \frac{3u\Delta_{ry}^2}{8} \right\} \qquad (6.176)$$

Normally only the first term of the above equation can be used to satisfy the phase correction requirement, so corresponding to each point target, a fixed phase term is multiplied on the echo signal of each pulse:

$$\exp\{-j2\pi f_c \cdot \varepsilon R/c\} \tag{6.177}$$

The phase error caused by the DPC approximation can be corrected. Theoretically, in the Vernier array scheme, the phase correction consists of two parts: the phase error caused by the Stop-and-Hop approximation and the DPC approximation. These two parts of the correction can be carried out simultaneously or separately.

4. Motion compensation of synthetic aperture sonar (SAS)

Synthetic aperture sonar (SAS) imaging assumes that the sonar moves along an ideal straight trajectory during the synthetic aperture process. However, in practice, the trajectory of the towed body always deviates from the ideal straight-line trajectory due to factors such as sea waves, sea winds, and the towing system, all of which can cause the towed body to deviate from the intended motion trajectory. This resulting deviation in the sonar's position leads to changes in the delay of the echo signals. If these errors are not compensated for, the imaging quality will be significantly degraded.

The motion compensation of synthetic aperture sonar SAS is generally divided into two methods: motion compensation based on inertial measurement system and motion compensation based on raw echo data. Since the synthetic aperture sonar SAS generally installs the measurement system on the towed body platform to record the position and attitude of the platform in real time, this chapter mainly introduces the motion compensation method based on the inertial measurement system.

1) Motion compensation system based on inertial measurement

(1) Influence factors of motion error

The motion errors of the synthetic aperture sonar mainly include: ① non-zero acceleration along the track direction; ② non-zero velocity in the horizontal and lead direction perpendicular to the track direction; ③ and the existence of three kinds of yaw, pitch, and roll motions around three coordinate axes. All these deviations introduce phase errors causing echo Doppler signal distortion and affecting the imaging quality. Among them, motion error item ① causes spatial non-equal spacing distribution of the emitted pulses along the trajectory direction, and items ② and ③ cause position errors in the line-of-sight direction.

(2) Tow body coordinate system and inertial coordinate system

Inertial measurement system involves two coordinate systems, namely, inertial coordinate system and drag body coordinate system. The inertial coordinate system, also known as the geographic coordinate system, is a right-angle coordinate system defined by X (north), Y (east) and Z (down) three orthogonal components. The origin of the drag body coordinate system is taken in the center of mass of the inertial guide of the measurement system, x'axis is taken as the forward direction of the longitudinal axis of the drag body, y' axis is perpendicular to the x' axis and points to the right of the drag body (when the drag body is placed normally on the ground), and z' axis is determined by the rule of right-handed helix. In general, the drag body coordinate system does not coincide with the inertial coordinate system. The inertial coordinate system is fixed, while the drag body coordinate system will change with the change of drag body motion attitude. In motion compensation, it is necessary to consider the position conversion from the drag body coordinate system to the inertial coordinate system to obtain the actual position of the phase center of each array element in the inertial coordinate system.

(3) Elimination of the "lever arm" effect

The so-called "lever arm" effect refers to the phenomenon where, due to a certain distance between the inertial navigation centroid and the equivalent phase center of the array elements, during the motion of the towed body, the velocity and displacement of each array element do not equate to those of the inertial navigation centroid. There exists a "lever arm" relationship between the two in a certain sense. Therefore, to obtain the position of the phase center of each array element, it is necessary to eliminate the influence of the "lever arm" effect. This effect can be eliminated by pre-measuring the position of the phase center of each array element in the towed body coordinate system.

(4) Determination of reference track and correction of installation error angle

Theoretically, any straight line can be selected as the ideal trajectory for reference, and here a straight line closest to the actual trajectory is selected as the ideal trajectory for reference. The mean value of the sailing direction (i.e. yaw angle heading) of the towing body is selected as the ideal reference track direction, and all the motion errors are calculated with this track direction as the reference.

After the reference ideal trajectory is determined, the difference between the heading and the mean value of the heading of the voyage is used as the yaw angle for solving the position of the phase center of the array element, so as to convert the x'-direction of the inertial coordinate system from the due north direction to the direction of the reference ideal trajectory, the y'-direction is perpendicular to the x'-direction horizontally to the right, and the z'-direction is perpendicular to the x'-direction vertically to the downward direction. In addition, due to the angular error of the inertial measurement system at the time of installation, the velocity data measured in the three directions are not completely orthogonal to the x, y and z directions of the trajectory coordinate system, resulting in the possibility that there may be a constant linear DC component of the velocity in the y and z directions, and since the displacement is obtained by integrating the velocity during positional solving. Therefore, even for a small velocity component, the displacement obtained after integration may be very different. This linear component cannot be removed by the coordinate transformation from the drag body coordinate system to the inertial coordinate system. If the displacement obtained by the velocity integral is ignored, the final phase center position may deviate from the actual position, thus affecting the motion compensation accuracy, so it is necessary to estimate and correct the mounting error angle.

After the reference trajectory is determined, the x-direction is the forward direction of the towed body, while the mean value of the velocity in the y-direction and the z-direction should be zero, and due to the angular error during installation, the obtained velocity in the y- and z-directions is coupled with the x-direction, which results in the mean value of the velocity in the y- and z-directions not being zero. Therefore, the installation error angle can be estimated based on the size of the mean value of the velocity data measured by the measurement system in the x, y and z directions, and the velocity can be corrected, and the corrected velocity can be used for the solving of the position to reduce the error caused by the installation angle.

2) Motion compensation algorithm based on inertial measurement system

(1) Point-by-point delayed summation imaging algorithm

The basic idea of the point-by-point delay summing imaging algorithm is to align the signals according to the delay from the target to each azimuthal sampling point, and then coherently superimpose them to obtain the imaging value of each imaging point. Traditional point-by-point time-delay summation algorithm in the two-dimensional imaging, for convenience, often ignoring the depth direction, but the final image through the distance projection mapped to a two-dimensional oblique plane coordinate system (Fig. 6.27). In the oblique plane coordinate system, the X-axis is the ideal trajectory direction, and the Y-axis is perpendicular to the direction of the trajectory of the oblique distance to the direction.

Assuming that the array element moves along the ideal horizontal straight-line trajectory direction, only the position in the azimuthal and diagonal coordinate directions are considered in imaging. According to the position of the phase center of the array element, the distance from each pixel to the center of the array element in the imaging area can be calculated, and then the sampling delay can be obtained, and according to the delay, the image luminance value of each point in the imaging area can be obtained. For the next sampling position, the same method is used to calculate the brightness value of each point in the imaging area and superimposed on the image, and finally the images of all sampling positions are superimposed together to form the final synthetic aperture sonar imaging map.

(2) Point-by-point compensation imaging algorithm based on the actual trajectory

In the point-by-point delay summation imaging algorithm, the delay calculation uses the delay between the ideal position of the phase center and the target. After the actual position of the phase center is obtained through the inertial measurement system, the delay between the actual position of the phase center and the target can be calculated directly, so as to achieve the compensation of the motion error. This is the basic idea of the point-by-point compensation imaging algorithm based on the actual trajectory.

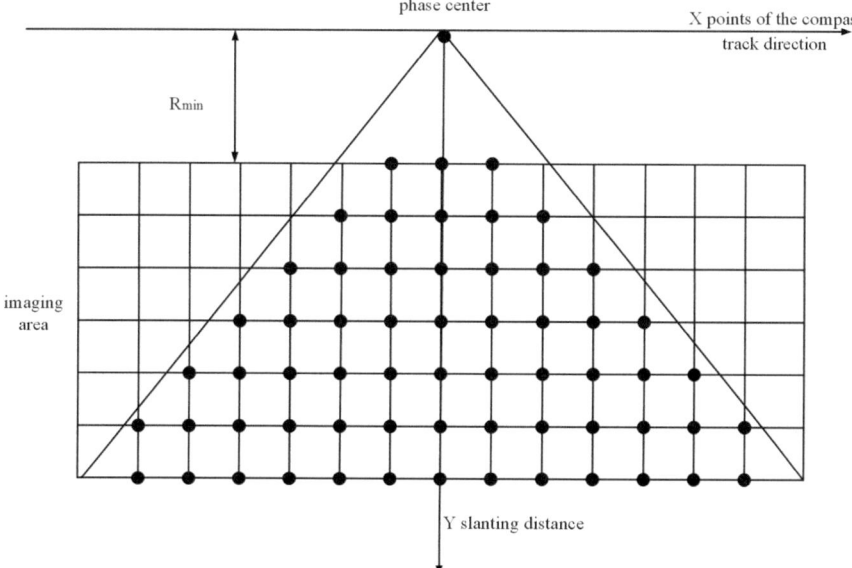

Fig. 6.27 Schematic of 2D oblique plane imaging

The point-by-point compensation imaging model based on the actual trajectory can achieve point-by-point imaging by replacing the ideal trajectory position with the actual position of the phase center of the array element obtained by the inertial measurement system. From the actual position of the phase center of the array element, the distance from each pixel to the center of the array element in the imaging area is calculated, and then the sampling delay can be obtained, according to which the image brightness value of each point in the imaging area can be obtained. For the next sampling position, the same method is used to calculate the brightness value of each point in the imaging area and superimposed on the image, and finally the images of all sampling positions are superimposed together to form the final synthetic aperture sonar imaging map. Therefore, as long as the attitude measurement data is accurate enough, such imaging processing will be able to compensate the effect of motion error very well.

References

1. Satour, A., S. Montresor, M. Bentahar, et al. 2013. Acoustic Emission Signal Denoising to Improve Damage Analysis in Glass Fibre-Reinforced Composites. *Nondestructive Testing and Evaluation* 29(1): 65–79.
2. Yang, Huigu, Ping Ju, Liang Hua, et al. 2017. Study on Denoising of Acoustic Emission Signals Based on Wavelets. *Modern Electronics Technique* (13): 78–80,84.

3. Huang, N.E., Z. Shen, S.R. Long, et al. 1998. The Empirical mode Decomposition and the Hilbert Spectrum for Nonlinear and Non-stationary Time Series Analyse. *Proceedings of the Royal Society of London. Series A: Mathematical, Physical and Engineering Sciences* 454(1971): 903–995.
4. Rui, Yu., Zhang Shouming, Bi. Guihong, et al. 2015. Recognition Method of Leakage Acoustic Emission Signal Based on EMD and SVM. *Computers and Applied Chemistry* 10: 1259–1264.
5. Feng, Xu., Liu Yunfei, and Song Jun. 2011. Feature Extraction of Acoustic Emission Signal Based on Median Filtering, SVD and EMD. *Journal of Instrumentation* 12: 279–2719.
6. Jianchi, Cui. 2016. *Research on Feature Extraction of Acoustic Emission Signal with Noise Based on Independent Component Analysis*. Shenyang: Shenyang University of Aeronautics and Astronautics.
7. Ju, L., C. Zhao, Z. Xin, et al. 2005. An Approach of Speech Enhancement by Sparse Code Shrinkage. In *International Conference on Neural Networks & Brain*.
8. Akaike, H. 1974. Markovian Representation of Stochastic Processes and Its Application to the Analysis of Autoregressive Moving Average Processes. *Annals of the Institute of Statistical Mathematics* 26 (1): 363–387.
9. Kitagawa, G., and H. Akaike. 1978. Aprocedure for the Modeling of Non-stationary Time Series. *Annals of the Institute of Statistical Mathematics* 30 (1): 351–363.
10. Maeda, N. 1985. A Method for Reading and Checking Phase Times in Autoprocessing System of Seismicdata. 38: 365–380.
11. Ohtsu, M. 2015. *Acoustic Emission (AE) and Related Non-destructive Evaluation (NDE) Techniques in the Fracture Mechanics of Concrete: Fundamentals and Applications*. Woodhead Publishing.
12. Nengjun, Yang, Yao Chunjiang, Yuan Xiaojing, et al. 2016. *Material Damage Detection Technology Based on Acoustic Emission*. Beijing: Beijing University of Aeronautics and Astronautics Press.
13. Lin, Huafang, and Mingyuan Xu. 2003. Beam Formation of Uniform Linear Array. *Information Technology* 27(5): 22–24.
14. Tian, Tan, Guozhi Liu, and Dajun Sun. 2000. *Sonar Technology*. Harbin: Harbin Engineering University Press.
15. Grant, H., and P. Andrew. 1995. *Simulation of Beamforming Techniques for the Linear Array of Transducers*. IEEE.
16. Zhu, Ye. 1990. *Principles of Active Sonar Detection Information*. Beijing: Ocean Press.
17. Murino, V., and A. Trucco. 2000. Three-Dimensional Image Generation and Processing in Underwater Acoustic Vision. *Proceedings of the IEEE* 88 (12): 1903–1948.
18. Carraraw, G., R.S. Goodman, and R. M. Majewski. 1995. *Spotlight Synthetic Aperture Radar Signal Processing Algorithms*. Boston, MA: Artech House.
19. Gough, P. T., and D. W. Hawkins. 1997. Unified Framework for Modern Synthetic Aperture Imaging Algorithms. *International Journal of Imaging Systems and Technology* 8: 343–358.

Chapter 7
Application of Acoustic Technology in Marine Engineering Inspection

7.1 Application of Acoustic Emission Technology

Offshore platforms are widely used in the exploration and development of marine oil and gas resources. Due to the combined effects of various random loads such as sea wind, waves, sea ice and tides over long periods, coupled with the marine environment corrosion, material aging and material defects and a variety of cumulative damage, reducing the overall resistance of the platform components, seriously affecting the safety and durability of the structure. Due to the harsh working environment of the marine platform and the variety of background noise, the traditional non-destructive testing methods are not ideal for monitoring the structure of the marine platform [1]. Acoustic emission (AE) technology has a broad application prospect in offshore platforms due to its high sensitivity, all-weather 24 h non-stop inspection and other characteristics.

During the service period of the offshore platform, it will be affected by various factors such as seawater and marine organisms, platform operation and life, and weather. In the AE monitoring of the platform, all kinds of damages, such as cracks and corrosion, are the focus of monitoring; while the impact and friction produced by personnel operation and marine environment on the platform will interfere with the normal fault signals, which should be weakened and eliminated in the monitoring process.

The key to the application of acoustic emission technology is the processing and analysis of AE signals, while the weakness of the AE acoustic emission signals of marine platforms and the diversity of interference noise increase the difficulty of acoustic emission signal processing and analysis. In recent years, advancements in the signal processing and analysis methods for acoustic emission technology, especially in the research on feature extraction and pattern recognition of acoustic emission signal, will further promote the application of acoustic emission technology in the ocean platform, detection and monitoring.

7.1.1 Fatigue Crack Signal Identification of Tube Nodes of Offshore Platforms

Tube nodes, as an important connection of conduit rack offshore platforms, are prone to damage phenomena such as cracks and extensions at the toe of the intersecting weld under complex loads, which seriously affects the stability of the structure. Therefore, exploring the acoustic emission signal characteristics of dynamic cracks is of great significance for the study of online health monitoring and damage assessment of pipe nodes of offshore structural components.

Acoustic emission technology has the ability to monitor the structural cracks of the conduit rack offshore platform, which can be monitored and localized at the early stage of crack development with high sensitivity, and is particularly suitable for online, continuous and remote monitoring. Acoustic emission monitoring adopts the traditional parametric analysis method for feature extraction, which cannot comprehensively, clearly and quickly express the characteristics of the acoustic emission signals, and it is difficult to realize the online monitoring of fatigue damage of the pipe nodes of the offshore platform and the assessment of the damage degree. Based on the crack extension theory, the wavelet distribution scale spectrum and wavelet energy coefficient method are used to extract the features of fatigue cracks in tube nodes of offshore platform structures, and the online monitoring of tube node cracks can be realized by establishing the relationship between wavelet energy coefficient and crack extension.

1. Wavelet energy coefficient method

Wavelet analysis is a commonly used analytical method in acoustic emission signal processing and analysis, which can simultaneously characterize the local characteristics of the signal from both time and frequency domains. The acoustic emission signal is decomposed at multiple scales using wavelet variations, if k layers of wavelet decomposition results in the decomposition of the original signal into k + 1 components in the frequency range, so the total energy of the signal can be expressed by the energy of the wavelet coefficients of each layer:

$$E_f = E_{ak} + \sum_{j=1}^{k} E_{dj} (j = 1, 2, \ldots, k, k \epsilon Z) \tag{7.1}$$

In the equation, E_f is the total energy of the signal; E_{ak} is the energy of the approximate wavelet coefficients when the scale decomposition; E_{dj} is the energy of the wavelet coefficients of the jth layer.

In order to better reflect the pattern of change of the signal, the coefficients after dimensionality normalization of the scale is called wavelet energy coefficients, i.e.

$$\gamma_a = \frac{E_{ak}}{E_f}, \quad \gamma_d = \frac{E_{dj}}{E_f} \quad (j = 1, 2, \ldots, k, k \epsilon Z) \tag{7.2}$$

7.1 Application of Acoustic Emission Technology

The wavelet energy coefficients represent the energy distribution of the signal within various frequency intervals, and different distributions result in different characteristics of the acoustic emission sources and different damage conditions. Therefore, by comparing the changes of wavelet coefficients of each layer, we can obtain the time–frequency characteristics of crack expansion at different stages, and establish the correspondence between wavelet energy coefficients and the crack expansion process, and the wavelet energy coefficients can effectively characterize the acoustic emission characteristics of the pipe node during the crack expansion process.

2. Fatigue crack test

In order to establish the correspondence between the wavelet energy coefficient and the fatigue crack extension of the tube node, Liu et al. [2] carried out a model test for the construction of the tube node of the offshore platform. The platform pipe node adopts the form of typical offshore platform pipe node, the material of pipe node selects high strength structural steel D36 with good comprehensive mechanical properties, and the weld dimensions of branch pipe and main pipe follow the geometric characteristic parameters of American Welding Society (AWS):

$$\alpha = \frac{2L}{D}, \beta = \frac{d}{D}, \gamma = \frac{D}{2T}, \tau = \frac{t}{T} \tag{7.3}$$

in the equation, α is the length coefficient of the chord tube; β is the diameter ratio of the branch chord tube; γ is the wall thickness coefficient of the chord tube; τ is the wall thickness ratio of the branch chord tube; d is the diameter of the branch chord tube; D is the diameter of the chord tube; L is the length of the chord tube; t is the thickness of the branch chord tube; T is the thickness of the chord tube.

A typical T-tube node geometry model and its geometric parameters are shown in Fig. 7.1.

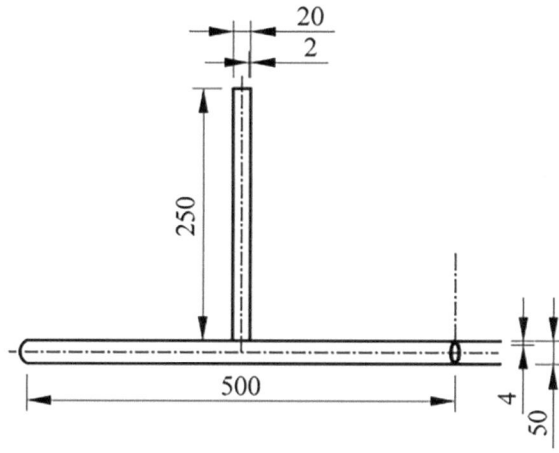

Fig. 7.1 T-pipe node geometry model (unit: mm)

The locations prone to fatigue cracks in tubular joints can be analyzed and determined using structural finite element analysis software (such as ANSYS, Midas, etc.). Pre-cracks are made at the locations where stress concentrations and cracks are likely to occur in T-type tubular joints. During the finite element analysis, a load of 10 kN is applied along the chord direction to determine that the most vulnerable location is 30.5 mm away from the chord axis and 2.5 mm away from the weld fusion line. A 0.18 mm molybdenum wire is used to pre-cut a 1 mm deep crack at this location along the radial direction of the branch pipe. A fatigue testing machine is used to conduct fatigue tests on the component, and acoustic emission signals are collected during the test.

3. Test acquisition parameter setting

The PRX15 acoustic emission (AE) sensors, with a resonant frequency of 150 kHz and sensitivity greater than 65 dB, are attached to both sides of the weld seam of the T-type tubular joint using a coupling agent to pick up AE signals from crack propagation in the joint. The parameters of the AE acquisition system are as follows: the frequency response of the acquisition card is 0.01–4 MHz, with a gain of $\pm$ 0.2 dB; the frequency bandwidth of the preamplifier is 15 kHz to 2 MHz, with a gain of 40 dB; the acquisition software can display AE signals and characteristic parameters in real-time full-waveform mode.

The AE detection instrument is set to a sampling frequency of 1024 kHz, with a single waveform sampling time of 16 ms and a software-triggered rising edge. Full-waveform acquisition is employed for AE signals from crack propagation, and the acquired signals are filtered to remove background noise such as experimental environmental noise, in order to accurately identify crack propagation signals. The specimen is loaded using a press, and when the load reaches 2 kN, the press indicates a load drop, indicating that the ultimate load of the specimen has been exceeded and crack propagation begins. When the load drops to 0.75 kN, the specimen fractures. The total sampling time for AE signals throughout the entire process is 160 s.

4. Wavelet analysis of acoustic emission signal

Since the collected acoustic emission signals contain very serious background noise such as environmental noise, resulting in the inability to accurately identify the acoustic emission signals excited by crack extension, it is necessary to use the method of wavelet analysis to decompose, reconstruct and filter out the low-frequency noise of the acoustic emission signals.

The acoustic emission signal data generated during crack expansion is relatively large, and it belongs to non-stationary signals. When selecting a wavelet basis, it is necessary to minimize or avoid data distortion and accurately achieve wavelet decomposition. Discrete wavelet transform (DWT) should be performed to minimize the computational complexity during wavelet transform. The wavelet transform at various scales should be able to well contain and reflect defect information. In order to highlight the characteristics of AE signals, a certain order of vanishing moments is required. The wavelet basis should have certain local analysis capabilities in both the time domain and the frequency domain, which means it should have compact support in the time domain and rapid decay characteristics in the frequency band of the frequency domain.

Considering the above requirements comprehensively and combining the characteristics of various wavelet functions, the Daubechies wavelet is selected as the wavelet basis for wavelet analysis, with the order of vanishing moments generally chosen between 4 and 7. In the experiment, the AE activity frequency distribution range is above 10 kHz, and the sampling frequency is set to 1024 kHz. The determination of the wavelet decomposition scale can be based on the sampling frequency and the frequency range of the AE signals. When determining the decomposition scale, it should be ensured that the identified frequency is not greater than the lowest AE signal frequency.

According to the basic theory of wavelet decomposition, the decomposition scale k should satisfy the following condition:

$$\frac{f}{2^{k+1}} \geq 10 \qquad (7.4)$$

The decomposition scale k for wavelet analysis of acoustic emission signals can be obtained by taking the common logarithm on both sides of Eq. (7.4) simultaneously:

$$k \leq \log_2 \frac{f}{20} = \log_2 \frac{1024}{20} = 5.678 \qquad (7.5)$$

According to the range of decomposition scale of wavelet analysis determined by the above formula, dB5 wavelet is selected to decompose the collected acoustic emission signals in 5-layer and 6-layer decomposition respectively, and the specific decomposition scale is determined according to the actual decomposition effect. When the collected acoustic emission signal is decomposed by 6-scale, the six-layer low-frequency coefficients mainly fluctuate slightly around 0, which is of little significance for signal analysis and energy extraction; while when the 5-scale decomposition is performed, the high-frequency wavelet coefficients and low-frequency approximation coefficients of each layer can effectively reflect the characteristics of the signal in this frequency band. Therefore, 5-scale wavelet decomposition is chosen to improve the aggregation of wavelet scale diagram, reduce noise interference, and establish the relationship between crack extension signal and wavelet energy coefficients clearly and quickly.

5. Relationship between crack extension and wavelet energy coefficient

Through the 5-scale decomposition and reconstruction of the acoustic emission signal wavelet in the tube node test, the fatigue crack extension is divided into three stages: plastic deformation stage, crack extension stage and instability fracture stage.

(1) Plastic deformation stage

Before loading, due to the influence of stress concentration caused by prefabricated cracks, the force required for the yielding of the specimen is small, and the plastic

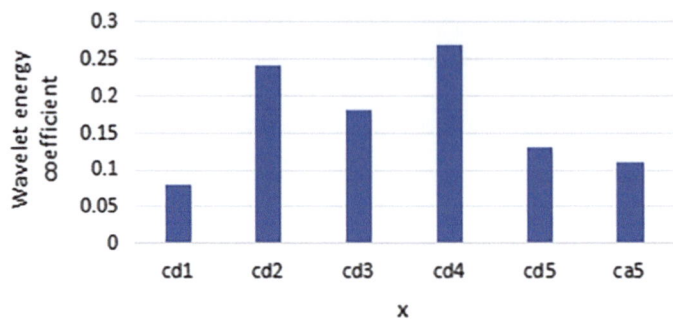

Fig. 7.2 Distribution of wavelet energy coefficients of acoustic emission signals in plastic deformation stage

deformation stage of crack initiation is relatively short, and the acoustic emission activity is relatively low at this time. With the continuous action of the external force, the shape of the specimen changes, dislocation and slip movements occur within the material, the stress concentration at the tip of the prefabricated crack is unloaded, and the acoustic emission signal is gradually strengthened, and after reaching yield, the acoustic emission signal will weaken.

Figure 7.2 shows the results of wavelet decomposition of the acoustic emission signal generated in the plastic deformation stage and the distribution of wavelet energy coefficients. In the plastic deformation zone, the acoustic emission signals are mainly concentrated in the second, third and fourth layers of the high-frequency part, accounting for 24.07, 18.07 and 27.77% of the total energy, respectively, while the acoustic emission signals distributed in the frequency range of 0–32, 32–64 and 256–512 kHz are relatively weak, with the proportions of 10.57, 11.50 and 8.02%.

(2) Acoustic emission signals in crack expansion stage

After passing through the plastic deformation, the specimen enters the crack extension stage, where the microscopic cracks have developed into a macroscopic crack under the action of the external force, the crack further extend, continuously generating new dislocation slips, the material continuously releases elastic waves. The acoustic emission becomes more active, with a significant increase in the intensity of the signal. Moreover, the signals evolve from continuous signals to burst signals, exhibiting a certain periodicity.

During continuous uniaxial loading, the acoustic emission signals generated by crack propagation undergo three stages of "weak - strong - weak". This is due to the process of crack tip blunting, propagation, and re-blunting that occurs under the action of loading after crack propagation begins. As the load further increases, these three processes repeat. In order to conduct a more specific and comprehensive analysis of the crack development process, the second stage is further subdivided into three processes: the initial stage of crack propagation, the stable stage of crack propagation, and the late stage of crack propagation. The wavelet energy coefficient distributions for these three stages are shown from Figs. 7.3, 7.4 and 7.5, and the

7.1 Application of Acoustic Emission Technology

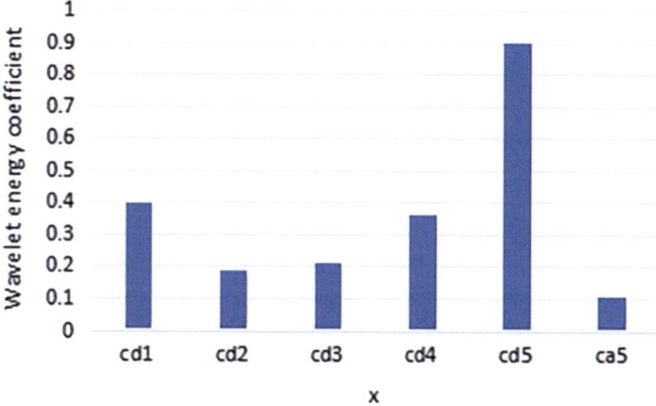

Fig. 7.3 Distribution of wavelet energy coefficients of acoustic emission signals during the initial stage of crack propagation

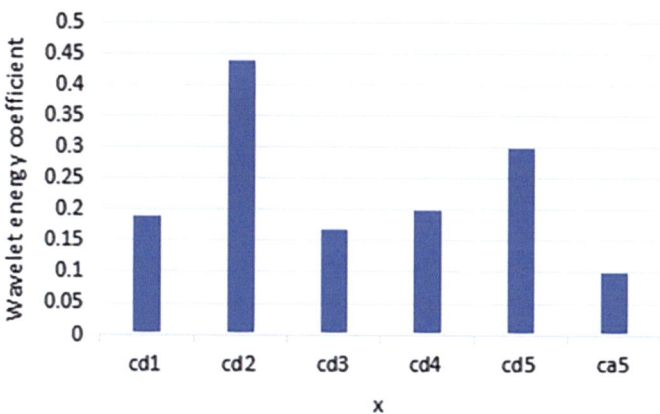

Fig. 7.4 Distribution of wavelet energy coefficients of acoustic emission signals during the stable stage of crack propagation

frequency ranges and energy coefficients of wavelet decomposition are listed in Table 7.1.

Combined with the results of wavelet decomposition of crack propagation and the distribution of wavelet energy coefficients, it can be seen that the high-frequency detail coefficients of the three stages change significantly. Compared with the early stage of crack extension, the energy proportion of cd1 increases by nearly 14% and the proportion of cd2 increases by 28%; however, after entering the later stage, the energy proportions of these two detail coefficients decrease by nearly 11% and 10%, respectively. Therefore, the acoustic emission signal characteristics during crack propagation can be effectively identified by cd1 and cd2.

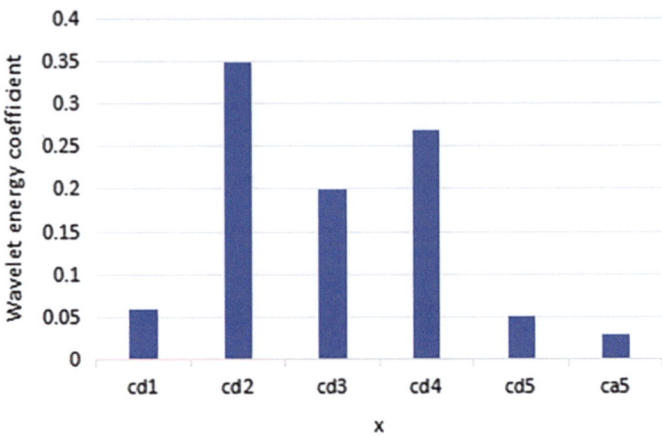

Fig. 7.5 Distribution of wavelet energy coefficients of acoustic emission signals at the late stage of crack expansion

Table 7.1 Distribution of wavelet energy coefficients during crack expansion

Wavelet coefficient	Frequency range (kHz)	Wavelet energy coefficient		
		Initial stage of crack propagation	Stable stage of crack propagation	Late stage of crack propagation
cd1	256512	0.0424	0.1801	0.0703
cd2	128–256	0.1791	0.4418	0.3473
cd3	64–128	0.2054	0.1492	0.2122
cd4	32–64	0.3688	0.1718	0.2791
cd5	13–32	0.0952	0.0401	0.0576
ca5	0–16	0.1091	0.0171	0.0335

(3) Unstable fracture stage

When the crack expands to the critical crack length, unstable fracture occurs, which produces a strong burst of acoustic emission signals at the moment of fracture. At the later stage of crack propagation, the acoustic emission phenomenon generated inside the material is significantly weakened and enters a short intermittent period, where the acoustic emission signal is basically undetectable, which is determined by the intermittent nature of crack propagation. Additionally, due to the application of external loads, the crack size gradually increases. After entering the unstable stage, the crack propagation rate rapidly increases, and fracture occurs instantaneously with the instantaneous release of energy. During this process, intense burst-type acoustic emission signals are generated and rapidly decay. Figure 7.6 shows the wavelet decomposition and energy coefficient distribution of the acoustic emission signals during unstable fracture of the specimen.

7.1 Application of Acoustic Emission Technology

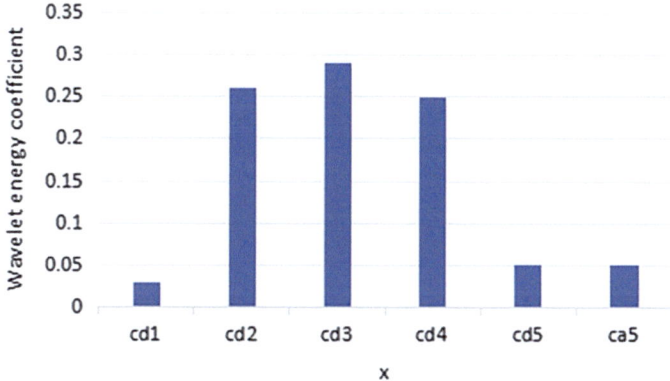

Fig. 7.6 Distribution of wavelet energy coefficients of crack unstable fracture acoustic emission signal

The acoustic emission signals at the instant of unstable fracture are mainly composed of signals from three frequency intervals, with the third-level wavelet coefficient cd3 occupying a dominant position. When the energy coefficient of the cd3 level is dominant, it can be determined as the state of unstable fracture.

Comprehensively, the acoustic emission signals excited during the crack development of pipe nodes are superimposed by a variety of high-frequency signals, and their signal characteristics will change significantly with the different crack states, while the wavelet energy coefficient can clearly reflect the change rule of signal characteristics. Therefore, through the wavelet energy coefficient analysis method, different stages of crack development can be effectively judged. The wavelet energy coefficient clearly reflects the changes in AE signal characteristics and can effectively identify the crack propagation state, providing a new method for health monitoring of offshore platforms and intelligent crack identification.

7.1.2 Corrosion Damage Signal Clustering Recognition of Tension Strand

As an important material for prestressed concrete members, the stress corrosion of steel strand is a cause of serious loss of prestress or even failure of the structure. When using acoustic emission technology to analyze the signal, due to the complexity of the stress corrosion mechanism, it is necessary to use intelligent cluster analysis method for effective identification. For the stress corrosion of steel strand, its development is not only a class of corrosion morphology degree of simple deepening process, or contains a variety of different corrosion morphology of complex evolution. During this period, different forms intertwine and overlap with each other, increasing and decreasing respectively, making it difficult to monitor and distinguish them purely throughout the entire process. Therefore, relying solely on overall signal analysis cannot achieve the goal of in-depth exploration of the evolution laws of the stress corrosion process. To fully leverage the advantages of acoustic emission testing technology and distinguish the types of acoustic emission signal sources corresponding to different corrosion forms, providing more detailed references for practical corrosion detection and mechanism research, technologies such as signal clustering and pattern recognition can be used to further process the acoustic emission signals collected during the corrosion process, thereby judging the source and characteristics of the signals during corrosion. For steel strand stress corrosion, principal component analysis can be used to extract the principal component features that contribute significantly to stress corrosion in acoustic emission signals as parameters, and a clustering analysis method improved by particle swarm optimization can be employed for pattern recognition of acoustic emission signals.

1. Principal component analysis method

In the parametric analysis of the data collected by the acoustic emission detection system, because the data often have many eigenvalues, some eigenvalues have a certain correlation, and some are relatively independent. Before performing data clustering analysis, it is necessary to compress the data from a high-dimensional feature space to a low-dimensional space. This not only reduces the complexity of the calculations but also allows for the selection of the principal component features with larger contribution rates.

Principal Component Analysis (PCA), as a dimensionality reduction tool, is capable of projecting multi-dimensional feature data onto a space of a few comprehensive feature values [3]. When dealing with high-dimensional data, it is important to avoid information overlap and redundancy, ensuring that the processed data have minimal correlation between different features. PCA adopts a dimensionality reduction method to find several comprehensive factors to represent the original data through projection transformation. These obtained comprehensive factors reflect the original data as much as possible while being relatively independent of each other.

For example, consider N two-dimensional data points with feature vectors x_1 and x_2. When these N two-dimensional data points are plotted on the same plane, as shown in Fig. 7.7, their distribution approximates an ellipse. From Fig. 7.7, we can observe that the coordinates x_1 and x_2 of the N points have a certain correlation. However, when the original coordinates are rotated by an angle θ, we take the coordinate axis y_1 along the length of the ellipse and y_2 along the short axis, as shown in Fig. 7.8. After the rotation, the functional relationship between x and y is as follows:

$$y_{1j} = x_{1j} \sin \theta + x_{2j} \sin \theta \qquad (7.6)$$

$$y_{2j} = x_{1j}(-\sin \theta) + x_{2j} \cos \theta \qquad (7.7)$$

Fig. 7.7 Distribution of sample characteristics

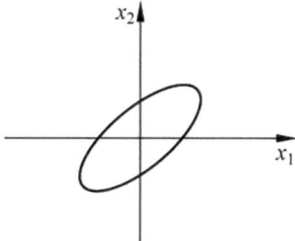

Fig. 7.8 Schematic diagram of principal components

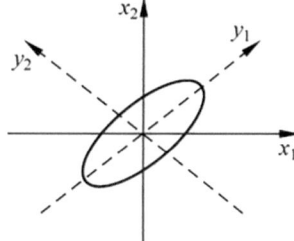

In the relation, $j = 1, 2, \ldots, N$. If written in matrix form, it can be expressed as

$$Y = \begin{bmatrix} y_{11} & y_{12} & \cdots & y_{1N} \\ y_{21} & y_{22} & \cdots & y_{2N} \end{bmatrix} = \begin{bmatrix} \cos\theta & \sin\theta \\ -\sin\theta & \cos\theta \end{bmatrix} \cdot \begin{bmatrix} x_{11} & x_{12} & \cdots & x_{1N} \\ x_{21} & x & \cdots & x_{2N} \end{bmatrix} = UX \quad (7.8)$$

where, U is a projection matrix and an orthogonal matrix, i.e., $U^T = U^{-1}, UU^T = I$.

Under the new coordinates, observing Fig. 7.8, we can find that the correlation between the coordinates y_1 and y_2 of the N points is almost 0. The distribution of the N points under the new coordinates shows that the variance along the y_1 axis is greater than that along the y_2 axis.

After the transformation, y_1 and y_2 are composite variables of the original variables x_1 and x_2. Comparatively, the variance of the N points on the y_1 axis is larger. If we represent the original two-dimensional features using a one-dimensional variable, we would choose the one-dimensional composite variable on the y_1 axis to replace the original variables. Here, y_1 is referred to as the first principal component, and y_2 as the second principal component.

Similarly, we can extend the application of principal component analysis to convert a set of data with N samples and each sample has n eigenvalues $(x_1, x_2, \cdots x_n)$ to get n composite variables, i.e.

$$\begin{cases} y_1 = c_{11}x_1 + c_{12}x_2 + \cdots + c_{1n}x_n \\ y_2 = c_{21}x_1 + c_{22}x_2 + \cdots + c_{2n}x_n \\ \vdots \\ y_n = c_{n1}x_1 + c_{n2}x_2 + \cdots + c_{nn}x_n \end{cases} \quad (7.9)$$

In the formula, the coefficient $\mathbf{C_{ij}}$ satisfies $C_{k1}^2 + C_{k2}^2 + \cdots + C_{kn}^2 = 1 (k = 1, 2, \cdots, n)$, where the coefficient $\mathbf{C_{ij}}$ need to meet the conditions:

1. x_1 and $y_j (i \neq j; j = 1, 2, \ldots, n)$ are independent of each other;
2. Assuming that $y_1, y_2, \ldots, y_n$ is a linear combination of all combinations of $x_1, x_2, \ldots, x_n$ in satisfying Eq. (7.9) and that the variance of $y_1, y_2, \ldots, y_n$ decreases sequentially, then $y_1, y_2, \ldots, y_n$ is referred to as the 1st, 2nd, ..., nth principal components of the original data variable, respectively.

Let $X = \begin{bmatrix} X_1 \\ X_2 \\ \vdots \\ X_n \end{bmatrix}$ be an n-dimensional vector, and the vector $Y = \begin{bmatrix} X_1 \\ X_2 \\ \vdots \\ X_n \end{bmatrix}$ can be transformed according to Eq. (7.9).

The formula can be expressed as $Y = CX$, C is an orthogonal matrix and satisfies $CC^T = I$, I is a unit matrix.

Y is a new orthogonal coordinate system obtained after the transformation of X. When represented in Y, the variance of the N points on the y_1-axis is the largest, followed by the y_2-axis, and similarly, the variance on the y_n-axis is the smallest. Meanwhile, the covariance between the different y_i-axis and y_j-axis (where $i \neq j$) of the N points and is 0, that is

$$YY^T = (CX)(CX)^T = CXX^TX^T = A \tag{7.10}$$

where, $A = \begin{bmatrix} \lambda_1 & & & \\ & \lambda_2 & & \\ & & \ddots & \\ & & & \lambda_n \end{bmatrix}$.

Assuming that X has undergone normalization, XX^T is the obtained correlation matrix. Let $R = XX^T$, then Eq. (7.10) is expressed as $CRC^T = A$. By left-multiplying the equation by C^T, we have

$$RC^T = C^T A \tag{7.1}$$

Written in algebraic form, it is expressed as:

$$\begin{bmatrix} r_{11} & r_{12} & \cdots & r_{1n} \\ r_{21} & r_{22} & \cdots & r_{2n} \\ \vdots & \vdots & & \vdots \\ r_{n1} & r_{n2} & \cdots & r_{nn} \end{bmatrix} \times \begin{bmatrix} c_{11} & c_{12} & \cdots & c_n \\ c_{21} & c_{22} & \cdots & c_{2n} \\ \vdots & \vdots & & \vdots \\ c_{n1} & c_{n2} & \cdots & c_n \end{bmatrix} = \begin{bmatrix} c_{11} & c_{12} & \cdots & c_n \\ c_{21} & c_{22} & \cdots & c_{2n} \\ \vdots & \vdots & & \vdots \\ c_{n1} & c_{n2} & \cdots & c_n \end{bmatrix} \times \begin{bmatrix} \lambda_1 & & & \\ & \lambda_k & & \\ & & \ddots & \\ & & & \lambda_k \end{bmatrix} \tag{7.12}$$

Expanding Eq. (7.12) results in n^2 equations, An analysis is conducted on the n equations obtained by expanding the first column.

7.1 Application of Acoustic Emission Technology

$$(r_{11} - \lambda_1)c_{11} + r_{12}c_{12} + \cdots + r_{1n}c_{1n} = 0$$
$$r_{21}c_{11} + (r_{22} - \lambda_1)c_{12} + \cdots + r_{2n}c_{1n} = 0$$
$$\vdots$$
$$r_{n1}c_{11} + r_{n2}c_{12} + \cdots + (r_{nn} - \lambda_1)c_{1n} = 0 \quad (7.13)$$

Solve for the eigenvalues of the coefficient matrix of the above system of equations:

$$\begin{vmatrix} r_{11} - \lambda_1 & r_{12} & \cdots & r_{1n} \\ r_{21} & r_{22} - \lambda_1 & \cdots & r_{2n} \\ \vdots & \vdots & \vdots & \vdots \\ r_{n1} & r_{n2} & \cdots & r_{nn} - \lambda_1 \end{vmatrix} \quad (7.14)$$

The determinant is expressed as $|R - \lambda I| = 0$. Similarly, for $\lambda_2, \lambda_3, \ldots, \lambda n$, $\lambda j (j = 1, 2, \ldots, n)$ are the roots of $|R - \lambda I| = 0$, where λ is the root of the characteristic equation, and the corresponding C_{ij} are the components of the eigenvector.

Suppose the N eigenvalues of R are $\lambda_1 > \lambda_2 > \cdots > \lambda_n \geq 0$, and the eigenvector corresponding to λ_1 is C_i, Let

$$C = \begin{bmatrix} c_{11} & c_{21} & \cdots & c_{n1} \\ c_{12} & c_{22} & \cdots & c_{n2} \\ \vdots & \vdots & & \vdots \\ c_{1n} & c_{2n} & \cdots & c_{nn} \end{bmatrix} = \begin{bmatrix} c_1 & c_2 & \cdots & c_n \end{bmatrix} \quad (7.15)$$

The equation relative to y_1 is:

$$\text{Var}(C_1 X) = C_1 X X^T C_1^T = C_1 R C_1^T = \lambda_1$$

Similarly, $\text{Var}(C_i X) = \lambda_i$, That is, y_1 has the largest variance, followed by y_1, and they also have covariance.

$$\text{Cov}(C_i^T, C_j X) = C_i^T R C_j \quad (7.16)$$

From Eq. (7.11) we get $R = \sum_{a=1}^{N} \lambda_a C_a C_a^T$, so Eq. (7.16) becomes

$$\text{Cov}(C_i^T C^T, C_j X) = C_i^T \sum_{a=1}^{N} \lambda_a C_a C_a^T C_j = \sum_{a=1}^{n} \lambda_a (C_i^T C_a)(C_a^T C_j) = 0 \quad (7.17)$$

$x_1, x_2, \ldots, x_n$ are projected to obtain new space vectors:

$$\mathbf{y}_1 = \mathbf{C}_1^T X$$
$$\mathbf{y}_2 = \mathbf{C}_2^T X$$
$$\mathbf{y}_n = \mathbf{C}_n^T X \tag{7.18}$$

$y_1, y_2, \ldots, y_n$ are orthogonal to each other, and the variance of y_i is λ_i, so $y_1, y_2, \ldots, y_n$ are called the 1st, 2nd, …, nth principal components, respectively. The contribution rate of the ith principal component is defined as $\lambda_i / \sum_{k=1}^{n} \lambda_k (i = 1, 2, \ldots, n)$, and the cumulative contribution rate of the first m principal components is defined as $\sum_{i=1}^{m} \lambda_i / \sum_{k=1}^{n} \lambda_k$, the first m (m < n) principal components are selected so that their cumulative contribution rate reaches the engineering requirements (95% is recommended), and thus the original data can be reduced from n-dimensional to m-dimension.

The diagram of dimensionality reduction using Principal Component Analysis (PCA) is shown in Fig. 7.9. The main functions of PCA are noise reduction and redundancy removal. Noise reduction through PCA involves eliminating principal components with relatively low contribution rates, which is achieved by setting the off-diagonal elements of the resulting covariance matrix to zero. Redundancy removal refers to discarding dimensions in the samples that exhibit insignificant variations. The primary computational steps of PCA include data normalization, calculating the covariance matrix of the samples, decomposing the covariance matrix into eigenvalues and eigenvectors, and obtaining a new matrix through dimensionality reduction. The algorithm flowchart is illustrated in Fig. 7.10.

2. Cluster analysis method for particle swarm improvement algorithm

Particle Swarm Optimization (PSO) is an effective global optimization algorithm [4] originally inspired by the research of Kennedy and Eberhart in the United States on

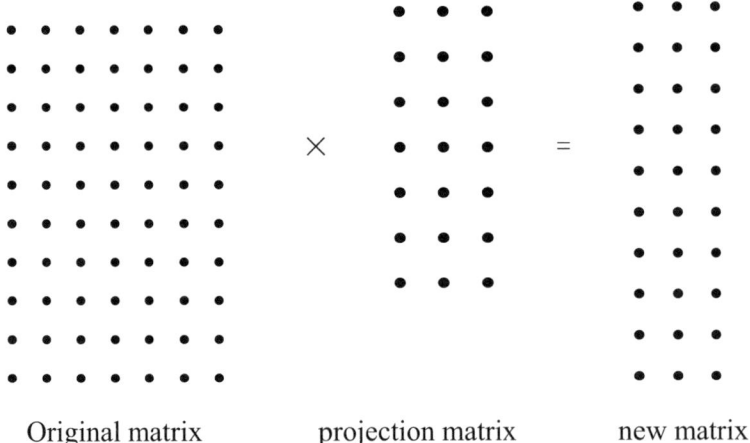

Fig. 7.9 Diagram of dimensionality reduction using principal component analysis

7.1 Application of Acoustic Emission Technology

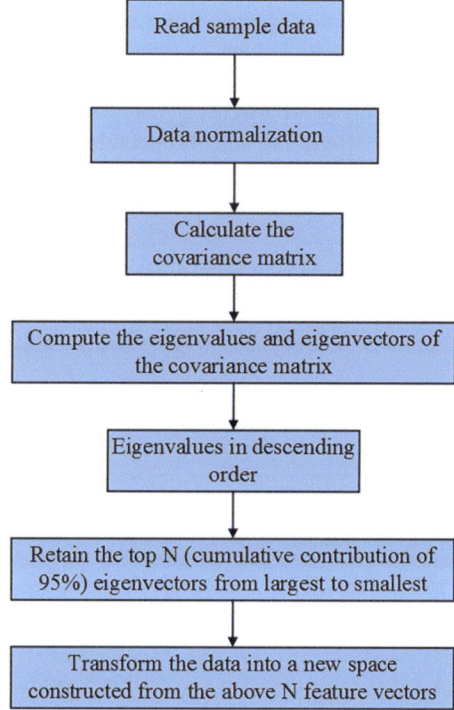

Fig. 7.10 Flow algorithm of principal component analysis

the foraging behavior of bird flocks, which was later applied to solve combinatorial optimization problems. PSO is a global optimization algorithm based on evolutionary ideas, utilizing cooperation and competition among particles in the swarm to ultimately generate collective intelligence for guiding the optimization search. PSO employs a global search strategy using a velocity-displacement model. During the computation, each generation of solutions in the swarm possesses the dual advantages of "self-learning" and "learning from others," allowing the algorithm to find optimal solutions with fewer iterations.

Particle Swarm Optimization achieves the search for optimal solutions in complex spaces through collaboration and competition among individuals, viewing each solution to an optimization problem as a particle in the search space. Initially, an initial particle swarm is generated, with each particle representing a feasible solution to the optimization problem being studied and assigned a fitness value based on the objective function. All particles search within the solution space, with their movement direction and distance determined by their velocity. In each iteration of the algorithm, particles update themselves based on two "extremes": one is the best solution found by the particle itself, and the other is the current best solution in the swarm.

In an n-dimensional space, assume there are m particles, denoted as $Z = \{Z_1, Z_2, \ldots, Z_m\}$, which represent solutions to a combinatorial optimization problem. The position coordinate of each particle is represented as $Z_i = \{Z_{i1}, Z_{i2}, \ldots, Z_{im}\}$, and its movement direction is denoted as $V_i = \{V_{i1}, V_{i2}, \ldots, V_{im}\}$. The entire particle swarm moves in the solution space, continuously adjusting its position through local extrema and global extrema to search for new solutions. During its movement, each particle can record the best solution it has found, denoted as P_{id} (local optimal extremum), and the best solution found by all particles is denoted as the global optimal extremum. Once both optimal solutions are found, the velocity-displacement update formula for each particle is as follows:

$$v_{id}(t+1) = w \cdot v_{id}(t) + \eta_1 \cdot \text{rand}() \cdot (P_{id} - z_{id}(t)) + \eta_2 \cdot (P_{gd} - z_{id}(t)) \quad (7.19)$$

$$z_{id}(t+1) = z_{id}(t) + v_{id}(t+1) \quad (7.20)$$

where, $v_{id}(t+1)$ represents the velocity of the ith particle in the d-th dimension during the $(t+1)$th iteration. To prevent the particle's velocity from becoming too large, a maximum velocity v_{max} can be set. That is, when $v_{id}(t+1) > v_{max}$, $v_{id}(t+1)$ is set to v_{max}; and when $v_{id}(t+1) < -v_{max}$, $v_{id}(t+1)$ is set to $-v_{max}$, as implemented by Eq. (7.21).

$$w = w_{max} - iter \times \frac{w_{max} - w_{min}}{iter_{max}} \quad (7.21)$$

where, iter represents the current iteration number, and $iter_{max}$ is the preset maximum number of iterations. w denotes the inertia weight, which is used to maintain the motion inertia of the particles. If $w = 0$, the particle's velocity has no memory, and the particle swarm will directly converge to the current global optimal position, losing the ability to search for better solutions. Typically, w is set to a random number between 0 and 1.

η_1 and η_2 represent acceleration constants, which are velocity adjustment parameters indicating the acceleration weights of particles moving towards the extreme points P_{id} and P_{gd}, respectively. If $\eta_1 = 0$, it means that the particle loses its "self-awareness" and only has "sociality". Although the particle will converge quickly, it is prone to falling into local extrema. Similarly, if $\eta_2 = 0$, the particle loses its sociality and only has "cognitive" ability, losing the collaboration and competition mechanism among particles. At this point, the algorithm is equivalent to m particles searching independently, unable to effectively find the global optimal solution. Generally, η_1 and η_2 are set to values around 2. rand() generates a random number between 0 and 1.

Observing the formula, we find that the particle's velocity update is mainly composed of three parts, and the update mechanism is shown in Fig. 7.11.

(1) The particle's own original velocity $v_{id}(t)$;
(2) The particle's direction $P_{id} - z_{id}(t)$ with respect to the best position experienced by itself;
(3) The direction $P_{gd} - z_{id}(t)$ of the best position experienced by the particle with all the swarm.

7.1 Application of Acoustic Emission Technology

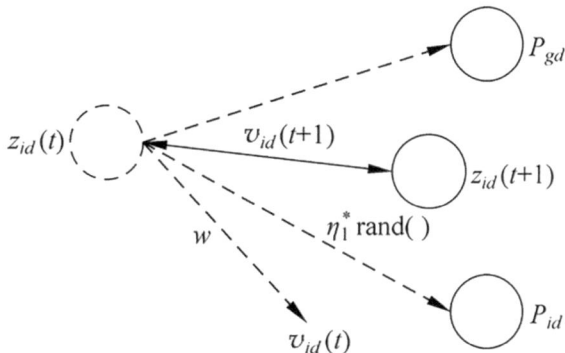

Fig. 7.11 Schematic diagram of particle update

The specific steps of the particle swarm algorithm are as follows.

(1) Initialize the particle swarm and randomly specify the initial position Z_i and initial velocity V_i of the swarm.
(2) Determine the new position of the particles according to the velocity and position.
(3) Calculate the fitness value fit of each particle.
(4) For each particle, compare its current fitness value with the fitness value at its best-known position. If the current fitness value is better, then update P_{id} accordingly.
(5) For each particle, compare its current fitness value with the fitness value of the best-known position among all particles in the swarm. If the current fitness value is better, then update the global optimal solution P_{gd}.
(6) Find P_{id} and P_{gd} and update the velocity and position of the particle according to Eqs. (7.19) and (7.20).
(7) Check if the termination condition is met. If it is, end the process; otherwise, proceed to step 3 for the next iteration.

The flow of particle swarm algorithm is shown in Fig. 7.12.
Translated with DeepL.com (free version).

3. Clustering analysis of acoustic emission signal of tension cable stress corrosion

Common acoustic emission clustering parameters include count, amplitude, energy, rise time, duration, average frequency and so on. Before applying the particle swarm damage algorithm, the first step is to determine the number of clusters, and the common evaluation indexes are DBI index and distance cost function. Figure 7.13 shows the clustering diagram of stress corrosion damage source of tension cable using DBI index, from which it can be seen that the stress corrosion damage source of bridge tension cable can be divided into 4 categories, namely plastic deformation, crack formation, crack propagation and fracture.

The clustering distribution of different acoustic emission characteristic parameters of bridge tension cable corrosion damage is shown in Fig. 7.14. According to

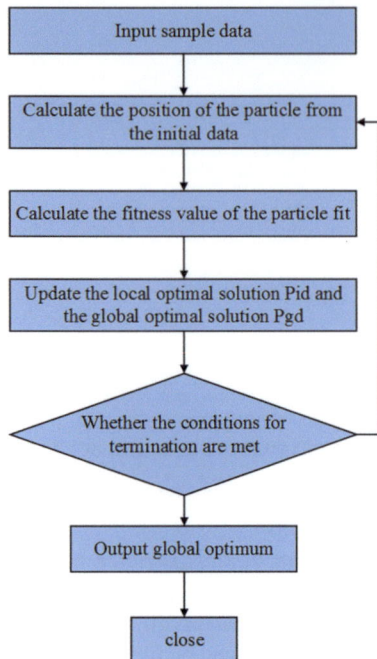

Fig. 7.12 Flowchart of the particle swarm optimization

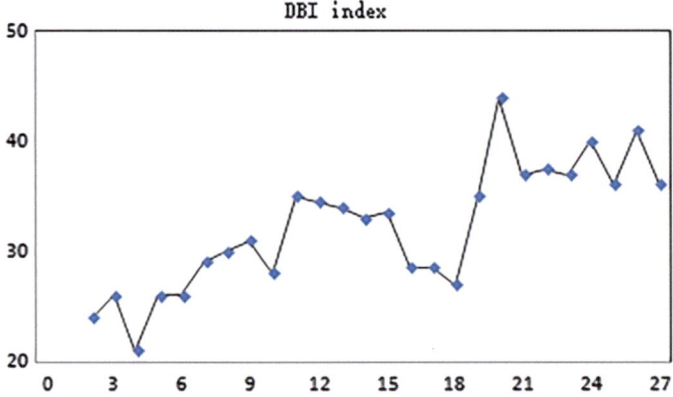

Fig. 7.13 Clustering number of stress corrosion damage sources of tension cable

Fig. 7.13 it can be seen that the bridge tension cable corrosion acoustic emission signals can be divided into four categories. Cluster 1 (pink) shows that the rise time, count, energy, duration, amplitude are in a relatively low state, while the average frequency distribution is more dispersed, which is mainly the electrochemical dissolution of the strand, the surface of hydrogen, bubble generation and rupture lead to

7.1 Application of Acoustic Emission Technology

more acoustic emission signal generation, this stage of corrosion is more active, the formation of pitting corrosion on the surface of the strand and the production of the corresponding corrosion products. Cluster 2 (green) more than cluster 1 is characterized by acoustic emission characteristics of the distribution of parameters is similar, only the two in the frequency range of the distinction between the more open, which is mainly corrosion products in the strand alternately attached and detached ultimately form a relatively stable attachment stacking structure, corrosion products in the accumulation of a certain degree will form a local occlusion of the original battery, resulting in the generation of local acidic corrosion and the generation of hydrogen; When the corrosion products fall off, hydrogen bubbles will be released at the same time local occlusion cell reaction termination, which formed a basic synchronous hydrogen bubble release floating process and corrosion products fall off the sinking process, from pitting corrosion into uniform corrosion. So cluster 1 and cluster 2 have more of the same information characteristics, the difference is that cluster 2 contains some high amplitude corrosion product shedding signals and lower frequency uniform corrosion signals. Cluster 3 (blue), which corresponds to the region with higher rise time, count, energy, duration, amplitude than the previous two signals, lower frequency, and the proportion of the corresponding time occurring at a later stage is greater than that of the former, mainly corresponds to the signals generated in the rapid development stage of stress corrosion cracking. Cluster 4 (red), compared with the other signals, the parameters are significantly larger than the other categories, characterized by the fact that each characteristic parameter is in a very high state, except for the frequency, and the corresponding time is mainly generated near the end of the experiment, mainly due to the destabilization and development of the strand cracks, and signals generated at the time of fracture.

4. Particle swarm clustering algorithm for acoustic emission of cables based on principal component analysis

The previous use of the Particle Swarm Optimization (PSO) clustering algorithm successfully classified the acoustic emission sources of stress corrosion in bridge cables. However, there are still many overlapping areas in the acoustic emission characteristic parameter diagrams, mainly due to the correlation among these parameters. To improve the effectiveness of clustering analysis, principal component analysis (PCA) is first conducted on the acoustic emission characteristic parameters. Based on the PCA results, the PSO clustering algorithm is then applied to the principal components with significant contributions to identify the stress corrosion damage sources in cables. PCA projects multi-dimensional feature data onto a few comprehensive feature spaces, minimizing the correlation between the processed data across different features. This not only reduces the computational complexity but also allows for the selection of principal component features with higher contribution rates. Through projection transformation, several comprehensive factors are identified to represent the original data, and these factors can reflect the original data as much as possible while remaining relatively independent of each other. Finally, the data after PCA is clustered using the PSO algorithm.

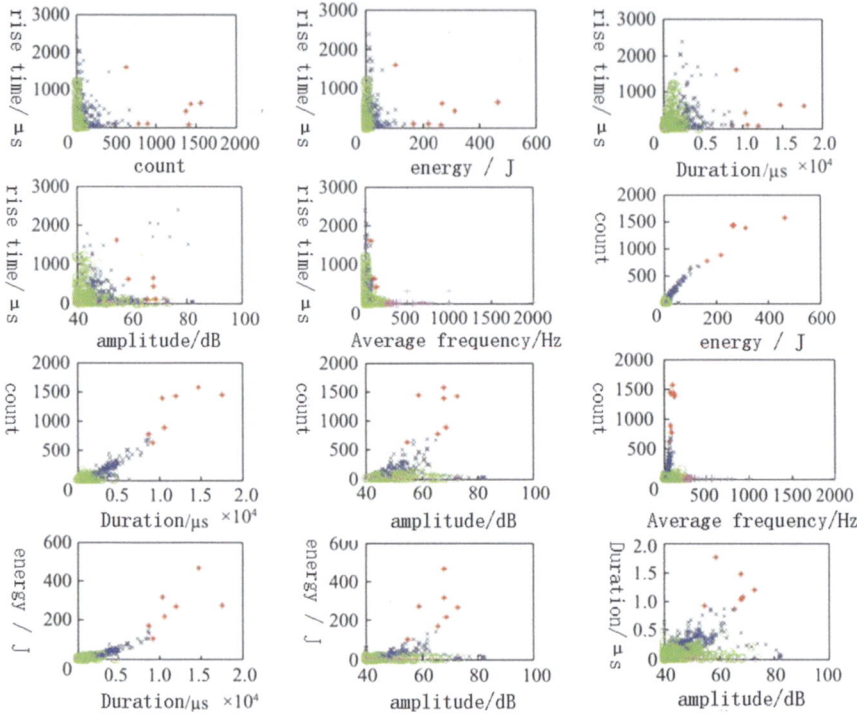

Fig. 7.14 Correlation diagram of the distribution of the characteristic parameters for stress corrosion of cables

This time, the first three orders of principal components whose contribution rate reaches 91% are selected for analysis, and the contribution rates of the first three orders of principal components are 51.7%, 23.9% and 16.4% respectively. The relationship of principal component 1 over time is shown in Fig. 7.15.

To further understand the distribution structure of the data, we can observe its distribution on the principal component space as shown in Fig. 7.16.

Fig. 7.15 Time-principal component 1 distribution

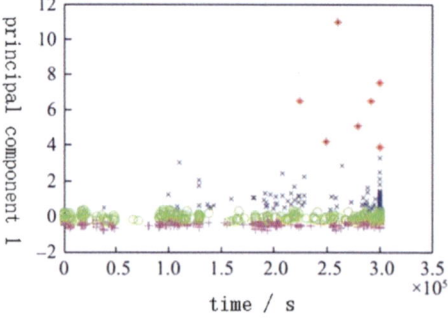

7.1 Application of Acoustic Emission Technology

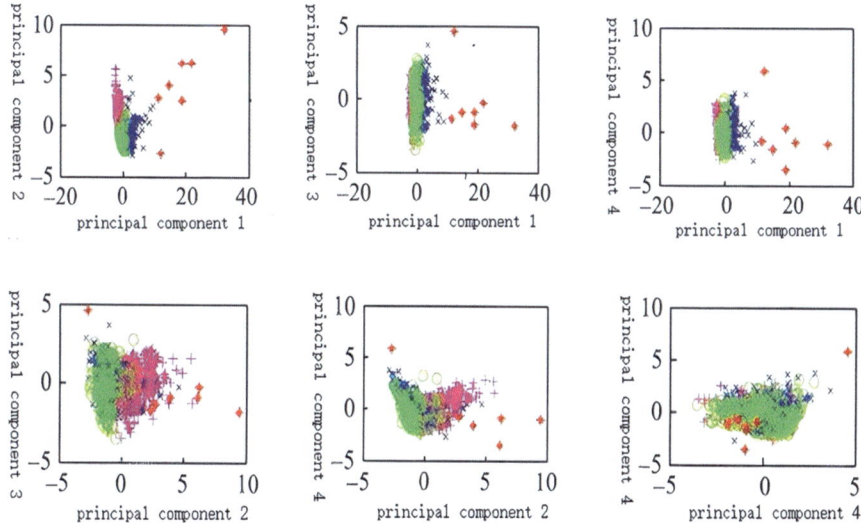

Fig. 7.16 Distribution structure of data on principal component space

As shown in Fig. 7.16, after conducting principal component analysis (PCA) followed by clustering analysis, the acoustic emission sources and damage evolution process of stress corrosion in bridge cables become clearer. The stress corrosion damage can be distinctly divided into three stages: crack initiation, crack propagation, and fracture of the bridge cables. Different damage stages exhibit typical acoustic emission characteristics. The graph of Principal Component 1 over time clearly shows the changes in the acoustic emission characteristic parameters of stress corrosion damage in steel strands and the moment of strand breaking. Further analysis combining the PCA results reveals that after clustering through PCA, the four types of damage sources are completely separated. The correlation diagram between Principal Component 1 and Principal Component 2 highlights the distinct spatial distribution of the four different types of damage sources. Cluster 1 and Cluster 2 share similar signal characteristics, occurring throughout the entire period with relatively low feature values and high quantities. Further verification of their characteristics includes the rupture and shedding of the passivation film on the steel strands, crack nucleation at pitting sites, hydrogen formation, and the bursting and release of bubbles. However, Cluster 1 has a larger component on the Principal Component 2 axis, mainly due to its broader distribution of average signal frequencies. As corrosion time increases, cracks at pitting sites continuously propagate under stress, and cracks converge, leading to a significant increase in acoustic emission characteristic parameters, with Cluster 3 signals dominating. In the final stage of corrosion, the steel strands fracture, generating a large number of acoustic emission signals (Cluster 4).

7.2 Application of Sonar Technology

7.2.1 Detection of Underwater Structure Engineering

Underwater structures in hydraulic engineering are constantly exposed to complex underwater flow environments throughout the year. Their defects are characterized by difficulties in detection and treatment, as well as sudden occurrences with severe consequences, making their inspection particularly challenging.

The underwater structures in hydraulic engineering mainly include the facing structures of the upstream slope below the normal water level, intake tower structures, and structures such as aprons, baseplates, stilling basins, and caissons. Common potential hazards include overall settlement of structures, cracks, joint gaps, or water stop damage on the surface of concrete structures, and corrosion of metal structures. Currently, the primary methods for inspecting underwater structures in hydraulic engineering include visual inspection, remotely operated vehicle (ROV) inspection, laser scanning, fan-scan sonar imaging, and so on.

1. The inspection of hydropower projects

A certain hydropower project is located on a foundation of deep, highly compressible, and fluid plastic silt. Therefore, the aprons before and after the sluice gate of the project were constructed using a prefabricated underwater lifting hollow box scheme. First, underwater dredging and leveling were conducted on the aprons before and after the sluice gate. After the prefabricated hollow box was self-floatingly transported to the predetermined location, it was positioned, filled with water, and submerged. A schematic plan is shown in Fig. 7.17.

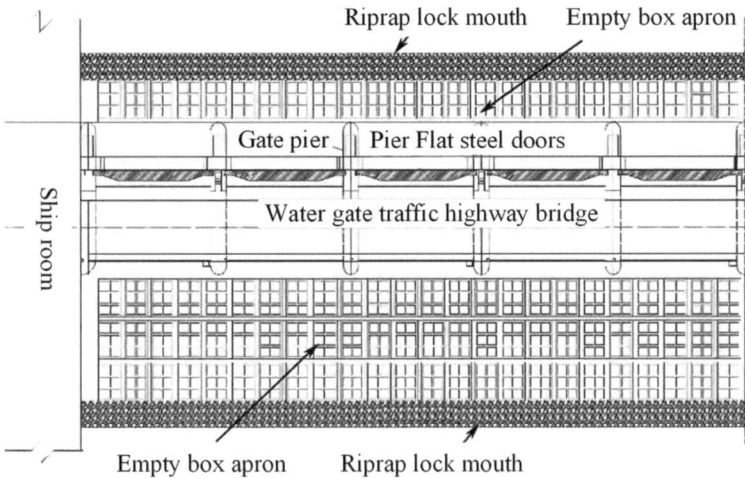

Fig. 7.17 Schematic plan of the hydropower project

7.2 Application of Sonar Technology 277

Based on the characteristics of hydraulic engineering, an underwater robot equipped with a sonar system was used to inspect the underwater structures and assess their condition. The underwater imaging was conducted using a multi-beam sonar system, which mainly consists of a sonar device, data acquisition device, equipment installation bracket, and connecting cables. The ideal coverage angle of the sonar device is 140°, with 256 beams and an ideal measurement depth of 100 m. The system can meet the requirements for most underwater structure and river channel measurements in hydraulic engineering.

The detection results from the multi-beam system indicate that the aprons inside and outside the sluice gate are presented relatively intact and clearly. The caisson structure is basically intact, with a clear outline and no damage. However, most of the caissons have become scattered and subsided under the long-term action of water flow, and some have been covered by silt (Fig. 7.18). The phenomenon of "sand running and stones sinking" is common in riverbeds with deep silt.

2. Dam Detection

Underwater robots are suitable for large-area and wide-range underwater inspections, providing intuitive and reliable results. However, they can only probe the surface conditions of hydraulic engineering and cannot conduct detailed inspections under conditions of turbid water, complex hydraulic structures, and complex flow patterns, making it difficult to achieve precise localization of defects. Nevertheless, underwater robots equipped with sonar systems can generate three-dimensional images of underwater structures through sonar imaging technology, enabling the inspection of underwater structures in combination with the robot's video and positioning systems.

Fig. 7.18 Detection results of the multi-beam imaging system

A reservoir dam was inspected using an underwater robot equipped with a two-dimensional sonar imaging system. The core detection system of the underwater robot includes a video camera system and a two-dimensional sonar imaging system, featuring high-resolution cameras (HD color zoom cameras and black-and-white low-light cameras) and high-brightness lighting. A real-life photograph of the reservoir dam is shown in Fig. 7.19, and the result of sonar imaging of the upstream faceplate of the dam is shown in Fig. 7.20. According to the inspection results: the surfaces of all faceplates have no obvious through cracks or large areas of missing concrete; the rubber water stops are basically intact, with locally cracked surfaces and some missing fixing nuts; a crack was found at a water depth of 1.2 m, which was verified to be a trace left by the removal of ribbons, bamboo strips, steel bars, or other strip-shaped objects that fell onto the concrete surface during concreting, and it is not a through crack.

Fig. 7.19 Photograph of the actual scene of the reservoir dam

Fig. 7.20 Sonar imaging of upstream panels of a dam

7.2.2 Application in Marine Platform Detection

1. Application of sonar technology in oil pipeline detection

Since the twenty-first century, the demand for oil has rapidly increased globally. With proven marine oil and gas reserves accounting for 34% of the world's total resources, the abundant oil and gas resources have attracted the attention of the entire world towards the ocean, which is considered a treasure trove of resources. Coastal countries worldwide have adopted the development of marine resources as a national development strategy. Subsea oil pipelines serve as the lifeline connecting offshore platforms and between offshore platforms and land, representing crucial facilities in offshore platforms.

During pipeline laying, to ensure that the subsea pipelines remain buried and avoid damage from various factors, the pipelines are generally buried to a certain depth within the seabed. However, due to the long-term impact of marine turbulent hydrodynamic forces near the pipeline, the buried pipeline gradually becomes exposed or suspended [5]. It is highly susceptible to damage from scouring, ship anchor lifting, or cyclic loading from ocean currents, leading to internal oil and gas leaks, causing significant economic losses, environmental pollution, and even threatening human safety. Therefore, detecting and assessing the burial depth, suspension, exposure, and even vertical or horizontal displacement of subsea pipelines is of great significance for ensuring safe pipeline operation.

Due to the effects of seabed turbulence and erosion near the pipeline, which may cause the pipeline to be in different states such as buried, suspended, or exposed, the use of a single sonar technology has already achieved good results in pipeline detection. To more effectively and reliably inspect pipelines between offshore platforms, it is necessary to integrate technologies such as multi-beam sonar, side-scan sonar, and sub-bottom profiler for detection.

The subsea pipeline detection data obtained through the multi-beam system can be processed and extracted using multi-beam data acquisition and post-processing software. It can display the beam situation of each Ping measurement section. Based on the monitoring data and mapping scale, point cloud data analysis can be conducted to obtain the underwater topographic map of the seabed near the pipeline and the two-dimensional detection results of the pipeline's cross-sectional and longitudinal profiles. For exposed and suspended subsea pipelines, side-scan sonar has advantages over multi-beam sonar in terms of high lateral resolution, detection efficiency, and lower detection costs.

Since both multi-beam and side-scan sonar acoustic detection equipment operate at frequencies in the hundreds of kHz range, the wavelength of the sound is too short. Therefore, they can only detect subsea pipelines that are exposed or suspended and cannot penetrate the seabed to detect buried subsea pipelines. The acoustic wave emission frequency of the sub-bottom profiler can be controlled within 1–5 kHz, enabling it to emit sound pulses with longer wavelengths, which can then penetrate the seabed to detect buried subsea pipelines.

In summary, during subsea pipeline inspection, the characteristics of multi-beam sonar, side-scan sonar, and sub-bottom profiler can be utilized for joint detection, improving the accuracy of pipeline inspection data and the reliability of results.

1) Multi-beam Detection of Submarine Pipelines

When using multi-beam sonar for pipeline detection, the acoustic wave reception zones intersect at the seabed, forming hundreds of beam footprints perpendicular to the track direction. By combining sound velocity profile data obtained from a sound velocity profiler, the phase and time of the acoustic signals within each beam footprint are estimated to obtain the water depth value for that beam footprint. Combining the water depth values measured from all beam footprints yields a high-density water depth profile perpendicular to the track direction, which can then be used to create a seabed topographic map in the track direction and a two-dimensional detection result for the pipeline's cross-sectional and longitudinal profiles.

When detecting subsea pipelines using multi-beam sonar, the point cloud data from the pipeline's cross-sectional and longitudinal profiles can represent the pipeline's occurrence state. The in-situ state of the pipeline can be judged based on the measured distance from the top of the pipeline to the mean sea level.

Figures 7.21 and 7.22 show the longitudinal and cross-sectional point cloud images of the pipeline, respectively, extracted from the two-dimensional imaging results of subsea pipelines using multi-beam sonar. Figure 7.21 indicates that the tested pipeline section is suspended above the seabed, with a maximum suspension height of 0.54 m. Analyzing the cross-section of the pipeline extracted in Fig. 7.22 reveals the distribution of seabed erosion along the cross-section and the suspension height of the pipeline. At this cross-section, the suspension height of the pipeline is 0.14 m.

2) Side-scan sonar detection of submarine pipeline

(1) Side-scan sonar detection of suspended pipelines

The side-scan sonar system has good detection effect on exposed and suspended submarine pipelines. Due to the complexity of the environment in which the pipeline is located, the degree of influence of the seabed currents and other environmental factors is different, and the pipeline located on the seabed surface is prone to be exposed and suspended. Side-scan sonar system is easy to read the acoustic image of these two situations.

The schematic diagram of side-scan sonar detection for exposed subsea pipelines is shown in Fig. 7.23. When using the side-scan sonar system to detect exposed subsea pipelines, a sonogram record of the detection is obtained. Due to the strong backscatter intensity of the exposed pipeline, it appears as a black strip in the side-scan sonar image. The seabed pipeline blocks the acoustic waves propagating towards the rear of the pipeline, resulting in no acoustic wave signals behind the pipeline. This is displayed as a white acoustic shadow area on the side-scan sonar sonogram image.

7.2 Application of Sonar Technology

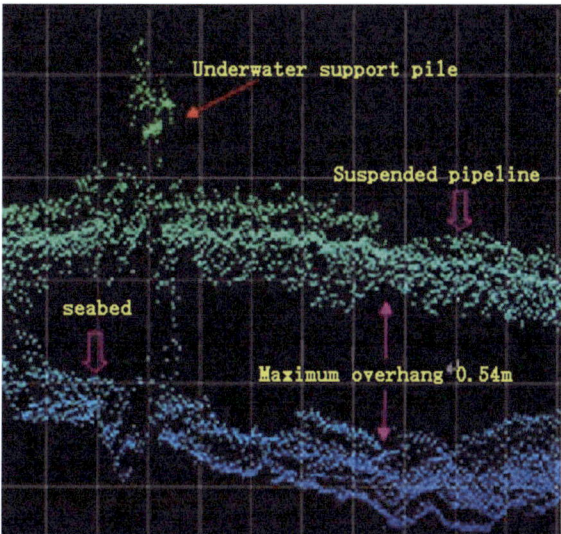

Fig. 7.21 Point cloud data map of longitudinal section of submarine pipeline

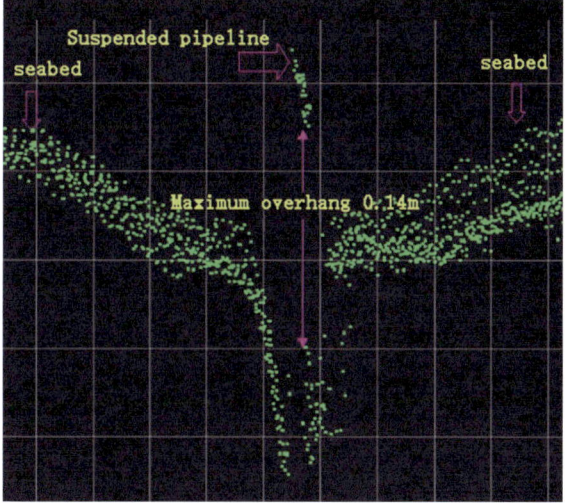

Fig. 7.22 Point cloud data map of cross section of submarine pipeline

When using the side-scan sonar system to detect pipelines suspended above the seabed, the detection result is shown in Fig. 7.24. Due to the strong scattering produced by the pipeline, it appears as a black strip target on the acoustic shadow image, while the seabed sediment image appears gray. However, the white acoustic shadow area does not appear immediately adjacent to the pipeline image but is separated from it by a gray seabed sediment area. This is because the seabed pipeline is in a suspended state, allowing acoustic waves to reach the seabed surface below the suspended pipeline. At the same time, since the distance between the area below the

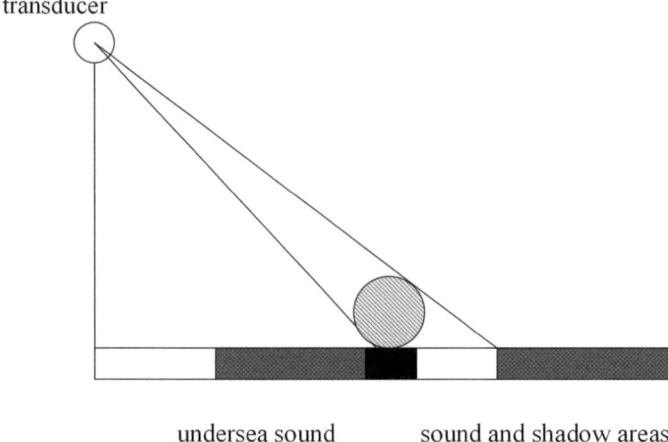

Fig. 7.23 Schematic of side-scan sonar detection of exposed pipelines

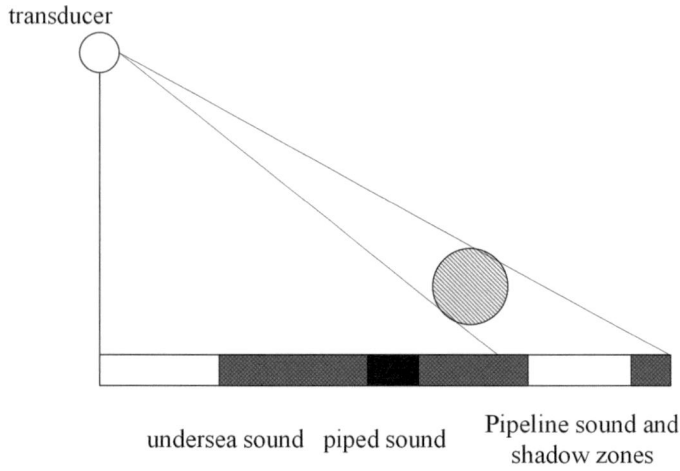

Fig. 7.24 Schematic diagram of side-scan sonar detection of submarine suspended pipeline

pipeline and the sonar transducer is greater than the distance between the pipeline and the transducer, there is a gray seabed sediment image between the pipeline image and the pipeline acoustic shadow area. This is a key feature for determining whether the pipeline is suspended based on side-scan sonar images. Therefore, the position and state of the subsea pipeline can be determined based on the acoustic shadow images detected by the side-scan sonar [6, 7].

7.2 Application of Sonar Technology

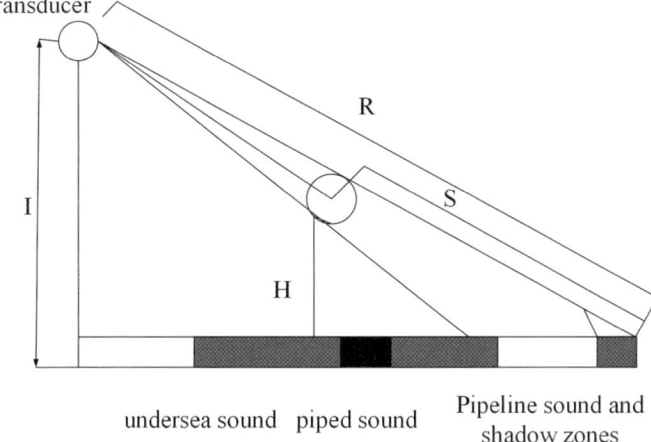

Fig. 7.25 Calculation sketch of the suspended height of submarine pipeline

(2) Calculation method of suspended height of submarine pipeline

When calculating the suspension height of a pipeline using side-scan sonar records for suspended pipelines, if the pipeline diameter is large, using a calculation formula that does not consider the pipeline diameter will result in significant errors. Therefore, the impact of large-diameter subsea pipelines on the calculation of suspension height cannot be ignored. Considering the pipeline diameter, the simplified diagram for calculating the suspension height of a subsea pipeline is shown in Fig. 7.25.

Assuming the seabed surface is a flat plane and ignoring the bending of sound rays during their propagation in seawater, based on the geometric relationships in Fig. 7.25, the formulas for calculating the suspension height h of the seabed pipeline and the grazing angle θ of the acoustic waves are as follows [8]:

$$h = \frac{HS}{R} - r\left(1 + \frac{\sqrt{R^2 - H^2}}{R}\right) \tag{7.22}$$

$$\theta = \arccos \frac{r}{\sqrt{(H-h-r)^2}} - \arccos \frac{H-h-r}{\sqrt{(H-h-r)^2}} \tag{7.23}$$

where H is the height of the towed fish above the seabed; S is the distance from the end of the pipeline's acoustic image to the end of its acoustic shadow area; R is the slant range from the transmission line to the end of the pipeline's acoustic shadow area; and r is the radius of the subsea pipeline (with a diameter of D). If the pipeline radius r is known, the height H of the towed fish above the seabed can be accurately measured automatically by the side-scan sonar system, while S and R need to be manually measured from the sonogram image of the side-scan sonar. Therefore, the accuracy of calculating the suspension height h of the subsea pipeline is mainly affected by the measurement errors of S and R in the detection sonogram image. From Eq. (7.22), it can be seen that the calculation of the pipeline suspension height is related to the four physical quantities H, S, R, and D. Based on this equation, the error formula for the pipeline suspension height can be derived. According to the error propagation formula [9], we can obtain:

$$\Delta h = \frac{\partial h}{\partial S}\Delta S + \frac{\partial h}{\partial R}\Delta R + \frac{\partial h}{\partial D}\Delta D + \frac{\partial h}{\partial H}\Delta H \qquad (7.24)$$

In the formula, since the radius of the pipe and the height of the towed fish can be measured accurately, their influence on the error of the overhanging height of the pipe is negligible. The above error calculation formula for overhanging height becomes.

$$\Delta h = \frac{\partial h}{\partial S}\Delta S + \frac{\partial h}{\partial R}\Delta R = \frac{H}{R}\left(1 - \frac{S}{R} \times \frac{\Delta R}{\Delta S} - \frac{DH}{2RL} \times \frac{\Delta R}{\Delta S}\right)\Delta S \qquad (7.25)$$

where ΔR and ΔS represent the systematic errors of R and S, respectively.

(3) Factors affecting the overhang height of submarine pipelines

The measurement errors ΔR and ΔS can be seen from the calculation formula for the suspension height of subsea pipelines. Based on the simplified diagram for calculating suspended pipelines and the side-scan sonar sonogram images of these pipelines, in order to accurately interpret and calculate the suspension height of subsea pipelines, the boundaries between the seabed image area and the white pipeline acoustic shadow area, as well as between the pipeline image area and the seabed image area, must be clear in the sonogram. Therefore, in addition to requiring a significant difference in backscatter intensity between the seabed sediment and the subsea pipeline, the testing of subsea pipeline suspension height also requires that the grazing angle of the acoustic waves during side-scan sonar detection fall within a reasonable range.

The grazing angle of the acoustic waves is determined by two factors: on the one hand, the boundary between the gray seabed image and the white pipeline acoustic shadow area in the sonogram must be clear, which requires a certain level of backscatter intensity from the seabed sediment. The backscatter intensity of the seabed sediment increases as the grazing angle θ increases, so the seabed sediment's sonogram image will be clearer when the grazing angle θ is greater than a certain lower limit during detection. On the other hand, the boundary between the black pipeline image and the gray seabed image in the sonogram must also be clear, which requires a certain difference in backscatter intensity between the pipeline and the seabed sediment. The difference in backscatter intensity between the subsea pipeline and the seabed sediment increases as the grazing angle θ increases, which determines that the accuracy of the detection results can be ensured only when the grazing angle θ is less than a certain upper limit. Combining these two factors, when the grazing angle of the acoustic waves from the side-scan sonar falls within an optimal range, the boundaries between the subsea pipeline, seabed sediment, and pipeline acoustic shadow area in the side-scan sonar sonogram image are the clearest, and the precision of calculating the pipeline suspension height is the highest. This range is defined as the optimal range for the grazing angle of acoustic waves.

In summary, the main factor affecting the testing accuracy is the difference between the backscatter intensity of the seabed substrate and the backscatter intensity of the seabed pipeline when the side-scan sonar is detected. In addition to the value of the acoustic sweep angle, the difference in the backscatter intensity is also affected by the acoustic wavelength of the side-scan sonar, the acoustic wavelength of the side-scan sonar, the beam opening angle, the radius of the submarine pipeline, the type of the submarine substrate, and other parameters. Therefore, in order to obtain high-quality detection of clear boundary acoustic image, for different types of side-scan sonar system and different types of sea environment and submarine pipeline, the optimal range of acoustic wave grazing angle may be slightly different.

3) Shallow stratigraphic profiler detection of submarine pipelines

Since both multi-beam and side-scan sonar acoustic detection equipment operate at frequencies in the hundreds of kHz range, the wavelength of the acoustic waves is too short. Therefore, these two sonar detection methods can only detect submerged and suspended subsea pipelines and cannot penetrate the seabed to detect buried subsea pipelines.

The acoustic wave emission frequency of the shallow subsurface profiler can be controlled within the range of 1–5 kHz, enabling it to emit acoustic pulses with longer wavelengths, which can penetrate the seabed to detect buried subsea pipelines. As shown in Fig. 7.26, the principle of detecting subsea pipelines using shallow subsurface profiling treats the acoustic wave propagation medium as a layered model. Seawater is considered the first propagation medium with a density of ρ_1 and a sound propagation speed of c_1. The underlying layers below the seawater are considered as the second, third, ..., and nth layers based on their sound propagation speeds, which are c_2, c_3, ..., and c_n, respectively, and their densities are ρ_2, ρ_3, ..., and ρ_n, respectively. Since subsea pipelines are generally made of steel or iron, the density of the pipeline and the speed of sound propagation within it are much higher than those of the layered seabed sediments.

Subsea pipelines differ from seawater and underwater subsurface structures in terms of their organizational structure and density ρ, which results in different sound propagation speeds c within them. When acoustic waves propagate downward from the seawater layer, reflection and refraction occur at the interfaces between layered materials. After processing the echo signals through certain means, distinct layering will be displayed on the display unit, and the shape of the subsea pipeline will appear as a clear arc. The detection of subsea pipelines by shallow subsurface profilers is based on this principle.

The degree of reflection and refraction of acoustic waves between two materials can be expressed using the reflection coefficient κ, which is a measure of the difference in wave impedance. Its definition formula is as follows:

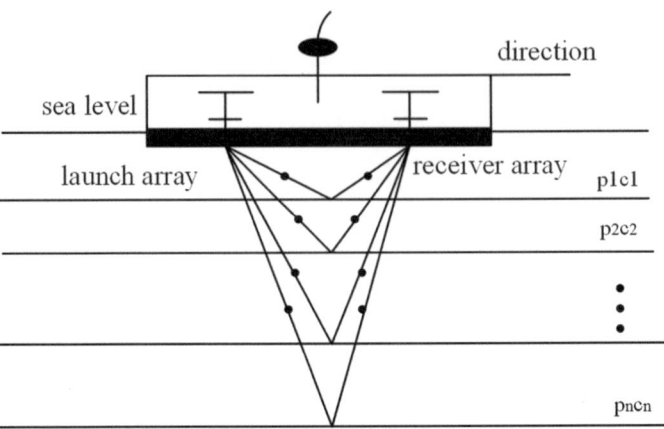

Fig. 7.26 Detection principle of shallow stratigraphic profiler

$$\kappa = \frac{\rho_2 c_2 - \rho_1 c_1}{\rho_2 c_2 + \rho_1 c_1} \tag{7.26}$$

In the formula, κ represents the reflection coefficient, also known as the wave impedance difference. ρ_1 and ρ_2 are the densities of the first and second propagation media, respectively, while c_1 and c_2 are the speeds of sound in the first and second propagation media, respectively. p_c is called the wave impedance, which characterizes the propagation properties of acoustic waves within the medium [10].

The clarity of the interface line between layered materials on the display unit is closely related to the wave impedances p_c of these two layers. If the wave impedances of the two materials differ greatly, the interface line between them on the display unit will be very clear; if the difference in wave impedances between the two materials is very small, the interface line on the display unit will be very blurry.

4) Example of submarine pipeline detection

Ding Jiandi [11] assisted Tianjin Luhai Surveying and Mapping Co., Ltd. in completing the inspection of the subsea pipeline between the ZH104 platform and Yuanhai yizhan platform. A comprehensive inspection of the pipeline was conducted using multi-beam sonar, side-scan sonar, and shallow subsurface profiler, and a special analysis of the pipeline inspection results was carried out. This section introduces the inspection results of the subsea pipeline between the ZH104 platform and Yuanhai yizhan platform using these three sonar technologies separately.

(1) Multi-beam sonar detection

Based on the seabed topographic map generated by multi-beam sonar and the varying roughness of the seabed surface, the seabed landforms can be classified into rough terrain (gullies), erosional platforms, spotted seabed, and smooth seabed. Through the bathymetric map generated from the multi-beam bathymetric data of the inspected area, the landforms surrounding the pipeline route can be visually interpreted. At the same time, a three-dimensional seabed DEM (Digital Elevation Model) can be established based on the underwater point cloud data obtained from different survey times. This model allows for an analysis of erosion and deposition in the seabed terrain surrounding the subsea pipeline.

(2) Side-scan sonar detection

Side-scan sonar was used to inspect the seabed landforms and pipelines between the ZH104 platform and Yuanhai yizhan platform. Combining the results of side-scan sonar and multi-beam testing, no seabed obstacles were found in the route area of the subsea oil pipeline from the ZH104 platform to Yuanhai yizhan. The seabed surface conditions were relatively complex, with erosion landforms and raised landforms formed by seabed pipeline maintenance structures dominating near the platforms. There were two obvious erosion areas, one near the ZH104 platform and the other near the Yuanhai yizhan. In the horizontal segment of the route, the seabed landforms were mainly rough seabed terrain and smooth seabed terrain. The developed seabed landforms were primarily rough terrain (gullies, seabed pipeline maintenance structures) and smooth seabed terrain. The presence of the ZH104 platform altered the hydrodynamic environment around it, leading to the formation of pits due to erosion around the platform base under the combined action of currents and waves. The terrain around these pits was relatively rugged and rough.

(3) Shallow stratigraphic profiler test results

Based on the shallow stratigraphic profiling results obtained from the shallow stratigraphic profiler, combined with historical data, a comprehensive comparative analysis was conducted on the amplitude, frequency, phase, continuity, and wave impedance combination relationships of the reflected waves. In the seabed pipeline route area, four distinct acoustic reflection interfaces could be identified within 18 m below the seabed surface, named R_0, R_1, R_2, and R_3, respectively. Among them, R_0 represents the seabed surface. Based on these acoustic impedance interfaces, the seabed strata can be divided into three layers from top to bottom, namely A, B, and C, as shown in Fig. 7.27.

According to the shallow stratigraphic profiling records of the pipeline route area, Layer A is located between the reflection interfaces R_0 and R_1. The interlayer reflection structure is mainly characterized by parallel bedding, with some local disorganization and strong reflection energy. The thickness of this layer ranges from 6.8 to 9.7 m, with a general thickness of approximately 8.9 m. The bottom interface R_1 of Layer A is clear, with strong reflection energy, and can be continuously traced with little undulation. Layer B is situated between the reflection interfaces R_1 and R_2, and its interlayer reflection structure is dominated by subparallel bedding and wavy bedding, exhibiting strong reflection energy. The thickness of this layer is between 3.3 and 4.4 m, with a general thickness of about 4.1 m. The bottom interface R_2 of Layer B is clear, with strong reflection energy, and can be continuously traced with minimal

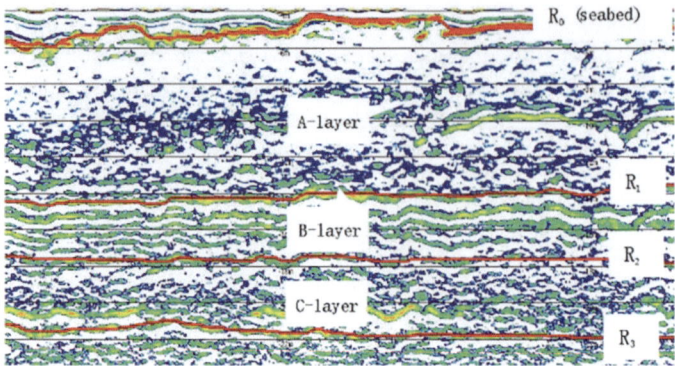

Fig. 7.27 Acoustic recording and interpretation of the centerline route from ZH104 platform to Yuanhai yizhan

undulation. The R_2 reflection interface represents an erosional unconformity. Layer C is located below the R_2 reflection interface, with medium-strong reflection energy in its upper part. Shallow stratigraphic records indicate that the upper part of this layer is mainly composed of wavy bedding. The thickness of this layer ranges from 3.4 to 4.2 m, with a general thickness of approximately 3.7 m. The bottom interface R_3 of Layer C is clear, with strong reflection energy, and can be continuously traced with little undulation.

(4) Pipeline comprehensive detection results

Based on the multi-beam full-coverage survey data, side-scan sonar data, and shallow stratigraphic detection data, the pipeline inspection results are analyzed according to the 1985 National Elevation Datum. The elevation of the submarine oil pipeline from the ZH104 platform to Yuanhai yizhan is relatively flat in the middle section of the route. Near the platform, the pipeline top elevation gradually increases, with the top elevation of the submarine pipeline ranging from − 6.36 m to 3.58 m. Additionally, the submarine pipeline section near the platform and the landing point of the route from the ZH104 platform to Yuanhai yizhan has undergone sand placement maintenance. When detecting the buried submarine pipeline using a shallow stratigraphic profiler, the image formed by the submarine pipeline is very similar to the image formed by the sand placement maintenance materials, making it basically impossible to distinguish the image of the submarine pipeline, as shown in Fig. 7.28.

The state of the pipeline under the seabed needs to be determined based on the detection results of side-scan sonar and shallow stratigraphic profiler. The planar position and suspension height of each route point of the pipeline are obtained through the detection results of the shallow stratigraphic profiler. The exposed and suspended pipeline segments, as well as their lengths, can be comprehensively identified through the detection results of side-scan sonar and multi-beam systems. Through the current

7.2 Application of Sonar Technology

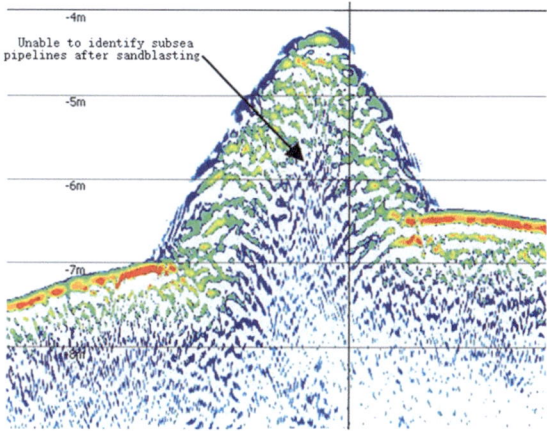

Fig. 7.28 Shallow profiling effect of sand blasted submarine pipeline

rerouting survey of the submarine pipeline, it was found that there are currently 26 exposed and suspended segments in the submarine oil pipeline from the ZH104 platform to Yuanhai yizhan. The bathymetric map obtained from multi-beam and side-scan sonar near a certain suspended pipeline segment is shown in Fig. 7.29.

2. 3D imaging sonar application in the detection of marine platforms

Offshore platforms in complex marine environments are jointly affected by various marine environmental factors such as wind, waves, currents, sea ice, storm tides, and earthquakes, leading to frequent damage incidents. Safety inspection and early warning for offshore platforms are crucial for their construction and operation.

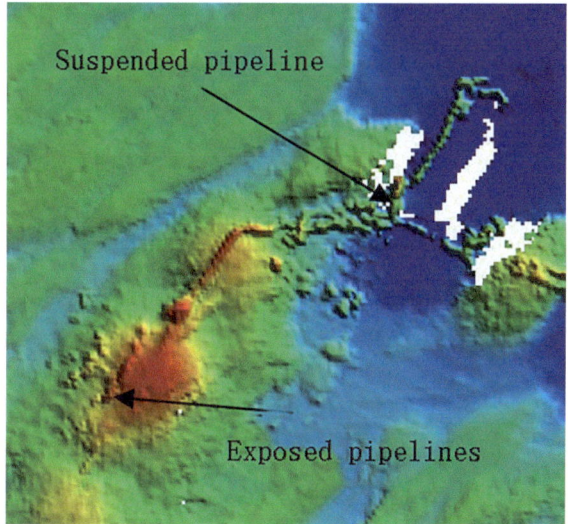

Fig. 7.29 Multibeam and side-scan sonar testing images of exposed, suspended pipelines

Common sonar technologies used for inspecting underwater offshore platforms include multi-beam sonar, side-scan sonar, and 3D imaging sonar. Existing multi-beam bathymetric systems face difficulties in detecting concrete defects on vertical surfaces when inspecting flow surfaces such as stilling basins and downstream aprons. Side-scan sonar can only obtain two-dimensional topographic sonograms, making it challenging to acquire dimensional parameters of erosion defects such as scouring. With the aid of 3D display technology, the 3D imaging sonar system can provide more detailed descriptions of the underwater target's outline and shape, representing an advanced means for detecting underwater detailed structures.

1) Overview of marine platform

An offshore platform in a shallow sea oilfield adopts the engineering model of artificial island-based offshore oil extraction on land. Three artificial islands have been successively constructed, including the target platform No.1, the landing platform for submarine pipelines (submarine pipeline landing platform), and the supporting platform for the submarine pipeline trestle on the target artificial island No.2. Due to the influence of offshore dynamic factors such as wind, waves, currents, and ice, there may be erosion around the pile foundations of the platforms, posing a serious threat to safe production in severe cases. Engineering designs and relevant standards have put forward requirements for regular inspections, which include bathymetric and seabed topographic surveying and mapping, pile foundation detection of platforms, pile foundation erosion, abandoned cables, and seabed obstacles.

Due to the shallow water depths (1–3 m) of each platform and their irregular shapes composed of cluster piles, it is difficult for survey vessels to approach for inspection. For many years, the pile foundations of each platform could only be qualitatively assessed through divers' exploration and touch, with no quantitative detection available. After technical comparisons, this project adopts 3D sonar scanning technology. The 3D sonar scanning equipment can display underwater topography in real-time and collect point cloud data, which can then be used to display 3D images of underwater objects and hydraulic structures in subsequent processing. 3D sonar has excellent visualization effects, capable of presenting the shapes of underwater objects at a 360° angle with a relative resolution distance of 4 cm, and can output 3D data of scanned objects. 3D sonar is very convenient to install and can be mounted on small boats, underwater excavators, survey vessels, underwater robots, etc. It is safe to operate and has a high safety factor.

2) Three-dimensional sonar imaging detection system

The testing equipment adopts Echoscope real-time three-dimensional sonar system and its auxiliary system, including the sonar head, computer terminal, power supply, measurement ship, shipboard relay station, inertial guidance system, test and analysis software, see Fig. 7.30. The instrumentation should be calibrated before the test, and the items of the calibration mainly include: transverse rocking (Roll) calibration, bowing (Yaw) calibration, Pitch calibration, X calibration, Y calibration.

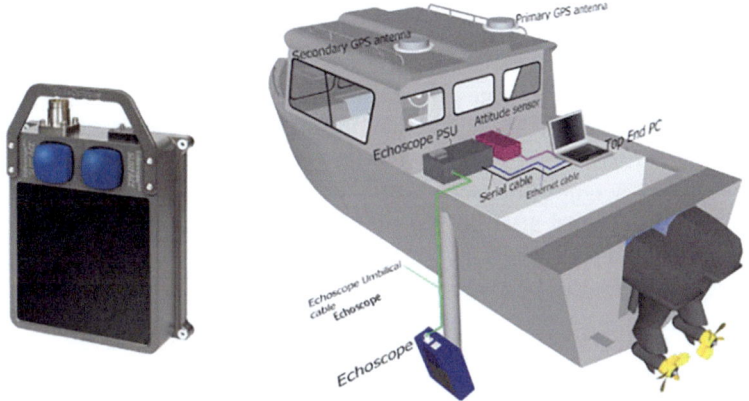

Fig. 7.30 Echoscope real-time 3D sonar system and its auxiliary system

3) Measurement method

Three-dimensional imaging sonar system transmits acoustic signals to the target area, and uses acoustic imaging methods to process the received echo signals. When the Echoscope sonar system works, an acoustic signal with a frequency of 375 kHz is emitted through the sonar probe to form a fan-shaped scanning area with a range of $50° \times 50°$, and each acoustic emission contains 128×128 wave beams arranged with the same spacing, and each acoustic spacing is $0.39°$; the system receives the echo signal and processes it to generate a two-dimensional image (frame). Each acoustic wave is spaced at $0.39°$; the system receives the echo signals and performs acoustic imaging processing to generate a two-dimensional image (frame); the system updates the data at a rate of 20 Hz, and then the series of frames are synthesized into a three-dimensional image by computer synthesis techniques. In order to ensure the measurement accuracy, attitude correction is carried out by an inertial navigation system to eliminate the effects of longitudinal and transverse swaying of the ship during navigation.

The horizontal datum adopts the Beijing 1954 Coordinate System with Gauss-Krüger projection; the projection parameters are a central meridian of $118° 30'$ E, an easting addition constant of 50×10^4 m, and a northing addition constant of 0 m. The vertical (depth) datum adopts the theoretical lowest astronomical tide. The water level control is based on the designed elevation of the landing points.

To meet the survey requirements of the waters surrounding the platform, a temporary benchmark station was established for this project. Temporary tidal gauge points were set up at each landing point, and manual tidal observations were conducted using the designed elevation of the landing points, with an accuracy of 1 cm. Measurements were taken and recorded every 10 min. Tidal observations began 10 min before each day's survey and ended 10 min after the bathymetric survey.

4) Detection results

The Echoscope 3D imaging system collects the echo signal of the surface pulse signal from the transmitter and generates 3D point cloud images in real time, and the dense point cloud data improves the resolution of the underwater structures. When the 3D scanning was carried out on platform 3 of the project, the scanning angle was 20°, and the scanning height of the underwater pile foundation was about 7 m. The 3D imaging results analyzed after the scanning on platform 3 are shown in Fig. 7.31, which show that there is a trench of 44 m in length and 14 m in width on the north side of the northwestern foundation pile, and the depth of the trench is 2.6 m at the deepest.

In order to further analyze the elevation distribution of the seabed near the pile foundation at the trench location, the software was used to generate the contour map at the target trench location of No. 3, see Fig. 7.32. The analysis of the trench location along the deepest part of the trench in Fig. 7.32 was carried out in section to extract the distribution of the seabed elevation along the section, see Fig. 7.33.

According to the results of Echoscope three-dimensional imaging system for the target of platform No. 3, the high-resolution three-dimensional point cloud generated by the sonar three-dimensional imaging system can clearly describe and display the details of the underwater structure, and through the software, it is possible to observe the landforms as well as the structure from any angle, cut the profile, and pick up the coordinates of any position point, which has a wide range of application prospects in the detection of the aerial ocean platform.

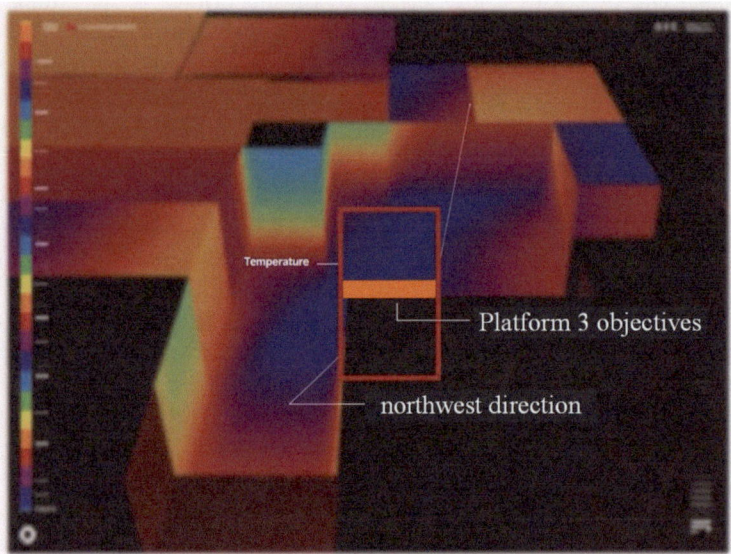

Fig. 7.31 Distribution of trench locations to the north-west of the platform 3 target

7.2 Application of Sonar Technology

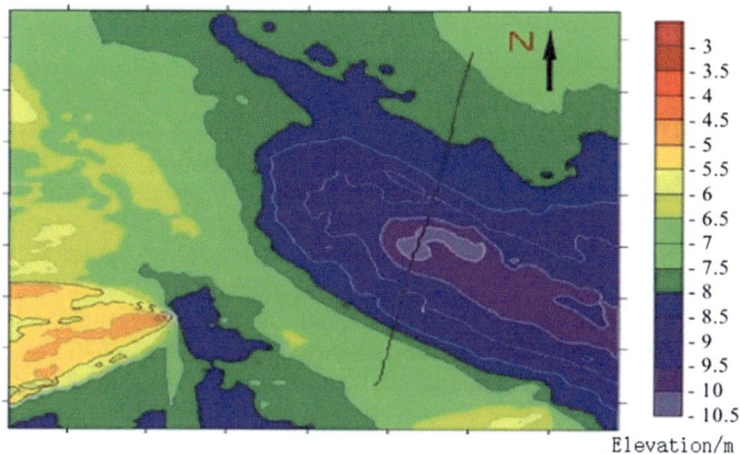

Fig. 7.32 Contour map at the target trench location of platform no. 3

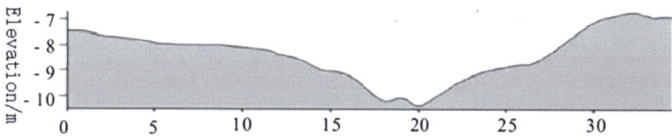

Fig. 7.33 Trench profile of platform no. 3

7.2.3 Bridge Underwater Structure Inspection

Bridge engineering is unanimously regarded as the lifeline project in transport engineering, and once the bridge collapses, it will cause significant losses to the traffic and social economy. Bridge structure inspection includes the inspection of bridge structure appearance, durability indicators, and load-bearing capacity. According to the results of the inspection, the technical condition, durability indicators and bearing capacity of the bridge will be evaluated, which will provide an important decision-making basis for the management and maintenance of the bridge.

Moreover, for some large-span or even super-large-span bridges crossing rivers, seas, and connecting islands, the main bridge foundations often adopt deep water foundations, with foundation depths sometimes exceeding 100 m. Deep water foundations mostly utilize pile foundations or deep-water caisson foundations, and the control of foundation construction quality during the construction process is crucial, such as the thickness of sediment at the bottom of pile foundations and the soil conditions within the caisson and around its cutting edge during its sinking process and when it is in place. Regarding the quality control of bridge deep water foundation construction, traditional methods are difficult to use for inspection, and sonar technology, due to its advantages in underwater measurement, is a promising measurement method.

1. Detection of topography of bridge deep-water foundation base

(1) Project Overview

Hutong Yangtze River Bridge is a control project of the new Hutong Railway. The bridge site is located 45 km downstream of the Jiangyin Yangtze River Highway Bridge and 40 km upstream of the Sutong Yangtze River Highway Bridge, with a total length of 11.072 km. It serves as a common corridor for the Hutong Railway, the Tongzhou-Suzhou-Jiaxing Intercity Railway, and the Xitong Highway. The upper level accommodates a six-lane highway, while the lower level has four railway tracks [12]. The main channel bridge of the Hutong Yangtze River Bridge is a double-tower continuous steel truss cable-stayed bridge with a span of (140 + 462 + 1092 + 462 + 140) meters, and its bridge layout is shown in Fig. 7.34. The main girder of the bridge adopts a three-main-truss structure, with diamond-shaped piers reaching a height of 325 m. The main piers from No. 26 to No. 31 all utilize open caisson foundations, with the upper part being reinforced concrete and the lower part being steel. Among them, piers 28 and 29 adopt rectangular open caisson foundations with rounded corners, with a top plan dimension of 86.9 m by 58.7 m, a rounding radius of 7.45 m, and a planar arrangement of 24 well holes, each measuring 12.8 m by 12.8 m [13]. The total heights of the open caissons for piers 28 and 29 are 105 m and 115 m respectively, with bottom elevations of − 97 m and − 107 m, as shown in Fig. 7.35 for pier 28.

(2) Methods of open caisson foundation inspection

The open caisson foundation of the main pier of the bridge has the following characteristics: the foundation base has a water entry depth from 100 to 110 m; the base area of the open caisson foundation is large, approximately 5100 m^2; there is a

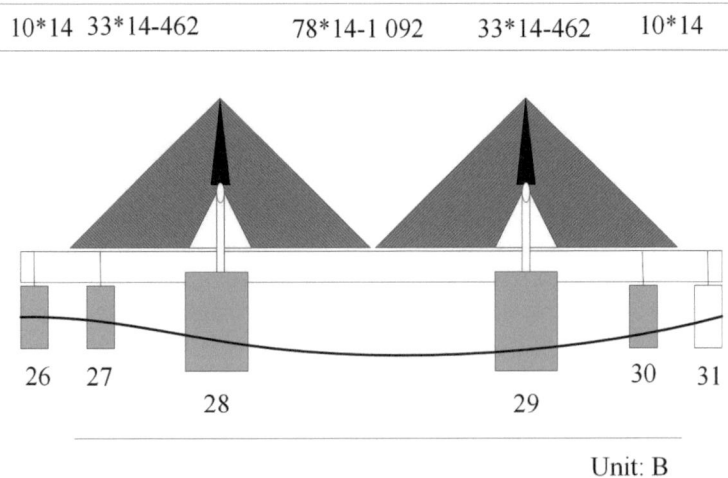

Fig. 7.34 Bridge arrangement of hutong Yangtze River bridge main channel bridge

7.2 Application of Sonar Technology

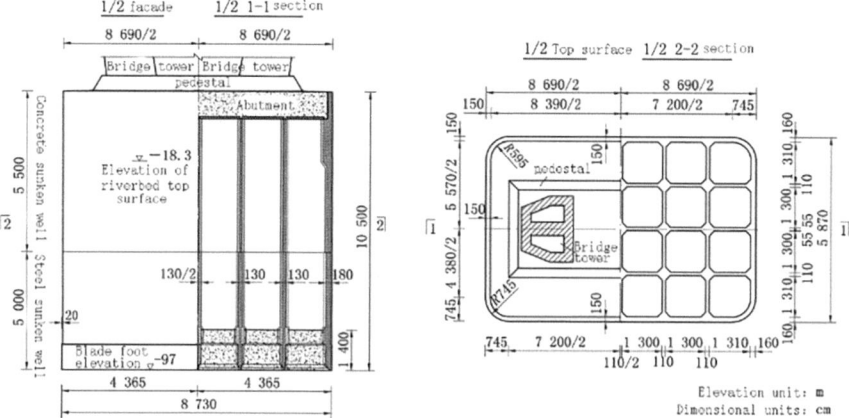

Fig. 7.35 Foundation structure of caisson of no. 28 main pier

significant height difference between the inner and outer walls of the open caisson, approximately 80 m; and the foundation is divided into 24 well holes, resulting in limited working space. Due to these characteristics, inspecting the open caisson is challenging, and conventional methods, such as those involving divers equipped with gear, large-scale mechanical equipment, and water-based inspection techniques, are difficult to implement.

Due to the higher requirements for the inspection of the open caisson foundation base set by the design of the Hutong Yangtze River Bridge, traditional foundation base inspection methods, which rely solely on qualitative analysis of inspection results and can only estimate measurement blind spots in the partition walls and cutting edges, cannot meet the required inspection accuracy. Instead, the "measuring rope survey combined with single-beam sonar detection verification" method [14] is adopted for the foundation base terrain. The measuring rope survey method involves lowering a measuring hammer with a rope from the water surface to the open caisson foundation. The surveyor relies on their experience to judge the moment when the measuring hammer begins to enter the base soil, measures the length of the rope that has been lowered, and determines the thickness of the base soil or sediment based on the corresponding elevation relationship. Since the foundation depth of this project is over 100 m underwater, the measuring rope survey method has a relatively large error, and thus a combination of the measuring rope and single-beam sonar method is used.

(3) Single-beam sonar inspection

The results of the measuring rope survey indicate that the top surface of the base soil layer is 1–3 m lower than the cutting edge, and the height difference within a single well hole is 0.7 m, indicating that the base is basically flat [14]. Based on the measuring rope survey, a single-beam sonar system is used for testing the base terrain. The single-beam sonar employed for the test is the SeaKing Hammerhead image sonar measurement system. During the inspection, the sonar emits acoustic waves and receives reflections from underwater objects, forming sonar images based on the different reflection speeds of the acoustic waves for base terrain verification. The No. 28 open caisson has a total of 24 well holes, with one sonar measurement point arranged in each well hole, totaling 24 measurement points. The single-beam sonar detection instrument is lowered to the bottom of the open caisson at the center of each well hole, and measurements are taken hole by hole to comprehensively grasp the terrain of the open caisson base, the burial depth of the cutting edge, and the void conditions of the partition walls.

In order to reflect the topography of the basement intuitively, the sonar is lifted up by 0.5, 1.0 and 1.5 m, and the results are analyzed below as an example of lifting up by 1 m. The single-beam sonar detection images of the basement and the upper 1 m of the basement are shown in Figs. 7.36 and 7.37. It can be seen from Fig. 7.36 that there is no soil in the middle of the borehole, the foot of the outer well wall is buried in the soil, and there is soil under the projection surface of the diaphragm wall, and the basement of each well borehole forms the form of a 'small pot', and the soil at the bottom of the foot of the outer well wall is higher than that under the diaphragm wall. From Fig. 7.37 it can be seen that the soil at the foot of the edge of the outer well wall is obviously reduced, and there is no soil under the diaphragm wall, in the borehole and in the middle of the entire sinkhole. The single-beam sonar detection results indicate that there is no large elevated mound of soil at the base, and the terrain is flat; the bottom of the diaphragm wall is not buried in the morass [14].

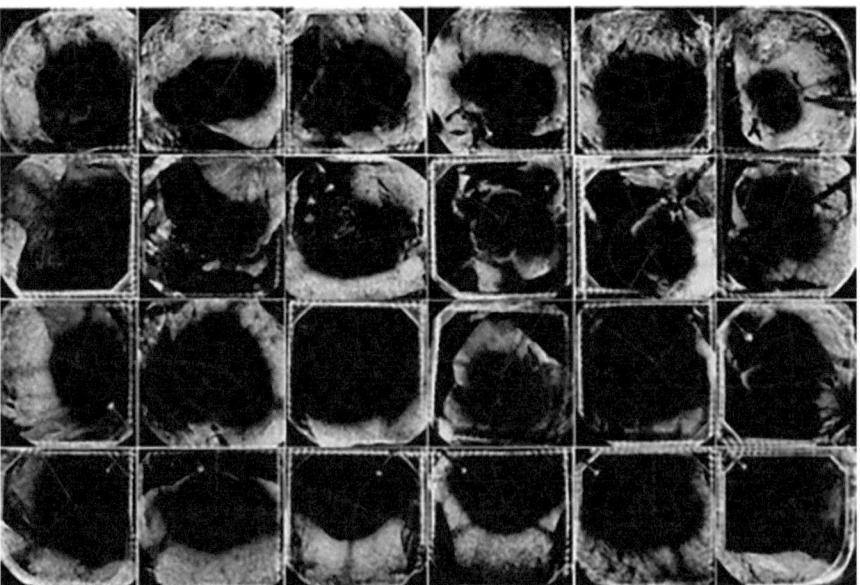

Fig. 7.36 Single-beam sonar imaging results of sinkhole foundation footing

7.2 Application of Sonar Technology

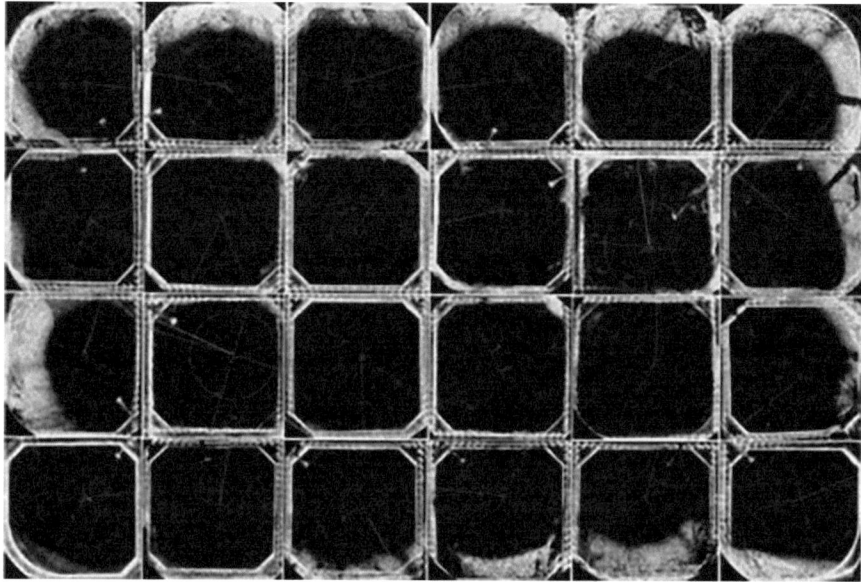

Fig. 7.37 Single-beam sonar imaging results at 1.0 m from the base of the foundation of the sinkhole

According to the sonar detection results combined with the results of rope measurement: the foundation of No.28 sinkhole meets the requirements of the design on the topography of the foundation, 'there should be no deep pits, steep canyons, and the entire surface of the foundation should be gentle and smooth'. The single-beam imaging map reflects the topography of the foundation intuitively, and the result can also be used as a reference basis for the burial depth of the outer rim of the sinkhole and whether the bottom of the sinkhole diaphragm wall is hollowed out or not.

Based on the sonar detection results combined with the measuring rope survey results, the foundation base of No. 28 open caisson meets the design requirements stating that "there should be no deep pits or steep scarps, and the entire base surface should be gentle and flat." The single-beam imaging map intuitively reflects the base terrain and the results can also provide a reference for the burial depth of the cutting edge of the outer well ring of the open caisson and whether there is voiding at the bottom of the partition walls of the open caisson.

2. Underwater foundation detection of existing bridge

The scouring and destruction of existing bridge foundation has been regarded as an important reason for bridge failure, and also a blind spot of underwater bridge detection in China. In the United States, bridge underwater detection began in the 1970s, and there are clear requirements for underwater inspections in the regulations. However, the current relevant specifications for highway bridge inspection in China do not stipulate the content of underwater inspection, and domestic underwater bridge inspection is currently in the exploratory stage. Some inspection units use divers or underwater robots equipped with video systems for underwater bridge inspection, but the effectiveness is not ideal due to factors such as turbid water quality and attachments on the structure's surface. With the development and application of sonar technology, especially the advancement of three-dimensional multi-beam real-time sonar imaging technology, relevant domestic units are also successively carrying out inspections of underwater bridge piers, foundations, and riverbed scouring.

The following is an introduction to the application of three-dimensional multi-beam real-time sonar imaging technology in the underwater foundation inspection of a railway bridge. The inspection utilizes the most advanced state-of-the-art 3DEchoscope three-dimensional multi-beam sonar system.

(1) 3D multi-beam sonar imaging system

3D Echoscope system is developed by the British CodaOctopus company, which is currently the most advanced international three-dimensional multi-beam real-time sonar observation system. The main hardware equipment of this system includes: the Echoscope multi-beam probe and control box, which updates "surface" true three-dimensional model data 12 times per second; the F180 system, equipped with GNSS-aided positioning and an Octans inertial attitude sensor, enabling precise measurements of position, heading, heave, and roll; the RTK-GPS system, which provides real-time corrections for centimeter-level positioning accuracy measurement data; a data acquisition computer power system and installation equipment, etc.

The software system, Underwater Survey Explorer, integrates data acquisition, processing and display, adopts the most advanced software development technology and patented mapping algorithms, which is capable of reproducing high-definition underwater scenes through three-dimensional point cloud data modes.

Compared to traditional multi-beam sonar detection, the working principle of the 3DEchoscope system upgrades the narrow "strip"-like cross-track detection signals and data stitching mode into a "surface"-like pulsed signal of a specific volume and performs data stitching. A single sonar signal pulse typically covers a surface area of $50° \times 50°$ (Fig. 7.38). The F180 system can accurately measure the position, heading, heave, pitch, and roll of the detection vessel, record the vessel's track and sway attitude information in real-time, and make real-time corrections to the depth and position information of the targets detected by each sonar signal. Meanwhile, based on the high-precision positioning data corrections from the RTK-GPS system, the detection data can be accurately stitched together to reflect the complete underwater environment of the survey area.

Therefore, the 3D Echoscope system can effectively avoid the data splicing blind area or error under the high water current operation environment, due to the lack of signal or distortion in the strip area caused by the sudden large-scale directional deviation and tilting of the vessel.

3D Echoscope system adopts a single high density of bathymetry data with the help of high-speed data transmission and processing, so that each frequency detected by the measured object echo can generate real-time three-dimensional point cloud image, with the acoustic data transmission update, three-dimensional image can be

7.2 Application of Sonar Technology

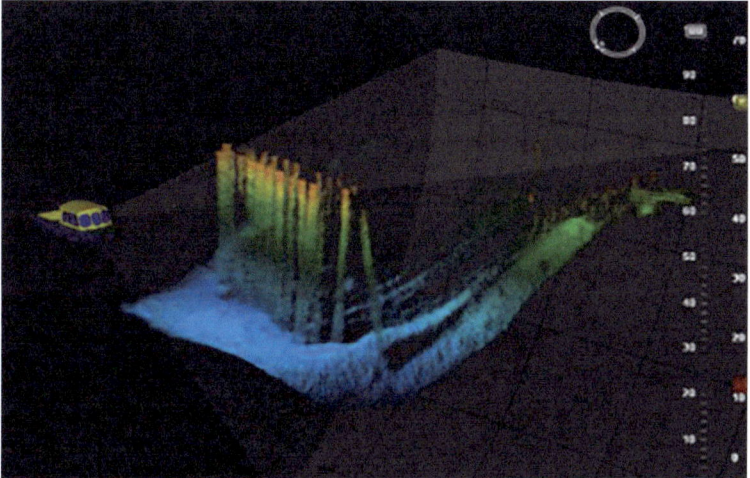

Fig. 7.38 Surface test mode of 3D multi-beam real-time sonar system

updated 12 times per second. Through 3D tracking and imaging, real-time monitoring is provided for underwater operations, enabling large-area scanning of the underwater environment. The display of images at various frequencies does not overlap but achieves instantaneous stitching, visualizing the entire scene in real-time. Additionally, when detecting underwater targets, all multi-beam points can be focused on the target according to detection requirements. The system automatically selects the optimal acoustic frequency and field of view angle for the sonar pulse based on the size of the target and its distance from the probe, while maintaining a constant number of beams, thereby enhancing the resolution and recognition capabilities for the target.

(2) Project Overview

A railway bridge, with a length of 351.74 m, was completed and opened to traffic in 1959. The river at the bridge site is narrow. At normal water levels, the river width is approximately 260 m and the water depth is about 10 m. The riverbed is covered with a 4 to 7.3 m thick layer of sand mixed with pebbles, underlying which is gray sandstone. During high water levels, the flow velocity reaches 4 m/s, causing severe scouring of the riverbed. The bridge has a total of 10 piers and abutments. The caisson foundations of the originally designed piers No. 5, No. 6, and No. 7 were changed to pile-column foundations. These pile-column foundations, located in the main river channel from the left bank to the right bank, involve sinking pile columns into the pebble layer. The pile columns have a diameter of 1.55 m, with each pier equipped with 9 pile columns ranging in length from 3 to 9 m, all embedded below the sandstone layer [15].

300   7 Application of Acoustic Technology in Marine Engineering Inspection

(3) Results of Bridge foundation inspection

This bridge adopts 3D Echoscope system to detect the foundation of Pier 5#, Pier 6#, Pier 7# and the nearby riverbed scouring respectively, and uses special analysis software to process and analyze the data to generate 3D point cloud images. The three-dimensional point cloud images of three pier foundations and riverbed scouring were inspected, and the exposure of pipe columns at localized positions of each pier are shown in Figs. 7.39, 7.40 and 7.41 [15].

(1) Analysis of the exposed situation of pipe columns

Through the analysis of the three-dimensional sonar image of each pier, we can find that the form of the pier and cofferdam is clearly visible, and the riverbed around the foundation of Piers No. 5, No. 6, and No. 7 have erosion phenomenon, with the

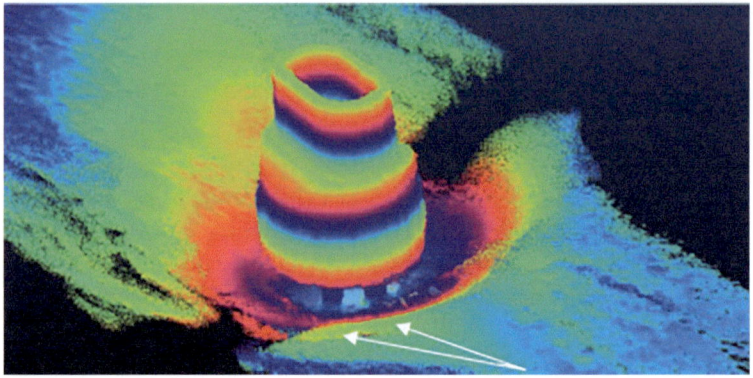

Fig. 7.39 3D sonar point cloud on the upstream side of Pier 5# foundation

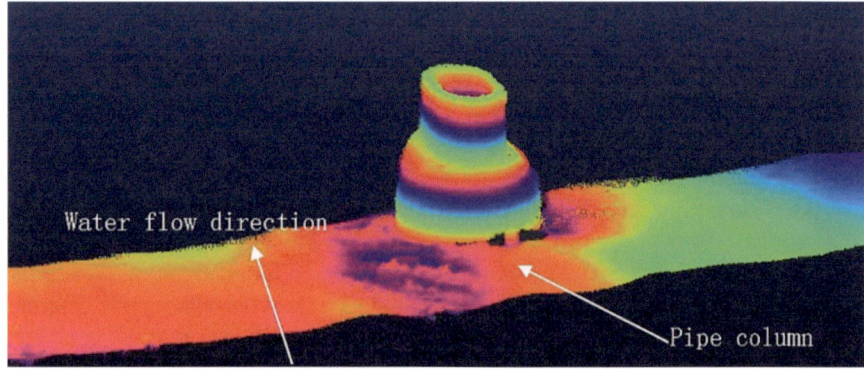

Fig. 7.40 3D sonar point cloud of the left side of Pier 6# foundation

7.2 Application of Sonar Technology

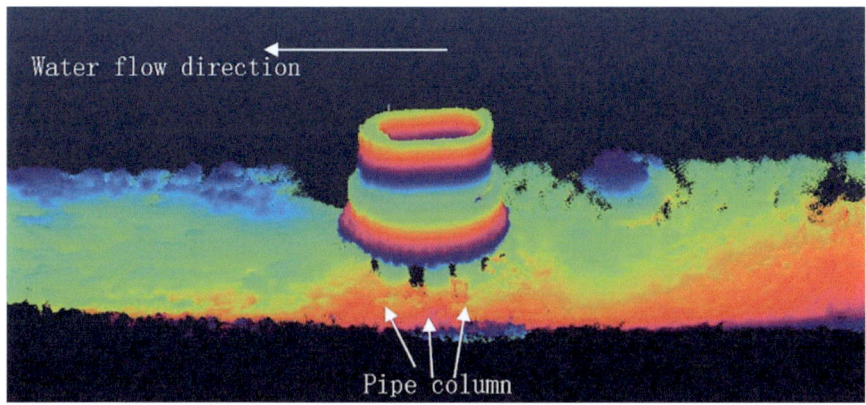

Fig. 7.41 3D sonar point cloud map of the left side of Pier 7# foundation

depth of erosion ranging from 3 to 5 m. The terrain around the piers was eroded, forming scour holes that were overall elliptical in shape along the river flow, with the upstream side having relatively deeper scour holes. Specifically, there were 4 exposed pipe columns on the riverbed at pier No. 5, 2 at pier No. 6, and 5 at pier No. 7. The exposed pipe columns were hollowed out, forming cavities between them. The distribution of the outer pipe columns of each pier is shown in Fig. 7.42.

(2) Analysis of riverbed scouring

Based on the results of the three-dimensional sonar imaging, the elevations of the riverbed cross-sections were extracted to create a longitudinal riverbed profile along the bridge. The profile was plotted along the line connecting the centers of piers No. 5, No. 6, and No. 7, as shown in Fig. 7.43.

Comparing the riverbed topography of the sonar profile with the riverbed topography and bedrock surface of the original design, it was found that the riverbed cover was mainly scoured most seriously around the bridge piers. Specifically, the cover layer around pier No. 7 was eroded to expose the bedrock surface. There was some erosion in the riverbed cover layer between piers No. 6 and No. 7, with a maximum erosion depth of approximately 1.7 m. On the other hand, there was a small amount

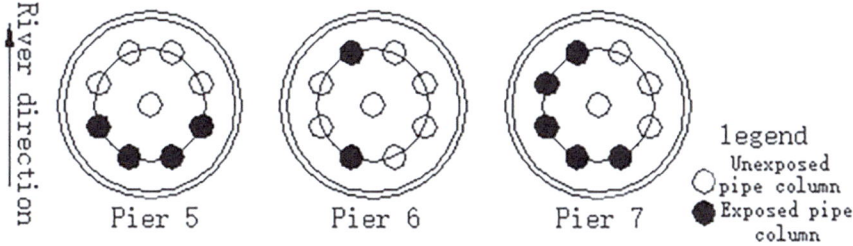

Fig. 7.42 Schematic plan distribution of exposed pipe columns at each bridge pier

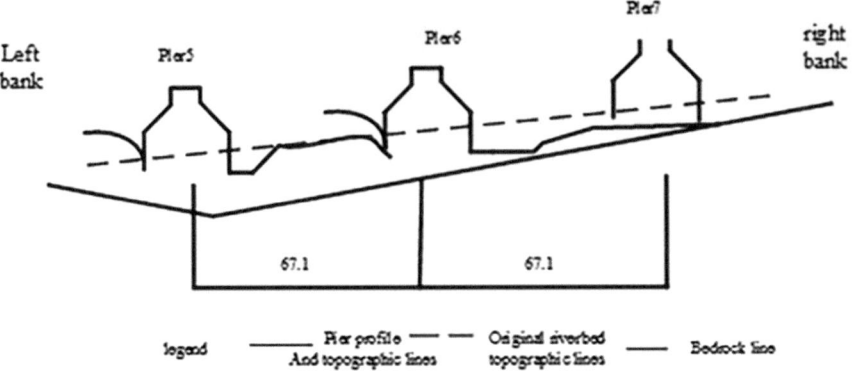

Fig. 7.43 Comparison between the present topographic line and the original riverbed topographic line

of deposition near the left bank side of the riverbed between piers No. 5 and No. 6, with a deposition thickness of approximately 0.4 m.

References

1. Li, Huajun, and Hezhen Yang. 2004. Research Progress in Structural Parameter Identification and Damage Diagnosis Technology for Offshore Platforms. *Engineering Mechanics* 21(Supplement 1): 113–117.
2. Liu, Guijie, Meng Xu, Sile Li, et al. 2013. AE Signal Recognition for Fatigue Crack Growth in Tubular Joints of Offshore Platforms Based on Wavelet Energy Coefficients. *Nondestructive Testing 35*(2): 1–7.
3. Li, Dongsheng, Wei Yang, and Yan Yu. 2017. *Acoustic Emission Monitoring and Evaluation of Civil Engineering Structure Damage—Theory, Methods, and Applications.* Beijing: Science Press.
4. Yang, Shuying, and Hua Zhang. 2015. *Pattern Recognition and Intelligent Computing—MATLAB Technical Implementation.* 3rd ed. Beijing: Electronics Industry Press.
5. Yang, Qiming, and E. Yang. 2013. Sensitivity Analysis of Vortex-Induced Vibration of Subsea Pipeline Spans. *Oil & Gas Storage and Transportation* (1): 8–11.
6. Ronghao, Wei, Chen Tiexin, and Guo Chen. 2014. Application of Side-Scan Sonar in the Survey of Suspended Subsea Pipelines. *Marine Surveying and Mapping* 34 (2): 63–65.
7. Wang, Lei, Xingping Xu, Xin Zhang, et al. 2013. Side-Scan Sonar Detection Method for Suspended Subsea Pipelines. *China Petroleum Machinery* 5(41): 50–52.
8. Lei, Tian. 2015. *Study on the Influence of Side-Scan Sonar Grazing Angle on Subsea Pipeline Detection.* Tianjin: Tianjin University.
9. Chunbao, Xiong, and Yin Xiaodong. 2010. *Surveying.* Tianjin: Tianjin University Press.
10. Li, Ping, and Jun Du. 2011. Overview of Shallow Stratigraphic Profile Detection. *Marine Science Bulletin* 30(3): 343–350.
11. Jiandi, Ding. 2016. *Research on Acoustic Detection Methods and Three-Dimensional Visualization Systems for Subsea Oil Pipelines.* Tianjin: Tianjin University.
12. Juntang, Li. 2015. Key Construction Technologies for the Open Caisson of the Main Channel Bridge of Hutong Yangtze River Bridge. *Bridge Construction* 45 (6): 9–17.

13. Chen, Tao. 2015. Analysis of the Floating Resistance of Steel Caisson for Pier No. 28 of Hutong Yangtze River Bridge. *World Bridges* 43(6): 79–82.
14. Zhang, Guizhong, and Xiaogui Ma. 2016. Underwater Detection Technology for Ultra-Deep Foundations of Giant Caissons of Hutong Yangtze River Bridge. *Bridge Construction* 46(6): 7–12.
15. Yang, Jianming, Yin Feng, Hongliang Sun, et al. 2017. Application of Real-Time 3D Multi-Beam Sonar System in Bridge Foundation Erosion Detection. In *Proceedings of the 15th National Academic Conference on Engineering Geophysical Exploration and Rock Engineering Testing*, 365–371.

Part III
Optoelectronics

Chapter 8
Overview of Optoelectronic Technology

8.1 Development History of Optoelectronic Technology

Photoelectric detection technology is based on the characteristics of the light waves radiated or reflected by the detected object to detect and identify the object of a kind of technology, this technology itself gives photoelectric technology in the military application of the four major advantages, that is, see more clearly, hit more accurately, faster response and survival ability.

Photoelectric detection technology is the core technology widely used in modern warfare, which includes photoelectric reconnaissance, night vision, navigation, guidance, homing, search, tracking and identification of a variety of functions. Optoelectronic detection includes the detection of light signals from ultraviolet light (0.2–0.4 μm), visible light (0.4–0.7 μm), infrared light (1–3 μm, 3–5 μm, 8–12 μm) and other bands.

The new generation of optoelectronic detection technology and its intelligence will enable the relevant weapons to obtain a longer range of action, stronger single target/multi-target detection and identification capabilities, so as to achieve more accurate strikes and rapid response, and to obtain the initiative of the war in the case of very small casualties. At the same time, the weaponry has a strong autonomous decision-making capability, which enhances the confrontation, anti-resistance and its own survivability. In fact, advanced photoelectric detection technology has become an important symbol of a country's military strength.

The distinguishing feature of modern high-tech war is first of all information warfare, and the primary task in information warfare is how to obtain information. Whoever gets more information, whoever gets the earliest information, who will grasp the initiative of information war. Optoelectronic detection is an important means of obtaining information. Microwave radar and photoelectric imaging equipment are often used together, complement each other's strengths and weaknesses, complement each other, can obtain more information, can obtain information earlier. The former has a long range of action and can work in all-weather conditions;

the latter has a high resolution and a strong ability to identify and resist interference. Reconnaissance satellites, early-warning satellites, early-warning aircraft and unmanned reconnaissance aircraft are often equipped with synthetic aperture radar and CCD cameras, infrared thermal imaging cameras or multispectral cameras at the same time. In order to improve the early warning capability of ballistic missiles, the space-based infrared system (SBIRS) being developed by the United States is proposed to use a dual-sensor programme, that is, a wide-field-of-view scanning short-wave infrared capture sensor and a narrow-field-of-view staring multicolour (mid-wave/long-wave infrared, long-wave infrared/visible) tracking sensor, which is capable of capturing and tracking ballistic missiles from the time of their launch to the time they are re-entered into the atmosphere. The CR-135S Cobra Sphere early warning aircraft, which has been equipped by the United States and is undergoing continuous improvement, uses visible light and mid-wave infrared cameras to accurately determine the launch of a missile at a distance of 420 km, determine the point at which the engine goes out, and calculate its trajectory and point of collision. Recently, a long-range laser range finder has been added to it, and its range is up to 400 km. The United States Navy is also developing an airborne optoelectronic sensor system for the theatre ballistic missile defence system, known as the "Door Police" system, which can carry out active/passive surveillance. It includes an infrared search and tracker (IRST), using a dual-band 6×960 elemental CdTe detector array, with a detection range of up to 800 km, and a range/tracker (LR/T), which tracks the target with a 128×128 elemental indium antimonide focal-plane array (about 5 μrad) and ranges the target with a laser (100–1000 km), so as to obtain real-time three-dimensional information about the target in the long range, and to win enough time to monitor the target. Real-time three-dimensional information of long-distance targets, and win enough time for early warning.

In other applications of optoelectronic technology, such as precision guidance, navigation, fire control, counter-weapons, communications, display and other aspects of a more important position.

8.1.1 Visible Light Detection

Visible light CCD and CMOS imagers due to its small size, light weight, low power consumption, long life, reliable and shock-resistant and many other features, is now widely used in military remote sensing, reconnaissance, aircraft navigation, missile and bomb guidance and other modern military equipment. Civilian is also extremely wide, such as security, monitoring, visual doorbell, video e-mail, video phones, video conferencing, digital cameras, and medical and biological sciences experiments recorded in the use of CCD and CMOS imagers.

Modern visible light imagers are digital and can be saved on floppy disks, hard disks and CD-ROMs and then read, displayed and printed out on computers. Such images can also be patched, cut and pasted and transmitted over long distances, which is one of the main elements of modern communications.

8.1 Development History of Optoelectronic Technology

The basic indicators of advanced image sensors are clarity (number of photosensitive elements), sensitivity (quantum efficiency), dynamic range (number of full-well charges), signal-to-noise ratio (dark current and other noise sources), etc., and often encountered with practical optical aperture, drag, halo, scintillation, image hysteresis, and other image properties related to the modern advanced technology is therefore to further improve these basic indicators and improve the image properties mentioned above. Efforts.

8.1.2 Infrared Detection

Since any object with a temperature above absolute zero radiates infrared light, the use of appropriate detectors sensitive enough to infrared light can detect the presence of an object even in the absence of light at night, and also obtain an image of its shape. The temperatures and peak wavelengths of radiation of some typical objects are shown in Table 8.1.

It can be seen that most of the infrared radiation from objects encountered in the war is between 1 and 12 μm.

However, in this band area of the signal is not always able to propagate far in the atmosphere, practice shows that there are only three band area of the signal can be propagated farther in the atmosphere, they are called: short-wave infrared (SWIR, 1–3 μm), medium-wave infrared (MWIR, 3–5 μm) and long-wave infrared (LWIR, 8–12 μm). Usually said military infrared technology, mainly for these three infrared bands, and also focus on the mid-wave and long-wave infrared.

For infrared detection equipment is the core of the infrared detector, in a sense the level of infrared detector determines the performance of infrared photoelectric detection equipment, the international general unit and multi-component known as the first generation of infrared devices, the focal plane lines and arrays known as the second generation of devices, dual (multi) band and intelligent focal plane devices known as the third generation of devices, and accordingly evolved into the sub-generation of infrared photoelectric detection equipment.

1. PtSi infrared detector

Table 8.1 Temperature and wavelength of infrared radiation for typical objects

Substance Name	Temperature (K)	Peak wavelength of radiation (um)
Tungsten light	2000	1.45
Boeing 707 aircraft nozzle	890	3.62
M-46 tank tail	473	6.13
F-16 Fighter Skin	333	8.70
Human body	310	9.66
Ice water (0 degrees)	273	10.6

This is the early infrared detector, working in the medium and short-wave, due to its relatively simple manufacturing process, the original homogeneity to do a good job, the cost is relatively inexpensive, and thus gained early military applications, such as early rattlesnake missiles using infrared guidance is the use of PtSi detectors, but due to its low quantum efficiency, low performance, affecting the performance of the weaponry to play, and was followed by InSb, HgCdTe detectors.

2. InSb infrared detector

InSb working band in the mid-wave is currently the most widely used, the most mature research, the military used to find the head of the often take 128×128 yuan gaze-type array, because there is a better performance/price ratio, the United States, Britain, Germany and Israel and other countries to develop a new type of air—air missiles are used in this specification. Requirements for precision, high-speed images or in high-value occasions often take 256×256, 640×480 or 512×512 InSb detector, the United States LockheedMartim production of "sniper" pods, Raytheon developed ATFLIR pods. NorthropGrumman and Israel Rafael company developed the LITENING pod and the United States of America forward-looking infrared systems company developed the AN/AAQ-22SAFIRE thermal imaging camera and the world's most advanced forward-looking, navigation and targeting equipment are used in the 640×480 yuan or similar size of the InSb arrays. 2000×2000 InSb and visible light combined with a low frame rate. 2000×2000 InSb combined with visible light to form low-frame-frequency, dual-colour cameras have been reported for battlefield and environmental surveillance.

3. HgCdTe infrared detectors

Due to the invention of HgCdTe infrared detectors, infrared detection of low-temperature targets (long-wave detection) is possible. From the principle can replace the first two types of infrared detectors, and thus this type of detector is the advanced countries in the West competing for the development of a class of detectors to the present is still the focus of the development of detectors, and focus on the second and third generation of infrared detectors, such detectors are divided into $4 (6) \times N$ line focal plane and array focal plane, the former is mainly the technology is more mature relative to the latter, and the use of parallel scanning technology can be achieved with the same number of elements of the number of array focal planes, more high performance, and the price is higher than the array focal planes. Plane higher performance, and the price is lower than the array focal plane, and thus the Western countries are also in the development of the list, the typical 4×288, $4 (6) \times 576$, 6×960, etc.; array focal plane typical varieties of 128×128, 256×256 (or 320×256), 512×512 (or 640×480), 1024×1024 and so on.

4. GaAs/AlGaAs quantum well infrared detector (QWIP)

Quantum well infrared focal plane detector in the 384×288, 640×512 scale above the large surface array and two-colour focal plane has application value, is currently mainly used in industrial and medical applications. There are also applications in the military field that allow long time integration, such as the German

8.1 Development History of Optoelectronic Technology

tank driver observation thermal imaging camera, using the 640 × 512 yuan long-wave quantum well infrared focal plane detector, the spectral response range of 8–9 μm. quantum well infrared focal plane detector is the most promising direction of the development of the future military applications in space, such as multicolour (4-colour), ultra-long-wavelength (14–16 μm) large surface arrays.

5. Non-cooled infrared focal plane detectors

Due to the cooling infrared focal plane detector power consumption, high cost, inconvenient operation and other unfavourable factors, countries are seeking to manufacture non-cooled infrared focal plane of the new technology, the current development of the technology is rapid. Non-cooled infrared focal plane detector has been developed to 320 × 240, 640 × 480 scale. Uncooled infrared detectors can be divided into two categories, namely, ferroelectric type and thermistor type. Vox and X-Si thermistor type non-cooling infrared detector development faster, the main technical focus is to improve the size of the array to reach 640 × 480 yuan order of magnitude, reduce the size of the photosensitive element to reach the centre distance of 25 μm, and further improve the noise-equivalent temperature difference, dynamic range, and other important performance indicators, and to further reduce the cost and ease of use and so on. Mainly used in industry, security, man-portable weapon observation and other fields, its price is relatively inexpensive, but also for performance requirements are not too high short-range missile guidance.

6. Third-generation infrared detectors

The United States has begun to develop the third generation of infrared detectors, and put forward the concept of the third generation of infrared cameras, mainly two-colour or three-colour high-performance, high-resolution, refrigeration-type thermal imaging cameras and intelligent focal plane array detectors, the U.S. Army Night Vision and Electronic Sensors Directorate that the development of the third-generation infrared sensors is the U.S. to maintain the advantages of the key to the night war. Therefore, the longer-term development trend of infrared detection technology is the development of third-generation infrared detectors, third-generation infrared focal plane detector main parameters are shown in Table 8.2.

Due to the continuous improvement of foreign infrared detector technology, it has become quite difficult to upgrade the technology on the detector chip. In order to further improve the performance of infrared detectors, people are now turning their attention to the infrared detector signal readout integrated circuit (ROIC). With the development of computer technology and integrated circuits, ROIC has made great progress, medium-scale infrared focal plane arrays and the corresponding readout circuit in the 1990s has formed a production scale, and now the advanced countries are being developed for large-scale focal plane arrays (three generations of devices) and its multiple functions of ROIC and intelligent focal plane arrays.

Intelligent focal plane array is also called the system on a chip, on-chip signal processing means in the photosensitive element on a chip or the region closest to the photosensitive element to mimic the function of the vertebrate retina, optical—electrical conversion of signals for the pre-processing, and then output the subsequent

Table 8.2 Third-generation infrared detectors require performance parameters

Focal plane array size	1000 × 1000, 1000 × 2000, 2000 × 2000
Photosensitive element area	18 × 18 μm²
Response Band (Wave)	At least two frequency bands
NETD	< 1 mkf/z (medium wave), < 5 mkf/z (long wave)
Dewar	High vacuum
Cooler	Mechanical or thermoelectric cooled 120–180 K
Target	Long range, strong clutter resistance
Spatial inhomogeneity	< 0.5 NETD
Electronic capacity	109
Frame rate	Can be added up to 480 Hz
Additional Functions	Non-uniformity correction and A/P circuits etc. included on-chip

data processing. Although this process does not belong to the process of receiving light signals directly, but the comprehensive performance of the photodetector has a great impact.

7. Ultraviolet detection technology

Ultraviolet detection technology in the field of national defence, national economy and scientific research has many applications, such as missile threat warning, interstellar communications, chemical and biological warfare agents detection and spectral measurement, engine and nuclear reactor monitoring, plant growth, radiation dosimetry, water purification, pollution monitoring (e.g., ozone) flame detection, gas (generator) furnace monitoring and ultraviolet astronomy.

In the last 20 years, three main types of ultraviolet (UV) detectors have been developed, namely photomultiplier tubes, imaging UV sensors, and AlGaN/GaN photodiode imaging arrays, referred to as first-, second-, and third-generation UV imaging sensors.

In many of these applications, it is desirable to detect only UV light and not be sensitive to visible and infrared radiation, especially sunlight, in order to reduce false detections and background flux. Therefore, in recent years in the field of short-wave ultraviolet detector research focused on the realization of "sun-blind" detector, that is, less than 280 nm above the photon insensitive detector, also known as the third generation of ultraviolet detectors.

Military applications of ultraviolet detection include missile guidance, incoming missile warning, detection of biological and chemical warfare agents, military meteorology and military short-range communications.

8.1.3 Microminiature Imaging Sensors

The main characteristics of micro-miniature imaging sensors are their significantly reduced size, high spatial resolution and low power consumption. The main application areas of microminiature imaging sensors are, for example, robotics, micro-vehicles, micro-aircraft, micro-spacecraft, unattended sensors and surveillance networks, and policing and law enforcement. Its application prospects are:

1. networked sensing measurement points in support of Network Centric Warfare (NCW);
2. portable surveillance and unattended networked surveillance for battlefield intelligence;
3. video control of UAVs and unmanned ground vehicles;
4. automated surveillance for facility policing;
5. target identification for dexterous weapon;
6. precision targeting systems for armoured vehicles and missiles;
7. robotic vision;
8. public security, border patrol, law enforcement and traffic surveillance.

The overall goal of videoconferencing over the next few years in terms of miniature sensors is to:

1. Demonstrate small-scale integrated networks of acoustic, seismic and infrared imaging miniature sensors;
2. Validate ultra-lightweight, low-cost, small-volume 3D integrated packages applied to compact designs with affordable overheads;
3. develop self-assembling networks to support classified communications of processed information, with tactical communications up to 10 km between the networked area of the micro-sensors and the warfighter.

8.1.4 Imaging Polarisation Detection

The purpose of imaging polarisation detection techniques is to improve the contrast of the target, suppress background clutter, provide information about the surface material of the target, and distinguish between natural and man-made objects. Therefore, imaging polarisation detection can improve the detection and identification of targets, and has the potential to detect camouflaged targets in isothermal objects.

At present, there are short-wave, medium-wave, long-wave and multi-spectral, hyper-spectral imaging polarisation detection test, the main test is mine detection, cloud measurement, and the Earth's ocean surface sea surface temperature, emissivity and sea wind vector detection, it should be said that it is still in the early stages of research.

8.1.5 Multi-spectral/Hyperspectral Imaging Technology

Optical remote sensing without condensation is an effective method for collecting target and background data. However, due to the limited optical spatial resolution, spatial information based on radiant intensity does not always provide sufficient target information, e.g., small targets at long distances or targets hidden under brighter background disturbances cannot be discriminated based on their radiant intensity characteristics alone. Therefore, multidimensional discriminative methods such as spectral, polarisation and temporal properties are used in remote sensing to identify targets and backgrounds and are becoming increasingly important. Spectral imaging is researched and developed under this concept, and spectral imaging technology can be roughly divided into three categories according to the number of bands and resolution: multispectral imaging, with 10–50 bands and a spectral resolution ($\Delta\lambda/\lambda$) of 0.1; hyperspectral imaging, with 50–1000 bands and a spectral resolution of 0.01; and polar spectral imaging with 7–100 bands and a spectral resolution of 0.001. At present, in addition to the spectral imaging, the spectral imaging technology can be used to identify the target and background. At present, in addition to polar spectral imaging technology is not used in military remote sensing, a variety of multi-spectral or hyperspectral imaging system has been equipped with remote sensing satellites, such as "IKONOS 2" (IKONOS) satellites and reconnaissance aircraft such as u-2 high-altitude reconnaissance aircraft, etc., the key civil applications are environmental monitoring and resource management.

Analysis of multispectral/hyperspectral imaging data has shown that the value of this unique data lies not in its ability to produce beautiful images, but in the information inherent in the unique spectral features obtained by multispectral/hyperspectral imagers, such as information on targets such as vehicles hidden under trees and buried mines.

The most used focal plane arrays for multispectral imagers are visible CCDs and infrared HgCdTe focal plane arrays, and in the future the mainstream of its development trend will remain CCDs and multi-colour infrared focal plane arrays. The technical characteristics of its development are: to improve the spectral resolution as much as possible; to make full use of all types of electromagnetic spectrum through the atmosphere; to the infrared, far-infrared and microwave expansion; will be divided into finer spectral bands. For example, the United States Landsat thematic mapper has seven spectral bands; the AVIRIS airborne visible and infrared multispectral imager is divided into 224 bands within the visible and infrared spectral bands; and China's airborne spectral imager has 72 bands, of which 32 are in the visible, 32 are in the short-wave infrared, and 8 are in the long-wave infrared.

Therefore, in the future, remote sensing technology will be towards the fusion of multi-spectral/hyperspectral imagers with multi-sensors, such as interferometric radar, passive radar and synthetic aperture radar, to allow for the simultaneous acquisition of multi-dimensional data in a sensor system that can, through advanced data fusion technology, obtain sufficient target information as needed, so that

8.1 Development History of Optoelectronic Technology   315

remote sensing technology will develop towards the goal of multi-scale, multi-band, all-weather, high-precision, high-efficiency and rapid development.

8.1.6 LIDAR Imaging Technology

Identification, classification, precision detection and accurate targeting of military targets are the goals pursued by LiDAR. LIDAR has become a high-sensitivity detection radar focusing on development due to its advantages such as anti-jamming and strong imaging capability.

Automatic Target Recognition (ATR) is required in many image processing, thus promoting the development of LiDAR. For example, in the detection of stationary targets in a terrain background, Doppler radar and visible or infrared thermal imaging systems have their difficult side, while LIDAR has the advantage of both high angular resolution for each image element and accurate distance data, with stable target and background characteristics, and thus can be accurately modelled in the ATR system. Of course, in some applications, LIDAR needs to work with infrared, visible, and millimetre-wave radar due to its narrow beam and limited scanning speed, which in turn improves system performance through data fusion. Laser imaging technology is currently the main scanning imaging, laser-illuminated distance-selective imaging, laser-illuminated single-shot imaging and coherent LiDAR.

Lidar and radio radar are the same in principle, the difference is the use of optical band laser transmitter and laser receiver adapted to it. Early use of CO_2 lasers for LIDAR transmitter technology has been quite mature, and has developed a variety of land-based and airborne prototypes, but due to the large size of CO_2 lasers, optical aperture, the detector needs to be cooled, and other factors constrain its mobile environment, especially the competitiveness of airborne tactical applications. With advances in laser diode pumping technology (DPSSL) and new solid-state laser material research, efficient, all-solid-state and human eye-safe compact solid-state LIDARs are being developed, and have been experimentally used in areas such as aberrant Doppler LIDAR, range imaging and obstacle avoidance. Semiconductor lasers have historically had great potential for LIDAR applications due to their small size, light weight, ruggedness, reliability, high efficiency and low cost, and in recent years the rapid development of high-power, small-beam-angle laser diodes, which are typically used for helicopter obstacle avoidance and ground-footprint detection, with the greatest application advantages before them in robotic vision systems and laser underwater target imaging detection.

It is precisely the development of focal plane array detector arrays suitable for detecting lasers has become the key to the development of floodlighting single-shot imaging.

8.1.7 Multi-sensor Data Fusion Technology

Modern detection technology are to the direction of multi-sensor fusion efforts to make up for some of the shortcomings of a single detection technology, so that the detected target information is as rich as possible, accurate, rapid, real-time, so that wartime to master the priority of information, the initiative, to win the valuable pre-emptive time, so as to win the war. Because of the multi-sensor fusion will inevitably use data fusion technology; in the current due to the emergence of new advanced sensors and advanced processing technology, as well as hardware and software improvements, so that real-time data fusion is more and more likely to be realised and get very fast development.

A single platform equipped with sensor types may include: radar, laser rangefinder/target indicator/tracker, forward-looking infrared system, TV (including laser TV), enemy identification, radar warning machine, missile approach warning receiver, laser warning receiver and other different types of fusion between sensors.

Multiple platforms equipped with different types of sensors can significantly expand the air, frequency and time domains of sensor detection through the use of increasingly sophisticated data-link technology.

8.2 Development Status of Foreign Airborne Photoelectric Equipment

Airborne ground-based photoelectric detection equipment is divided into three types according to the installation method: buried (photoelectric radar), semi-buried (photoelectric turret form) and external (pod). Comparatively speaking, buried small size, light weight, does not affect the aerodynamic performance of the flight, the disadvantage is the shape of the structure, affecting the photoelectric detection field of view, the search of airspace is small, can only be observed locally; external pod form, easy to mount different aircraft, the use of mobile and flexible, and the field of view, the search range, the disadvantage of occupying the hanging point; semi-buried turret can be equipped with navigation pods and targeting pods and other functions, the field of view, Search range, the disadvantage is that the aerodynamic resistance is large, can only be equipped with subsonic aircraft, used in subsonic conditions. In the past 10 years, Russia, the United States, Europe and other military powers continue to increase the investment in airborne photoelectric detection equipment, has formed a series of equipment spectrum. The following will introduce several typical foreign airborne photoelectric detection equipment.

Fig. 8.1 Sniper optical targeting pods

8.2.1 Sniper Optical Targeting Pods

Sniper optoelectronic targeting pod is the only targeting pod equipped by USAF for all ground attack fighters and bombers, as shown in Fig. 8.1. Sensor and functionality upgrades have been made to the USAF-defined ATP-SE status, which allows the pod to be mounted on six types of USAF fighter and bomber platforms (F-15, F-16, and F/A-18 fighters, A-10 attack aircraft, B-1 bombers, and B-52 bombers).

The Sniper targeting pod is 300 mm in diameter and adopts a common aperture optical system design, which is the smallest multi-functional pod in terms of weight and volume among the representatives of foreign advanced targeting pods. The head adopts wedge-shaped sapphire splicing light window, the internal configuration of 640×512 yuan indium antimonide focal plane medium-wave gaze detector, the use of micro-scanning, through the 4 times continuous electronic zoom, the formation of $4°$ and $1°$ dual-field of view, 649×494 resolution CCD TV, infrared indicator can be compatible with the night-vision goggles; the laser adopts the dual-band of 1.06 m and 1.57 m, respectively, for combat and training use and the installation of a laser spot tracker. Installed with laser spot tracker; Sniper targeting pod installed with inertial measurement device to achieve automatic target calibration, and can guide GPS-guided weapons; with graphic data link to achieve reconnaissance and collaboration, and digital video recording function.

8.2.2 Litening Optical Targeting Pods (LITENING)

Litening photoelectric targeting pod series models include: LiteningII, LiteningIII, LiteningER, LiteningAT and LiteningG4, mounted on the F-15, F-16, F/A-18 and other aircraft types.

The original LiteningII electro-optical targeting pod was an improvement on the LANTIRN, a dual pod system with two pods, a navigation pod and a target indicator pod. The LiteningII achieves both functions with a single pod and weighs about 360 lbs less than the LANTIRN, generating 20% less air resistance, and the laser uses a 1.06 m and 1.57 m dual bands for operational and training use respectively. Now the Litening pod has been developed to the fourth generation LITENINGG4, which meets the USAF Advanced Targeting Pod Sensor Enhancement (ATP-SE) programme standards, with the main subsystems being a 1k x 1k triple-field-of-view mid-wave infrared thermal camera, a 1k x 1k daytime CCD sensor, a laser pointer, a laser spot tracker, a laser marker, an inertial measurement device, and a bi-directional

Fig. 8.2 Litening optical targeting pods

Fig. 8.3 ATFLIR optical targeting pods

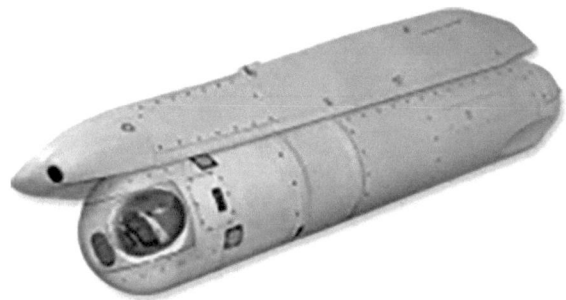

map transmission datalink, and the first introduction of the use of laser target imaging on a targeting pod. Figure 8.2 shows the Litening optronic targeting pod.

8.2.3 ATFLIR Optical Targeting Pods

The ATFLIR optronic targeting pod, developed by Raytheon, is the only targeting pod approved for carrier operations in the U.S., and has been equipped with the F/A-18A, C, D, and F/A-18F Super Hornet fighters. The optical path configuration of the ATFLIR pod is similar to that of the Sniper pod, and it adopts a common-aperture optics system design including mid-wave infrared (MWIR), CCD sensors, and lasers. The infrared field of view is available in 6° (wide), 2.8° (medium) and 0.7° (narrow), with a 640 × 480 elemental indium antimonide focal plane array in the 3.7–5 m band, and has a forward-looking infrared navigation module supplied by BAEsystem. According to public information, the recognition distance of ATFLIR pod can reach 68 km, and 117 km under ideal atmospheric conditions. Figure 8.3 shows the schematic diagram of ATFLIR optoelectronic targeting pod.

8.2.4 Damocles Electro-Optical Targeting Pod (DOT)

The Damocles optronic targeting pod was developed by Thales in France, and has evolved into the Damocles XF multi-role version, which is primarily mounted on

Fig. 8.4 Damocles optical targeting pods

Mirage 2000 and Rafale fighters. The Damocles pods are also equipped with high-resolution infrared and TV sensors, with long-range automatic target identification and damage assessment capabilities. It is fitted with modules such as a laser pointer, laser spot tracker, inertial measurement unit and a mapping datalink. There are two infrared detectors, one is the forward-looking infrared detector, which adopts a medium-wave 640 × 512 gaze focal plane with a large field of view of 24° × 18°, and the other is the target aiming detector, which adopts a third-generation medium-wave infrared thermal camera with two fields of view of 4.0° × 3.0° (wide) and 1.0° × 0.75° (narrow). The forward-looking infrared is not only used to assist navigation, but also can conduct a wide range of search for ground targets, realise guidance instructions for ground targets, help aiming pods to quickly find targets, and change the limitation that only known targets can be struck at night. The ground image detected by FIR and the targeting information are displayed on the pilot's helmet display, while the cockpit multi-function display can also be displayed simultaneously. Damocles optoelectronic targeting pod can be used for guiding all kinds of image-guided weapons and laser-guided bombs, Fig. 8.4 is the schematic diagram of the Damocles optoelectronic targeting pod.

8.2.5 OLS-35 Optronic Radar

It was learnt from the airshow that the OLS-35 photoelectric radar developed by the Russian Research Institute of Precision Systems (NIIPP) is used on the Russian 4++generation fighter SU-35, and it can also be upgraded for the SU-27/SU-30 fighters. The optoelectronic radar has the following features: scanning in airspace in air-to-air mode; detecting, locking on and tracking air, ground and surface targets; scanning surface terrain; target identification; irradiating ground targets with a laser illuminator; detecting and tracking targets irradiated by a coded laser; target angular co-ordinates, distance, and angular and linear velocities; and outputting TV, IR, and TV plus IR video information. and TV plus infrared video information output to the cockpit multi-function display; cross-linked with the onboard targeting and guidance integrated unit; with autonomous functions and radio silence mode. The product adopts medium-wave 640 × 512 pixel indium antimonide infrared focal

Fig. 8.5 OLS-35 optical radar

plane detector, which can track 4 targets simultaneously, with azimuth scanning range of not less than ± 90, vertical scanning range of not less than − 15 to +60, scanning time of not less than or equal to 4s, dimensions of 766 mm × 540 mm × 763 mm, and weight of 71 kg. the system can track 4 targets on a typical SU-30 aircraft. Target pursuit detection distance of 90km, head-on detection distance of 30 km; air-to-air laser ranging distance of 0.2–20 km, ground laser ranging distance of 30 km, Fig. 8.5 for the OLS-35 photoelectric radar.

8.2.6 T-50 Optronic Integrated System

The 101KC optronic integrated system developed by the Russian Ural Optical Machinery Plant for the T-50 fighter, as shown in Fig. 8.6, consists of the 101KC-B optical radar station for airborne targets, the 101KC-H multi-channel search-and-targeting optronic pod for ground targets, the 101KC-y airborne and ground situational information assurance subsystem, and the 101KC-O onboard protection station. Since the T-50 is currently a test model of Russia's fifth-generation fighter, few specific performance indicators have been made public, and the integrated optronic system was only demonstrated at the 2011 Moscow Airshow.

101KC-B optical radar station arranged in the T-50 fighter pilot cockpit windshield to the right, mainly used to detect, identify and track airborne targets and measure their coordinates, the working band for the infrared channel 3–5 m, visible—near infrared TV channel 0.4–0.9 m, laser channel 1.064 m (irradiation), 1.54 m (ranging), the search azimuth angle of ± 90, pitch angle Not less than − 15 to + 55, detection distance 100 km (tail), 40km (head-on) (for SU-30 fighter). 101KC-H multi-channel search and targeting photoelectric pods are suspended under the belly of the T-50

8.2 Development Status of Foreign Airborne Photoelectric Equipment

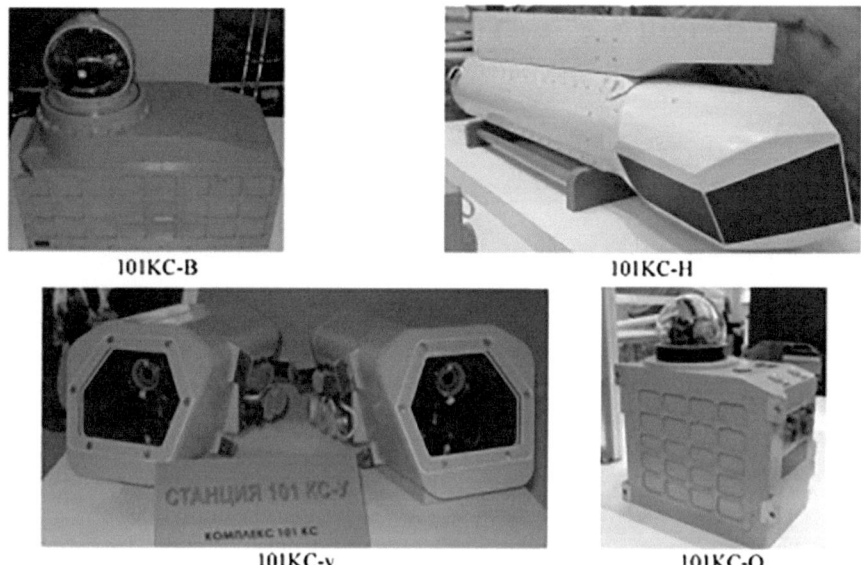

Fig. 8.6 T-50 optoelectronic integrated system

fighter for detecting, identifying, tracking and measuring the coordinates of ground targets and emitting lasers for testing and targeting instructions.

The 101KC-y Air and Ground Scenario Information Assurance System (ASIS) is equipped with a full complement of ultraviolet sensors for detecting air-to-air and surface-to-air missile launch scenarios. The 101KC-O Optical Protection Station (OPPS) is used to jam infrared-guided missiles, and is arranged on the fuselage surface of the pilot's cockpit, as shown in Fig. 8.6.

8.2.7 F-35 Electro-Optical Targeting System

The integrated electro-optical targeting system of the U.S. fourth-generation fighter F-35 is jointly developed by Lockheed Martin and Northrop Grumman, with integrated air-to-air/air-to- surface detection and targeting capabilities. The system consists of two parts: the first part is the Electro-Optical Targeting System (EOTS) developed by Lockheed Martin, as shown in Fig. 8.7; the second part is the Distributed Aperture Sensor System (DAS) developed by Northrop Grumman.

The EOTS is mounted in the forward section of the aircraft, and it integrates forward-looking infrared (FLIR), dual-mode laser, CCDTV, laser tracking, and laser designator. The EOTS provides air-to-ground high-resolution imaging and automated target tracking, air-to-air infrared search and tracking, laser-directed ranging, and laser spot tracking. The low-drag, stealthy EOTS is integrated into the F35's forward

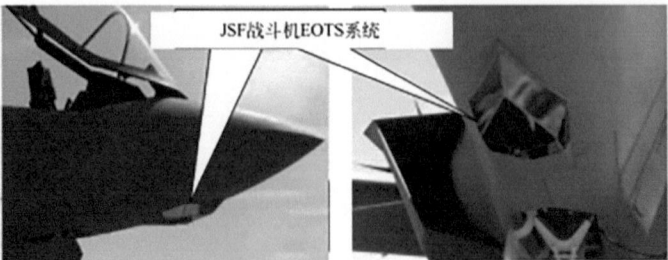

Fig. 8.7 JSF fighter EOTS system

fuselage through a rugged blue window and is connected to the aircraft's integrated control computer via a high-speed fibre-optic interface.

The DAS system consists of six identical sensors, each with a field of view of 90° × 90°, and the six sensors cover the aircraft's forward, aft, left, right, up, and down airspace, and the image data obtained by the sensors is transmitted via fibre optics to the integrated processor, which processes the image data from the six sensors and processes the overlapping portions of the field of view according to certain algorithms to obtain a global image, so that the whole system covers the whole airspace in 4π stereo angle without overlapping or gaps. DAS is used for situational awareness around the aircraft, missile warning and aiding navigation.

8.3 Development Trend of Domestic Airborne Photoelectric Detection Equipment

8.3.1 Airborne Earth-to-Ground Photoelectric Detection Equipment Form Development Trend

1. Equipment structure and devices

Equipment devices will use higher resolution mid-wave infrared focal plane detector, a new type of CCD TV devices and semiconductor pump lasers, to improve the target identification and recognition of the role of the distance, the structure of the use of advanced common aperture optical technology, so that the pod calibre effectively reduced.

2. Equipment configuration and size

Stealth has become the main performance indicators of modern combat aircraft, so the future of airborne ground electro-optical detection equipment will ensure the field of view at the same time, in the system's configuration and dimensions continue to modularisation and miniaturisation, the use of fuselage internal or conformal mounting instead of the original form of pods, which is conducive to reducing the radar reflective cross-section, but also reduces the flight resistance to meet the future needs of supersonic flight aircraft.

8.3.2 Technology Development Trend of Airborne Ground-to-Ground Photoelectric Detection Equipment

1. Multi-spectrum detection capability

Infrared, television and laser is still an important part of the future airborne electro-optical detection system, television detection of high resolution, image clarity, real, intuitive, conducive to the pilot to quickly find the target; infrared images can be detected in the night and daytime visibility is slightly worse than the case of detecting the target [4]; laser can be on the air/ground target ranging, irradiation to guide the laser-guided weapon aimed at attacking the target [5]. With the development of optoelectronic detection technology, the future of airborne optoelectronic detection system will be infrared, TV, laser single optical band to the direction of multi-spectral detection, while meeting the fire control detection, reconnaissance detection, alarm detection, optoelectronic countermeasures and other operational tasks.

2. Time-sensitive target rapid positioning capability

The enemy battlefield deep defence area is filled with a large number of tanks, armoured vehicles, surface-to-air missile positions and other time-sensitive targets. These time-sensitive targets are a big threat, the emergence of a short window of time, if we want to implement a rapid and accurate combat need to shorten the kill chain, especially rapid and accurate positioning. In the future, the ground-based photoelectric detection system will be able to conduct high-resolution imaging of the ground target and background, rapidly identify the target and locate it accurately, and guide the airborne weapons to strike precisely.

3. Air surface multi-target tracking capability

Air surface single target attack technology has been unable to meet the future combat needs of air-to-ground strike, air surface multi-target attack has become a key means of air-to-ground strike. Air-surface multi-target attack refers to the pilot manoeuvring an aircraft, in an entry process, the airborne weapon fire control system to multiple weapons to allocate multiple different targets, while accurately attacking multiple ground/sea surface targets. In air-to-ground support operations, the need to conduct precision strikes against a large number of time-sensitive targets (e.g., tank clusters, armoured convoys, missile positions, etc.) and to avoid the near-, medium-, and long-range echelon defensive fires of the air defence system requires that the aircraft have the capability to attack multiple targets in a single mission. For bombers with large bomb loads, there is a greater need to distribute various types of weapons to attack multiple targets.

4. Air-air and air-ground integrated detection capability

At present, the western countries to the air detection of photoelectric equipment is mainly photoelectric radar and infrared search and tracking system, to the ground detection of photoelectric equipment is mainly targeting pods and photoelectric turret. The U.S. F-35 aircraft electro-optical targeting system (EOTS) is mainly ground detection, taking into account the function of airborne detection. EOTS has high-resolution imaging, automatic infrared search and tracking, laser indication and ranging, laser spot tracking. In the future, the airborne ground electro-optical detection system will be integrated with the airborne electro-optical detection system to provide integrated detection capability for airborne stealth targets and ground time-sensitive targets.

5. Alarm and detection capabilities

Alarm detection requires large coverage of detection airspace, high pointing accuracy and low false alarm rate; ground-based photoelectric detection requires clear images and high targeting accuracy; airborne detection requires long distance, high angular accuracy and low false alarm rate. The future airborne photoelectric detection system design through reasonable air, time and frequency domain management measures, should be able to achieve a comprehensive alarm and detection capability for incoming threat targets.

6. Multi-sensor fusion detection capability

As the number and types of airborne sensors continue to increase [6], optoelectronic detection system and fire control radar, electronic warfare, data chain systems and other combat information grows rapidly, multi-sensor fusion detection can be used to meet the needs of different combat missions. In the single aircraft avionics system, optoelectronic sensors can be based on stealth needs with radio frequency sensors, respectively, active and passive collaborative search, tracking and identification, to achieve integrated detection of a single aircraft; multi-machine can be through the data chain will be more than one optoelectronic sensors networking, fusion of target tracking and identification, to achieve multi-machine networked detection.

7. Application-oriented cost control

According to the operational mission requirements, the overall programme design of the system is reasonable, leaving a margin of hardware and software for future upgrades, but also reasonably controlling the configuration of the payload, and developing useful products that are affordable to the military.

Airborne ground-based photoelectric detection equipment is a highly integrated photoelectric product that integrates optical, mechanical and electrical technologies. With the continuous development and breakthroughs in optoelectronic component technology and computer processing technology, its field of application will continue to expand, and it can be foreseen that the airborne ground-based photoelectric detection equipment will play a more and more important role in future aerial combat.

References

1. Strojnik, M. 1999. Distributed-Aperture Infrared Sensor Systems. In *Proceedings of SPIE-The International Society for Optical Engineering*, vol. 3698, 58–66.
2. O'Neil, W.F. 1997. Processing Requirements for The First Electro-optic System of the Twenty-First Century. In *Digital Avionics Systems Conference*, 5.1–15–22. IEEE.
3. Lu, Jianming, and Yi Cai. 2015. Military Application of Soviet/Russian Infrared Technology, 163–175. Beijing: The Publishing House of Ordnance Industry.
4. Liu, Xingyun. 2001. Research on Airborne Infrared Search and Tracking Technology. *Laser and Infrared* 31(5): 273–276.
5. Shen, Yang, and Mingwen, Tang. 2003. A Review of Airborne Infrared Search and Tracking System (IRST). *Infrared Technology* 3(1):13–18
6. Shu, Jinlong, Liangyu Chen, and Zhenfu Zhu. 2003. Development Status and Development Trend of Foreign Infrared Search and Tracking Systems. *Modern Defence Technology* 31(4): 38–41.
7. Chen, Miaohai. 2003. Application and Development of Airborne Optoelectronic Navigation Targeting System. *Electro-optic and Control*, 10(4): 42–46.

Chapter 9
Principles of Photovoltaics

9.1 Principle of UAV TV Camera and Tracking and Positioning

9.1.1 Overview

TV camera is a technology that applies electronic technology to convert, record, transmit and reproduce the activities of scenery or people, and it is also an important way and means to record sound and moving images. UAV TV camera system is composed of five major parts, such as TV image ingestion (also including infrared image ingestion), recording, transmission, reception and reproduction, and control, which is used to complete the continuous reconnaissance of a certain area on the ground by the UAV, as well as the rapid positioning of the ground target and other functions.

TV camera is one of the important sensors for UAV to take images, as early as 1925, the British successfully tested the mechanical scanning TV, at the same time, the Americans also invented the electronic scanning system and photoelectric camera tubes. around 1930, the British, Russian and other countries started the mechanical black and white TV broadcasting, and after that, in 1951, the U.S. test-broadcasted a kind of colour TV signals which are incompatible with black and white TVs; and in 1953, the U.S. adopted the mechanical black and white TV broadcasting system. In 1953, the United States adopted the NTSC colour television system to achieve compatibility with the black-and-white television system. It should be said that from the 1950s onwards, along with the TV reconnaissance technology began to be gradually applied to the field of military intelligence acquisition, the aviation and aerospace reconnaissance technology has also been developed rapidly, and the accuracy and timeliness of its intelligence acquisition has also been greatly improved.

(1) Signal acquisition

Television signals are mainly ingested by the television camera, the camera's task is to decompose and convert the light image of natural scenery into electrical signals represented by voltage or current. When the scene of the reflected or scattered light into the lens of the camera, first in the camera device (such as solid-state camera device CCD surface) to form a two-dimensional charge image corresponding to the light image of the scene. This charge image using charge coupling transfer method, the formation of a one-dimensional function of the time-varying electrical signal. This process is carried out continuously to produce a continuous image signal for the purpose of transmitting an image. Directly from the photographic element on the electrical signal is very weak, and with a lot of noise, distortion and other defects, so the camera is also set up in a variety of processing circuits, used to amplify the signal, de-emphasis of noise, a variety of corrections, compensation, conversion, and a series of processes, and finally the output of an ideal and standards-compliant analogue or all-digital television signals.

(2) Editing and recording

The signal obtained by UAV TV camera system is mainly divided into two ways, one way is directly transmitted to the ground through wireless link, and the other way is directly transmitted to the on-board recording equipment. Therefore, there are two main ways of editing and recording of UAV: one is to record the TV image on the on-board electronic disc (or video recorder), so that the complete TV image can be obtained when the radio is interfered with; one is to directly record the image transmitted to the ground through the wireless link on the media such as magnetic tape, and under this condition, the parameter information such as the photoelectric rotary table of the on-board TV camera can also be in the form of character superimposed on the TV image for later editing, observation and analysis. When the UAV TV reconnaissance mission is over, the TV signals are then collected from the recording equipment, and the reconnaissance images of important geographical areas or seasons are edited and processed, so as to inform or share the reconnaissance intelligence information with the demand units.

According to the requirements of the UAV flight time and the size of the recording space of the electronic disc, in order to ensure that the electronic disc has enough space to record the TV signals, under the condition of not affecting the observation and analysis of the TV images, the on-board recording equipment can compress the TV images by a certain proportion.

(3) Transmission

The sending transmission of the UAV TV camera system mainly relies on the wireless communication link, i.e. using microwave technology to send the TV images under airborne conditions to the ground through the airborne wireless transmitter, and the ground data display (control) terminal decodes the received TV signals and displays or records them in real time on the data display (control) terminal. For small UAVs, since their radio link only completes the flight control function of the UAV, the TV images need to be transmitted using specialised wireless map transmission equipment.

(4) Receiving Reproduction

After solving the transmission problem, the TV image is transmitted to the ground control terminal in the form of an electrical signal, and the receiving equipment is used to receive the signal for reproduction.

Receiving equipment is a TV monitor or video capture card, TV monitor can directly display TV images on the screen, while the video capture card also needs to use video playback software for display.

(5) Control and processing

TV camera system to achieve real-time reconnaissance and tracking and positioning tasks, there must be a corresponding TV camera control system and TV image data processing system. The photoelectric turntable (PTZ) and its control mechanism, TV image tracking and processing are the important components of the UAV TV camera system. The optoelectronic turntable can automatically adjust the optical axis pointing of the TV camera according to the direction, pitch and other angular parameters sent by the ground station (or remote control), so as to ensure that the optical centre of the TV camera is always pointing at the target to be detected, thus realizing the tracking and positioning function of the target; the on-board control processing equipment can automatically adjust the TV image's field of view, image contrast, and image data processing according to the TV camera's focal length size, image correction and other commands sent by the ground. The on-board control and processing equipment can automatically adjust the TV image's field of view size, image contrast and sharpness according to the commands sent from the ground.

9.1.2 CCD Structure and Principles

In 1969, CCD (Charged Coupled Device) was invented by W.S. Boyle and G.E. Smith of the Bell Institute and published the following year. Because CCDs have the function of storing and transmitting signal charges, they are widely used in memory, displays, and delay elements. The key application of the CCD image sensor, using a simple structure called the frame transfer (frametransfer) method (FT-CCD), was also invented by the Bell Institute in 1971.

With the publication of the 250,000-pixel high-resolution CCD image sensor for civilian cameras in 1985, the practical application of CCD image sensors was officially announced; after that, the era of CCD image sensors finally came to an end, and after its official application, many basic technologies were developed, including the functions of improving the image quality and expanding the size of the image

CCDs are functionally integrated devices that combine a sensing part with a scanning part, and are used to replace the TV camera tubes to make the camera more compact. The CCD is a functional device that combines a sensing section and a scanning section, and is used to replace the TV camera tube to convert the optical image taken by the camera lens into an electronic signal, and to perform tasks such as optical/electrical conversion, information storage, and scanning and reading. Among them, scanning and reading is the use of the charge transfer method to achieve.

Due to the continuous development and application of CCD technology, the two types of image sensors published in 1966 and 1967, respectively, were gradually phased out. With the deep application of CCD image sensors in the field of cameras, in 1993 for still images of all-pixel readout method of CCD was successfully developed, the technology to improve the resolution of the camera at the same time, but also greatly improve the development of digital cameras.

Another widely used image sensor is the CMOS device, this kind of image sensor can be divided into two categories: one is its pixels do not have signal charge amplification, called PPS (Passive Pixel Sensor, passive pixel sensors); the other is the pixel with signal amplification, called APS (Active Pixel Sensor, active pixel sensors). CMOS image sensors are mostly APS, so the creation of the APS type image sensor in 1966 marked the birth of CMOS. Subsequently, APSs using photodiodes and MOS transistors were published in 1968 and 1969, respectively.

1. CCD basic structure

CCD that charge-coupled device, also known as CCD image sensor or image controller. It is a semiconductor device, CCD has a number of neatly arranged photodiodes, can sense the light, and optical signals into electrical signals, amplified by the external sampling and analogue-to-digital conversion circuit, and then realize the analogue image to the digital image of the conversion. CCD implanted on the tiny light-sensitive material is called a pixel (Pixel), a CCD contains more pixels, the more the resolution of its picture to provide. The more pixels a CCD contains, the higher the resolution it can provide.

The basic unit of a CCD is a MOS capacitor, which is capable of storing electric charge, and its structure is shown in Fig. 9.1. P-type silicon, for example, in the P-type silicon substrate by oxidation on the surface of the SiO_2 layer, and then in the SiO_2 precipitation layer of metal as a gate, P-type silicon in the majority of carriers are positively charged holes, a minority of carriers are negatively charged electrons, when the metal electrodes on the application of a positive voltage, the electric field can be through the SiO_2 insulating layer of the carriers to be repelled or attracted. So the positively charged holes are repelled away from the electrode, the remaining negatively charged minority carriers in the immediate vicinity of the SiO_2 layer to form a negatively charged layer (depletion layer), once the electrons enter the electric field due to the role of the can not come back out, it is also known as the electron potential well.

When the device is subjected to light (light can be shot from the gap between the electrodes through the SiO_2 layer, or through the substrate of the thin P-type silicon

9.1 Principle of UAV TV Camera and Tracking and Positioning

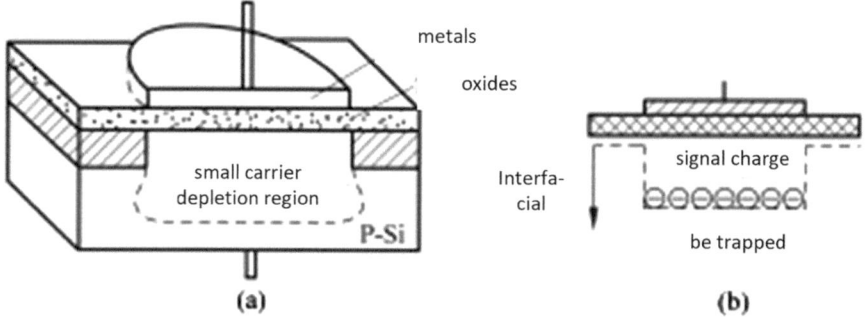

Fig. 9.1 CCD structure and operating principle diagrams. **a** Cross-section of a MOS capacitor used as a minority carrier storage unit. **b** Potential well with signal charge, represented by the liquid at the bottom of the well.

shooter), the energy of the photon is absorbed by the semiconductor, resulting in an electron—hole pairs, which appeared in the electrons are attracted to be stored in the potential wells, which can be conducive to these electrons. The stronger the light, the more electrons collected in the potential well, the weaker the light is vice versa, so that the strength of the light into the number of charges, to achieve the conversion of light and electricity, and the potential well of electrons collected in the storage state even if the light stops for a certain period of time will not be lost, which achieves the memory of the light.

In short, the above structure is essentially a tiny MOS capacitor, which constitutes a pixel, both "light-sensitive", but also can leave "latent shadow", light-sensitive role is to rely on the accumulation of electronic charge generated by the light intensity, latent shadow is the pixel to stay in each capacitor charge is not equal to the formation, if we can manage to put the potential well collected electrons in the storage state even if the light is stopped for a certain period of time will not be lost, which will achieve the memory of the light. If you can manage to transfer the charge in each capacitor to the output in sequence, and then form rows and frames and go through the "development" to achieve the transmission of the image.

2. CCD working principle

CCD image sensor work process mainly includes four steps:

(1) Photoelectric conversion, that is, light is converted into a signal charge;
(2) Storage of charge, storage of signal charge;
(3) The transfer of charge, the transfer of signal charge;
(4) The detection of charge, the signal charge is converted into an electrical signal.

1) Photoelectric conversion

Photoelectric conversion is based on the intensity of light irradiated to the photographic surface of the charge, that is, the presence of electrons in the material from the light to obtain energy to change the state, as long as the application of a small amount of electric field electrons present the phenomenon of free movement. Physically, photoelectric conversion can be divided into two state changes: one is the external photoelectric effect, the other is the internal photoelectric effect.

The external photoelectric effect refers to the electrons on the surface of the solid, accept the photon (photon) energy is released into the vacuum phenomenon. At this time, the need for the valence band and the vacuum energy levels between the energy difference, this energy difference is called the work function (Work Function). The internal photoelectric effect refers to the phenomenon in which electrons in several energy levels of a solid are excited into higher energy electrons by the energy of a photon.

Specifically, in Si single crystal, one of the semiconductors, the energy of electron orbitals of atoms is distributed in the form of bands with the periodicity of the crystalline lattice, and the energy levels of electrons can be divided into valence band and conduction band. When in the state of the energy level, that is, in the energy band of the valence band of electrons can accept the energy of light, the phenomenon of excitation to the conduction band, known as the internal photoelectric effect.

Since the electrons excited to the conduction band can move as long as a small voltage (electric field) is applied, the signal charge can be taken out according to the strength of the light. CCDs that use semiconductors use the internal photoelectric effect to obtain the signal charge generated by photoelectric conversion.

When a semiconductor absorbs light, it converts the energy of photons into the energy of electrons. In the conversion process, photons with the required energy to excite electrons from the valence band to the conduction band, photoelectric conversion is called the basic absorption. However, for the absorption of light, the light wavelength sensitivity brings significant impact and is a very important condition.

2) Charge storage

Charge storage is to collect the signal charge from photoelectric conversion until the output of the storage action. In a typical CCD image sensor, the signal charge generated by photoelectric conversion is stored in the photodiode.

Most CCD image sensors make use of the property that negatively charged electrons can be absorbed by a high potential to store the signal charge centrally. The state of potential distribution in which the charge is stored is called a potential well.

How to create a high potential well above the surrounding potential in a Si single crystal to store charge is described here as an example of a surface-type MOS capacitor.

The MOS capacitor shown in Fig. 9.2 has a substrate formed from P-type Si. First, the back of the substrate (semiconductor end) of the MOS capacitor is grounded, and a positive voltage is applied to the surface motor (metal end). In this way, the entire substrate is approximately at the ground potential before the voltage is applied at the electrode. Under the influence of the voltage applied to the electrode, the potential distribution changes, and the potential on the surface of the Si substrate located under the electrode rises. In this state, since the surface of the Si substrate closest to the electrode is surrounded by the ground potential, a potential well is formed in the part

9.1 Principle of UAV TV Camera and Tracking and Positioning

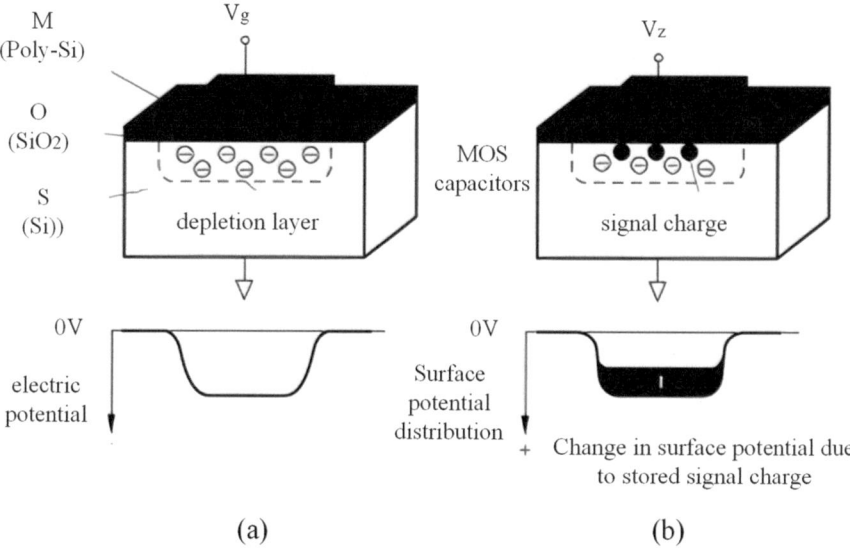

Fig. 9.2 MOS capacitors with surface potentials. **a** without signal charge; **b** with signal charge

of the substrate having the highest potential, and negatively charged electrons can be stored there.

Since there is a Si oxide insulator (SiO_2) between the electrode and the Si substrate, electrons cannot flow to the electrode. If the electrons of the signal charge are once stored in this potential well, the surface potential decreases as the number of charges changes. For a qualitative explanation of signal charge storage, the surface potential distribution shown in Fig. 9.2 can be viewed as a bucket, with the electrons of the signal charge stored in the bucket like water.

3) Charge transfer

The charge transfer process between four electrodes in close proximity to each other is represented in the CCD shown in Fig. 9.3. It is assumed that at the beginning there is some charge stored in the deep potential well below the first electrode with a bias of 10 V, and that the other electrodes are loaded with lower voltages (e.g., 2 V) greater than the threshold. Let Fig. 9.3a be the zero moment (initial moment). After n moments, the voltage on each electrode becomes as shown in Fig. 9.3b, c. If the voltage on the electrodes at this time becomes as shown in Fig. 9.3d, with the first electrode voltage changing from 10 to 2 V and the second electrode voltage remaining at 10 V, the shared charge is transferred to the potential well below the second electrode, as shown in Fig. 9.3e. It can be seen that the deep potential well and the charge packet move one position to the right.

By applying a certain regularly varying voltage to each of the CCD electrodes, the charge packets under the electrodes can move in a certain direction along the semiconductor surface. The CCD electrodes are usually divided into groups, each

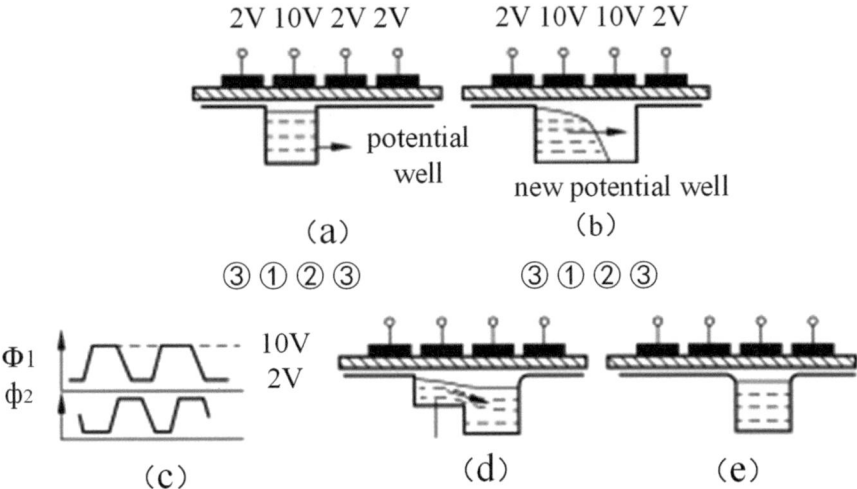

Fig. 9.3 Charge transfer process in a three-phase CCD

group is called a phase, and the same clock pulse is applied. The internal structure of the CCD determines the number of phases required to make it work properly. The structure shown in Fig. 9.3 requires three-phase clock pulse, the waveform shown in Fig. 9.3c, such a CCD is called a three-phase CCD. three-phase CCD charge coupling (transfer) mode must be in the three-phase overlapping pulses in order to be able to a certain direction of the transfer of the unit-by-unit. In addition, it must be stressed that the CCD electrode gap must be very small for the charge to be transferred unimpeded from one electrode to the neighbouring electrode. If the electrode gap is relatively large, the potential well between two neighbouring electrodes is separated by the potential barrier, can not be merged, the charge can not be from one electrode to the other electrode to completely transfer, CCD will not be able to work under the action of external pulses in the positive range.

The maximum gap that can produce full coupling conditions is generally determined by the specific electrode structure, surface state density and other factors. Theoretical calculations and experiments have confirmed that, in order not to prevent the interface below the electrode gap to impede the charge transfer barrier, the length of the gap should be less than 3 μm, which is roughly the same conditions under the same conditions of the semiconductor surface width of the deep depletion zone size. Of course, if the thickness of the oxide layer and the density of surface states are different, the results will be different. But for the vast majority of CCDs, a gap length of 1 μm is small enough.

CCDs with electrons as the signal charge are called N-type channel CCDs, referred to as N-type CCDs, while CCDs with holes as the signal charge are called P-type channel CCDs, referred to as P-type CCDs. due to the mobility of the electrons (the speed of movement per unit of field strength) is much greater than the mobility of

9.1 Principle of UAV TV Camera and Tracking and Positioning

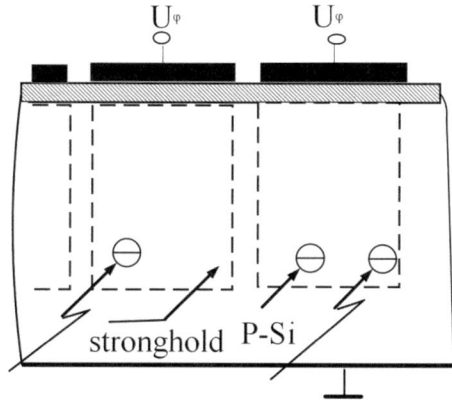

Fig. 9.4 Backside Illumination Light Injection

the holes, so the N-type CCDs have a much higher frequency of operation than the P-type CCDs.

4) Charge injection and detection

(1) Charge injection

In the CCD, there are many methods of charge injection, in summary, can be divided into two categories of optical and electrical injection.

Light injection refers to the light irradiation to the CCD wafer, in the semiconductor body near the gate to generate electron—hole pairs, the majority of its carriers are discharged by the gate voltage, and the minority carriers are collected in the potential well to form a signal charge. The light injection method can be further divided into front-illumination type and back-illumination type. Figure 9.4 shows the schematic diagram of back-illuminated light injection.

Electrical injection means that the CCD samples the signal voltage or current through the input structure, and then converts the signal voltage or current into signal charge. There are many methods of electrical injection, the main two commonly used are current injection and voltage injection.

(2) Detection of charge

Effective collection and detection of charge is an important issue in CCDs. One of the important characteristics of CCDs is that the signal charge is transferred without any capacitive coupling to the clock pulse, which is unavoidable at the output. Therefore, the choice of appropriate output circuit can minimise the extent to which the clock pulse is capacitively fed into the output circuit. The current CCD output methods mainly include current output, floating diffusion amplifier output and floating gate amplifier output (Fig. 9.5).

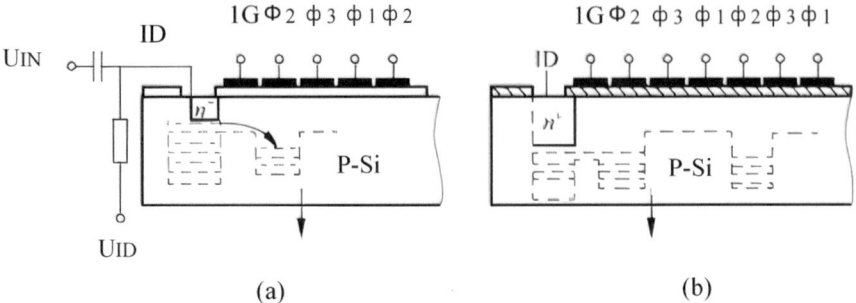

Fig. 9.5 Electrical injection method. **a** Current injection; **b** voltage injection

9.1.3 CCD Classification and Characteristics

1. Classification of CCD

Charge-coupled camera devices, also referred to as ICCD, its function is to convert the two-dimensional optical image signal into a one-dimensional video signal output. there are two major types of ICCD linear and surface. Both need to use the optical imaging system will be the scene image imaging in the CCD image-sensitive surface. The image-sensitive surface will be illuminated on each image-sensitive unit of the image illumination signal into a small number of carrier density signal stored in the image-sensitive unit (MOS capacitor). Then, it is transferred to the CCD's shift register (potential well under the transfer electrode) and sequentially shifted out of the device under the action of a drive pulse to become a video signal.

For linear devices, it can directly receive one-dimensional optical information, but not directly transform the two-dimensional image into a video signal output. In order to get the video signal of the whole two-dimensional image, it is necessary to use the scanning method to achieve.

(1) Line CCD camera device of two basic forms

1) Single-channel linear CCD

The structure of a three-phase single-channel linear CCD is shown in Fig. 9.6. As can be seen from the figure, the photosensitive array is separate from the transfer—shift register, and the shift register is blocked. In the light integration cycle, the grating electrode voltage of this device is high, and the photosensitive region generates a charge under the action of light stored in the photosensitive MOS capacitive potential well. When the transfer pulse arrives, the signal charge in the potential well of the line array photosensitive array is transferred in parallel to the CCD shift register, and finally shifted out of the device one bit at a time under the action of the clock pulse to form a video pulse signal.

This structure of the CCD transfer more times, low efficiency, modulation transfer function MTF is poor, only for like less sensitive unit of the camera device.

9.1 Principle of UAV TV Camera and Tracking and Positioning

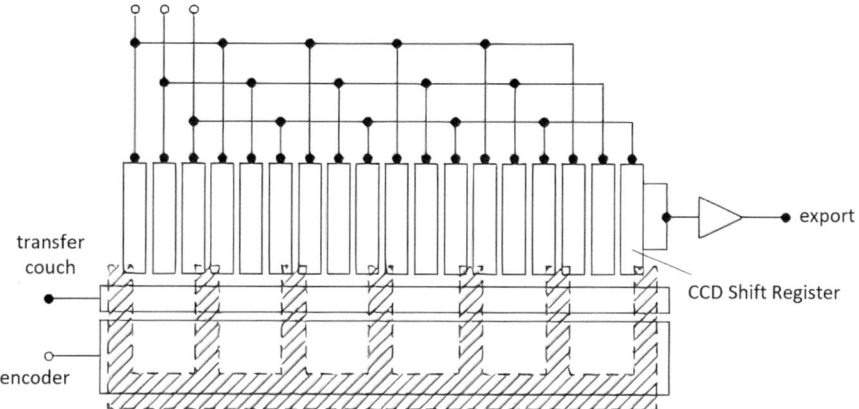

Fig. 9.6 Single-channel linear CCD structure

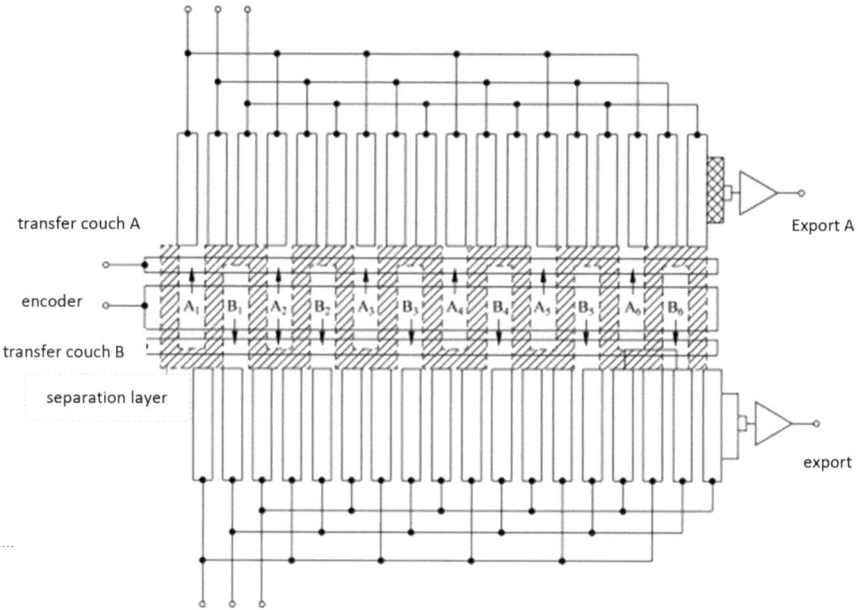

Fig. 9.7 Dual channel linear CCD structure

2) Dual-channel line array CCD

Figure 9.7 shows a double channel line camera device. It has two rows of CCD shift registers A and B, divided on both sides of the image-sensitive array. When the transfer gate A and B for high potential (for N-channel devices), the optical integration array signal charge packet at the same time in the direction of the arrow

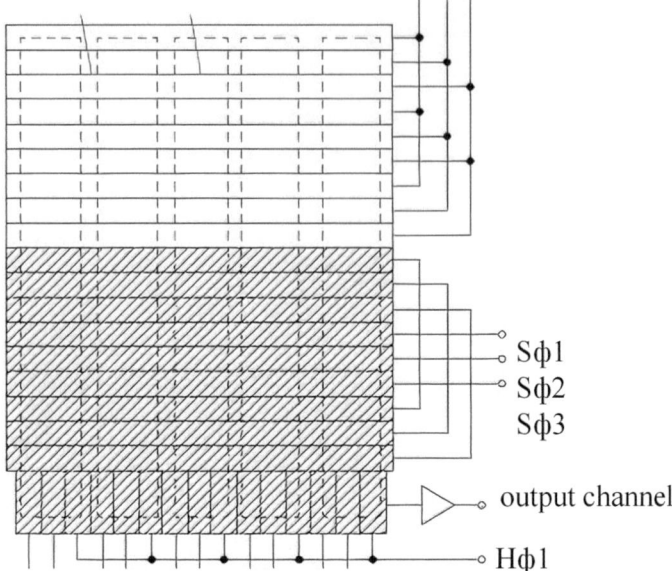

Fig. 9.8 Three-phase frame transfer surface array CCD structure

to the corresponding shift register, and then under the action of the drive pulse were transferred to the right, and finally to the video signal output. Obviously, the same like the sensitive unit of the dual-channel line array CCD than a single-channel line array CCD transfer less than half the number of times, its total transfer efficiency is also greatly improved, so the general higher than 256-bit line array CCD are dual-channel.

(2) Surface-array CCDs

According to a certain way will be a one-dimensional line CCD photosensitive units and shift registers arranged into a two-dimensional array, that can constitute a two-dimensional surface array CCD. due to the arrangement of different ways, the surface array CCD is often a frame transfer, column transfer, line transfer and full-frame transfer and other ways.

1) FT frame transfer mode CCD

As shown in Fig. 9.8 frame transfer mode CCD structure by the light-sensitive part (imaging area), storage (temporary storage) and horizontal displacement (horizontal readout) registers and other three parts. The imaging area consists of a number of charge-coupled channels arranged in parallel (dashed box in the figure), with each channel separated by a trench barrier and horizontal electrodes running across each channel. Assuming that there are M transfer channels and N imaging units per channel, there are a total of units in the entire imaging zone. The structure and number of cells in the staging area are the same as those in the imaging area. Both the staging area and the horizontal readout register are masked.

Each frame of signal charge accumulated by optical/electrical conversion in the photoreceptor section is quickly transferred to the memory section during field blanking, and the photoreceptor section re-enters the state of charge accumulation. The signal charge of a frame transferred to the storage section transfers a signal equivalent to one scanning line to the horizontal displacement register during each line blanking period. The signal charge entering the horizontal shift register is transferred to the output at a standard line scanning speed during the line positive range, and a standard television signal is finally obtained. Since the signal charge generated by the photoreceptor section is transferred to the memory section at one time, the memory section must have the same number of pixels and area as the photoreceptor section, and the difference is only that a metal shading layer is added to the surface of the memory section to prevent light exposure. In this frame transfer method, the electrode structure of the CCD is simple, and the decomposition power and sensitivity are high. However, the total area of the device is large because the area of the light-sensitive imaging area and the charge storage area are the same. In addition, in the light-sensitive area due to the accumulation of charge and transfer in the area is not separated, in the transfer process there is still charge accumulation, thus generating vertical trailing phenomenon. And because the transparent electrode has a certain degree of absorption of light, especially on the blue light absorption is more, and cause the colour picture bias. Because of the above disadvantages, frame transfer method in the early CCD camera used outside, has been rarely used. At present, FT-CCD has been greatly improved.

2) Interline transfer method CCD

Interline transfer mode CCD structure shown in Fig. 9.9, it is the light-sensitive part and the transfer of the Department to do the horizontal inter-arrangement, pairs of close together, below the horizontal displacement register. In addition to the light-sensitive part of the other parts covered with a metal cover layer, blocking the light into the CCD interior, the transfer section is also a temporary part of the charge. Each unit (pixel) of the light-sensitive section is one-to-one coupled with each unit of the transfer section, and the two are controlled by a transfer control gate. The signal charge generated by the optical/electrical conversion in the photoreceptor section (all pixels) is quickly transferred to the staging section by the control gate during the field fading period, so that the transfer of all the pixels can be completed in only one transfer, and the photoreceptor section returns to the state of charge accumulation again. The signal charge which is transferred to each temporary storage part works the same as the storage part of the CCD of the frame transfer method, and during the line blanking period, one scanning line is shifted out at a time and sent to the horizontal displacement register below, and then transferred to the output terminal during the line scanning period in order to obtain the standard TV signal.

(3) FIT frame-line transfer method CCD

Frame-line transfer method CCD is a combination of frame transfer method and line transfer method, its structure is shown in Fig. 9.10. It includes light-sensitive part,

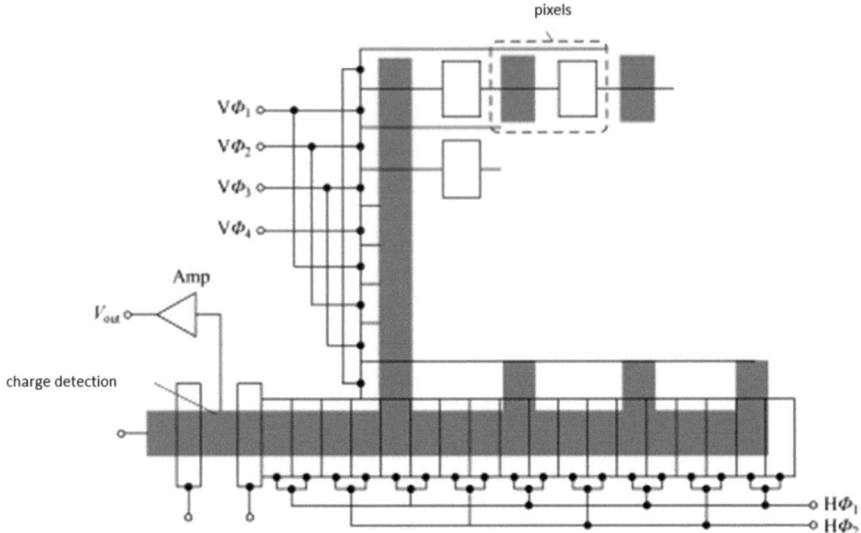

Fig. 9.9 Interline transfer method CCD structure

transfer part, storage part and horizontal displacement register and other parts. As can be seen from the figure, the upper part of the FIT is the same as the interline transfer method, while the lower part is the same as the frame transfer method. The operating process of the CCD in the frame-to-interline transfer method is as follows: the signal charge is accumulated by the photoreceptor section during the field positive range, passed through the control gate during the field blanking period, and at the same time transferred to the corresponding potential well of the transfer section at one time, which is the same as that of the interline transfer method. The signal charge temporarily stored in the transfer section is then quickly transferred to the storage section below. During the field positive travelling period, under the control of the line frequency clock pulse, the pixel charge is shifted into the horizontal displacement register line by line during the blanking period of each line. And then, during the line positive travelling period, the horizontal shift register moves each pixel to the output to produce a video signal. The process of transmitting later in the memory section is the same as the frame transfer method. This frame-to-row transfer method has the advantages of both the frame transfer and the inter-row transfer methods because it is a combination of the two methods, and in particular, the vertical trailing in the case of high brightness is significantly improved. However, the CCD of this transfer method has a large chip area and high manufacturing cost due to the additional memory section, which is suitable for high-grade TV cameras.

2. CCD characteristics

CCD performance is to determine the performance of the entire camera is an important criterion, in a sense, is also an important embodiment of the camera performance indicators.

9.1 Principle of UAV TV Camera and Tracking and Positioning

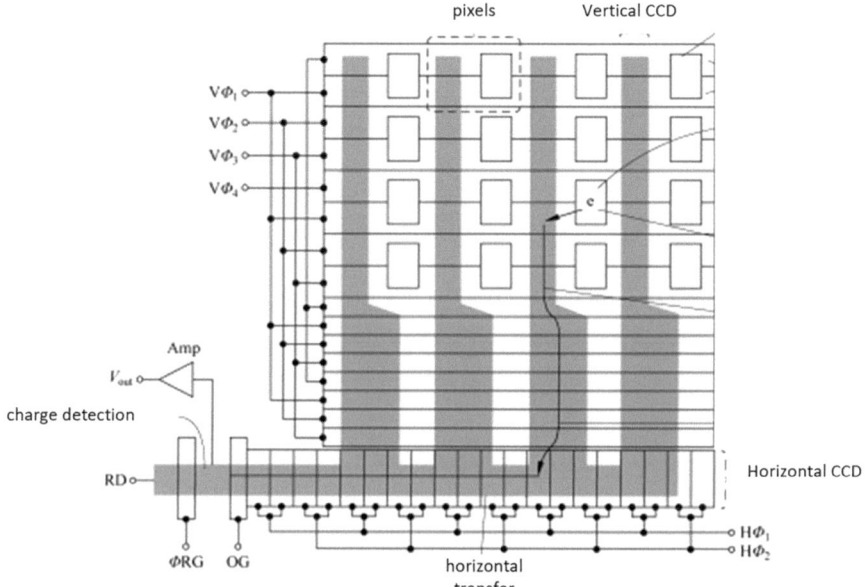

Fig. 9.10 Frame-to-line transfer method CCD structure

(1) Decomposition power

In the CCD on the light-sensitive units, that is, the number of pixels is a certain amount, if the number is too small, or if the size is too large, it will affect the picture resolution of the degree of precision—when the object is extremely fine details, the CCD can not be distinguished, then there will be aliasing interference. Therefore, the number, size, and density of light-sensitive units on the CCD will have a significant impact on image quality.

The signal decomposition power can be improved by increasing the number of pixels, adding an optical low-pass filter in the optical path to suppress the high-frequency components of the image, and using spatial biasing techniques to make the base colour signals of aliasing interferences in opposite phases and cancel each other out.

The decomposition power is commonly evaluated by the modulation transfer function. Figure 9.11 plots the M7T curves of a line array CCD under broadband light source and narrowband light source illumination.

Line array CCD solid-state camera device to more bits of photosensitive unit development, there are now 256×1, 1024×1, 2048×1, 5000×1, $10,550 \times 3$ and so on. The higher the number of pixels, the higher the resolution of the device, especially for the measurement of object size, the higher the number of pixels, the higher the accuracy of the measurement can be obtained by using the high number of photosensitive units of the line array CCD device. In addition, when a mechanical scanning device is used, it is also possible to obtain a video signal of a two-dimensional image

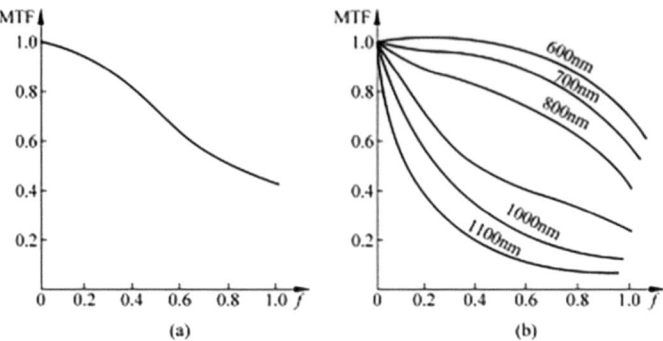

Fig. 9.11 MTF curve of a line array CCD. **a** 2856 K incandescent light source, **b** monochromatic light source illumination

with a line array CCD camera. The resolution of the second dimension obtained by scanning depends on factors such as the scanning speed and the height of the CCD photosensitive unit.

The output signal of a two-dimensional surface array CCD generally adheres to the scanning method of a television system. Its resolution in the horizontal and vertical directions is different, and the horizontal resolution is higher than the vertical resolution. When evaluating the resolution of an array CCD, only its horizontal resolution is evaluated, and the evaluation method of the number of TV lines, which is the evaluation method of the TV system for image resolution, is utilised. The TV line evaluation method shows that the number of black and white bars that can be distinguished in the horizontal direction on an image is its resolution. Horizontal resolution and the number of pixels in the horizontal direction of the CCD, the more the number of pixels, the higher the resolution, the number of pixels of the existing surface array CCD has been developed to 512 × 500, 795 × 596, 1024 × 1024, 2048 × 2048, 5000 × 5000 and so on, the resolution of the more and more high.

(2) Sensitivity

CCD sensitivity to the luminous flux of 1 lumen (lm) of the 3200 K colour temperature of light projected onto it by the size of the current intensity generated to define the CCD sensitivity is higher, indicating that the CCD photoelectric conversion capability is stronger.

CCD spectral sensitivity depends on quantum efficiency, wavelength, integration time and other parameters. Quantum efficiency characterises the CCD chip for different wavelengths of light signals photoelectric conversion capabilities. CCD chip made of different processes, its quantum efficiency is different. Sensitivity is also related to the way the light, back-illuminated CCD quantum efficiency, the spectral response curve without ups and downs, positive illumination CCD due to reflection and absorption losses, the spectral response curve there are a number of peaks and valleys.

9.1 Principle of UAV TV Camera and Tracking and Positioning

(3) Dynamic range

Dynamic range is the CCD can reproduce the picture of the relative ratio between the bright and dark parts. The higher the better this range, the dynamic range of the CCD determines the CCD conversion of the maximum charge and the influence of clutter interference is limited to a certain amplitude of the minimum charge of the difference. The improvement of the former depends on the improvement of material technology, the rationality of the CCD structure design and the size of the voltage on the electrode and other factors, while the latter is also related to the circuit structure and so on.

(4) Dark current

CCD dark current is caused by internal thermal excitation carriers. Under normal operating conditions, the CCD in the low frame rate operation, can be a few seconds or thousands of seconds of accumulation (exposure) time to collect low-brightness images, MOS capacitance is not saturated non-equilibrium state. Over time, however, dark currents fill the potential well with hot electrons before photoelectrons are formed, allowing the few carriers generated due to thermal excitation to bring the system into equilibrium. As a result, undesired dark currents can exist even in the absence of light or other means of charge injection into the device. Due to defects in the lattice dot pattern, the dark current can vary greatly from pixel to pixel. On images with long exposure times, a star-like pattern of stationary noise is produced. This effect is due to a small number of pixels having an anomalously large dark current, which can generally be subtracted from the image after recording unless the dark current has saturated the electrons in the potential well.

In addition, the dark current is temperature dependent. The higher the temperature, the more carriers are produced by thermal excitation, and thus, the larger the dark current. It is calculated that the dark current can be reduced by 1/2 for every 10 °C decrease in temperature.

(5) Noise

There are several sources of noise in the CCD as follows:

1) Noise caused by charge injection into the device;
2) The charge transfer process, the noise caused by changes in the amount of charge;
3) Noise generated by the detection.

The average noise value of the CCD is shown in Table 9.1, and the noise associated with the CCD sensor is shown in Table 9.2.

a) Photon noise

Since photon emission is a random process and hence the optical charge collected in the potential well is also random, this becomes a source of noise. Since photon emission is a random process, and hence the optical charge collected in the potential

Table 9.1 CCD noise

Types of Noise		Noise level (number of electrons)
Output Noise		400
Transfer Noise	SCCD	1000
	BCCD	100
Output Noise		400
Captured Noise	SCCD	1150
	BCCD	570

Table 9.2 Noise associated with CCD sensors

Noise sources	Size	Representative values (rms carriers)	
Photon noise	N_s	100, $N_s = 10^{-4}$	
Dark current noise		1000, $N_s = 10^{fi}$	
Optical fat-zero noise	iV_{nc}	100, $N_{OC} = N_{smax}$	
Electronic fat-zero noise	iV_{FZ}	300, $N_{fz} == 10\% N_{Sm} \ll x$	
Capture noise	$400 C_w$	100, $C_w = 0.1\text{PF}$ ($N_{Sm} = 10^5$)	
Output noise	See Table 9.1	10^3, SCCD	2000 transfers
Noise Sources		10^2, BCCD	
Photon noise	400	200, $C_{out} = 0.25\text{pF}$	

Note Ns is the size of the charge packet

well is also random, this becomes a noise source. Since the noise source is independent of the CCD sensor and depends on the nature of the photons, it becomes the fundamental limiting factor of the camera device. This noise mainly has an effect on the camera at low light intensity.

b) Dark current noise

Like photon emission, dark current is also a random process, and thus also become a source of noise. Moreover, if the dark current of each CCD unit is not the same, it will produce graphic noise.

c) Fat-zero noise

Including optical fat-zero noise and electronic fat-zero noise, optical fat-zero noise by the use of the size of the bias light decision, electronic fat-zero noise by the electronic injection of fat-zero body decision.

d) Capture noise

In SCCD, it is caused by interface defects, and in BCCD, it is caused by body defects, but the capture noise in BCCD is small.

e) Output noise

This noise is caused by the thermal noise generated during the reset of the output circuit. This noise can be compared with the noise of CCD if converted to the root mean square value.

In addition, the device's cell size or different spacing can also be a source of noise, but this source of noise can be reduced by improving photolithography.

(6) Colour reproduction characteristics

Colour reproduction depends on the uniformity of the CCD sensitivity response to different wavelengths of light. Only in its spectral response to the wavelengths of light is relatively uniform and consistent under the premise of the colour can be restored to the normal picture. In the current technical conditions, the CCD colour reproduction characteristics can fully meet the visual requirements of the human eye.

(7) Black spot and trailing

Due to thermal excitation effect, CCD sometimes in the absence of incident light may also produce a number of charges, so that when the input screen is black, the output screen is shown as a certain dark outline display, called black spot. By controlling the accuracy of the CCD material, improving the internal structure of the CCD, and adding a black spot correction circuit to the circuit, the effect of the black spot can be reduced to a level where it is not noticeable.

Due to the accumulation of certain light-sensitive unit charge in high brightness under the number of too much overflow leakage to some neighbouring positions of the light-sensitive unit, resulting in the screen to produce a vertical bright band of the phenomenon is called tail dragging. In the CCD using overflow leakage settings can eliminate tailing, get good high brightness characteristics.

(8) CCD chip pixel defects

Pixel defects: For illumination in the 50% linear range, if the pixel response deviates from its neighbouring pixels by more than 30%, it is a pixel defect.

Cluster Defects: The number of defects exceeds 5 pixels in a range of 3×3 pixels.

Column defects: more than 8 pixels of defects in a column within a range of 1×12 pixels.

Row defects: more than 8 pixels of defects in rows within a set of horizontal pixels.

9.1.4 Optical Stabilisation Platform Principle and Characteristics

1. Stabilised Platform Principles and Characteristics

The UAV optoelectronic platform system usually includes sub-systems such as sensors, optical systems, carrier platforms and data memory. The sensor can be classified into visible light, infrared and laser sensors according to the working band; the optical system can be classified into fixed focus, multi-stop switchable fixed focus and continuous zoom lenses according to the focal length; the carrier platform can be classified into two-axis, three-axis or multi-axis stabilised platforms according to the number of stabilised axes; and the data memory can be classified into ordinary and massive types according to the capacity.

From the above classification, the photoelectric platform is more varied, which does not include the laser detection and distance-selective imaging under development, solid-state 3D LIDAR and other new photoelectric loads. However, looking at the development process of its technology and application, the optoelectronic platform has experienced several stages of single-load optoelectronic platform, dual-load optoelectronic platform and multi-load optoelectronic platform, and the specific application depends on the demand.

From the point of view of the actual equipment of UAV optoelectronic platforms in various countries, single-load platforms are still in use to a certain extent due to their lightweight, easy maintenance, easier input into the battlefield, and technological advancement, but the trend of their being replaced by dual-load and multi-load optoelectronic platforms is irreversible, and multi-load optoelectronic platforms are gradually being developed into mainstream configurations.

From the perspective of specific optoelectronic loads, optical cameras are being replaced by television cameras; infrared line scanners are being replaced by forward-looking infrared; and low-light cameras are rarely used.

From the technical and application perspectives, the UAV-carried optoelectronic platform is developing in the direction of high resolution, high sensitivity, high precision, multi-functionality, small size, light weight, long service life, high reliability, and resistance to shock and vibration. Overall, UAV optoelectronic platforms are ushering in a new round of development boom driven by demand traction and technology.

The basic idea of direct positioning of UAV video images is to realise the conversion process from the image coordinates of the target to the ground coordinates through the comprehensive solving of the aircraft's positional attitude parameters, rotary table angle parameters, camera parameters and so on. In the coordinate conversion process, it involves five coordinate systems: image coordinate system, camera coordinate system, turntable coordinate system, aircraft coordinate system and earth coordinate system.

The airborne optoelectronic stabilisation platform is a gyro-stabilised platform installed on a motion carrier such as a UAV. A ground detection device is installed on the platform to search or track ground targets based on visual axis stabilisation

9.1 Principle of UAV TV Camera and Tracking and Positioning

technology. In recent years, optoelectronic stabilisation platforms have been rapidly developed and are widely mounted on various carrier platforms such as helicopters, fixed-wing aircraft, UAVs, ships and vehicles, which can be used in military fields such as reconnaissance, targeting, missile guidance, etc., as well as in non-military fields such as search and rescue, anti-smuggling, security, environmental monitoring and forest fire prevention.

The optoelectronic stabilisation platform is one of the key equipments to achieve UAV video image tracking and positioning processing. It adopts gyroscope as the feedback element to isolate the angular perturbation of the moving base to the load, so that the load is stabilised in the fixed inertial space of the rotating platform, using the characteristics of the gyroscope to keep the platform platform azimuthally stable; it is also used to measure the attitude of the moving carrier and to establish a reference coordinate system for the measurement of the linear acceleration of the carrier, which is also referred to as gyroscope platform and inertial platform.

Optical stabilisation platforms are usually equipped with a variety of optical sensors such as infrared thermal cameras, visible light camera systems and laser range finders. When the UAV carries out a mission, the optoelectronic stabilisation platform can isolate the carrier's interfering motion and keep the optical sensor's aiming line stable in the inertial space angle, thus ensuring that the detection equipment on the platform can always point at the target.

2. Typical Optically Stabilised Platforms

The development of optoelectronic stabilised platforms has evolved with the evolution of gyroscopes. As early as 1936, ball bearing type power gyro stabilized platform appeared, used as a stabilizer for rangefinder on warships. After the successful development of three-axis gyro stabilized platform XN-U in 1950 in the U.S.A., in the inertial guidance system of missiles and carrier rockets, there appeared one after another hydrostatic air-floating gyro platforms, dynamically pressurized air-floating free-rotor gyro platforms, liquid-floating gyro platforms and so on, due to the gyro platforms using these As the gyro platform adopts these gyroscopes with buoyant support and reduced friction torque, its accuracy has been improved. The United States were applied in the "Panshin I" missile, "Saturn" launch vehicle, "Minuteman I", "Minuteman n "missiles and" Polaris "missiles." In the late 1960s, the United States developed a simple structure, high precision, low-cost flexible gyroscope as a sensitive element of the flexure gyro platform, which in the "Trident I" submarine ground Missile, "Tomahawk" cruise missiles, as well as "Panshin n" missiles on the application. With the research and development of gyro platform technology, in 1973 the United States developed a floating ball platform without frame, that is, advanced inertial reference ball platform. In order to further improve the guidance accuracy and reliability, the support system and temperature control system of the floating ball platform has been improved. Gyro-stabilised platforms have been developed from frame gyro platforms to floating ball platforms, and the mass of gyro platforms has been developed from tens of kilograms to only 0.8 kg, and the outer dimensions have been developed from more than 0.5 m to only 0.08 m for small gyro platforms.

StarSafire series of turrets developed by FLIR. The StarSafireID has been selected by the US Coast Guard as the standard turret for use with the Hawkeye UAV. It has a diameter of 38.1 cm, and its sensors include an infrared camera, a colour zoom camera, a high-resolution colour reconnaissance camera (capable of penetrating fog), a laser illuminator, and a laser rangefinder. One of the infrared imaging devices is a large mid-wave infrared focal plane array of 640X480 arrays, which produces fairly clear infrared images. Thermal imagers are available in four fields of view: a wide field of view with a field of view range of 25°, a medium field of view with a field of view range of 5.4°, a narrow field of view with a field of view range of 1.4°, and so on. Colour daytime cameras are also available with wide, medium and narrow fields of view. Colour reconnaissance cameras have an extremely narrow field of view of 0.29° for ultra-long distance reconnaissance in addition to the medium and ultra-narrow fields of view. It should be noted that the selection of the field of view of the lens involves the relationship between the scope of the picture and the magnification of the target. The narrower the field of view, the higher the magnification of the target; conversely, the wider the field of view, the wider the coverage, but not able to see the details of the target.

The BriteStar turret is slightly larger, with a diameter of 40.6 cm and a mass of 54.43 kg, and is currently mainly used with the US Navy's "Fire Scout" unmanned helicopter. In addition to the visible/infrared sensors, laser ranging and illumination systems of the SafireIE turret, the turret is also equipped with a laser targeting system to guide laser-guided weapons. The turret has an in-built targeting module that allows the imager and the photocamera to automatically target the point of illumination of the laser pointer, ensuring that the two types of sensors act in unison and facilitating the controller's ability to quickly switch between different spectral images.

The latest addition to the Safire family is the StarSafireHD, the first all-digital HD SteadyShot airborne sensor turret to use a 1500 mm focal length lens, all sensor-generated images are fully digital, avoiding the loss of resolution caused by analogue-to-digital conversions, and high-bandwidth digital video transmission ensures that all scene details are transmitted from the sensors back to the platform in full. The turret has seven sets of built-in imaging sensors or laser sensors, which can be controlled via just one fibre optic. The U.S. Army has used the SafireHD and StarSafireIE turret systems in the Rapid Airborne Initial Deployment (RAID) programme. The plan for small tethered balloons equipped with high-performance sensors, in about 100 m altitude to implement surveillance, early warning, for the deployment of U.S. troops in Iraq and Afghanistan to provide vigilance protection.

The MX series turret is manufactured by WESCAM in the United States. It was initially used on various manned aircraft, such as the MX-20 turret for the United States Navy's P-3C maritime patrol aircraft and the Coast Guard's C-130 reconnaissance aircraft, and the smaller MX-15 turret for the Coast Guard's HU-25 Falcon aircraft. The MX-15 turret was further modified with an improved version of the Mosquito drone produced by General Atomics Aeronautical Systems, which was used in Iraq for one and a half years. The sensor system consists of a colour zoom camera, a colour or monochrome reconnaissance camera with an ultra-long focal length lens, an infrared zoom camera, a laser rangefinder, and a laser illumination

9.1 Principle of UAV TV Camera and Tracking and Positioning

unit. The infrared camera has four fields of view: wide, medium, narrow and ultra-narrow, and the colour zoom camera has three fields of view: wide, medium and narrow. And with the ultra-telephoto reconnaissance camera, it is possible to observe the details of the target.

WESCAM has also developed the MX-15D, based on the MX-15, which incorporates a laser target designation system for guiding laser-guided weapons into the turret. The MX-15D model has the longest optoelectronic/infrared identification range of any turret in the 15in (~ 38.1 cm) diameter class. There is also the more compact MX-12 turret, with a diameter of 30.5 cm and a mass of 24 kg, which can be equipped with any four of the three levels of infrared zoom cameras, colour zoom cameras, high-magnification reconnaissance cameras, laser rangefinders, and laser irradiators, as required. In order to capture clear images and locate the target precisely, it is necessary to avoid sensor jitter and eliminate the influence of UAV flight on sensor imaging; otherwise, even the longest focal length lens and the highest resolution imaging element can only obtain blurred images. For this reason, the gimbal of all MX series turrets is equipped with an inertial guidance system (IMU) consisting of a solid-state fibre-optic gyroscope and accelerometers, which is used to accurately determine the platform's attitude and form control signals, so that the turret's sensors are always stably pointed at the target.

"The Multi-spectral Target Acquisition System (MTS) turret was developed by Raytheon Company in the United States, and is mainly used for the Predator of the United States Air Force and the Predator B, which has stronger flight performance." The MTS-A turret, used to equip the Predator, has a diameter of 43.2 cm and incorporates an electro-optical/infrared sensor and laser indication system. The MTS-B turret for the Predator B is heavier and has a longer sensor range than the A model, with a diameter of 53.9 cm and a weight of 103.5 kg. 7 fields of view are available, including 2 ultra-narrow fields of view (0.08° × 0.11°), 1 narrow field of view, 3 medium fields of view and 1 wide field of view (35° × 45°). × 45°). At ultra-narrow field of view (highest optical magnification), the IR and photoelectric cameras have 2× and 4× electronic zoom ratios, respectively, to further magnify the image.

The MTS series of turrets, like the MX series of turrets, also houses an IMU unit. the most outstanding technical feature of the MTS series of turrets is the image fusion capability as standard, which Raytheon has been working on for more than 10 years. The system systematically analyses the signal-to-noise ratio of 300,000 pixel (640 × 480) infrared and visible light images, then selects the sharpest pixels of the two images and superimposes them to create a single image, in a fully automated process. Raytheon will also make further improvements to the MTS turret, including increasing the resolution and clarity of the optoelectronic/infrared system, developing a short-wave camera that can contain more detail, adopting smaller diode-pumped lasers, and developing a compressed video format, in order to improve the performance of the turret system so that the information is transmitted faster and occupies less bandwidth.

3. Optoelectronic stabilisation platform development trend

(1) High-performance sensors and integration technology

Optoelectronic sensors include front-view infrared, TV camera, laser range finder, laser illuminator, etc. Among them, TV camera has high resolution and colour image, but it is only applicable to daytime; front-view infrared is suitable for nighttime reconnaissance; laser range finder is used to measure the distance to the target, and laser illuminator is used to indicate the target for precision-guided weapons.

With the development of optoelectronics and microelectronics technology, optoelectronic sensors are also being updated and some new types of airborne optoelectronic sensors have appeared. The use of large-aspect/micro-optical CCD camera greatly improves the resolution and reception sensitivity; high-definition television video technology, from interlaced scanning to progressive scanning and image processing, can eliminate image skew, improve the speed of imaging and image clarity, has become the U.S. Department of Defense tactical needs and the video of the UAV with the industry standard; the application of digital cameras can be taken into account the camcorder and the camera's The application of digital cameras can take into account the different needs of video cameras and camcorders.

Sensors are constantly improving their performance while developing in the direction of integration to adapt to different operational needs. One is the integration of multiple sensors on the same platform to achieve complementary functions and performance, such as the U.S. Army's UAVs equipped with optoelectronic platforms usually integrating 4–7 types of sensors to meet the demand for reconnaissance and strike integration. Another is that the same sensor integrates multiple functions, such as the quantum well infrared detector with high integration of 1–4 million pixels and high sensitivity of mk level, which can detect 2–3 spectral bands at the same time; and the laser rangefinder/target indicator is developed in the direction of fusion, so as to be able to both measure the distance to the target and indicate the target to be attacked for the weapon system at the same time.

(2) High-precision Stable Targeting and Tracking Technology

High-precision stable aiming and target tracking is one of the core performances of UAV optoelectronic platforms, and its precision determines whether the sensor can accurately detect and accurately judge the target, which affects the combat effectiveness of UAVs.

1) High-precision composite stable aiming technology. When equipped with high-definition dynamic video source data, the stability of the image is particularly important in order to obtain useful information. The commonly used forms of stabilising platforms are reflector stabilisation and platform overall stabilisation. Secondary composite stabilisation in the platform overall stabilisation to achieve coarse stability on the basis of the organic combination of reflector fine stabilisation technology. This technology puts forward higher requirements for platform control technology, and also requires the support of multi-spectral common

9.1 Principle of UAV TV Camera and Tracking and Positioning

optical path technology. "The MTS-A and MTS-B optoelectronic platforms equipped with Predator UAVs have already realised the practical application of this technology.

2) Multi-spectral common optical path technology. Visible light television, infrared thermal camera, micro-optical television, CCD camera, laser range/illuminator, spot tracker, laser illuminator, etc. are commonly used sensors in airborne optoelectronic systems. Visible light, infrared, and laser work in different bands and require corresponding optical systems and windows, which results in larger volume and quality of the optical system, more difficult calibration of the optical axis, and difficult to improve the accuracy of steady aiming. Multi-sensor using a common optical path structure, integrated design can effectively reduce the quality of the system, reduce the volume, increase the aperture of the optical system, improve the resolution, sensitivity and accuracy of steady aiming. Raytheon company in the MTS-A type and MTS-B type photoelectric platform through the use of reflective optical system to make this technology has been applied.

3) Airborne self-calibration technology. The multi-sensor integration of the optoelectronic platform makes it a very critical issue for the sensors to keep the optical axis parallel, which has a great impact on the target positioning accuracy. The traditional practice is to carry out periodic calibration on the ground, but due to the ground, air environment differences, such as temperature changes, external vibration, stress release and other factors, will cause the calibrated optical axis in the air to reoccur deviation, so it is necessary to take the air real-time calibration of the approach to calibrate the optical axis of the parallelism difference. This eliminates the carrier vibration, impact and temperature and air pressure changes on the optical axis, and can be calibrated at any time when needed to improve positioning accuracy, better meet the operational requirements.

(3) New imaging technology

Visible light, infrared is the traditional imaging means, the application of new imaging technology can break through the limitations of the environmental conditions and the limitations of the sensor itself, to provide richer and more detailed information about the target, to provide the basis for accurate judgement of the commander.

(1) Multi-spectral/Ultra-spectral imaging technology. Multi-spectral (tens of spectral bands) and hyper-spectral (hundreds of spectral bands) imaging is a technology that uses sensors to perform multiple spectral imaging of the target, which can achieve a high recognition rate and ultra-fine observation of target information. Hyperspectral imaging technology has strong anti-camouflage, anti-concealment and anti-spoofing capabilities, which facilitates the comprehensive mastery of target information in the theatre of operations. In addition, passive hyperspectral imaging through aerosol clouds can provide early warning of non-traditional attacks, as well as detection and identification of biochemical warfare particles. At present, the first generation of airborne hyperspectral sensors have been successfully applied to the U.S. Predator and Pioneer UAVs,

while the multi-spectral data and information products of commercial satellites have become the backbone of commercial applications, with a resolution of several metres.
(2) Light Detection and Distance Imaging Technology. Light detection and distance imaging technology, i.e., active illumination and distance-selective imaging detection technology, is a means of detecting and identifying low-observable targets using the high spatial resolution imaging characteristics of laser beams. Under moderate cloud cover, dusty and smoky environments, imaging is achieved through the use of precise, short-pulse lasers that capture reflected photons; high-resolution imaging in low visibility at night can also be achieved, resulting in a clearer image and a longer range of action, achieving long-range character recognition capability.

(4) Reconnaissance information processing technology

With the development of sensor technology, the information reconnaissance through the photoelectric platform will be more and more rich, these massive reconnaissance information needs to be processed comprehensively in order to provide effective decision-making support for the commanders. Reconnaissance information processing technology to meet the operational needs for the direct purpose, at present, mainly presents the following aspects of the development trend.

(1) Target information integration technology. Modern weapon system information combat requirements reconnaissance system can provide more target reconnaissance information, it is necessary to obtain the target reconnaissance video at the same time to further develop the target geolocation, passive ranging, motion state estimation, image enhancement, splicing reconnaissance, geographic registration and other technologies to improve the efficiency of the target search and capture, and to achieve the first enemy discovery, the first enemy precision strike purpose.
(2) Image intelligent processing technology. The development of image/signal processing technology and network technology makes image intelligent processing possible, which will greatly reduce the burden on personnel and even keep people away from the battlefield. For example, through the active target prompter, the sensor can automatically search for targets that match the characteristics of the target library, or prompt the operator to focus on the targets that have changed since the last observation or targets that are significantly different from the environment. Currently, Raytheon has adopted image auto optimisation technology on the MTS-A platform, which can maximise the information displayed on the image and enhance situational awareness and remote monitoring capabilities.
(3) Image fusion technology. Different types of sensors have different focuses on target detection, each with its own strengths and weaknesses, and it is necessary to fuse images from different sources to maximise the use of the resulting image information.

9.1.5 TV Image System Tracking and Localisation

1. UAV TV image target positioning methods and processes

Currently, there are three main methods for target positioning using UAV TV reconnaissance images.

(1) Non-real-time positioning based on image matching pattern

This method mainly makes use of accessible multi-source image resources, and under the condition of establishing a pre-baseline image, matches the digitally processed and geometrically corrected UAV TV image with the pre-baseline image with high accuracy, thus realising the precise positioning of the target of interest. This method has outstanding advantages such as high target positioning accuracy and multi-point simultaneous positioning, but the non-real-time working mode restricts the scope of its application.

(2) Real-time positioning based on UAV telemetry data

This method directly infuses the position information and attitude information of the UAV at the instant of target positioning, as well as the angle information and ranging information of the reconnaissance rotary table into the positioning solution model, so that the target coordinates can be quickly solved. This method is adopted by all active UAS because of its outstanding advantage of good real-time performance. However, when using TV (visible/infrared) reconnaissance equipment to track and locate ground targets, the position and attitude errors of the UAV, the angle of the reconnaissance rotary table and the ranging errors, etc. inevitably affect the target positioning accuracy, and thus the target positioning accuracy is lower than that of the first method.

(3) Target positioning based on spatial rendezvous

This method is essentially an expansion of the second method. In the process of TV reconnaissance performed by the UAV, after discovering the target of interest, it enters the tracking state, while the laser ranging equipment continuously fires the laser to measure the distance from the TV camera to the target, and collects the flight telemetry data and image data after tracking. These telemetry data include three attitude angles of the aircraft, two angles of the camera turntable, camera focal length, aircraft position and laser distance; then the telemetry data are combined to construct a rendezvous model of multiple positions in space to the same target on the ground, and the rendezvous model is used for the levelling calculation to achieve the target localisation; the multi-point spatial rendezvous solving process has a better rejection and suppression of the error, so that a higher positioning accuracy can be achieved. positioning accuracy.

The target positioning process of UAV TV image is carried out by the TV camera mounted on the photoelectric stabilised platform, which reconnaissance ingests the ground image and displays it on the display terminal of the ground control platform. When the target of interest appears on the screen, the marker generated by the tracker is used to cover the target, and the tracker outputs the target's coordinates on the screen to the positioning solving terminal; the solving terminal calculates the coordinates to control the rotating of photoelectric stabilised platform according to the position of the target and the relevant parameters sent by telemetry, and then it can calculate the target positioning accuracy. According to the position of the target and the relevant parameters sent by telemetry, the solution terminal calculates the signal for controlling the rotation of the photoelectric stabilized platform, which is sent to the aircraft through the remote control system and sent to the photoelectric stabilized platform after processing by the flight control terminal on the aircraft, and the photoelectric stabilized platform is rotated under the action of the signal to make the target move towards the centre of the screen, and the control signal is also changed after the target has been moved; when the target moves to the vicinity of the screen centre, the camera focus (telephoto) will be adjusted to enlarge the image of the ground scene in order to determine the centre position of the target more accurately when positioning and calculate the geographic coordinates of the target. Adjust the camera focal length to short focus, can increase the shelter area of the TV camera, when the target to focus on reconnaissance appears on the screen, quickly adjust the camera focal length, enlarge the image of the target area, and then click on the target with the mouse, you can give the target's screen coordinates, and transmit it to the ground target positioning solution terminal, so as to give the actual geographic coordinates of the target.

In order to facilitate the completion of the target positioning task, the position of the target and the aircraft, the focal length, the altitude, and the attitude angle of the aircraft are also transmitted to the ground in real time and displayed on the screen as required.

2. Definition and Composition of Imaging Tracking

The so-called imaging tracking is to use the image characteristics of the scene to achieve the tracking of the target. The tracking device is usually composed of a detection system and a servo mechanism. The detection system provides measurement information and the servo mechanism completes the tracking of the target. The overall performance of the tracking system mainly refers to the tracking speed, the spatial range of tracking, and the frequency range of tracking. These overall tracking performances are largely dependent on the sensitivity and accuracy of the detection system, which are the two main performances.

The imaging tracking method has the following advantages:

1) In the case of natural or artificial interference, imaging tracking can suppress the influence of interference based on its rich information content to improve detection and tracking accuracy;
2) It can provide a richer amount of information than point tracking;

9.1 Principle of UAV TV Camera and Tracking and Positioning

3) has an image recognition function that can be used to recognise targets and their types from complex backgrounds;
4) have high tracking accuracy.

For the imaging system, the scanning method is usually used to divide the object space for sequential sampling, and then composite imaging; recent point detection system also uses scanning method to improve the performance of the detection system.

In the scanning mode of detection system in the detection of the target, in the case of the target distance is far away, detected only the target of the image of the point; when the target distance becomes smaller, it will gradually show the target like to. The detection system working with scanning mode is a kind of extensive imaging detection system.

Imaging tracking system consists of a camera, image monitor, image signal processing circuit, servo mechanism and other components. Usually in the imaging tracking system also contains an image recognition part, the structure of the imaging tracking system is shown in Fig. 9.12. The functions of each part are specified as follows.

(1) Camera

Camera is a device for videotaping the scene. According to the different radiation sources, it is divided into millimetre wave camera, infrared camera, visible light camera, laser camera and other types. The function of the camera is to provide the tracking system with information about the state of the target, and the requirements of the observation system are not the same as those of the imaging tracking system, which does not require much information about the texture of the scene.

(2) Image Monitor

The image monitor is used to display images of the scene for observation, and it uses analogue signal processing for imaging. Provide tracking and identification of

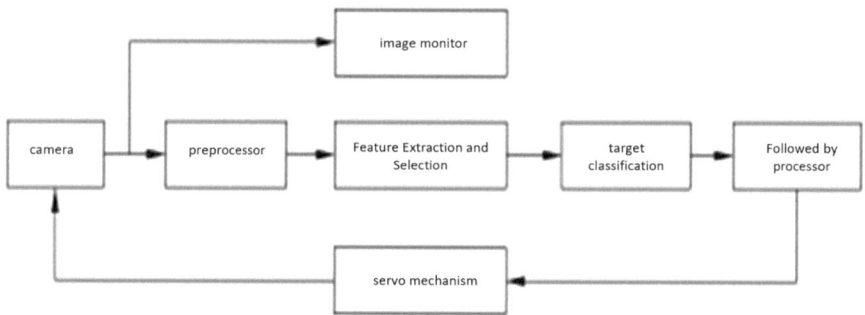

Fig. 9.12 Image tracking system structure

signals are used in digital signal processing, so it is necessary to be taken by the camera through the A/D converter will be converted to digital signals in the time sequence of analogue signals processed after processing, the capacity and speed of the A/D converter by the camera's total number of pixels per frame, the total number of grey levels and the speed of the frame to determine.

(3) Pre-processor

The function of the pre-processor is to pre-process the image signal to improve the image quality and reduce the amount of operations. The contents of image preprocessing are as follows.

1) denoising: the spatial domain of the neighbourhood average, median filter, matched filter, Kalman filter, gradient weighted average and frequency domain low-pass filtering.
2) Image correction: geometric correction of images, normalisation of image signal quantisation, etc.
3) Data compression: hierarchical search, grey scale compression, image projection, magnitude sorting, Hoffman coding, transform coding, etc. Neighbourhood averaging and filtering is also a means of data compression.
4) Image enhancement and compensation:There are overall image enhancement, high-frequency compensation, histogram equalisation, logarithmic transformation and so on.

For specific systems, which preprocessing steps to take depends on the quality of the ingested image signal and system requirements.

(4) Feature extraction and feature selection

Feature extraction: from the scene of the original grey-scale image to extract the image depiction features is an important step in image processing, depiction of the extraction of features known as feature extraction. Considering the method of feature depiction, it is divided into two categories: line feature depiction and area feature depiction. From these depiction features, the shape features of the target and the moment features of the target can be obtained. The shape features mainly include area, perimeter, aspect ratio, roundness, density, etc.; the moment features include form centre, higher order moments and invariant moments.

The above features can be obtained at the same time in the process of image processing, and then according to the needs of image recognition and tracking, in accordance with the principle of feature selection in the above features to select some useful features for further operations to achieve the purpose of compression of dimensions and simplification of operations.

(5) Target classification

After feature selection, the target features are classified according to certain classification criteria to identify the target. These criteria such as: minimum distance method, minimum average loss method, tree classification method, etc..

For imaging tracking systems, the target is usually determined on the basis of shape without having to know more details about the scene image.

(6) Tracking processor

This is the key part about imaging tracking. Operations related to tracking patterns, tracking state estimation, and filter prediction are all included in the tracking processor. The signal output from the tracking processor is the amount of error signal of the tracking system relative to the target state.

(7) Servo Mechanism

The target error signal sent by the tracking processor is first passed through the control processor of the servo mechanism in order to derive the required control signal to control and adjust the working state of the whole tracking system and drive the tracking mechanism to track the target.

3. Real-time correlation tracking

Real-time correlation tracking is the most widely used tracking method for tracking moving targets in thermal imaging systems and TV imaging systems. It can implement fast tracking of moving targets according to the real-time measurement of the target position.

Composition and principle: the scene signal is converted into a video signal by the camera, the A/D converter quantifies the video signal, the resulting digital image is stored in the image memory, and the D/A converter converts the image stored in the image memory into a corresponding analogue video signal for display on the monitor. Before the relevant detection, a reference image of the tracked target must exist in the tracking system. The reference image is typically acquired in real time by human intervention before the system performs tracking. In order to adapt to the tracking system and the target distance change and time delay caused by the target brightness, geometry, relative position relationship and other characteristics of the change, the system must have a base image update item. The function of the preprocessor is to preprocess the real-time image acquired from the image memory to make it more realistic and more adaptable to the relevant detectors for detection processing, as shown in Fig. 9.13.

At the same time, according to the relevant processing results and the benchmark image update guidelines to determine the next relevant detection of the benchmark image; repeat the above process, the real-time tracking of the target is achieved.

4. Imaging tracking mode and image matching

There are two important research aspects in imaging tracking technology, namely: motion analysis of sequence images and structural design of imaging system. The motion analysis of sequence images is the basis of imaging tracking, and the structural design of imaging tracking system involves two aspects of imaging detection and tracking. The basic point of the imaging system research is the three aspects of tracking accuracy, intelligence and image recognition function.

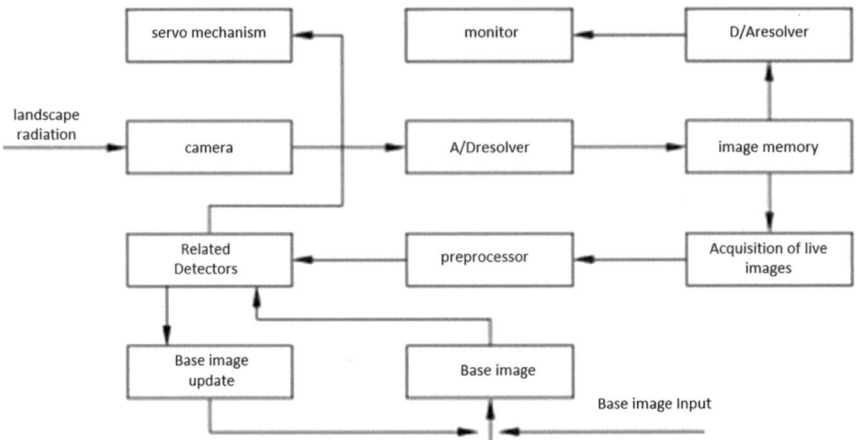

Fig. 9.13 Real-time image correlation tracking system composition schematic diagram

When the tracking system works, it needs to extract the position information of the target from the image of the target, and then form the tracking error signal to drive the servo mechanism to track the target.

The size, shape, grey scale and its distribution of the target image, as well as the resolution of the image system and other factors with the structure of the imaging detection system, imaging tracking system on the target in the working state, environmental conditions and other different and have a time-varying nature. The tracking mode adopted by the tracking system on the target should be changed accordingly with the above image parameters in order to get the best tracking performance.

Image matching: Identifying homonymous points between two (or more) images is the core problem of computer vision. Image matching can be used for the localisation of a single active target or for the detection, classification and localisation of a larger region of the scene. From the tracking point of view, the probability of correct interception and the localisation accuracy are the main performance indicators of image matching; from the system structure aspect, the real-time computing of the image matching system is the key, which depends on the performance of the dedicated image processing hardware.

(1) Correct interception probability and localisation accuracy

In the presence of other errors such as noise in the image, the resulting correlation function may have several peaks or valleys with random ups and downs. These random peaks or valleys may affect the correct matching of the image. The higher the probability of correct matching, the higher the reliability of the search interception process and the higher the correct interception rate.

There will also be errors between the match points obtained under the effect of noise and other errors and the true match points, and the variance of the matching error is often used to describe the positioning accuracy of the matching and positioning system.

9.1 Principle of UAV TV Camera and Tracking and Positioning

(2) The impact of various errors on the matching performance

Real-time ingested scene image compared with the reference image, often due to the image ingested when the geometric conditions and radiation conditions of the differences and other reasons and there are a variety of errors.

1) Geometric distortion. The errors belonging to this category include image rotation, changes in image scale, changes in perspective direction, and so on. All these errors increase the overlapping area of the two image mismatches, thus reducing the matching probability and decreasing the positioning accuracy.
2) Selecting a smaller real-time map size (or correlation window) is necessary to reduce the impact of such errors.
3) Gray scale distortion. The causes of grey scale distortion are roughly irradiation conditions and changes in the reflectivity or emissivity of the scene itself when ingesting the real-time image, differences in camera type and sensitivity between the real-time image and the reference image, and changes in environmental conditions.

The whole grey scale distortion of the real-time image relative to the reference image has no effect on the Prod algorithm, but it has a greater effect on the MAD algorithm, reducing the probability of correct interception.

(3) Matching algorithms

There are many matching algorithms, taking into account the matching speed and matching accuracy, usually use the common correlation algorithm, feature matching algorithm, hybrid algorithms, here is only a brief introduction, please refer to the relevant information.

1) Ordinary correlation algorithm. This algorithm is based on the overall characteristics of the whole image, and the preprocessing is also completed as a whole. Through the calculation of the correlation function between the real-time image and the reference image, the offset of the real-time image relative to the reference image is obtained by the best matching.

Advantages: the operation is carried out in the case of a smaller image signal-to-noise ratio, with the effect of suppressing the inherent noise, and this algorithm is more effective in the occasions where there are only local errors.

Disadvantages: When the image within the correlation window has a large uniform area with little detail, the correlation function value may be flatter and it is difficult to detect peaks. Differences in scale size, geometric distortion, image rotation, and radiometric brightness between the real-time and reference images can also cause difficulties in matching.

2) Feature matching method. This method is the first reference map and real-time map for feature extraction, in the feature matching algorithm, such features are usually graphic edges, boundary line segments and their vertices, but also for their symbols and structural descriptive features; and then match the features of

the two images for the matching operation; matching operation in the measure is often taken between the corresponding feature quantity of the Euclidean distance, as a mismatch error measure, and then establish the feature quantity distance measure Then the feature distance measurement matrix is established, and the real-time map is shifted relative to the reference map in order to get the value of the measurement matrix for the best match, and finally the positioning is carried out according to the best match.

Advantages: When there is only a regional error in the image, it will not have a significant effect on the extraction of features, so it is more suitable to use the feature matching algorithm, and the algorithm of segmenting the whole image into a number of homogeneous zones and then matching the features can make the correlation significantly steeper, so it can improve the positioning accuracy.

Disadvantage: Because image segmentation and feature extraction requires a certain signal-to-noise ratio, so the feature matching algorithm can only be used in the case of a larger signal-to-noise ratio.

3) Hybrid Algorithm. A hybrid algorithm is only the reference map is divided into a number of homogeneous areas, and then according to the reference map of the homogeneous areas of each image offset position of the real-time map for segmentation, the homogeneous areas of the two images in the regional correlation value, and finally the regional correlation value will be added to get the total value of correlation; another hybrid algorithm for the observation of the scene of the radiation, the characteristics of the camera and the nature of the error and the integrated use of correlation algorithms and feature matching algorithm. The other hybrid algorithm is a combination of correlation algorithm and feature matching algorithm for the observed scene radiation, camera characteristics and error nature.

9.2 Principle of Infrared Imaging for UAV

The origin of infrared reconnaissance technology can be traced back to 1800, when the British astronomer W. Herschl (W. Herschl) in the process of looking for ways to protect the eyes when observing the sun discovered the infrared radiation, he will be placed in the prismatic dispersion of mercury thermometer in different positions of the solar spectrum, to observe the thermal effect of the sun's spectrum of various parts of the thermal effect, found that the largest thermal effect is outside the red end of the visible spectrum, and called this light "invisible rays", later called "infrared". He placed mercury thermometers in different positions of the solar spectrum dispersed by prisms to observe the thermal effect of each part of the solar spectrum, and found that the location with the greatest thermal effect was outside the red end of the visible spectrum, and called this light "invisible rays", which was later referred to as "infrared rays". Later, the French physicist Michel Blanc called it "infrared radiation".

9.2 Principle of Infrared Imaging for UAV

In the 1930s, active infrared night vision began to appear, Germany in the Second World War, the first to apply it to the war; the next 30 years, active infrared reconnaissance instrument has been the main military reconnaissance equipment at night. Until the 1970s, due to the rapid development of passive infrared reconnaissance technology, the dominance of active infrared reconnaissance technology was broken.

Infrared reconnaissance technology is the use of target reflection or radiation of infrared characteristics of the difference to detect the target, access to information technology. It is based on the intensity and wavelength difference of infrared radiation of the object, the use of reconnaissance devices will be invisible to the naked eye infrared radiation difference into the naked eye can see the image or data, so as to extract useful information process. Infrared imaging reconnaissance devices from a technical point of view can be divided into two categories: direct and indirect infrared imaging devices.

Direct infrared imaging is the target reflected infrared through the infrared tube and infrared film directly into the visible image, this imaging technology is generally only 1–3 um below the near infrared radiation response.

Indirect infrared imaging breaks through this limitation, it uses optical scanning and the mid- and far-infrared radiation sensitive solid semiconductor materials, as well as the scene itself the difference between the various parts of the radiation, the target and the background of the infrared radiation energy into electrical signals, the electrical signal processing and amplification, and then through the display device into a visible image, it can be the wavelength of the mid-red radiation of 3–5 μm and 8–14 μm of the It can respond to mid-red radiation with a wavelength of 3–5 μm and far infrared radiation with a wavelength of 8–14 μm. Since these wavelengths of infrared radiation are also known as thermal radiation, indirect infrared imaging can also be called thermal imaging, i.e., the reproduction of the fine temperature distribution map of an object. This imaging technique overcomes both the disadvantages of active infrared night vision, which requires an artificial source of infrared radiation and the consequent vulnerability to self-exposure, and passive microlight night vision, which relies entirely on ambient natural light. The infrared imaging system has the ability to penetrate smoke, fog, haze, snow and other restrictions as well as to identify camouflage, is not blinded by bright light or flash interference on the battlefield, and can achieve long-range, all-weather observation.

9.2.1 Physical Basis of Infrared

The phenomenon of an object radiating energy due to its temperature is called thermal radiation. Thermal radiation is a universal phenomenon in nature, all objects, as long as their temperature is higher than the thermodynamic temperature zero (− 273.15 °C) will produce radiation.

Usually, the energy propagated in the form of electromagnetic waves is called radiant energy. Radiant energy can represent both the total electromagnetic energy emitted by a radiation source over a defined period of time and the energy received by the surface of a blocked object. However, since most of the detectors used are not cumulative, they respond not to the total energy transmitted, but to the rate of transmission of the radiant energy, i.e., the radiant power. Therefore, the radiant power and, by derivation, several physical quantities in radiometry belong to the fundamental radiative quantities.

9.2.2 Infrared Imaging Technology

Since the 1980s, infrared imaging devices and their system technology has been developing rapidly, and the equipment involved mainly includes infrared viewers, infrared scopes, periscopic infrared thermal imaging cameras, fire control thermal imaging cameras, infrared tracking systems, forward-looking infrared systems and infrared cameras, etc., which are mainly used in night reconnaissance and surveillance, targeting and shooting, guidance and air defence, navigation, search of manned and unmanned aircraft, tracking, identification, capture, observation and fire control and other fields. At present, the United States, Britain, France, Germany and Russia and other countries in the field of this technology development and application of the world's leading position.

Infrared thermal imaging technology can be divided into two types: cooled and uncooled. The former has the first, second and third generations, and the latter can be divided into two types: pyroelectric camera tubes and thermoelectric detector arrays.

1. The first generation of infrared imaging technology

The first generation of infrared detection technology mainly consists of infrared detectors, optical machine scanners, signal processing circuits and video displays. The infrared detector is the core device of the system and determines the main performance of the system. Infrared detectors have indium antimonide (lnSb) and mercury cadmium telluride (HgCdTe or CMT) devices. Currently widely developed is a high-performance multivariate HgCdTe detector, the device has been as high as 60, 120 and 180 yuan. the early 1980s, a device called SPRITE detector (or swept product detector) in the United Kingdom, which is composed of a few narrow strips of the aspect ratio of more than 10:1 of the photoconductive HgCdTe elements, working under positive bias voltage. The SPRITE detector, in addition to its signal detection function, enables the delay and integration of the signal within the device, reducing the number of device leads and thermal load. Compared to multivariate detectors, Dewar bottles have a simpler structure, reduced process difficulty, and greatly improved reliability. An 8-bar SPRITE detector is equivalent to the performance of a $120 HgCdTe detector, but requires only 8 signal channels. In order to facilitate the organisation of mass production, reduce the cost of thermal imaging cameras, eliminating the need to repeat the design and development costs, easy to repair, maintain and effectively equip the troops, the United States, Britain, France and other countries have implemented a thermal imaging of the universal components. The United States thermal imaging universal components using multiple HgCdTe detector, and

9.2 Principle of Infrared Imaging for UAV

scanning system; the United Kingdom uses SPRITE detector, series, and scanning system. These two thermal imaging system temperature resolution can be less than 0.1 °C, image clarity can be comparable with the image enhancement technology.

2. The second generation of infrared imaging technology

The second generation of infrared imaging technology uses an infrared focal plane detector array (IRFPA), thus eliminating the need for an optical scanning mechanism. This focal plane array with the help of integrated circuits, the detector installed in the same chip and has the function of signal processing, the use of a very small number of leads on each chip thousands of detector signals read out to the signal processor. Because of the removal of optical machine scanning, this sensor with large-scale focal plane imaging is also known as a gaze sensor. It is small, lightweight and highly reliable. There can be arrays of detectors with hundreds of elements or more in the pitch direction, which gives a larger tensile field of view, and a special scanning mechanism can be used to complete a full 360° scan to maintain high sensitivity using a much slower scanning speed than a general-purpose thermal imaging camera. These devices mainly include InSbIRFPA, HgCdTeIRFPA, SBDFPA, uncooled IRFPA and multi-quantum well IRFPA.

3. Third-generation infrared imaging technology

The number of infrared focal plane detector units used in third-generation infrared imaging technology has reached 320 × 240 or higher, and its performance has improved by nearly three orders of magnitude. At present, the unit sensitivity of 3–5 f × m focal plane detectors is 2 ~ 3 times higher than that of 8–14 pm detectors. As a result, the overall performance of the 320 × 240-based medium-wave and long-wave thermal imaging cameras is not much different, so the 3–5 fxm focal plane detectors are particularly important in the third generation of focal plane imaging technology. In the long run, HgCdTe focal plane detectors with high quantum efficiency, high sensitivity, and coverage of the mid- and long-wave phases are still the first choice for focal plane device development.

4. Uncooled infrared imaging technology

Due to the refrigeration type infrared detector material is expensive, the detector of the finished product rate is very low, resulting in a refrigeration type infrared imaging system is expensive; at the same time, the refrigeration type infrared imaging system needs a set of refrigeration equipment, increasing the system cost, reducing the reliability of the system; in addition, the refrigeration type infrared imaging system power consumption, large volume, bulky and heavy, and it is difficult to achieve miniaturisation, which limits the refrigeration type infrared imaging system of the Widely used.

Uncooled infrared focal plane detector arrays have the advantages of room temperature operation, no refrigeration, wavelength-independent spectral response, relatively simple preparation process, low cost, compact size, easy to use, maintenance and reliability, etc. Therefore, a new vitality development direction has been formed, which aims at obtaining excellent infrared imaging performance with lower cost, smaller size and lighter mass. In recent years, three different types of uncooled infrared focal plane detector arrays have been successfully developed, and the physical mechanisms of these three different types of uncooled infrared focal plane detector arrays are as follows.

(1) thermopile: according to Seebeck (Seebeck) effect detects the temperature gradient between the hot end and the cold end, the signal form is voltage.
(2) radiometric thermometer: detection of temperature changes caused by carrier concentration and mobility changes, the signal form is resistance.
(3) Pyroelectric: detects changes in dielectric constant and spontaneous polarisation strength caused by temperature changes, with the signal in the form of a charge.

Of these three different types of uncooled infrared focal plane detector array devices, the radiometric thermometer array has seen the most rapid development and impressive success. It is fabricated using silicon micromachining techniques similar to the silicon process, and is generally constructed as a bridge in order to achieve effective thermal insulation. The detector is electrically interconnected to the silicon readout circuitry by two support legs. The sensitivity of a radiometric thermometer depends mainly on its thermal insulation from the surrounding medium, i.e. the thermal resistance–the higher the thermal resistance, the higher the obtainable sensitivity. Currently, the temperature resolution of the radiometric thermometer array can reach 0.1 K. In 2000, the French company Sofradir produced the first uncooled focal plane infrared detector, the detector array size of 320×240 yuan, the pixel centre distance of 45 pm, the fill factor is greater than 80%, the noise equivalent temperature difference (NETD) to 0.1 K (typical), the device's performance indicators to reach today's world advanced level. The performance index of the device has reached the advanced level in today's world.

9.2.3 Infrared Detectors

Infrared detector is the core component of infrared system, thermal imaging system, infrared detector research has always been the core of infrared physics and infrared technology development. At present, the use of solid radiation irradiation and the electrical properties of the photoelectric effect made of photon detector sensitivity range has been extended to more than 30 μm band, short-, medium- and long-wavelength infrared unit detector performance has reached or close to the theoretical level of the background limit. The second generation of image sensitivity element in thousands of pixels and even tens of thousands of pixels below the line array and surface array detector performance has also reached or close to the background limit, the uniformity of the device and the rate of finished products has also been significantly improved. Especially in the use of CCD readout circuitry to successfully solve the focal plane photon detector array output signal integration, delay and multiplexing problems, the third generation (more than 100,000 pixels) of the focal

9.2 Principle of Infrared Imaging for UAV

plane array detector has begun to enter the practical stage, the signal-to-noise ratio and the information rate has been substantially improved, from the structure of the infrared imaging system to bring about fundamental changes.

1. Classification of infrared detectors

There are many types of infrared detectors, classification methods are also many. For example, according to the wavelength can be divided into near-infrared (short-wave), in the infrared (mid-wave) and far-infrared (long-wave) detector; according to the operating temperature, can be divided into low-temperature, medium temperature and room-temperature detector; according to the use and structure, can also be divided into a unit, multiple and gaze-type array detector. Infrared detector in the photoelectric imaging system, mainly used to complete the infrared incident radiation to the conversion of electrical signals, so it can be imaging type, can also be non-imaging type. Therefore, from the theory of the general work of the conversion mechanism to be classified. In terms of its working mechanism, generally can be divided into thermal detectors and photon detectors (or photodetectors) two categories.

(1) Thermal detectors

Thermal detectors absorb infrared radiation, resulting in a temperature rise, accompanied by a temperature rise and the occurrence of certain physical property changes. Such as the generation of temperature difference electric potential, resistivity changes, spontaneous polarisation intensity changes, gas volume and pressure changes. Measurement of these changes can measure the energy and power of infrared radiation they absorb. The above four are common physical changes, the use of one of the physical changes can be made of a type of infrared detector. Such as the use of temperature difference electrical effect made of thermocouples; the use of resistivity changes in the thermistor or resistance measurement of the radio thermometer; the use of gas pressure changes in the gas detector (Gaurai box) and so on. Here mainly introduces can be used for thermal imaging pyroelectric detector and microbolometer and so on.

1) Pyroelectric detector

The working principle of pyroelectric detector with pyroelectric camera tube target works the same principle, just in the area size and signal readout methods and other aspects of the larger differences. Pyroelectric detectors and CCD devices mixed to provide the prospect of work without refrigeration. Due to the differential properties of pyroelectrics, the modulation of human radiation is required when used for imaging of gaze arrays, but of course it can also be used for scanning arrays.

The structure of the infrared charge-coupled device of the pyroelectric CCD hybrid is to make a pyroelectric film between the channel and the metal gate of the MOS field effect transistor, i.e., in series with the gate to form an infrared charge-coupled device, and to modulate the depth of the potential well of the MOS structure by the voltage generated by the pyroelectric detector. In this way, the signal charge is transferred due to the variation of the potential well depth. When the voltage is a constant and large enough, the potential well depth can be several kT, and the charge makes the potential well essentially full, with the drain on top of the N-barrier and into the CCD channel, and the voltage is regulated until the modulation of the scene is no longer attenuated by the dark current.

Two factors affect the detection rate: (1) the responsivity of the pyroelectric detector. This means that in the case of direct coupling, the noise of the CCD will dominate, so amplification is required between the detector and the CCD; ② the pyroelectric detector has to incur heat dissipation losses at the wafer interface. with a TGS layer thickness of 20 pm, the signal drops to about 1/30 at a modulation frequency of 20 Hz.

Typical infrared charge-coupled devices made with an oxide layer of 6.0×10^{-8} thickness and a TGS of 1(T5 cm^2 area have a minimum resolvable temperature difference of 0.2 K when operating at 20 fps in an 8 to 12 pm window. It should be noted here that in order for the actual device to achieve the desired performance, higher thermal insulation is required to avoid substrate thermal loading and crosstalk interference between image elements.

2) Microbolometer

Microbolometer (Microbolometer) is a detector material to absorb incident radiation to make its own temperature changes, which in turn makes the detector of other physical properties (such as resistance, capacitance, etc.) changes in the principle of thermal detector array. Commonly used microbolometer: ① thermistor microbolometer, its sintered semiconductor film as a light-sensitive element; ② metal film microbolometer, the use of resistance temperature coefficient of the metal as the material made into a thin film, the surface coated with black as a light-sensitive element; ③ dielectric microbolometer, which is the use of dielectric materials, the parameters of the principle of change with the temperature of the device.

Microbolometer offers the prospect of working without refrigeration. Microbolometer is in the IC-CMOS silicon chip using precipitation technology, with Si3N, supporting a high resistance temperature coefficient and high resistivity thermistor material VCh or polycrystalline silicon made of micro-bridge structure of the device (monolithic FPA), which receives thermal radiation caused by temperature changes and changes in the resistance value of the DC coupling without the need for a chopper, only a semiconductor cooler to maintain its stable operating temperature. Compared with pyroelectric UFPA, microbolometer can use silicon integration process, low manufacturing cost, good linear response and high dynamic range, good insulation between image elements and low crosstalk and image blurring, low noise as well as high frame rate and potentially high sensitivity. However, its bias power is limited by power dissipation and large noise bandwidth is difficult to compare with pyroelectric.

3) Microbolometer thermopile

Microbolometer thermopile is a number of radiation thermocouples connected in series to form a thermal detector device, the principle of the use of the temperature difference electrical effect. That is, when two different materials of metals or

9.2 Principle of Infrared Imaging for UAV

semiconductors form a closed circuit to form a thermocouple, if the two junction junction in a temperature increase by incident radiation irradiation, and the other junction is not exposed to incident radiation irradiation and the temperature remains unchanged, then due to the two junctions are at different temperatures and make the closed circuit to produce a temperature difference in electric potential, the measurement of the temperature difference in electrical potential can be measured to get to be measured by the size of the radiant energy or power. The measurement of the temperature difference electric potential can get the amount of radiant energy or power to be measured. Series connection of thermocouples accumulates the temperature difference potential generated at each junction improves the response rate, in addition, the series connection also makes the resistance of the radiation thermocouple increases and is easy to cooperate with the amplifier, but also reduces its response time.

Microbolometer thermopile is usually made of thin film technology into a thin film, its advantages are high response rate, stable performance, solid structure, can have a uniform response in a wide range of wavelengths, the use of cooling is not required. Microbolometer thermopile is widely used in spectrometer calibration, etc. In recent years, it has been successfully applied to the field of thermal imaging technology.

(2) Photon detectors (photodetectors)

Certain solids are irradiated by infrared radiation, where the electrons directly absorbed infrared radiation and produce a change in the state of motion, resulting in a change in some of the electrical parameters of the solid, this change in electrical properties collectively referred to as the solid photoelectric effect (within the photoelectric effect). According to the size of the photoelectric effect, the number of absorbed photons can be measured. Infrared detectors made using the photoelectric effect are also known as photon detectors or photodetectors. These detectors rely on the direct absorption of infrared radiation by internal electrons, without the intermediate process of heating the object, and therefore have a fast response. In addition, the structure of these detectors are relatively robust, can work in relatively harsh conditions, and thus photoelectric detectors are the fastest growing and most commonly used infrared detectors today. Commonly used photoelectric detectors are as follows.

1) Photoelectron Emission Detector

When light is irradiated onto the surface of certain metals, oxides or semiconductors, if the photon energy is sufficiently large, it is possible to cause the surface to emit electrons, a phenomenon known as photoelectron emission (external photoelectric effect). The use of photoelectron emission made of visible light detectors and infrared radiation detectors, collectively known as photoelectron emission detectors. Such as variable image tube, image intensifier and part of the camera tube are such devices. In addition, photoelectric tubes, photomultiplier tubes, etc. are also such devices. The time constant of these devices is very short, only a few milliseconds. Therefore, in laser communication, often using special photomultiplier tube.

Most photoelectron emission detectors work only with visible light. The only photocathodes that can be used in the near-infrared are the silver-oxo-caesium photocathode S-1, the polybasic photocathode series, and the negative electron affinity potential photocathode. Therefore, the development of new infrared photocathode is also one of the urgent tasks of infrared technology.

2) Photoconductive detector

When infrared radiation is projected to the semiconductor device, will make the body of some electrons and holes from the original non-conductive bound state into a free state that can conduct electricity, thereby increasing the conductivity of the semiconductor, a phenomenon known as the photoconductivity effect. The use of photoelectric conductivity effect in the manufacture of infrared detectors called photoelectric conductivity detector (referred to as PC devices). This type of device has a simple structure, the largest variety, and the widest range of applications.

3) Photovoltaic detectors

In the semiconductor P–N junction and its nearby regions to absorb photons with sufficiently large energy, in the junction area and the vicinity of the junction to release a few carriers (electrons or holes), they are in the vicinity of the junction area by diffusion into the junction area, in the junction area by the role of the built-in field, the electron drifts to the N area, the hole drifts to the P area. If the P–N junction open circuit, then the two ends will produce a voltage, a phenomenon known as the photovoltaic effect. The use of this effect made of infrared detectors called photovoltaic detectors (referred to as PV devices).

Photovoltaic detector response speed is generally faster than the photoconductive detector, is conducive to high-speed detection, it can be used for both direct detection, can also be used for differential reception. Photovoltaic device structure is conducive to the arrangement of two-dimensional arrays, people are interested in it and CCD devices coupled to form a focal plane array of infrared detectors.

4) Photomagnetoelectric detector

When the infrared light irradiated to the semiconductor surface, if there is an external magnetic field, then generated near the semiconductor surface of the electron - hole pairs in the process of diffusion to the semiconductor interior, the movement of electrons and holes in the magnetic field will be biased to the side of the action of the semiconductor on both sides of the potential difference. This phenomenon is called the photomagnetoelectric effect. The use of this effect made of infrared detectors called photomagnetoelectric detectors (referred to as PEM devices). Early had a photomagnetoelectric InSb detector commodity, but with the improvement of the quality of semiconductor materials, coupled with the photomagnetoelectric detector needs to bring a magnet is very inconvenient, this device has been rarely used. At present, the optical magnetoelectric effect is sometimes used in conjunction with photoconductors to measure the carrier lifetime, in order to avoid the troublesome radiation calibration work, but also can be measured to a lower carrier lifetime.

9.2 Principle of Infrared Imaging for UAV

In addition to the several types of devices introduced above, there is the use of photon traction effect detector devices, infrared conversion devices and quantum well devices.

2. Infrared detector performance parameters

Infrared detector performance can be described by many parameters, but the most basic are three parameters: the detector's ability to detect infrared radiation, wavelength response range and response speed. The detection ability contains two aspects: that is, the unit of radiant power people shoot to the detector produced by the size of the signal and the detector to identify the ability of weak signals.

(1) Responsivity

Response is a description of the incident to the detector units of radiant power produced by the signal size of the ability to performance parameters. It is defined as: infrared radiation incident perpendicular to the detector photosensitive element, the detector output signal voltage of the root-mean-square value of R and the incident radiant power of the root-mean-square value of the ruler of the ratio. The responsivity of the detector is related to the wavelength and modulation frequency of the incident radiation.

(2) Noise equivalent power

The detection capability of an infrared detector depends not only on the responsivity, but also on the noise level of the detector itself. A detector with higher responsivity and lower noise will be able to detect signals with weaker radiated power. Therefore, any detector has a detectable radiated power threshold determined by its own noise level.

(3) Response time (or time constant)

Response time is the relaxation time for a detector to convert incident radiation into electrical output, and is a quantitative parameter indicating the speed at which the detector operates. Because infrared detectors are inert, the response to infrared radiation is not instantaneous, but rather fast or slow depending on the detector material.

Some IR detectors have two response times. This is because there is one response time for one wavelength of radiation and another response time for another wavelength of radiation. In fact, in the operating frequency range, responsivity, detection rate IT are frequency-dependent, in addition to special needs should try to avoid the use of detectors with two response times.

In addition to the above introduction of the responsivity, noise equivalent power, detection rate and response time and other parameters, in the use of detectors should also meet the following conditions:

1) There is a linear relationship between detector responsivity and radiation intensity;

2) The responsivity of the detector on the receiving area is uniform;
3) Detector and optical system matching, the detector's receiving area should be the same size as the optical system into the optical image;
4) When the detector is connected with the preamplifier, the internal resistance of the detector should match the impedance of the amplifier.

3. Infrared focal plane array detectors

Currently, most thermal imaging systems still use a single unit or a simple multivariate detector that uses optical-mechanical scanning in both vertical and horizontal directions to obtain a two-dimensional image. Working with this method not only requires a large bandwidth of the system, but also makes the whole system large and complex due to the optical-mechanical scanning system, and the reliability is not ideal. Although the series-scanning, parallel-scanning and series–parallel-scanning systems used in recent years have been able to increase the system sensitivity as well as reduce the system's bandwidth, they have not been able to cause a breakthrough change in the structure after all. Therefore, there has been a quest for gaze detector arrays that can have a considerable number of cells in two dimensions. In this case, the spatial sampling of the infrared image is such that each scene element corresponds to a focal plane array element, and the whole system has no moving parts. The two-dimensional focal plane gaze array is scanned by a two-dimensional multiplexer, and should also include a uniform correction and calibration part of the focal plane array. Due to the use of focal plane gazing array, can make the thermal imaging system to overcome the shortcomings of the optical—mechanical scanning system, at the same time, because the gazing array can use almost all the incident infrared photons, which improves the thermal sensitivity of the system, theoretically estimated, the smallest temperature resolution up to a few milli degrees Celsius.

Charge-coupled devices applied to the infrared detector, successfully solved the delayed integration of the output signal of the infrared detector array on the focal plane and the multiplexing problem, making the infrared focal plane gaze array completely practical, the information rate and the signal-to-noise ratio is also greatly improved, so that the structure of the thermal imaging system has undergone a fundamental change. Since such devices can be fabricated using an integrated circuit type fabrication process, they can in principle be produced in large quantities, making it possible to obtain less expensive infrared focal plane array devices. Especially for the mixed structure of the focal plane array, the detection element and the information processing element can be tested separately, are up to standard before interconnection, can be expected to have a higher yield. Due to foresee the significance of infrared focal plane arrays in military applications, some developed countries, governments and military departments are to give a huge amount of funding, so the development rate is extremely fast, especially in the late 1980s, the development rate is even more amazing.

Because of the special nature of infrared light imaging, and can not be used directly for visible light CCD infrared light imaging. Because the detection of infrared images to be subject to a variety of limitations. This is mainly a thermal target around the background radiation is too strong, the target is too low on the contrast between the background. High background and low contrast severely limits the accumulation time of the device. A long accumulation time not only makes the background governing radiation exceed the loading capacity of the CCD, but also the dark current of the device will overwhelm the weak signal. In addition to the background subtraction procedure before signal injection, the background effect can be reduced if a low-efficiency, long-term accumulation CCD is used, but it is important to ensure that there is a response to the temperature difference of the CCD.

9.2.4 Infrared Detector Cooling

In order to reduce the noise of the infrared detector and obtain a high signal-to-noise ratio, the detector needs to be cooled so that it operates at a low temperature. Since the detector takes up very little space in the thermal imaging system, the cooler assembly, consisting of a Dewar bottle and a cooler, is usually required to be miniaturised. Because of the complexity of the manufacturing process of microchillers, it has been the technical key to the development and production of thermal imaging systems.

In the cooler assembly of an infrared detector, a Dewar's bottle is a heat-insulating container that prevents radiation, convection and conduction. According to the materials used, Dewar bottles are mainly divided into glass Dewar bottles and metal Dewar bottles. Small glass Dewar bottles commonly used in thermal imaging systems consist of an inner and outer wall, a lead, an infrared window, and other parts. The outer surface of the inner wall and the inner surface of the outer wall are coated with a reflective layer, and the inner and outer walls are pumped into a vacuum to form an adiabatic layer.

The method of obtaining low temperature is roughly physical and chemical two kinds, in the infrared detector refrigeration commonly used physical methods. Due to the use of different occasions and requirements of the refrigeration temperature, different principles can be used to make a suitable refrigerator.

1. Liquid nitrogen cooler

Phase change refrigeration principle: the material phase change refers to its aggregation state change, the material phase change, need to absorb or release heat, this heat is called phase change latent heat. The use of refrigeration working substance phase change heat absorption effect, such as solid working substance melting heat absorption or sublimation heat absorption, liquid gasification heat absorption and so on to achieve refrigeration.

Liquid nitrogen refrigerant is injected directly into the cold liquid chamber of the Dewar's bottle, constituting a liquid nitrogen refrigerator. The detector is inside the vacuum layer of the Dewar's flask, and a cold shield is used to limit the background radiation received by the detector from the surrounding area. Since the boiling point of liquid nitrogen is 77 K, the required cooling temperature of the detector is maintained. Dewar's bottle cooler has the advantages of simple structure, stable cooling temperature, cold enough.

2. Gas throttling cooler

Joule–Thomson effect: when the temperature of the high-pressure gas is lower than its own conversion temperature, and through a very small throttling hole, due to the expansion of the gas and make the temperature drop. If the low-temperature gas after throttling returns to cool the incoming high-pressure gas, so that the high-pressure gas is throttled at lower and lower temperatures, and this process is repeated continuously, the required low temperature can be obtained to achieve the purpose of refrigeration.

Gas throttling refrigerator is based on the Joule–Thomson effect made, also known as Joule–Thomson refrigerator. Figure 9.14 shows the flow chart of Joule–Thomson effect refrigeration, the refrigeration working material is high-pressure nitrogen. High-pressure nitrogen from the inlet into the heat exchanger, through the throttle hole throttling expansion and cooling; cooling nitrogen through the circuit back to the heat exchanger, and high-temperature high-pressure nitrogen heat exchange, so that the temperature of high-pressure nitrogen before throttling is lowered, and then discharged through the exhaust port. Thus, the high-pressure nitrogen is throttled and expanded at a lower temperature, and the temperature is further reduced. This process continues so that the high-pressure nitrogen is throttled and expanded at lower and lower temperatures, and the temperature after expansion is lower and lower, and ultimately a portion of the nitrogen can be liquefied in the refrigeration chamber to obtain a low temperature of nearly 77 K.

Jiao—soup cooler is currently one of the more mature cooler, with refrigeration components of small size, light weight, no moving parts, mechanical noise, easy to use and so on; but the availability of gas source is poor, high-pressure cylinders are heavier, the purity of the working gas requirements are harsh, the general content of impurities shall not be higher than 0.01%, otherwise caused by the throttle hole clogging and stop working.

Jiao—soup refrigerator including open and closed loop type two. The open type refers to the refrigeration medium, throttling expansion after discharge, no longer recycling, generally used in the requirements of refrigeration time is short in the device. Closed cycle refrigerator refers to the refrigeration of high-pressure gas from the compressor continuous supply, throttling expansion recovery, by the compressor and then compressed into a high-pressure gas, and then used for throttling expansion of refrigeration, refrigeration workmass recycling, mostly used in the requirements of long-term continuous operation of the system.

In order to obtain a lower refrigeration temperature, two coke—soup cooler can be coupled together to form a two-stage coke—soup cooler. The refrigerator with two kinds of work material: one is used to obtain the pre-cooling stage temperature; the other is used to obtain the final temperature. Such as nitrogen—neon two-stage coke—soup refrigerator, with nitrogen for the pre-cooling stage to obtain a low

9.2 Principle of Infrared Imaging for UAV

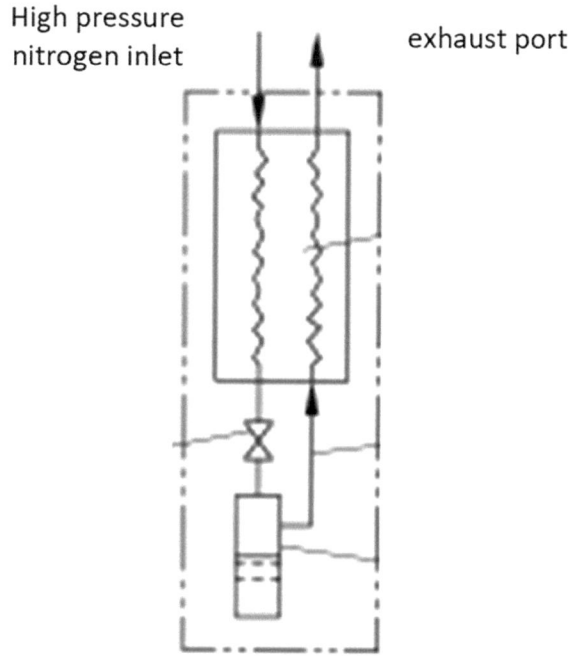

Fig. 9.14 Flow of Joule–Thomson effect refrigeration

temperature of 77 K, with neon to obtain the final low temperature of 30 K. General use of closed-loop refrigeration system, the need for two compressors at the same time the supply of two refrigerant medium, so the high cost of the refrigerator, large volume, large mass, applicable to ground stations in the infrared system.

3. Stirling cycle cooler

Due to the isentropic expansion of gas, not only through the expansion of the piston outward output of mechanical work, and expansion, the internal potential energy of the gas should be increased, which should be consumed by the gas itself to compensate for the internal function of the gas itself, so the gas isentropic expansion of the temperature will be significantly reduced.

Stirling cycle refrigerator using gas isentropic expansion principle and work, it is by the compression chamber, cooler, regenerator and refrigeration expansion chamber and other parts. In the compression chamber, there is a compression piston; in the refrigeration expansion chamber there is an expansion piston. In order to make the structure compact, reduce the boundaries of the size of the regenerator installed in the expansion piston, the regenerator packing is a large heat capacity at low temperatures, not £ steel mesh or lead particles. The regenerator connects the compression chamber and the refrigeration expansion chamber, and the refrigeration medium (nitrogen or hydrogen) can circulate freely, constituting a closed loop system.

In the actual working process, the two pistons are mounted on the same crankshaft through their respective connecting rods, and there is a fixed phase angle difference between the two connecting rods, which moves continuously according to the sinusoidal law. The crankshaft speed is very high, generally more than 500r/min, so the approximate continuous compression and refrigeration expansion, refrigeration efficiency is high.

Stirling cycle refrigerator (Stirling-CycleRotaryCoolers) is a kind of refrigerator with wide use and long life, with the advantages of compact structure, small volume, light weight, wide range of refrigeration temperature (77 ~ 10 K), short start-up time, high efficiency, long service life, simple operation, and can be used for long term continuous work; however, due to the high speed movement of piston at the cold head, it requires high processing technology, otherwise it may be used as the cold head. Processing technology requirements are high, otherwise it may produce large mechanical vibration, resulting in an increase in device noise, therefore, the price is more expensive.

For this reason, people developed a split Stirling cycle refrigerator. In this refrigerator, the compression part and the expansion part of the separation, which is connected to a gas pipe to reciprocating motor instead of the original crank linkage mechanism rotary motor drive. Split Stirling refrigeration not only maintains the overall Stirling refrigeration high efficiency strengths, but also vibration, wear and tear and work contamination, leakage is greatly reduced, life and reliability greatly improved; but also allows for a larger and heavier compressor is installed in a more suitable location, with the optical system is more convenient (Fig. 9.15).

Advantages of the semiconductor cooler is a simple structure, long life, high reliability, small size, light weight and no mechanical vibration and impact noise, easy maintenance, only consume electricity.

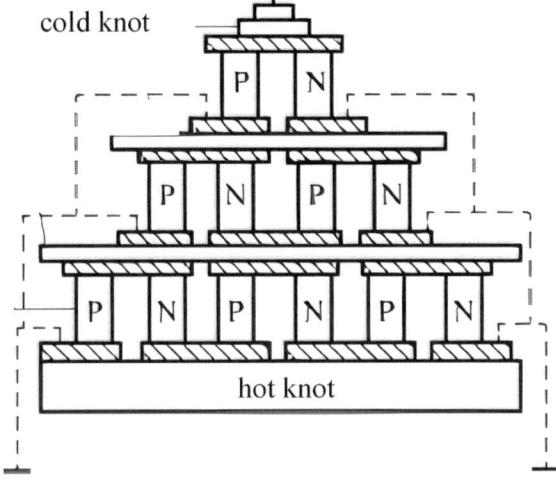

Fig. 9.15 Three-stage semiconductor cooler

9.2 Principle of Infrared Imaging for UAV

5. Radiation cooler

Radiation heat transfer: if the two objects have different temperatures, the high temperature object has to radiate energy, the temperature decreases; while the low temperature object absorbs the radiation energy, the temperature rises. Due to the cosmic space in a high vacuum, deep cryogenic state, in this special environment, the object can be and the surrounding deep cold (about 3 K) space for radiation heat exchange, so that the hot objects continue to cool down, to achieve the purpose of refrigeration.

Radiation cooler consists of cold sheet, radiator, brim, multi-layer insulation layer and outer shielding. In order to obtain different cooling temperatures, can be one, two or three or more radiators of different sizes in series to form a single-stage, two-stage or three-stage refrigerator, Fig. 9.16 shown in the European ESA satellite on the radiation cooler, it can be infrared detectors cooled to 95 K.

Radiation cooler has the advantage of long service life, no additional refrigeration power, there is no moving parts, so there will be no vibration, impact noise, high reliability; disadvantage is the requirement of the satellite's operating orbit and attitude control, to ensure that the radiation cooler is always aligned with ultra-low-temperature cosmic space, does not allow the sun or the Earth, such as infrared radiation directly to the refrigerant in the radiator on the radiators.

The cooler is essential to ensure that the infrared detector obtains the best working performance, which requires a reasonable choice of appropriate coolers according to

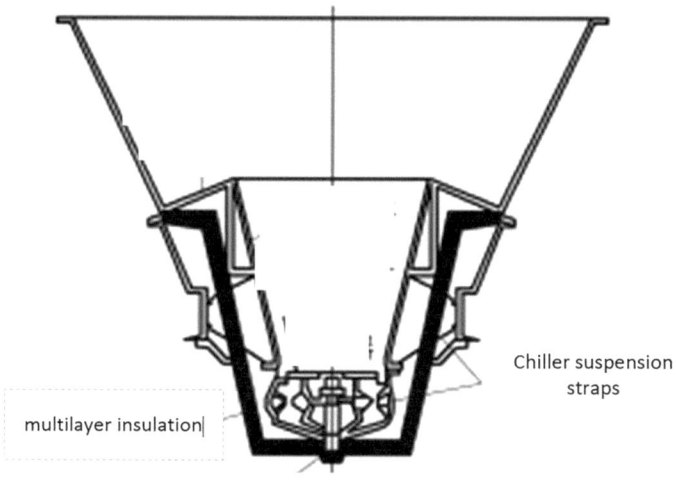

Fig. 9.16 Radiation chiller on the European ESA satellite

the working conditions and requirements of the infrared imaging system. The main indicators that characterise the performance of the cooler are the cooling temperature, the time of cold going down, power consumption, decomposability, boundary size, service life and repairability.

9.2.5 Infrared Thermal Imaging System

Infrared thermal imaging system can transform the infrared radiation naturally emitted by an object into a visible thermal image, thus extending the visual range of the human eye to the mid-wave/long-wave infrared band. In recent years, advances in related technical fields have led to the rapid development and wide application of thermal imaging technology. The quality of thermal images has reached the level of black-and-white analogue TV signals, and still images are comparable to high-quality black-and-white photographs.

This section discusses the components of a thermal imaging system and their functions and structures from the imaging principle of the thermal imaging system.

1. Thermal imaging principle

All objects in nature, as long as its temperature is higher than absolute zero, is always constantly emitting radiant energy. Therefore, in principle, as long as the collection and detection of these radiant energy, through the detector signal acquisition and processing to form a thermal image corresponding to the radiation distribution of the scene. This thermal image reproduces the radiation rise and fall of each part of the scene and can show the characteristics of the scene.

Figure 9.17 illustrates how a thermal imaging system converts the temperature and radiant emissivity differences of a scene into a visible thermal image with the simplest unit detector optical scan. The infrared optical system focuses the flux distribution of infrared radiation emitted by the scene onto the photosensitive surface of a detector located at the focal plane of the optical system; the optical machine scanner located between the focusing optical system and the detector consists of two sets of scanning mirrors, both vertical and horizontal, and when the scanner is in operation, the beam of light arriving at the detector from the scene is moved in response to the beam of light, thus sweeping out a raster in the object space like that of a television; when the scanner is made to sweep the detector over the scene in the form of a TV raster, the detector is swept over the object. When the scanner in the form of a TV grating makes the detector sweep through the scene, the detector will be point by point received by the scene radiation into the corresponding electrical signal sequence, or, in other words, the optical machine scanner constitutes the scene image swept through the detector sequentially, the detector sequentially to the scene of each part of the infrared radiation is converted into an electrical signal, after the video processing of the signal in the synchronous scanning of the monitor to show the scene of the thermal image.

9.2 Principle of Infrared Imaging for UAV

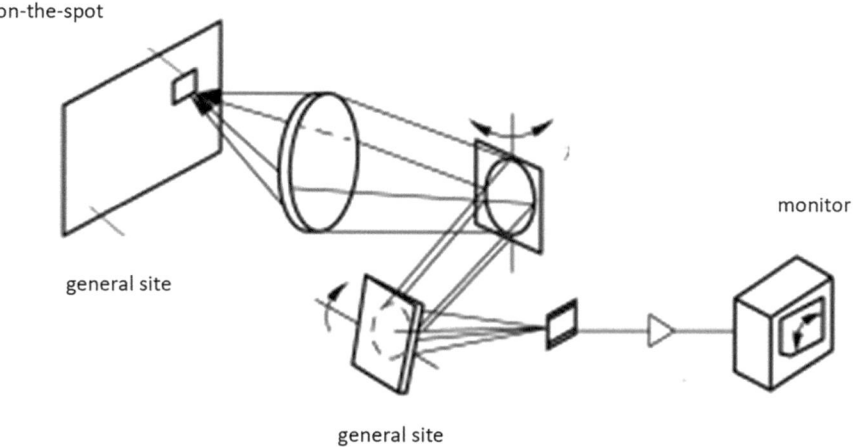

Fig. 9.17 Principle of T. operation of a unit-optical machine scanning thermography system

2. Types and Composition of Thermal Imaging Systems

According to the principle of infrared detectors, thermal imaging systems can be divided into cooled and uncooled types. According to the imaging method, the thermal imaging system can be divided into optical scanning type and gaze type thermal imaging system.

Figure 9.18 shows the block diagram of the optical machine scanning type thermal imaging system, the whole system mainly includes infrared optical system, infrared detector and cooler, electronic signal processing system and display system four components. Optical machine scanner so that the unit or multiple array detectors sequentially swept through the field of view of the scene, the formation of a two-dimensional image of the scene. In an optical-mechanical scanning thermal imaging system, the detector converts the received radiation signal into an electrical signal, and the background radiation is eliminated from the scene electrical signal by means of a DC-isolating circuit to obtain a thermal image with good contrast. Optical machine scanning type thermal imaging system due to the presence of optical machine scanner, the system structure is complex, larger, less reliable, the cost is also higher, but due to the relatively low requirements of the detector performance, the technical difficulty is relatively low, become the main international practical thermal imaging type after the 1970s, and there are still some important applications.

Gaze-type thermal imaging system using the focal plane detector array, so that each unit in the detector corresponds to a microfacet in the scene, Fig. 9.19 shows the block diagram of the gaze-type thermal imaging system, compared with Fig. 9.18, the gaze focal plane thermal imaging system eliminates the optical scanning system, while the detector pre-amplifier circuitry and the detector, integrated in the detector array located in the focal plane of the optical system, which is the so-called "focal

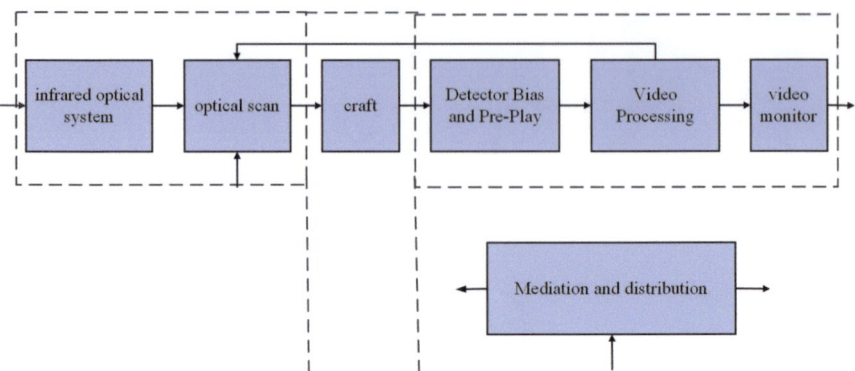

Fig. 9.18 Block diagram of an optical scanning thermal imaging system

plane" meaning. This is what is meant by "focal plane". In recent years, the development of gaze focal plane thermal imaging technology is very rapid, PtSi focal plane detector, 512 × 512, 640 × 480 and 320 × 240, 256 × 256 image elements of the refrigerated InSb, HgCdTe detector as well as non-cooled focal plane detector have made important breakthroughs, the formation of a series of products. The current development and application of scanning focal plane detector is also very rapid, and its difference with Fig. 9.18 is mainly in the detector preamplification and detector integration.

Pyroelectric infrared imaging system (also known as thermal TV) also belongs to the gaze-type thermal imaging system, which uses pyroelectric material as a rake surface, made of pyroelectric camera tubes, without the need for optical scanning, the direct use of the electron beam scanning and the corresponding processing circuitry, composed of a TV camera-type thermal imaging camera. Due to the simplified structure, no refrigeration, low cost, although the performance is not as good as the optical scanning thermal imaging system, but there are still some market applications.

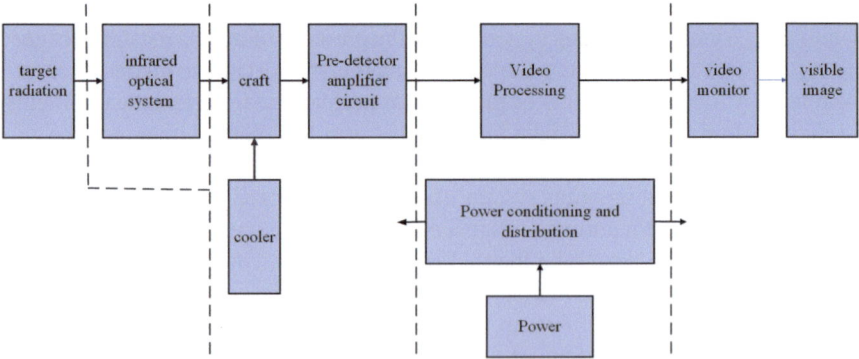

Fig. 9.19 Block diagram of a gaze focal plane thermal imaging system

Currently, the most common method of dividing thermal imaging into generations is to refer to optical-mechanical scanning systems based on separated arrays of single or multiple detectors as first-generation thermal imaging systems, and systems based on focal-plane detectors as second-generation thermal imaging technology. A new, more nuanced method of dividing thermal imaging has been proposed by the United States Army in 1997.This method of dividing thermal imaging systems into generations is called the first-generation thermal imaging system:

(1) Thermal imaging systems consisting of an optical-mechanical scanner with a single or multiple detector are referred to as first-generation thermal imaging systems;
(2) Scanning thermal imaging cameras are referred to as second-generation thermal imaging cameras, which have a 1900-detector element level and a feature size of approximately 30 pm; and
(3) Gaze-type thermal imaging cameras are referred to as third-generation thermal imaging cameras with a 307,000-detector-element level and a reduced detector size of 20 pm.
(4) Dexterous focal plane arrays with advanced signal processing and operating bands covering the visible, near-infrared, mid-infrared and far-infrared regions are referred to as fourth-generation thermal imaging systems.

9.3 Principle of Laser Scanning Technology

Three-dimensional laser scanning technology is a new technology that has emerged in recent years and has attracted more and more attention in the research field in China. It makes use of the principle of laser ranging, and by recording the three-dimensional coordinates, reflectivity and texture of a large number of densely packed points on the surface of the object to be measured, the three-dimensional model of the target to be measured as well as the data of various drawings such as lines, surfaces and bodies can be quickly reproduced. As the 3D laser scanning system can obtain the data points of the target object in a large number in a dense manner, so compared with the traditional single-point measurement, the 3D laser scanning technology is also known as a revolutionary technological breakthrough from the evolution of single-point measurement to surface measurement. The technology has also been tried, applied and explored in many fields such as cultural relics and monuments preservation, architecture, planning, civil engineering, factory renovation, interior design, building monitoring, traffic accidents handling, legal evidence collection, disaster assessment, ship design, digital city, military analysis and so on.

Three-dimensional laser scanning only measurement technology relative to the traditional topographic measurement technology, its high efficiency, strong expressive power, fine measurement details, topography, geomorphology measurement quickly, automatically get the DEM number and image model, and the results of a variety of forms, to meet the needs of different people on the exploration data and other characteristics. Underwater 3D laser scanning technology usually exists in the process of data acquisition of large-scale topographic maps in the field conditions of the many limitations of the shortcomings. Because of the complexity of the underwater terrain, and the water will cause refraction, resulting in data accuracy is not very high; due to the wave on the scanning results of some impact, resulting in complex post-data processing, the need for a longer period of time for data processing, delaying the progress of the follow-up work of the personnel input; three-dimensional laser attributed to the scanner's price is expensive, in the enterprise production cost minimisation, maximisation of the benefits of the social requirements, is not very suitable.

As the measurement principle of the three-dimensional laser scanner is the laser irradiation of the surface of the target to be measured, so that the surface of the formation of light spots, the object to be measured after the reflection of the light, the formation of scattered light by the sensor to analyse the object to be measured from the distance of the measurement station. Because the light from one medium into another medium, the light refraction, the optical path becomes curved, so that the ordinary three-dimensional laser scanner in the underwater spot imaging will appear offset, resulting in measurement errors. According to the principle of optical triangular diffuse reflection to study the laser ranging sensor, this sensor directly into the water, the light will be in the process of propagation, first through the water after the outside into the air, refraction occurs, through the lens in the detector imaging, you can more accurately measure the range of the distance to be scanned, and to overcome the impact of water fluctuations on the data, to get more accurate data, to establish a more realistic three-dimensional virtual reality. The 3D laser scanner can be used underwater to measure the distance to be scanned more accurately.

When the 3D laser scanner is underwater, through the on-site observatory, it records the real-time wave conditions in real time, measures and counts the wind speed, current speed, current force direction, effective wave height and period of the buoy waves in each time period, and in the process of data processing, the RTK tide adopts the integer multiples of the half period of the wave to weaken the influence of the wave on the results of the 3D laser scanning, and to improve the efficiency of the data processing in later stages as well as the accuracy of the data in the measurement of the underwater topography. accuracy in underwater terrain surveying.

9.3.1 Basic Concept

Three-dimensional laser scanning technology, also known as real-life reproduction technology, as a high-tech in the mid-1990s, is another technological revolution in the field of surveying and mapping following the GPS technology. Through the method of high-speed laser scanning and measurement, large-area, high-resolution and rapid access to the surface of the object at each point of the (x.y.z.) coordinates, reflectance, (R.G.B.) colour and other information, and by these A large number of dense point information can be quickly reconstructed 1:1 true-colour three-dimensional point cloud model, for the subsequent internal processing, data analysis and other work to provide an accurate basis. It has the characteristics of rapidity, high efficiency, non-contact, penetration, dynamics, initiative, high density, high accuracy, digitalisation, automation, real-time, etc., which is a good solution to the bottleneck of real-time and accuracy in the development of space information technology. It breaks through the traditional single-point measurement method and has the unique advantages of high efficiency and high precision. Three-dimensional laser scanning technology can provide three-dimensional point cloud data on the surface of the scanned object, so it can be used to obtain high-precision and high-resolution digital terrain model, mainly through the method of high-speed laser scanning and measurement, large-area high-resolution rapid access to the surface of the object to be measured three-dimensional coordinate data, a large number of spatial point information, is the rapid establishment of the object's three-dimensional image model of a new technical means.

Three-dimensional laser scanning technology makes the application of engineering big data possible in many industries. Such as the reverse engineering of industrial measurement, contrast detection; construction engineering in the completion of acceptance, alteration and expansion design; measurement engineering in the displacement monitoring, topographic mapping; archaeological projects in the data archiving and restoration projects and so on.

Three-dimensional laser scanning system includes the hardware part of data acquisition and the software part of data processing. According to the different carriers, three-dimensional laser scanning system can be divided into airborne, vehicle-mounted, ground and handheld types. Apply scanning technology to measure the size and shape of the workpiece and other principles to work. Mainly used in reverse engineering, responsible for surface scanning, three-dimensional measurement of the workpiece, for the existing three-dimensional objects (samples or models) in the absence of technical documentation, you can quickly measure the object's contour of the collection of data, and be constructed, edited, and modified to generate a common output format of the surface digital model.

9.3.2 Technical Principle

Three-dimensional laser scanner is based on the laser's coherence, directionality, monochromaticity and high brightness and other characteristics, while focusing on the ease of operation and measurement speed, so as to ensure the comprehensive accuracy of the measurement, and its principle of measurement is mainly divided into four aspects of ranging, scanning, angular measurement, orientation.

1. Three-dimensional laser scanner ranging principle

As laser ranging is a very important part of laser scanning technology, for laser scanning positioning and access to spatial three-dimensional information is a very important role. At this stage, the ranging methods are mainly: phase method, triangular method, pulse method.

Ranging methods have their advantages and disadvantages, but mainly focus on the relationship between range and accuracy, pulse measurement of the longest distance, but the accuracy will be reduced with the increase in distance. The phase method is used for medium-range measurement, which has a relatively high measurement accuracy, but it can only get the distance value through two indirect measurements. Triangulation has the shortest measuring range, but the highest accuracy, which is suitable for close range and indoor measurement.

2. Angle measurement principle of 3D laser scanner

Differing from the conventional instrument dial angle measurement, the laser scanner is by changing the laser light path and get the scanning angle. Two stepper motors are mounted together with the scanning prism, thus achieving horizontal and vertical scanning respectively. Stepping motor is also a kind of control micro-motor which converts the electrical pulse signal into angular displacement, and it can realise the precise positioning of the laser scanner.

3. Scanning principle of 3D laser scanner

For three-dimensional laser scanner is through the built-in servo-driven motor system precision control of the multi-faceted scanning prism rotation, determines the direction of the laser beam, can let the pulse laser beam can be along the horizontal axis direction and vertical axis direction for rapid scanning. The scanning is controlled by an oscillating scanning mirror and a rotating orthopolyhedral scanning mirror. The oscillating mirror is a plane mirror, which is driven by a motor to oscillate back and forth. This type of ranging is a kind of indirect ranging method, which detects the phase difference between the transmitted and received signals, so as to obtain the distance of the target to be measured. Ranging accuracy is high, mainly used in precision measurement and medical research, the accuracy can reach millimetre level.

4. Orientation principle of 3D laser scanner

Three-dimensional laser scanner scanned point cloud data are in its custom scanning coordinate system, but the data post-processing requirements are under the geodetic coordinate system, which requires the data under the scanning coordinate system to be converted to the geodetic coordinate system, this process is known as three-dimensional laser scanner orientation.

Three-dimensional laser scanner using the principle of laser ranging, through high-speed measurement and recording of a large number of dense points on the surface of the object to be measured three-dimensional coordinates, reflectivity and texture and other information, can be quickly reproduced by the target of the measurement of the three-dimensional model and lines, surfaces, bodies and other kinds of graphic data. As the 3D laser scanning system can obtain the data points of the target object in a dense and large number of points, so compared with the traditional single-point measurement, 3D laser scanning technology is also known as the evolution

9.3 Principle of Laser Scanning Technology

from the single-point measurement to the surface measurement of the revolutionary technological breakthrough.

This technology adopts non-contact high-speed laser measurement to obtain the geometric data and image data of terrain or complex objects, and finally through the post-processing software to process and analyse the collected point cloud data and image data, convert them into three-dimensional spatial position coordinates in the absolute coordinate system or establish the three-dimensional visualization model of the structurally complex and irregular scene, which saves time and effort, while the point cloud can also be outputted in At the same time, the point cloud can also be exported to many different data formats, which can be used as the data source of the spatial database and meet the needs of different applications (Fig. 9.20).

(1) short-range laser scanner: its longest scanning distance of no more than 3 m, generally the best scanning distance of 0.6–1.2 m, usually this type of scanner is suitable for small mould measurement, fast scanning speed and high accuracy, can be up to three hundred thousand points, accuracy to earth 0.018 mm. eg: Minolta's VIVID 910, handheld three-dimensional data scanner FastScan etc., belong to this category.
(2) in the distance laser scanner: the longest scanning distance is less than 30 m, which is mostly used for large moulds or indoor space measurement.
(3) long-distance laser scanner: scanning distance greater than 30 m, which is mainly used in buildings, mines, dams, large civil engineering and other

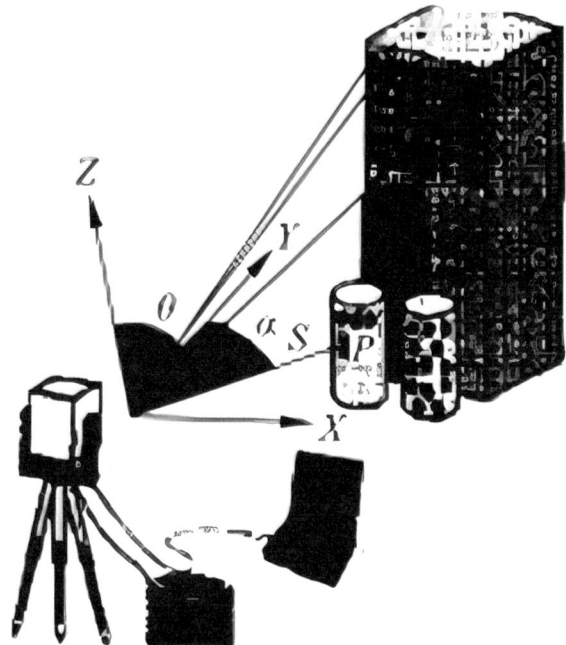

Fig. 9.20 Terrestrial laser scanner system composition and coordinate system

measurements. For example: Austria Rieglcompany's LMS Z420i three-dimensional laser scanner and the Swiss Leica company's C10 laser scanner, etc., belong to this category.

(4) aviation laser scanner: the longest scanning distance is usually greater than 1 km, and need to be equipped with accurate navigation and positioning system, which can be used for a wide range of terrain scanning measurement.

The reason for this classification is that the effective distance of laser measurement is an important condition for the application range of 3D laser scanners, especially for the observation of large features or scenes, or inaccessible features, etc., must take into account the actual measuring distance of the scanner. In addition, the farther away the measured object is, the relatively poorer the accuracy of the feature observation. Therefore, to ensure the accuracy of the scanned data, it is necessary to use the scanner within the standard range specified for the corresponding type of scanner.

Regardless of the type of scanner, the construction principles are basically similar. The main construction of a 3D laser scanner consists of a high-speed, accurate laser rangefinder with a set of reflective prisms that guide the laser and scan at a uniform angular velocity. The laser rangefinder actively emits laser light, and at the same time receives signals reflected from the surface of the natural object so that the distance can be measured, for each scanning point can be measured from the station to the scanning point of the slant distance, together with the scanning of the horizontal and vertical direction angle, you can get the spatial coordinates of each scanning point and the station. If the spatial coordinates of the station are known, the three-dimensional coordinates of each scanning point can be obtained.

9.3.3 Data Processing

1. Typical Data Processing Flow

Data processing mainly includes point cloud pre-processing, point cloud splicing, overall point cloud processing, and map production.

(1) Degree requirements and control

Firstly, the degree of the 3D overall point cloud model is consistent with the degree of the original point cloud; secondly, in each step of the 3D point cloud model establishment, the point cloud simplification and triangular network establishment algorithms are improved to ensure that the accuracy of the overall 3D point cloud model meets the requirements.

(2) Cloud Data Preprocessing

Point cloud data preprocessing is an indispensable step in the processing of 3D laser scanning data, point cloud data model is the basic data for the subsequent processing work in the industry, and the results of the point cloud data directly affect the quality

9.3 Principle of Laser Scanning Technology

Fig. 9.21 Point cloud preprocessing flowchart

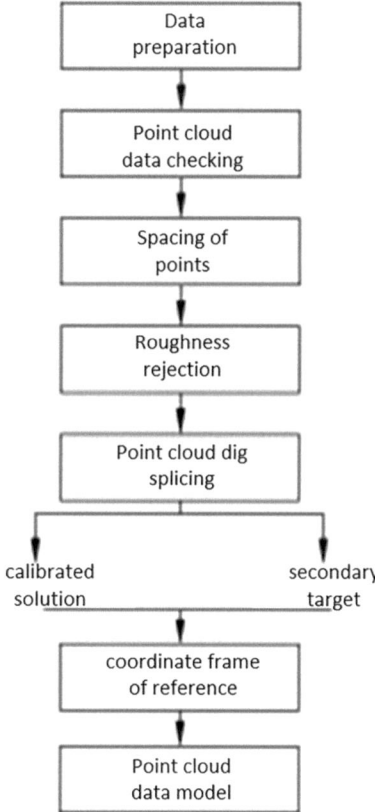

of the subsequent model establishment and the use of the point cloud model for the analysis of orthophoto maps and the related graphics production effect. The pre-processing process is shown in Fig. 9.21.

1) Data preparation

Prepare 3D point cloud raw data.

2) Cloud data checking

During the scanning process, due to the mutual occlusion between the scanned objects, it is inevitable to cause missing data. At the same time, scanning external conditions may also cause data stratification. Therefore, point cloud data checking is an essential step in point cloud data preprocessing. If the delamination is serious, the data is regarded as unqualified data; if the delamination is less or there is no delamination, the data is regarded as qualified, and a small amount of delamination can be dealt with by the subsequent point cloud roughness rejection and denoising operation.

3) Point spacing measurement

Point spacing measurement refers to the use of 3D software to manually sample the shortest spatial distance between two neighbouring points in the point cloud. Through the point spacing measurement operation, it can check whether the density of the acquired point cloud meets the degree requirement, and provide help for the data simplification step in the subsequent three-dimensional model building. The spacing should meet the accuracy requirements.

4) Roughness Rejection

Due to some abnormal vibration of the instrument and specular reflection during the measurement process, the real data points are often mixed with unreasonable noise points, containing a large number of roughness, errors and irrelevant information, so it is necessary to denoise the data and smooth the filtering process.

5) Cloud data splicing

Using professional software, automatically splicing the 3D point cloud data of each station, and checking the splicing accuracy of the station and the station, and dealing with the problems in time. Extract the target point cloud, delete the redundant point cloud, and export the point cloud data in pieces.

6) Reference system unification

Due to the inadequacy of the structure of the laser measuring machine or coordinate measuring machine itself, as well as the needs of the data processing process, in some cases it is necessary to carry out a coordinate transformation of the data point cloud to unify all the station data into an independent coordinate system. According to the coordinates of the target points measured by the field, the point cloud data model can be realised to be transformed, which is convenient for subsequent processing.

2. Point cloud processing theory

Compared with traditional measurement methods, the advantages of 3D laser scanning data acquisition speed and high sampling frequency lead to point cloud data with high redundancy, non-linear error distribution, incompleteness and other characteristics, which brings great difficulties to the intelligent processing of massive 3D point cloud: ① It is difficult to effectively integrate the point cloud data from multiple viewpoints, platforms and sources, which restricts the advantageous complementarity between the data and results in incomplete description of the complex scene; ② Difficulty in expressing the structure and semantic features of complex object models, and serious limitations in model availability, which greatly restrict the accurate perception and cognition of complex scenes. In recent years, scholars at home and abroad have conducted in-depth research on point cloud processing theory and methods for data quality improvement, automated fusion, point cloud classification and target extraction, and on-demand multilevel expression, and more progress has been made.

9.3 Principle of Laser Scanning Technology

(1) Generalised point cloud model theory and methods

Aiming at the serious defects of fusion difficulty, target extraction difficulty and 3D adaptive expression difficulty of multi-source and multi-platform point cloud data, some literatures have proposed the scientific concept and theoretical research framework system of generalised point cloud. Generalised point cloud refers to bringing together multi-source and multi-platform spatial data such as laser scanning, photogrammetry, and multi-source acquisition, and through cleaning, alignment and integration, realising the composite model that is based on point cloud, with a unified datum, and integrated with data, structure, and function, from multi-angle and visually relevant to omni-directional and visually irrelevant.

Point cloud data quality improvement includes geometric correction and intensity correction. On the one hand, the geometric position of the point cloud has errors and its distribution is uncertain due to the influence of the ranging system, the environment and the positioning and stance. The geometric position correction of the point cloud using the calibration field and known control points can improve the positional accuracy and usability of the scanned point cloud. On the other hand, the reflection intensity of the laser point cloud reflects the physical properties of the feature to a certain extent, and plays a key supporting role for the fine classification of the feature, however, the reflection intensity of the point cloud is not only related to the physical properties of the feature surface, but also affected by the scanning distance, incidence angle and other factors. Therefore, it is necessary to establish a point cloud intensity correction model for correction, in order to correct the influence of laser incidence angle, the distance of the feature from the laser scanner and other factors on the reflection intensity of the point cloud.

(2) Multi-source, multi-platform 3D point cloud fusion

Due to the limited observation range and inconsistent spatial reference of single viewpoint and single platform, in order to obtain all-round spatial information of the target area, it is not only necessary to carry out point cloud fusion between stations/strips, but also point cloud fusion of multiple platforms (e.g., airborne, vehicle-mounted, and ground stations, etc.), so as to make up for the lack of data brought by a single viewpoint and a single platform, and to realise a complete and fine description of digital reality for a wide range of scenes. In addition, due to the limited ability of the laser point cloud and its intensity information to portray the target, it is necessary to fuse the laser point cloud with the image data, so that the point cloud not only has high-precision 3D coordinate information, but also has richer spectral information.

The fusion of different data (e.g., laser point clouds of different sites/strips, laser point clouds of different platforms, laser point clouds and images) requires the association of homonymous features. To address the shortcomings of the traditional manual alignment method of low efficiency and high cost, scholars at home and abroad study the statistical analysis method based on the correlation of geometrical or textural features, but due to the differences in imaging mechanisms, dimensions, scales, accuracies, viewpoints, and so on, between different platforms and sensor data, there are still problems with its universality and robustness, and it is also necessary to break through the following bottlenecks: robust and discriminative homonymous feature extraction, global optimised alignment model building and anti-differential solving.

(3) Fine Classification and Target Extraction of 3D Point Cloud

Fine classification of 3D point clouds is the process of identifying and extracting artificial and natural feature elements from disordered point clouds, which is the basis for subsequent applications such as digital ground model generation and 3D reconstruction of complex scenes. However, laser point cloud classification on different platforms focuses on different topics. Airborne laser point cloud classification mainly focuses on targets such as large-scale ground, building roofs, vegetation, roads, etc. Vehicle-mounted laser point cloud classification focuses on targets such as roads and road facilities on both sides, vegetation, building facades, etc., while ground station laser point cloud classification focuses on the detailed interpretation of specific target areas. Among them, the point cloud scene has various targets, complex morphological structure, target occlusion and overlapping, and very different spatial density differences, which is a common problem for automatic fine classification of 3D point clouds. Accordingly, many scholars at home and abroad have carried out in-depth research and made some progress, using point-by-point classification method or segmentation clustering classification method on the basis of feature computation to identify the point cloud and extract the targets. However, due to the insufficient feature description capability, the quality of classification and target extraction cannot meet the application requirements, which greatly limits the use value of 3D point cloud. At present, the deep learning method that simulates the human brain breaks through the difficulty of over-reliance on manually defined features in the traditional classification method, and has shown great potential in the classification and interpretation of 2D scenes, but in the fine classification of 3D point cloud scenes, it still faces many difficult problems: the establishment of the sample library of massive 3D datasets, the construction of neural network models suitable for 3D structural feature learning and its interpretation of 3D data for big scene and its application in the interpretation of large scene 3D data. In summary, the semantic understanding of targets and their structures, the learning of global and local features of 3D targets at multiple scales, and the classification and extraction of multi-targets guided by a priori knowledge or third-party auxiliary data are important research directions in the future.

(4) On-demand multi-level representation of 3D scenes

After large-scale point cloud scene classification and target extraction, the target point cloud is still discrete, disorderly and highly redundant, which cannot explicitly express the target structure and the spatial topological relationship between the structures, and it is difficult to effectively meet the application requirements of 3D scenes. Therefore, it is necessary to convert the discrete and disordered point cloud into a geometric primitive combination model with topological relationships through scene 3D representation, and two types of methods, data-driven and model-driven, are commonly used, in which the main problems and challenges include: automatic repair of 3D models to overcome the impact of local data loss on model incompleteness; automated and robust reconstruction of targets with complex shapes and structures; and development of the 3D reconstruction from visualisation-based to computable reconstruction. 3D reconstruction to computable analysis-centred 3D reconstruction to improve the usability and goodness of results. In addition, different application themes require different levels of detail for different types of targets in the scene, and the 3D representation of the scene needs to strengthen the adaptive multi-scale 3D reconstruction methods for various types of 3D targets, and to establish a multilevel representation model of scene-target-element with correct mapping of semantics and structure.

3. Challenges in point cloud processing

In recent years, the extensive use of star, air, and ground scanning as well as portable ubiquitous sensors (e.g., RGB-D depth cameras) has not only improved the timeliness, granularity, and coverage of point cloud acquisition, but also brought new characteristics of multi-temporal phases, streaming, and diverse attributes of point clouds, thus generating multidimensional point cloud data. Multi-dimensional point cloud is essentially a multi-dimensional intensive sampling of three-dimensional geometric, physical and even biochemical characteristics of geographical objects/phenomena in the physical world, which not only records the three-dimensional spatial structure of the features, but also records the physical characteristics of the feature targets (e.g., waveforms, reflective intensity, etc.). Digging deeper into the intrinsic characteristics of the multidimensional point cloud is crucial for enhancing the intelligence of multidimensional point cloud processing and revealing the changing law of the complex dynamic three-dimensional scene. Although good research results have been achieved in point cloud processing, the intelligent processing of multi-dimensional point cloud still faces the following great challenges:

(1) Scale conversion of multi-dimensional point cloud geometry and attribute synergy

Explore the error distribution law of point cloud acquired by different platforms, establish the scale-dependent feature point quality assessment model; study the feature point cluster aggregation and hierarchical method that integrates the physical characteristics of point cloud; establish the multiscale integration method of multi-dimensional point cloud based on feature hierarchy, so as to realise the automatic unification of the spatial and temporal benchmarks of multidimensional point cloud.

(2) Discovery and Classification of Multi-dimensional Point Cloud Changes

Establish the change discovery and extraction method of multi-dimensional point cloud under the unified spatio-temporal reference framework, study the association method between multi-dimensional point cloud and 3D model of features based on time window, extract the geometric and attribute changes of spatial elements of features, and study the visualisation and analysis method of changes in spatial structure of features, so as to provide scientific tools for revealing the changing law of spatial elements.

(3) Accurate understanding of complex 3D dynamic scenes

Based on machine learning, artificial intelligence and other advanced theories and methods, we will explore the theory and methods of structured modelling and analysis of multi-dimensional point clouds, study the establishment of accurate positioning, classification and semantic modeling of polymorphic targets in complex three-dimensional dynamic scenes, establish multi-dimensional point cloud oriented feature description, classification and modelling methods for all kinds of elements in the three-dimensional dynamic scenes, and build bridges between the multi-dimensional point clouds and geo-computing models.

The breakthroughs in the above key challenges will form a complete theoretical and methodological system for the full three-dimensional (full coverage, full elements, full relationships) modelling of generalized point clouds, thus realizing the leap from "static, visual and quantitative" to "dynamic, simulation and analysis" in point cloud processing.

4. 3D Laser Scanning and Point Cloud Processing Development Trends and Prospects

In recent years, the development of sensors, communication and positioning technology, artificial intelligence, deep learning, virtual/augmented reality and other fields of advanced technology important progress to promote the digital reality (digital reality) era. Laser scanning and point cloud intelligent processing will comply with the needs of the digital reality era towards the following aspects of development.

(1) Three-dimensional laser scanning equipment will be single waveform, multi-waveform to single photon and even quantum radar, in the acquisition of data from the current geometric data to the geometry of the data to the geometry, physics, and even biochemical characteristics of the integrated collection.
(2) Three-dimensional laser scanning platform will also be a single platform based on the main change to a multi-source, crowdsourcing-based air-ground flexible platform, so that the target for a full range of data acquisition, the current national key research and development plan focus on special projects: domestic air-ground holographic three-dimensional remote sensing system and industrialisation (No. 2016YFF0103500) has supported the relevant research.
(3) The key issues such as feature description, semantic understanding, relationship expression, target semantic model, multi-dimensional visualisation, etc. of point cloud will be rapidly developed in the direction of automation and intelligence driven by advanced technologies such as artificial intelligence, deep learning, etc., and the point cloud will become a new type of model after the traditional vector and raster models in mapping and geo-information, which will strongly enhance the degree of automation of feature target cognition and extraction and the ability of knowledge-based service. Knowledge-based services.

(4) The development of virtual/augmented reality and Internet of Things + will promote the expansion of 3D laser scanning products from specialised applications to popular, consumer-level applications to meet the demand for networked multi-dimensional dynamic geographic information services.

9.3.4 Typical Products

3D laser scanning technology has a wide range of applications. The symmetrical application of forward modelling technology (e.g. CATIA, UG, CAD, etc., operated by man) is called reverse modelling technology (e.g. direct model reduction from solid or real-world view). Reverse modelling can reconstruct the content of changes in the process of design, production, experimentation, use, etc., and then carry out a variety of structural characteristics analysis (such as deformation, stress, efficiency, process, technology, attitude, prediction, etc.), testing, simulation, emulation, CIMS, CMMS, virtual reality, flexible manufacturing, virtual manufacturing, virtual assembly, etc., which is very important for the Finite Element Analysis, Engineering Mechanics Analysis, Fluid Dynamics Analysis and other software is very important, for the accuracy of the work can also be suitable for post-processing mapping, measurement and so on.

There are several kinds of reverse data acquisition technology applied at present, for example, the French MENSI three-dimensional laser scanning technology is the latest application technology in the evolutionary chain of three-dimensional reconstruction technology. It is different from the traditional technical means. Previously available traditional techniques include:

1. Discrete single-point acquisition of three-dimensional coordinates, such as CMM, CMM tracker, CMM, latitude and longitude, and so on. The shortcoming is that there are difficulties in data acquisition for complex structural surfaces, bodies, etc. that require massive point cloud acquisition.
2. Based on two-dimensional optical photographic principle, and then use three-dimensional software to simulate three-dimensional model (i.e., from two-dimensional to three-dimensional), such as close-up photogrammetry. Its shortcomings are: the existence of optical inherent deformation errors, depth of field is not enough, the physical surface pre-processing, the reference point set after the triangular plane misalignment, two-dimensional photographic conversion and indirect data uncertainty and other difficulties.

MENSI three-dimensional laser scanning technology can really do directly from the physical object for rapid reverse three-dimensional data acquisition and model reconstruction, that is, from three-dimensional to three-dimensional panoramic three-dimensional measured data reconstruction. It does not need to do any surface treatment of the physical object, and the depth of field is very long, avoiding the error caused by optical deformation factors, and each 3D data in its laser point cloud is the real data of the target directly collected, thus making the post-processed data real and reliable, so people take it as an effective means of rapid acquisition of spatial data. It is mainly for large and medium-sized target entities or real scenes, and can directly reflect the real-time, changing and real morphological characteristics of objective things. Since MENSI 3D laser scanning technology is supported by precision automatic sensing technology, CCD technology and remote sensing tracking technology, the quality of 3D point cloud data for acquiring physical objects or real scenes is very high, which ensures the authenticity, uniformity, real-time, operability, integrity, wide-area, monitorability and maintainability of the data.

In short, spatial data is a complex, interlaced, changing attributes, surface structure is only one of the attributes, and the task of reverse engineering will also be extended with the application of environmental quantification, virtual manufacturing, flexible manufacturing, tooling process, workpiece combination, digital factory, process operations, visual simulation, virtual reality, etc., and the social horizontal application surface will be further expanded.

Bibliography

1. Hongda, Chen, and Zuo Chao. 2004. *Very Short Distance Optical Transmission Technology*. Beijing: Science Press.
2. Hongyu, Ren. 2018. Application of Three-Dimensional Laser Scanning in the Integrated Measurement of Water and Underwater. *Science and Technology Wind* 31: 81.

Chapter 10
Optical Key Technology

10.1 Three-Dimensional Laser Scanning Technology

Three-dimensional laser scanning technology is a new technology that has emerged in recent years and has attracted more and more attention in the research field in China. It makes use of the principle of laser ranging, and by recording the three-dimensional coordinates, reflectivity and texture of a large number of densely packed points on the surface of the object to be measured, the three-dimensional model of the target to be measured as well as the data of various drawings such as lines, surfaces and bodies can be quickly reproduced. As the 3D laser scanning system can densely obtain a large number of data points of the target object, so compared with the traditional single-point measurement, 3D laser scanning technology is also known as the evolution from single-point measurement to the surface measurement of the revolutionary technological breakthrough.

Airborne laser 3D radar system (LightDetectionAndRanging, abbreviated as LiDAR) is a kind of integrated opto-mechanical system that combines laser scanner, global positioning system (GPS) and inertial navigation system (INS) as well as high-resolution digital camera, etc. It is used to obtain the laser point cloud data and to generate accurate Digital Elevation Model (DEM), Digital Surface Model (DSM), and at the same time obtain the Digital Orthophoto (DOM) information of the object, and through the processing of the laser point cloud data, a real three-dimensional scene map can be obtained.

Laser ranging technology is one of the main technologies of 3D laser scanner. The principle of laser ranging is mainly based on four types: pulse distance method, phase ranging method, laser triangulation method, and pulse-a-phase type. At present, the three laser scanners used in the field of surveying and mapping are mainly based on the pulse ranging method of ranging, and the close-range three-dimensional laser scanners mainly use the phase interference method of ranging and laser triangulation method of ranging. The types of laser ranging technology are described as follows:

1. Pulse ranging method

Pulse ranging method is a high-speed laser time ranging technology. Pulsed scanner in the scanning laser emits a single point of laser, recording the laser echo signal, by calculating the laser time of flight (TimeofFlight, abbreviated as TOF), using the speed of light to calculate the distance between the target point and the scanner. This principle of distance measuring system can reach a distance of a few hundred metres to thousands of metres. The laser ranging system mainly consists of a transmitter, a receiver, a time counter, and a microcomputer.

Pulse ranging method is also known as pulse time-of-flight difference ranging, due to the use of a pulsed laser source, suitable for ultra-long distance measurement, the measurement accuracy is mainly limited by the pulse counter operating frequency and the pulse width of the laser source, the accuracy can reach the order of metres.

2. Phase ranging method

Phase type scanner is to emit an uninterrupted beam of integer wavelength laser, through the calculation of the phase difference of the laser wave reflected back from the object to calculate and record the distance of the target object. The phase-based measurement principle is mainly used in scanning measurement systems that carry out medium distances. The scanning range is usually within 100 m and its accuracy can reach millimetre orders of magnitude.

Phase-based scanner due to the use of a continuous light source, the power is generally lower, so the measurement range is also smaller, the measurement accuracy is mainly limited by the accuracy of the phase comparator and the frequency of the modulation signal, increase the frequency of the modulation signal can improve the accuracy, but the measurement range is also smaller, so in order to improve the measurement accuracy without affecting the measurement range of the premise, generally set more than one frequency modulation frequency.

3. Laser triangulation

Laser triangulation is the use of triangular geometric relationships to find the distance. First by the scanner to emit laser light to the surface of the object, the use of the CCD camera at the other end of the baseline to receive the reflected signal from the object, record the angle between the incident light and the reflected light, known as the laser light source and the baseline length of the CCD between the triangular geometric relationship deduced from the distance between the scanner and the object. In order to ensure the integrity of the scanned information, many scanners scan only a few metres to tens of metres. This type of 3D laser scanning system is mainly used in industrial measurement and reverse engineering reconstruction. It can achieve sub-millimetre accuracy.

4. Pulse-phase ranging method

Combining the two methods of pulse ranging and phase ranging produces a new method of ranging: pulse-phase ranging, which uses pulse ranging to achieve a rough measurement of the distance and phase ranging to achieve a precise measurement of the distance.

The 3D laser scanner is mainly composed of a ranging system and an angular measurement system as well as other auxiliary function systems, such as a built-in camera and a dual-axis compensator. The principle of operation is to obtain the distance from the scanner to the object to be measured through the ranging system. Then the goniometric system obtains the horizontal and vertical angles from the scanner to the object to be measured, and then calculates the three-dimensional coordinate information of the object to be measured. In the scanning process and then use its own vertical and horizontal motors and other transmission devices to complete a full range of scanning of the object, so that continuous scanning and measurement of space with a certain sampling density, you can get the target object to be measured by the dense three-dimensional coloured scattered data, known as the point cloud.

10.2 Ultra-High Resolution Photoelectric Imaging Technology

High resolution has been the development of aerial camera pursued by one of the key indicators, but the aerial camera work in dynamic imaging, and thus many factors shadow high-resolution imaging, such as forward motion, vibration, environmental changes (temperature, pressure, and photographic distance) caused by aerial imaging fuzzy, and therefore to take the corresponding compensation techniques to ensure high-resolution imaging is its key problem.

1. Optical system

When the size of the imaging device CCD is determined, the optical system determines the performance, volume and weight of the aerial camera. In order to obtain high-resolution images require optical system focal length is very long, but due to the limitations of the calibre, the relative aperture is generally not allowed to be very large, mainly because of the aviation technology on the volume and weight of the special requirements, its volume and weight reduction means that the flight speed, flight time increases. According to the requirements of focal length, relative aperture, field of view and wavelength band, the optical system of the camera can be refractive, refractive-reflective and reflective in various forms.

Refractive system correction of the secondary spectrum is an important work, generally used with special dispersion of optical materials such as CaF_2 crystals, FK glass, etc., but the refractive index temperature coefficient of this special dispersion of optical materials is negative and very large, the environmental temperature change caused by the displacement of the image plane is very large, in order to reduce the impact of the temperature environment and barometric pressure on the displacement of the image plane, if you use the ordinary glass to correct the secondary spectrum, its relative Dispersion is very close, in order to eliminate the advanced aberration, the structure of the optical system is necessarily complex, the weight increases. In

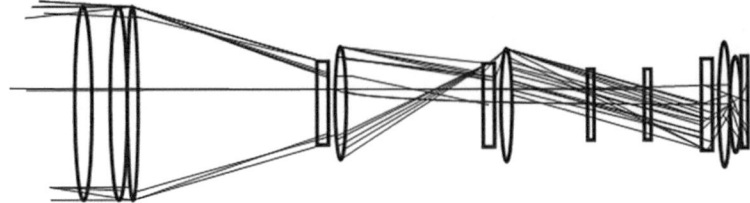

Fig. 10.1 Diagram of an aerial camera refraction system

addition, the refractive system is quite sensitive to changes in air pressure. Therefore, the refractive system is more suitable for the system with smaller aperture, larger field of view, and narrower wavelength band. Figure 10.1 shows the refractive system of an aerial camera.

Advantages of refractive-reflective systems optical focus is almost generated by the reflecting surface, which does not generate chromatic aberration, so the secondary spectrum is very small; and before and after the reflecting surface are air, so air pressure changes reflector has no effect on the displacement of the image plane, while the refractive element is very small optical focus, so refractive-reflective systems are insensitive to changes in ambient air pressure; refractive-reflective systems are also much simpler than refractive systems in the structure of the refractive-reflective system. The disadvantage is that there is a centre blocking, which loses energy and reduces the modulation transfer function (MTF). Commonly used refractive system for the Cassegrain system, the field of view is relatively small, generally within 2, the aberration is sometimes not well-corrected, so the secondary mirror and the correction group of a piece of the lens using aspherical to properly increase the field of view and correction of advanced aberration. Figure 10.2 shows the structure of the Cassegrain system.

Fig. 10.2 Cassegrain system structure

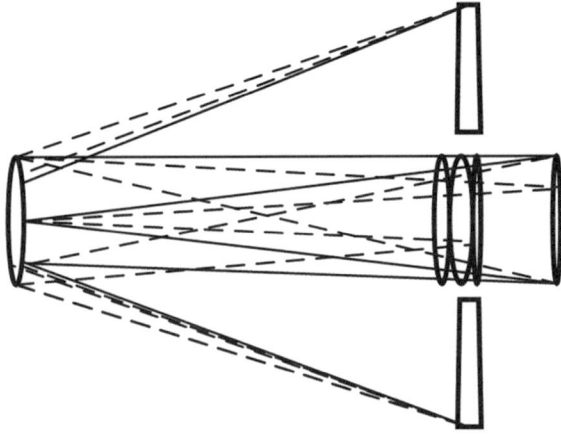

Fig. 10.3 Reflection system

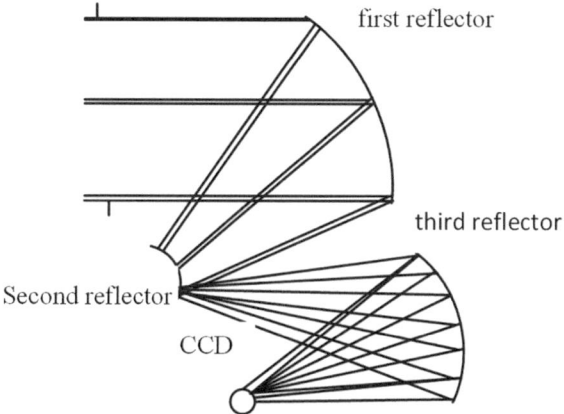

The advantage of the total reflection system is that the wave band is not limited by optical materials, so it is especially suitable for ultraviolet and infrared work; there is no chromatic aberration and the structure is simple. Disadvantage is the need to use aspherical surface, optical detection is difficult, difficult to install and adjust; field of view angle is small, especially suitable for aerospace earth observation and astronomical telescopes, the Hubble telescope primary and secondary mirrors are hyperbolic, the field of view is only 28'. Figure 10.3 for the three-reflection system, the system has a greater spatial correction of aberration, but mounting is more difficult to ensure.

Therefore, the specific use of which form of structure depends on the actual technical requirements. For aerial cameras in the long focal length, medium and high altitude is more often used folding reverse, the field of view is generally smaller, but also to meet a certain coverage to improve the efficiency of reconnaissance.

2. Like shift compensation technology

Aircraft flight process in the exposure time, the target to be photographed like the presence of relative motion between the light-sensitive media (film or CCD, etc.), that is, like shift, like the existence of the shift will lead to imaging fuzzy, contrast deterioration, clarity degradation, and ultimately affect the quality of imaging. Like shift compensation technology is to ensure a high level of image resolution is an important technical means. When the camera hardware conditions. The main factors affecting image resolution are light time and flight speed. It is generally believed that the amount of like shift does not exceed 1/3–1/2 pixels can be considered not to cause image blurring, do not need to compensate for the device.

3. Vibration control technology

In many high-resolution aviation photoelectric imaging system, despite the use of high-quality sensors, but the image quality is not ideal, limiting the main factors of high-resolution imaging is often not due to the electronics or optical system caused by the image fuzzy degradation is mainly caused by aviation vibration. When the photoelectric equipment and the carrier connected to the amplitude and phase of each point is inconsistent, resulting in an angular displacement of the optical visual axis, seriously affecting the clarity and resolution of optical imaging.

Vibration control technology mainly includes electronic image stabilisation, optical image stabilisation, passive vibration isolation and active control technology. Because passive vibration isolation has the advantages of high reliability, no energy required, simple structure and economic practicality, aviation optoelectronic equipment often uses passive vibration isolation technology to isolate high-frequency vibration.

According to the relevant literature and theoretical analysis, the angular displacement on the image quality of the impact is much larger than the line displacement, so in the design of the system should be as far as possible to control the carrier from the angular displacement, and inhibit the platform seat frame substrate line vibration into the angular displacement. According to the principle of no angular displacement can be divided into three-way equal stiffness and double parallelogram structure.

4. Automatic focusing technology

Usually designed optical systems only consider the material properties at room temperature and pressure, and the assessment of imaging is also based on this. However, the optical system working at 30 km altitude is subjected to very large changes in ambient temperature, typically up to -50 °C, pressure is 0.1 Pa. Due to the thermal instability of optical and structural materials, when the temperature changes will cause the system to be out of focus, and the imaging quality will deteriorate. In the many environmental changes affecting the optical system imaging factors, temperature-induced changes in the refractive index of optical materials is the most important factor, the impact of this effect is more obvious especially in the infrared system, and now the high-resolution aerial camera is more to long focal lengths, high altitude oblique vision, long-distance target photography as a feature, when the longer the focal length, the more pronounced out of focus. In order to meet the requirements of the aviation camera high resolution must be through a special design or certain compensation techniques, so that the focal plane is unchanged or changes are very small.

Compensation methods can be generally divided into three categories according to different situations: mechanical heatless technology, electromechanical heatless technology, optical heatless technology. However, the high degree of uncertainty in aerial reconnaissance photography and the complexity of environmental changes make it difficult to use a certain method to get complete compensation. Therefore, it is necessary to adopt auto-focusing technology according to different environments. Aerial camera automatic focusing system method is as follows: first determine the infinity height corresponding to the position of the image surface as a focusing reference, focusing on different photographic heights before, first of all, to compensate for changes in atmospheric pressure and temperature caused by changes in the reference plane, that is, self-collimating focusing; secondly, as a basis for calculating the different photographic heights caused by changes in the position of the image point,

10.3 Offshore High Humidity, Salt Spray and Temperature Protection …

which is determined by the position of the focusing mirror. The above key technology is to eliminate changes in the base caused by changes in atmospheric pressure and temperature.

For aerial CCD camera using image processing method to determine the focusing reference is a more ideal method, the method through the known target image quality image processing to achieve automatic focusing. The image processing of the image quality is realised by a circuit system with high accuracy.

Shipboard optoelectronic equipment working environment is harsh, the environment of the three defence design in the development must pay attention to. Combined with the actual corrosion problems, from the application of materials, structural optimisation and process protection and other aspects of research.

Shipboard photoelectric equipment, especially the cabin equipment in the long-term maritime climate, its high temperature, high humidity and corrosive substances in the air, salt spray and a variety of mould on the equipment has great destructive properties. And these climatic factors and environmental conditions are very complex, can affect each other, but also interact with each other to accelerate the damage to the equipment. Equipment leakage surface if directly exposed to the humid atmosphere, usually adsorbed a layer of water film, water film containing salts in the formation of electrochemical corrosion of metal surfaces necessary electrolyte film, this electrolyte film on the bare metal surface has a strong corrosive. Higher temperatures at lower latitudes increase the rate of galvanic corrosion. Humidity and heat is not only on the metal with erosion, the PCBA coating on the printed circuit board is also more destructive, can cause the circuit board blistering, short circuits between the lines and so on.

Three-proof test can refer to the national military standard, such as damp heat (alternating) and salt spray combination of test methods are as follows: test temperature: high-temperature stage temperature is 40–2 °C, the relative humidity is 93 ± 3%, the room temperature stage temperature is 25 ± 2 °C, the relative humidity is 95% (Note: the cabin equipment for the 93% of the 3%). Salt concentration requirements for 5% (Note: the cabin equipment for 2 min. test to 24 h for a cycle cycle, each cycle is divided into warming, high temperature and high humidity, cooling and high humidity at room temperature 4 stages. The number of test cycles is 2, 4, 6. Before and after the experiment should be twice on the test product appearance inspection, electrical performance inspection and mechanical performance inspection.

10.3 Offshore High Humidity, Salt Spray and Temperature Protection Technology

Shipboard optoelectronic equipment working environment is harsh, the environment of the three defence design in the development must pay attention to. Combined with the actual corrosion problems, from the application of materials, structural optimisation and process protection and other aspects of research.

Shipboard photoelectric equipment, especially the cabin equipment in the long-term maritime climate, its high temperature, high humidity and corrosive substances in the air, salt spray and a variety of mould on the equipment has great destructive properties. And these climatic factors and environmental conditions are very complex, can affect each other, but also interact with each other to accelerate the damage to the equipment. Equipment leakage surface if directly exposed to the humid atmosphere, usually adsorbed a layer of water film, water film containing salts in the formation of electrochemical corrosion of metal surfaces necessary electrolyte film, this electrolyte film on the bare metal surface has a strong corrosive. Higher temperatures at lower latitudes increase the rate of galvanic corrosion. Humidity and heat is not only on the metal with erosion, the PCBA coating on the printed circuit board is also more destructive, can cause the circuit board blistering, short circuits between the lines and so on.

Three-proof test can refer to the national military standard, such as damp heat (alternating) and salt spray combination of test methods are as follows: test temperature: high-temperature stage temperature is 40–2 °C, the relative humidity is 93 ± 3%, the room temperature stage temperature is 25 ± 2 °C, the relative humidity is 95% (Note: the cabin equipment for the 93% of the 3%). Salt concentration requirements for 5% (Note: the cabin equipment for 2 min. test to 24 h for a cycle cycle, each cycle is divided into warming, high temperature and high humidity, cooling and high humidity at room temperature 4 stages. The number of test cycles is 2, 4, 6. Before and after the experiment should be twice on the test product appearance inspection, electrical performance inspection and mechanical performance inspection.

10.4 High-Precision Stable Platform Technology

With the development of the intensity of optoelectronic confrontation, target stealth technology, weapon precision and range in modern war, it is required that the optoelectronic payload has a longer range, a wider spectral perception range, and a higher aiming and tracking precision. To achieve long-distance, high-precision target detection and targeting, the stabilised platform must have high-precision image stabilisation capability, and the high-precision stabilised platform is the basis and guarantee for airborne optoelectronic payloads to perform the combat mission. At present, the stabilisation accuracy of advanced foreign airborne optoelectronic payloads has reached the sub-pixel level.

In the design of high-precision stabilisation platform, through the optimisation of the system axis frame composition, structural layout optimisation, material and control component selection, etc., to improve the system structural rigidity, reduce the axis coupling and friction torque, and improve the carrier disturbance torque isolation ability, the control system through the use of new technology and new methods to improve the control loop bandwidth and gain, and to improve the system visual axis stability. For example, SniperAT stabilisation platform adopts flexible optical base design technology, AN/AAQ-30 adopts 5-axis stabilisation platform technology,

BRITEStar II adopts 6-axis image stabilisation technology, MX-25 adopts 5-axis active image stabilisation and 6-axis passive vibration damping technology.

The coarse-precise combination stabilisation system is an effective technical way to improve the stabilisation accuracy of the system. On the basis of the universal stabilisation platform, the high-precision FSM component is added to eliminate the residual error of coarse stabilisation through precision compensation, so as to improve the aiming line stabilisation accuracy, and at the same time, due to its small rotational inertia, it can greatly increase the resonance frequency and improve the tracking bandwidth and response speed of the system. FSM is the key technology in the coarse-precision combination stabilisation system, and the use of this stabilisation technology can make the stabilisation accuracy of the optoelectronic system reach the micro-arc level or even the nano-arc level, and achieve the sub-pixel level stabilisation of the image. The two-degree-of-freedom high-precision FSM mirror technology can not only compensate for the residual aiming line stabilisation and improve the stabilisation accuracy, but also can be used to compensate for image motion blur to achieve wide-area search and reconnaissance applications of the "stepping gaze", and at the same time can be used for the infrared imaging system of the "micro-scan", to achieve sub-pixel super-stability. At the same time, it can be used in the "micro-scan" of infrared imaging system to achieve sub-pixel super-resolution infrared imaging.

10.5 High-Precision Target Recognition Tracking and Positioning Technology

In order to achieve "wide-area search, accurate positioning, rapid destruction, real-time assessment", as well as networked collaborative warfare capabilities, the use of satellite positioning, inertial measurement and gyro-stabilized GPS + IMU + STA combination of technologies, to achieve high-precision target search and positioning, tracking and targeting, is currently an important development direction of advanced airborne optoelectronic payload systems. development direction. Using GPS + IMU + STA combination technology, the system has the following functions and features ① Reduce the installation error and dynamic error due to the system damper, significantly improve the target guidance and positioning accuracy, ② Realise the weapon system high-precision automatic calibration of the axes and automatic calibration of the optoelectronic payload, solve the optical axis error due to the material, assembly, environmental changes, etc., and improve the target positioning and aiming accuracy, ③ By the target position, motion speed and motion direction, the target position and tracking and aiming accuracy can be achieved. By measuring the characteristics of target position, movement speed and movement direction, combined with video tracking, it improves the anti-interference, memory tracking and multi-target tracking capability of automatic target tracking of optoelectronic payload, ④ It improves the auto-focusing capability of optoelectronic sensors, ⑤ The system has strong versatility, which improves the adaptability of the aircraft-carrying platform.

10.6 Photoelectric Information Processing Technology

Along with the development of optoelectronic payload equipment, improving its target detection capability has always been an important part of research. In order to improve the target detection and identification distance and long-range reconnaissance and surveillance capability of the system, in addition to the improvement of detector performance and the continuous development of the new detection concept system, advanced image enhancement processing technology is an effective way to enhance the performance of optoelectronic payloads, which has been widely used in foreign optoelectronic payloads and has significantly improved the performance of the system. For example, AN/AAQ-30 adopts advanced range-extension technology based on local image enhancement, which increases the target recognition distance by 60% MTS-A/B adopts enhancement processing technology based on automatic image detail optimisation, which effectively enhances the scene perception and remote reconnaissance and surveillance capability. Image enhancement processing technology has always been a hot research topic and a key technology in the development of airborne optoelectronic payloads.

The infrared image enhancement processing algorithm can be divided into spatial domain and frequency domain processing according to the processing domain. According to the method of algorithm implementation can be divided into grey scale correction, image smoothing, image sharpening, image enhancement, colour processing and so on. With the development and application of various mathematical tools such as artificial neural networks, genetic algorithms, wavelet variations, fuzzy theory and mathematical forms, new algorithms continue to appear. In recent years, with the improvement of infrared focal plane detector performance, infrared image detail enhancement technology has received extensive attention from researchers. For the characteristics of infrared imaging, research has proposed a variety of infrared image detail enhancement processing algorithms, through the enhancement of the infrared scene in the target and the background of the grey scale contrast and the target's own structural features contrast, to solve the problem of recognising low-contrast targets in the scene of the high dynamic range. digital detail enhancement (DDE) technology proposed by the FLIR company is a very good solution to the problem at present.

Visible light camera sensor is an important imaging sensor of the photoelectric load, but in poor meteorological conditions such as hazy weather, due to the role of suspended particles in the atmosphere on the scattering of reflected light from the target, so that the colour distortion and lightening, contrast reduction and other image degradation, which seriously affects the visual effect of the image and the performance of the target detection and identification, the use of computer image processing technology for the visible image dehazing is a good solution to improve the image quality. The dehazing of visible light images using computer image processing techniques is an effective technical way to improve visible light imaging quality and target detection performance under hazy weather. Comprehensively analysing the existing image enhancement algorithms, there is room for further improvement in the algorithm's computing power, scene adaptability, large dynamic weak contrast small target recognition, and human eye perception matching. With the development of sensor technology, imaging mode and imaging system, the development of image enhancement processing technology will have the following characteristics:

(1) With the development and application of new imaging technologies such as hyperspectral imaging, three-dimensional imaging, polarisation imaging, image enhancement processing technology will develop in the direction of multi-feature based, multi-dimensional space, depth, time, spectral, polarisation;
(2) With the development of compressed perception theory, adaptive coding aperture imaging and other computational hybrid imaging techniques, image enhancement from post-processing to the development of imaging-processing integration, the comprehensive use of optical systems, sampling and image reconstruction processing techniques to achieve a large field of view, high-resolution infrared imaging;
(3) With the development of distributed aperture omnidirectional detection, multi-spectral sensors of the same loading coordinated detection applications, image enhancement technology will be developed to multi-source, heterogeneous, multi-spectral image fusion enhancement processing direction;
(4) The research on image enhancement processing algorithms pays more attention to the visual characteristics of the human eye, and develops in the direction of visual perception-based.

Bibliography

1. Miller, D.A.B. 1990. *OptandQuantive Electronics* 22, s61–s98.
2. Cheng, Xiaowei, Ying Che, and Chang Xue. 2009. Key Technology and Development of High Resolution Imaging of CCD Digital Aerial Camera. *Electro-Optics and Control* 4.
3. Zhang, Weiguo, Yukun Wang, and Bin Wang. 2008. Protection Technology and Design of Shipboard Optoelectronic Equipment. *Infrared Technology* 30(4).
4. Jian, Qiao, Cao Lihua, and Shi Long. 2012. Three-Proof Design of Shipboard Optoelectronic Equipment. *Applied Optics*.
5. Dong, Meng, Zhang Dong, Meng Kai, Wang Fang. 2018. Research on Three-Proof Design of Shipboard Optoelectronic Equipment. *Science and Technology Innovation and Application* 19.
6. Shupeng, Ji. 2017. Development and Key Technology of Airborne Optoelectronic Payload Equipment. *Aviation Weapon* 06.

Chapter 11
Application of Optical Technology in Marine Engineering

11.1 Application of Airborne Marine Lidar in Offshore Engineering

Airborne marine LiDAR is an advanced technology that uses airborne blue-green laser transmitting and receiving equipment to detect underwater targets in the ocean by transmitting high-power narrow-pulse lasers. It mainly solves the following problems in marine investigation: underwater topographic survey; river enclosure, harbour sedimentation changes; underwater site quality disaster; underwater resource survey; coastal zone engineering construction, etc.

Compared with sonar technology, although the detection distance of airborne marine LiDAR is small, its search efficiency and detection density are much higher than that of sonar, in addition, it also has many advantages such as strong mobility, low operating cost and easy to operate. Therefore, airborne marine LiDAR can be widely used in seawater hydrographic survey (including shallow water depth, seabed site mapping, seawater optical parameters of telemetry, etc.), underwater submarine detection, mine detection, fish detection, marine environmental pollution monitoring and many other fields. At present, sonar is still the only major technology in deep-water detection, while in shallow-water detection, airborne marine LiDAR has shown stronger competitiveness than sonar, and is a tempting new device.

11.1.1 Application of Red Tide and Pollution Monitoring

In recent years, the scale of red tide has shown an increasing trend. 1998–2003, in the Bohai Sea, the East China Sea have occurred in an area of several thousand square kilometres of mega red tide, which is very rare in the international arena. Due to the complexity of the formation mechanism of red tide, there is no very effective method to prevent the occurrence of red tide, but only through the means of monitoring and forecasting to reduce the losses caused by red tide. Commonly used methods are based on shipborne water quality monitoring and buoy station fixed company automatic monitoring methods. Shipboard optical instrument measurement method needs fixed sampling, chemical analysis and manual processing, there are problems such as slow measurement speed, very low efficiency and high cost, which can not meet the requirements of rapid access to large areas of water quality parameter occasions, and at the same time, also seriously affects the response time of disaster prediction. In recent years, the monitoring of red tide working platform from the traditional shipboard platform measurements, changing to more and more use of aviation, satellite to carry out detection. Satellite platform measurement requires more complex equipment and higher cost, while satellite visible remote sensing also has its own shortcomings, for example, it can not work around the clock, all-weather, rainy weather and night can not monitor the red tide, in addition, due to the low spatial resolution, it is very difficult to monitor the red tide on a small scale.

At present, the application of new technologies such as airborne-based aerial oceanographic sensing detection in the field of red tide monitoring and forecasting has attracted the attention of more and more countries. Since 1985, China has been carrying out red tide aerial cruise surveillance, emergency response, tracking and monitoring work, using China Marine Surveillance (CMS) aircraft as the aerial work platform, using infrared spectral region (0.7–0.9 μm) and ultraviolet spectral region (0.3–0.4 μm), real-time detection of seawater temperature and its changes, and according to the characteristics of the red tide sea area, the temperature of the red tide sea area should be higher than the normal seawater temperature, in order to carry out the monitoring and forecasting of red tides. Using the on-board marine LiDAR, we introduced the concept of scattering coefficient of red tide algae for the optical physical phenomena presented in the process of red tide, and realized the prediction and detection of red tide in the process of red tide by detecting the backward scattering signals of the laser, and monitoring the change of the density of the red tide algae in the process of red tide, which is an additional method for the monitoring of red tide based on the on-board marine LiDAR.

11.1.2 Marine Sounding Applications

Airborne laser bathymetry system using repetition frequency 200 Hz modulated Q frequency doubling Nd: YAG laser, aircraft height 500 m, flight speed 70 m/s, it is easy to achieve 50km2/h coverage speed. The study of the U.S. Navy shows that an aircraft flying 200 h a year to complete the measurement task, a conventional survey ship takes 13 years to complete, and the ratio of the cost of airborne and survey ship is 1:5 (including the cost of data processing). The International Hydrographic Association (IHA) requires that the bathymetric error within 30 m of water depth should be no more than 0.3 m, and the relative error of water depth greater than 30 m should be no more than 1%, and the airborne laser bathymetry method can meet this requirement. Airborne blue-green laser ocean bathymetry does not replace the traditional acoustic and multispectral imaging methods, and sonar technology should still be used in deep-sea areas; multispectral imaging technology is used as a census method. However, airborne blue-green laser is undoubtedly the most effective bathymetric means on the continental shelf.

11.1.3 Application of Three-Dimensional Landscape Simulation in Coastal Zone

Coastal zone is the ground bomb unit that connects, compounds and crosses the ocean system and land system, which is the most active natural area on the surface of the earth, and also the human activity area with the most favourable conditions of resources and environment, and has the closest relationship with the survival and development of human beings. In addition, the coastal zone is the most sensitive to global change, and is subject to strong land and sea effects, making it an ecologically fragile zone in the transition between land and sea and a sensitive area for environmental change. In recent years, with the development of the marine economy, there has been an increasing number of various development activities in the coastal zone, which has led to increasing pressure on the coastal zone, unprecedented changes in resources and the environment, and the emergence of a number of problems that hinder sustainable development. Comprehensive management and monitoring of coastal zones is a key element in achieving sustainable development of the marine economy and the marine ecosystem, and has attracted great attention and concern from the State.

It is clearly proposed in "Digital Ocean" that, with the support of high-performance computers and advanced visualisation equipment, the use of scientific visual computing, 3S (a collective term for three technologies: remote sensing, geographic information system, and global positioning system), three-dimensional visualisation, virtual reality, simulation, interoperability and other technologies, based on the "Digital Ocean Based on the "digital ocean" spatial data framework, powerful model support and three-dimensional visual information expression, it realizes the digital reproduction and prediction of the seabed, water body, sea and coast with full information, establishes the three-dimensional digital sea surface and coastal landscape model including the natural landscape, human elements, natural environment, marine facilities, etc., and establishes the visual expression reflecting the process of change of marine resources and environmental elements. model, and realise the visualisation of the dynamic changes of the sea. When traditional aerial photography technology is used for three-dimensional visualisation of the coastal zone, the measurement and DEM (Digital Elevation Model) editing and processing of the external control is time-consuming, costly and long; the use of satellite remote sensing stereo image pairs to obtain the DEM and combined with the image can also be generated three-dimensional landscapes, but will be subject to the many limitations of the optical sensors, and the accuracy is not high. The use of LiDAR technology for airborne laser scanning and trapping can quickly obtain high-density, high-precision three-dimensional coordinates of the target. Model construction, texture mapping and ortho-correction of the cloud data with software support can facilitate the construction of large-area 3D models. At present, improving the filtering method of laser point cloud and enhancing the fusion effect of laser point cloud with image and feature model are the hot and difficult points of research in the field of visualisation. Through the processing of marine and coastal zone LiDAR data, the reconstruction of three-dimensional landscapes can be realised, which can better demonstrate the spatial distribution of various types of objects in the coastal zone, and provide technical support for the comprehensive management of the ocean and coastal zone and the dynamic monitoring of its development activities.

11.1.4 Underwater Military Target Detection

1. Detection of submarines

In the Second World War, the submarine played an extremely important role, and after the war, all countries pay more attention to the development of submarines. The biggest advantage of the submarine is good concealment and mobility, nuclear power propulsion in the submarine on the practical application of the submarine, the development of the submarine has entered a new stage. Nuclear submarines not only have better concealment and manoeuvrability, but also have much better underwater speed and other performance than conventional submarines. At present, 43 countries in the world have submarines, a total of nearly 1000, of which 37% are nuclear-powered submarines.

The strategic thinking of the United States and Russia and other countries in the naval forces of nuclear submarines as a key item of naval equipment. The United States has 38 active ballistic-guided nuclear submarines, 95 attack nuclear submarines, and only four conventional submarines. Soviet Navy at the end of the Second World War, only a weak "coastal defence" force, but later developed into "oceanic expansion type" super sea power, 359 active submarines, including 64 nuclear-powered ballistic missile submarines, conventional and cruise missile submarines 68, nuclear submarines and cruise missile submarines. Cruise missile submarines 68, nuclear-powered attack submarines 67.

Submarine in modern warfare occupies an important position, at the same time, anti-submarine and submarine detection also has important military significance.

The concept of airborne lidar oceanographic (ALH) detection systems was initiated in the mid-1960s. At that time, military experts wanted to search for submarines under the sea using a new invention—lasers. Since blue-green lasers have a strong seawater penetration capacity than other wavelengths of electromagnetic radiation, they could be utilised to search for submarines. Airborne transmitting system on the sea surface laser scanning, by the submarine reflected back by the laser airborne receiving system detected, or due to the submarine surface absorption of laser is stronger than the surrounding seawater, submarines in the seawater to form a "black hole" and be detected. An airborne LIDAR oceanographic system with high-precision scanning, positioning, data processing and imaging can complete the task of detecting submarines. Laser dive detection has the advantages of high positioning accuracy and geometric resolution, dive search continuity and good mobility. The traditional means of airborne hydroacoustic dive detection, still occupies a dominant position in the current, but it has its own shortcomings, and face the enemy submarine to use noise reduction and sound against new technologies such as serious challenges: many of the same family in the development of acoustic submarine sound detection at the same time, has long begun the feasibility of laser submarine sounding, 1963 to 1967, the United States of America, the University of Ohio for the Naval Avionics Laboratory for the blue-green laser sounding of the feasibility of the submarine. In 1970, the U.S. Navy and the National Atmospheric and Space Administration (NASA) with the first airborne LIDAR oceanographic detection system, successfully carried out on-site tests, especially on the "threadfin" submarine detection experiments. Its detection of submarine results, by L.M. Ott, H.K. ○ Umb ○ ltz et al. in the United States on the basis of the academic report on the Eighth Military Ocean Mapping Conference, the United States Defense Advanced Research Projects Agency in 1988 to invest more funds, airborne LIDAR submarine exploration research and testing topics. For example, Lockheed Sander was given the task by the United States Navy's PMA-264 office of exploring the use of higher-frequency electro-optical systems—LIDAR technology and SH-60 helicopter platforms—to carry out research and experiments on the detection of submarines. LIDAR submarine detection was also carried out earlier in the Soviet Union, where it was reported in 1986 that the Soviet Union had been able to detect underwater targets using LIDAR oceanographic technology from a low-altitude aircraft travelling at a speed of 160 km/h. The LIDAR technology was used to detect submarines from a depth of 130ni by passing a laser beam through a submarine. The use of laser beams through the 130ni deep seawater and silt, found sunken in the seabed for many years of the ship, and show its outline. Sweden has mounted LIDAR detection systems on submarine-hunting aircraft to detect submarines in territorial waters.

Submarine underwater laser imaging systems developed in the 1990s can be mounted on the bottom of a submarine or surface ship, or in an unmanned submersible vehicle, which can perform underwater reconnaissance or complete the task of dive detection and anti-submarine warfare. Underwater optical imaging systems commonly utilise blue-green lasers. It encounters serious laser backscattering problems, which are currently overcome mainly by the use of simultaneous scanning and distance-selective techniques.

(1) Submarine simultaneous scanning underwater imaging technology

The basic feature of this technology is that the laser beam reflected back from the underwater obstacles, received by the receiving mirror and reflected back to the receiving optical system, and then by the lens focused on the detector that the receiving mirror and the scanning mirror of the laser emission scanning at certain intervals, but to synchronise the operation. Since the optical path of the receiving laser does not use the same optical path as the transmitting laser, the effect of backward scattering of the transmitting laser on imaging is avoided. The video signal generated by the photodetector can show that the laser beam of this imaging method is very narrow, and the beam brightness is high, and thus higher quality images can be obtained at longer underwater distances.

(2) Submarine distance selective underwater imaging technology

This imaging system uses a high repetition frequency pulsed laser to illuminate the underwater scene. As mentioned in the previous chapter, to avoid the effects of laser backscattering on imaging, a distance-selective technique is used. This technique is characterised by the fact that the receiver remains 'off' until the illuminating laser is reflected back to the receiver by the target. As long as a certain distance from the target reflected light reaches the receiver at the moment, the receiver will open the selector gate, so that the reflected light into the receiver. This technique eliminates most of the backscattered light.

2. Detection of mines

The mines laid in seawater are the biggest obstacle to the amphibious combat troops in war, and at the same time, they pose a great threat to the navigation of ships. Airborne LIDAR mine detection is especially effective for tethered mines in shallow waters near the coast and in wave conditions, which not only makes up for the shortcomings of sonar detection technology, but also greatly improves the mobility and detection capability.

In 1987, the US Navy discovered from two British mine detection devices that a brass plate MagkLantern projector was available for mine detection in some US night vision optics, and the first experimental flight of the system was in 1988 on the KanrnnSH-2 Siren helicopter. According to director Mustin, the experiment was unsuccessful for a number of reasons, such as the fact that aircraft vibration impinged on the system's mirrors, causing them to deviate from collimation, which was a prominent problem. Another disadvantage of the early MagkLantern system was its limited scanning width, insufficient automatic target recognition and the need for a human to observe the TV in operation to provide real-time detection and classification. Despite this, the system worked well, providing imaging of fish swimming under seawater. Since then, the company's engineers have overcome the vibration problem by re-engineering the laser/camera pod to move key optical elements in the pod away from their resonance points. This resulted in more accurate and stable imaging. Another improvement consists of a more powerful laser and a high-capacity desktop computer. Reinforcement of the pod to withstand flight and the ocean environment greatly improved the system. Many better results were obtained in the 1989 experiments.

The advantage of the laser mine detection system is that it operates well in areas where acoustic detection systems fail. Moored mines close to shore or in shallow waters, due to the movement of the sea, which may contain waves, generate noise that may mask the sonar detector signals. Electro-optical detection systems are unaffected by such noise at depths of less than 20 m.

11.1 Application of Airborne Marine Lidar in Offshore Engineering

(1) MagicLantern system. The MagicLantern system, developed by Kaman, is mounted on the SH-2F helicopter, where a blue-green laser emits pulses into the sea at a frequency-doubled output wavelength of 532 nm with nanosecond pulse widths. A 20W laser outputs 100 mJ pulses of energy at a repetition rate of 40 Hz, and an enhanced charge-coupled device (CCD) camera captures the reflected laser energy and clears the reflected light from the surface of the water at regular intervals using the camera's shutter. The rough shape of the suspected mine object and its relative position in the water are displayed through computer processing.

(2) Detection. The system was tested under war conditions and produced excellent results. A large number of moored mines that were missed by conventional mine detection methods were detected by the MagicLantern system.

MagicLantern is capable of operating at sea altitudes of 140~320 m. Flying at low altitudes improves resolution and provides a better signal-to-noise ratio, but the field of view is limited. Higher altitude flights allow for a wider field of view and increased detection rates, since at the same speed, a higher field of view is possible compared to low altitude flights.

At the same speed, a larger area is covered compared to flights at lower altitudes. However, this reduces sensitivity and resolution. Changes in helicopter altitude have little effect on the detection depth of the system. The latest device also incorporates automatic target recognition, an improved set of optical sensors that provide higher data rates to identify targets from real-time imagery. The target recognition process consists of three parts: detection, classification and localisation.

When mine signals are detected, the system warns the operator in two ways: either by quickly collecting and incorporating the series of target signals generated, including imaging and other relevant data, and displaying them on the display; or by featuring a computer-generated tactical map showing the position of the mine object in relation to the helicopter's bearing. This simplifies the joint detection of any system on the helicopter and on the ship.

(3) Sabotage. The MagicLantern system is configured with a terminal sabotage device, the Terminator, which is a small underwater moving warhead that can be controlled. The operator commands the MagicLantern to target a specific mine. After the helicopter in which the MagicLantern is mounted has moved to a point where it is likely to explode at a right angle to the ship's keel, the operator controls the terminal through the helicopter's sonobuoy launching system. The MagicLantern images both the mine target and the terminal simultaneously, and the system's computers automatically steer the terminal's warhead to the target. The signal is emitted by the same laser used in the detection system. The terminal, which joins the detonation, then turns to the MagicLantern and enters another complete survey, detection and destruction run.

3. Underwater imaging detection technology for unmanned submersibles

Underwater laser imaging systems (LVIS) can be mounted on unmanned submersibles to identify buried, anchored and submerged mines in amphibious theatres of operation. The U.S. Navy developed in the mid-1990s a visual LVIS mine identification system that can be mounted on an unmanned submersible. The U.S. Navy developed a visual LVIS mine recognition system in the mid-1990s that could be installed in unmanned submersibles. After comparing various options, line-scan imaging was selected for the system.

11.1.5 Fish Detection

1. Traditional methods

Seabirds often congregate in the vicinity of fish close to the surface, which provides conditions for fishermen to observe fish. Usually the fishing fleet drives to areas where fish are present and observation detection is carried out by specialised fish observation ring guides. During the day, fish can be seen directly near the surface. At night, it is possible to observe the bioluminescence of the fish as they swim by. Visual observation allows for the recognition of certain fish profiles and the estimation of the size of some schooling areas. In some cases, fisheries managers also use observation methods. For example, in the Californian Sea, reports of fish observations have been collected over a 30-year period, providing a large number of timetables for the arrival of fish, such as mackerel and sardine schools, in the area, which serve as a standing index and reference for subsequent annual estimates of fish abundance.

In addition, traditional direct fish detection methods include fish sampling, trawling and sonar detection. Human detection ranges and rates are limited due to vessel speed constraints, and human estimates are quite inaccurate. Swimming schools of fish also avoid marine vessels, rendering them undetectable. The use of photographic records makes it possible to evaluate the information provided by the observer. However, taking photographs underwater is highly dependent on lighting conditions, the state of the water, and the skills of the operator. For these reasons, after much research and testing, it has been verified that airborne LiDAR marine detection systems can be an important tool for marine fisheries plants.

2. Airborne LiDAR Fish Detection

(1) Fish Detection Lidar System

Replacing visual observation with LiDAR will greatly increase the effectiveness of aerial observation, thus improving the detection capability. In addition, aerial detection of fish schools can solve the key catch (ETP) problem in the yellowfin tuna fishery. In the ETP, tuna are first caught by searching for schools of dolphins. This approach produces unfavourable results, including the fate of the dolphins having to die. Airborne LiDAR is capable of detecting and tracking larger schools of tuna without the aid of dolphins, thus providing an alternative detection method currently available.

Currently, the first is the simplest radiometer lidar, with no scanning system and the detector being the only unit detector. Laser pulses are emitted from the aircraft in

11.1 Application of Airborne Marine Lidar in Offshore Engineering

a fixed direction, usually just off the zenith angle. However, each laser pulse provides a profile of the seawater depth echo signal. Because the aircraft is always travelling forward, the system provides a two-dimensional image of the fish, one of axial depth and the other of the density of targets concentrated on the amplitude of the lidar cut in flight. The second type is the imaging radar, which produces horizontal imaging at a depth set by a distance selector. These individual pulses are combined into a hybrid image produced by the movement of the aircraft. The third type is the volumetric lidar, which utilises a scanning system and a single detector; each pulse provides a profile at one depth, which is configured into an integrated or three-dimensional image from the echoes of a number of individual pulses using the scanning system.

As mentioned earlier, volumetric lidars have been used in ocean bathymetric applications, and fish stocks can be observed during ocean surveys. Experts have evaluated airborne lidar systems for fish detection. Commercial equipment (e.g. Fishery) has been provided by KamanAerospace to parameterise several types of fish-detecting lidar systems.

(2) Airborne Laser Radar for Fish Detection

In 1981, Squire and Knimktz were the first to use LIDAR for fish detection. The system they used was a helicopter-mounted Naval Jurisdictional Lidar (NJLR), which took off from New Jersey for the voyage, and since 1982 the Russian Institute of Atmospheric Optics (RIAO) has been using airborne radiometric Lidar to detect fish in the ocean. The LIDAR system provides additional information about the scattering target: clear-water echoes; and a model of the model containing fish at a depth of 11 m. The clear water echo intensity decreases with depth, while the depolarisation ratio increases with concentration. Fish can be identified by the anomaly of increasing receiver strength and decreasing depolarisation ratio at a depth of 11 m.

Research on the Osprey lidar began in the UK in 1990. This device is a helicopter-borne radiation lidar for detecting tuna. Trials of this system were conducted from 25 September to 20 October 1992, when it was mounted on a CMS helicopter and used for turtle detection while guiding the Captain Vinecru Garm retracting grid vessel to fish. It basically operated 16 h per day in the eastern Pacific Ocean.

The National Oceanic and Atmospheric Administration (NOAA) Experimental Fishfinder Lidar of the Ocean (FLOE), also a radiometer lidar, 7 was developed specifically for aerial detection of bioluminescence in striped bass schools. Although it was mounted and flown in a small aeroplane, the actual operation was carried out on the sounding vessel R/TDavidStarrJerdan, which was also equipped with detectors for sonar and acoustic echoes.

In 1995, mackerel data were collected off the east coast of Vancouver Island, Canada, using the scanning lidar LARSEN500.The FishEyr lidar was also originally designed as an imaging system for fish detection. It images the contours of large fish or mammals or schools of fish. Distinguishing fish relies on seawater turbidity, with a typical resolution of a few tens of centimetres. This resolution provides information that can help identify fish. However, it is not favourable for estimating the size of schools of fish, as calculating the size of a school requires the thickness of the school.

LiDAR has become an important tool in the detection of turtle populations in several photosynthetic zones and in fisheries management. Optimal detection should include complementary or combined detection by shipboard efforts with sonar detection and direct sampling. Small aircraft (with a capacity of six passengers) can be equipped with radiation LIDAR, small infrared radiometers and colour TVs, which can study in detail the interactions of fish with sea surface temperature and ocean colour to the extent that the extent of the encircled fish can be determined more accurately from satellite imagery. Ideally, the pilot would be a professional fish observer, able to operate both the flight and the visual observations to provide reliable information on fish stocks and species.

3. Researching the age and origin of fish

The use of optics to produce Fourier transforms of fish data can help fisheries researchers tell the age and origin of fish. Experiments at the Virgilian Institute of Integral Technology and the State University of Blacksburg have shown that the Fourier transform is able to sort patterns containing fish size data according to the age of the fish and the area in which they lived when they were young.

11.1.6 Reef Detection and Distress Surveys

On 6 September 1998, a Swissair aircraft crashed off NovaScotca and Canadian aviation researchers with sonar imaging systems quickly arrived at the crash site. Sonar imaging can indicate large areas of concentrated damage, but it does not provide divers with images of sufficient resolution to search for debris that may be clues to the crash.

U.S. Navy Mine Warfare Laboratory with laser detection methods for mine countermeasures research and detection experiments require the following equipment: 21-inch imager, low-resolution forward scanning sonar detector; higher resolution side scanning sonar detector; high-resolution laser line scanning imager.

All parts are mounted in a torpedo-shaped "fish trawler" that is sunk into the sea and towed by an underwater vessel. In mine detection, the sonar is used first, and when a target is found, it is turned towards the target and then imaged with a laser beam scanner.

After the Swiss flight disaster, the Canadians had already done the scanning of the sonar, and later used the laser detector of the British Naval Coastal Systems Station to carry out laser scanning and imaging, and as the "fish trawler" device passes through the water, the imager moves forwards to establish two-dimensional imaging, thus finding some aircraft debris to be imaged.

As the "towfish" device passed through the water, the imager moved forward to create a two-dimensional image, which led to the imaging of some aircraft debris, providing an important basis for many clues to the cause of the crash.

11.1.7 Detection of Marine Underwater Resources

The Airborne Laser Fluorescence (ALF) sensor, a system for detecting oil layers at sea, developed jointly by BP Surveys of the United Kingdom and World Geoscience of Australia, is searching for oil fields far out at sea, more simply and cost-effectively than some current methods.

About 75% of the world's known oil fields are deeply buried, with very little oil spilling directly to the Earth's surface. Consequently, most offshore fields seep oil, and the detection of seepage indicates the presence of oil accumulations. However, in many production sites, seeping oil forms a thin film that is difficult to detect with the naked eye and other passive airborne or satellite-mounted detectors. On board an F-27 aircraft at an altitude of l00m, the ALE system emits 50 Hz laser pulses at the sea surface from a solid-state laser, and the laser beam, on any fresh oil film, produces a fluorescence, which is intercepted by the system's receiving telescope, and then split into the colours of the various constituents by a spectrometer with a wavelength operating range of 20–700 mm. The output signal from the spectrometer, fed back to a 500-channel diode array detector, whose information is recorded on channel 176 and stored along with information on navigation, aeromagnetic components and the environment. Television cameras simultaneously capture imaging in the field.3 This imaging can be used in data processing plants to help discern anomalies in the detection. Pollution and other substances can also cause fluorescence, and it is important to exclude this information from the sounding data, retaining the information on the oil film from which it can be sorted out as to the type of oil contained in the film: accumulations, regular oil, or heavy oil. The system was flight-tested in a number of sea areas where oil fields were known to exist, and it was demonstrated that the expected oil films did occur there.

The application of the ALF system to S-marks is not only on oilfields, it will also provide an opportunity to study the mysteries surrounding the worst maritime disasters. In the Second World War, the Australian battleships HMASSydney and HMASkormoran were lost to enemy fire and it is unlikely that they will ever return to the west coast of Australia, but traces of oil may still seep from their wrecks, making the system the perfect tool to find them. Similar airborne laser fluorescence sensors, SLEAF and SLEAF2, have been studied by Environment and Minerals Canada and are being used in marine oil field detection studies.

11.1.8 Detection of Marine Plankton

Phytoplankton in the oceans are responsible for more than half of the world's light synthesis. Measurement of the plankton population is of great significance, as it enables us to estimate the amount of carbon dioxide fixation, analyse the effects of the El Niño phenomenon and other changes in marine meteorology, and contribute to in-depth research on global environmental issues. Until now, observations of plankton have been made by sampling and analysing seawater in order to obtain the types of plankton. Even with the help of artificial satellites, only the surface layer of the ocean can be observed, but it is not possible to measure plankton and its distribution in the deep sea, nor is it possible to carry out remote control measurements.

In order to solve this problem, the Lithuanian Maritime Research Centre has developed an observation device that can measure phytoplankton growing at a depth of 30 to 50 m by transmitting laser light from a ship at sea. The device uses a green laser as a light source, and the distribution of zooplankton is determined by measuring the fluorescence and reflected light emitted by the laser. In the case of zooplankton growing at depths of up to 50 m, it is also possible to analyse the distribution of planktonic organisms by measuring the scattered light produced by cells irradiated by the laser.

11.2 Application of Infrared Detection in Offshore Engineering

Infrared detection system has a wide range, all the use of external objects themselves emit thermal radiation as a radiation source for detection can be called infrared detection system. The system is characterised by the fact that no light is required and detection is carried out in complete darkness. This kind of detection is passive, can be covertly observed and monitored, due to the use of wavelengths longer than visible light 10 to 20 times, in the smoke penetration ability is much larger, so in foggy days, smoke environment, you can observe a considerable distance. In order to improve the maintenance of maritime security of the war celluloid ability to defend the national territorial waters and maritime rights and interests. Infrared probe system can meet the demand for long-distance observation, search, surveillance, navigation and other functions at night and in poor visibility.

Can be applied in the sea infrared probe system has infrared front vision, infrared search tracker, incoming missile infrared alarm, infrared guided missiles and so on. Surface ships in order to attack the enemy, the need for real-time observation or visual air or sea surface posture, the type of enemy ships and activities: surface ships in order to their own safety, the need for real-time prevention of enemy attacks. Especially to prevent such as sea-skimming missiles, sea-skimming aircraft and other threats to the most serious attacks, but also to prevent attacks from high altitude. Infrared detection can fulfil the task of observing the scene and searching for threats, and it is also compatible with radar search and detection systems. When the radar system is subjected to electromagnetic interference or suppression, the infrared system can still work independently, so that the entire weapon system from failure.

1. Infrared forward-looking

Infrared forward-looking thermal imaging observation system, the output of the thermal distribution of external objects in grey scale form, can be used around the clock, and even in poor visibility in the case of normal work. Thermal imaging

11.2 Application of Infrared Detection in Offshore Engineering

observation equipment output is a video image, usually the same as the television system, can be observed in real time the activities of the target, will be installed on the ship, a wide range of uses.

(1) Navigation at night and in fog

The observation distance ranges from tens of metres to thousands of metres, and the observation targets include nearby ships, bridge piers, obstacles in the channel, docks, etc. The purpose of use is to ensure the safety of navigation and to prevent collisions. This type of thermal imaging camera requires a wide field of view of 15°–50°. Due to the large field of view, but also in order to reduce costs, generally without stabilisation and rotation. Thermal imaging cameras are small in size and usually use uncooled long-wave thermal detectors, which are convenient and easy to promote their use.

(2) Medium range observation

Observation distance of about 5 km, for night or fog monitoring, tracking, landing, counter-terrorism, search and rescue. Observation targets are mainly small and medium-sized ships, landing beaches, personnel activities. The thermal imaging camera can use an uncooled detector with a recommended focal length of 33.3 to 100 mm continuous zoom optical system, with a detector of size 384×288 metrics and intermetric spacing of 35 µm, and the field of view of the thermal imaging camera will be continuously variable between $22.8° \times 17.1°$ and $7.7° \times 5.8°$. The recognition range for a small 30 m long boat is over 6 km and the detection range for a person is about 1.9 km. for all round searching or observing of the target, the thermal imaging camera can be mounted on a slewing head, and if image stabilisation of the observation is required in case of rocking of the boat, the observation system needs to be mounted on a stabilising platform.

(3) Long-range observation

Observation distance of more than 8 km, for night long-distance observation, surveillance, search, tracking, etc., the observation target is mainly large ships and a variety of coveted ships, mastering the night sea ship activity situation thermal imaging camera can choose to use refrigeration-type long-wave or medium-wave detector. The size of the detector with at least 320×256 yuan, 640×480 yuan better; optical system focal length from 150–600 mm can be used; in order to meet the surveillance, search, tracking and targeting of the different requirements, can be designed into a two-stop or three-stop zoom form, the specific design based on the nature of the task of observing the requirements of the clarity of the target image and the size of the thermal camera and other factors.

The thermal imaging camera itself is relatively simple, generally consisting of an optical system, detector components, power supply, image formation circuitry and corresponding structural components. If it is required that the target can be identified and tracked or alarmed, the target identification and tracking circuits need to be added. In order to eliminate the effect of the swaying of ocean-going vessels on the observed image, the thermal imaging equipment should be mounted on a platform or in a sealed pod with stabilising features. Depending on the requirements thermal imaging cameras can also be integrated with visible cameras and laser range finders to form spherical turrets.

2. Infrared search and tracker

Infrared search tracker is used for searching, intercepting, tracking and aiming at distant targets, such as aircrafts, ships, flying missiles, sea-skimming missiles, etc., at a distance of generally more than 5 km. It is characterised by a large search field of view, often requiring a 180°–360° spatial range. The target is far away, capable of detecting and intercepting aircraft 20–30 km away. Such a long distance, it is impossible to require to see the shape of the target, even less than a pixel, which is different from the thermal image observation system, the target information to be obtained is not the details of the shape of the target, but the total number of the target, the spatial orientation, and the direction of flight. The angular velocity of the target is small, and it is not necessary to collect information at a high frequency, usually with a frame rate of 0.1 to 2.0 Hz.

Infrared search and tracker is usually used with a stable function of the rotary table for a wide range of search, with a higher frequency of the pendulum mirror for a small range of scanning, scanning and tracking.

The necessity of installing infrared search and tracker on ships is to be able to detect intruding aircraft and sea-skimming missiles at ultra-low altitude in a timely manner, as this area is a dead zone for radar operation. The IR search tracker can be mounted on the ship alone, or in most cases hinged on the search radar, as an auxiliary detection device to the radar system. The system can also be hinged with anti-aircraft guns or air-to-air missile launchers. When the search for a target is turned to tracking, the weapon launching system is aligned and continuously tracks the target, and can immediately attack if necessary, therefore, the infrared search tracker is an important part of the fire control system.

3. Infrared guided missile

Ships equipped with infrared guided missiles are mainly used to deal with airborne threats, such as ultra-low altitude aircraft, sea-skimming missiles, cruise missiles, etc., can also be used to attack the sea on the ship class targets, the role of the distance is generally up to 10 km. available guidance modulation disc unit detector guide and multi-guidance guided guided guide to improve the ability of anti-jamming at present the development of thermal imaging guided guided guided guide to the use of infrared guided missiles, simple equipment, fast operation, suitable for infrared guided missiles The use of infrared-guided missiles, simple equipment, fast operation, suitable for rapid response and attack on close-range threats.

At present, infrared guided missile technology has been quite mature, there are a variety of domestic infrared guided missiles for air and ship. Infrared guidance system more than the use of mid-wave infrared wavelengths, the main detection of the target nozzle, exhaust, boat smoke tube radiation and heat exhaust radiation, the most advanced guidance system can be cooled gaze focal plane array detector to achieve thermal imaging guidance, capture the field of view of 3°~5°. The role of the distance

11.2 Application of Infrared Detection in Offshore Engineering

and the type of target, the state has a great relationship, usually close combat type more requirements in about 10 km can reliably intercept the target. Stable tracking mechanism of many structural forms, the best breakthrough orthogonal dual-frame form, with polar coordinates three-frame platform structure, to achieve ± 90° large tracking field. Infrared guided missiles for marine vessels, can be used as ship-to-air and ship-to-ship near-type missiles. Infrared guidance can also be used for the end of the missile guidance stage, the realization of ships and other fixed targets in the medium and long-distance strikes, the end of the type of guidance is to be reached by the missile from the target at an appropriate distance, the guidance system to quickly search for and capture the target, and then into the automatic tracking state, until the target is destroyed.

4. Infrared alarm for incoming missiles (aircraft)

The infrared alarm of incoming missiles is mainly used to detect missiles and aircrafts and give timely alarms to facilitate the adoption of effective measures, or to cast interference, or to attack the enemy to ensure their own safety.

Incoming missile (aircraft) alarm also requires a large spatial range of detection, detection distance is farther, which is similar to the search and tracking device, but due to the threat of the target is relatively close to the alarm must be given in a timely manner, do not need to be as complex as the structure of the search and tracking device, does not require a stable platform, and even more do not need to be wide-range search and tracking mechanism, but only requires real-time monitoring of a very wide spatial range, once detected incoming missiles or aircraft, can immediately give the alarm signal Once an incoming missile or aircraft is detected, an alarm signal can be sent out immediately so that the operator can make further judgement on the danger and take the necessary countermeasures. Missile alarm structure is relatively simple, compact, can cover the field of view is usually up to 90° × 65°, if you want to search the hemispherical field of view in the azimuthal direction, it is necessary to install two, a larger field of view to install more than one device. The design should take into account the spatial area that poses the greatest threat to itself. Detection distance should normally be able to reach more than 5 km, according to the speed of approaching missiles, the detection of the target should have enough time to take countermeasures.

Incoming missile alarms are radar, ultraviolet and infrared types. Radar-type structure is complex, and is active, not conducive to conceal themselves; UV-type can only detect the radiation emitted when the missile engine work, once the engine work is over, will not detect the target, and the general working time of small missile engines is shorter, is not conducive to the full range of detection of incoming missiles; infrared-type in addition to detecting the radiation of the engine work, but also to detect the formation of high-speed missiles aerodynamically heated radiation, which can be used to detect the incoming missiles, and can be used to detect the target. The infrared type can also detect the radiation formed by the aerodynamic heating of high-speed missiles in addition to the radiation during engine operation, which can detect the whole process of incoming missiles and ensure timely detection of threatening targets.

If the ship is equipped with infrared search and tracker, and can monitor the full range of space, it can play the function of incoming missile alarm, can not be configured with a special alarm. If there is a dead space for air surveillance or no search tracker, it is very necessary to set up the alarm device for incoming missiles (including aircraft) to ensure their own safety.

5. Oil Leakage Detection

Oil leakage from ships is a major problem in maritime operations. Since crude oil or diesel fuel usually floats above the water column after a spillage due to its low density. Although the oil spill will be distributed and form a well-defined film. However, there is not a very sharp visual contrast between the oil film and the water surface before, and visual observation is still difficult.

Particularly in the direction of lower incidence angles, both the water column and the oil film will appear dark black. However, in rough or undulating waters, oil spill detection becomes even more difficult due to the ever-changing reflection of the water surface off the sunlight or sky, with the water surface being darker and brighter at times, which further strengthens the masking effect on the low-contrast oil film area.

Under different sea conditions and light conditions, the multi-band range of the spectrum brings hope for increasing the contrast between petrochemical products and the water surface, and infrared thermal imaging cameras are of great significance for oil spill detection by scanning the invisible light bands.

6. Emergency Response

Thermal imaging cameras are widely used in maritime search and rescue operations around the world. When a person falls overboard, the head of the person overboard is often a visible part of the body. Shown on video as a white or red 'head-shaped' image against a black or dark grey background, the head is clearly visible as it emits much more heat than water. This effect is achieved day and night, making it truly responsive 24 h a day, 7 days a week.

Thermal imaging does not require external illumination, making it difficult to conceal even the best camouflaged bits from view, and is able to see human-sized targets with a longer detection range than competitive TV cameras. Another advantage of using thermal cameras is that the heat signatures of people, boats, cruise ship docks, boat launches, piles and other things detected by maritime thermal cameras are usually more contrasty at night than during the day. As long as there is a small temperature difference between the target and its background, then the observer can easily detect it.

7. Safe navigation

Navigating a cruise ship in open water is very challenging. Navigating on ice-covered water is even more difficult and dangerous. Shipping companies are increasingly concerned about the safety of ice-class tankers when navigating in waters where the temperature is below 0 °C. The safety of ice-class tankers is a major concern for shipping companies. Thermal imaging cameras can detect ice in the sea at long distances, thus avoiding collisions.

Bibliography

1. Zhizhong, Li. 1998. Airborne LiDAR System and Its Application Prospect in Marine Survey. *Geological Frontiers* 2: 77.
2. Qiyang, Xu, Yang Kuntao, Wang Xinbing, and Xu Desheng (eds.). 2002. *Blue-Green Lidar Ocean Detection*. National Defence Industry Press.
3. Xuerong, Lu., Lv. Yuhang, and Cheng Mingyang. 2011. Application of Infrared Detection System in Military Sea. *Marine Information* 1: 5–7.

Part IV
Radar

Chapter 12
Overview of Radar Technology

5000 words Radar is a sensor device commonly used in marine unmanned platforms, and the marine unmanned platforms that we often mention, such as: unmanned boats, unmanned aircraft, etc., are basically installed with radar as the main information sensing device. In unmanned boats, navigation radar is generally used as a necessary device to ensure the safety of navigation. In unmanned aircraft platforms, synthetic aperture radar is often installed as a means of detection. So how exactly does radar enable detection? And what are the characteristics of radar on marine unmanned platforms? This chapter will focus on this aspect.

Radar is the Chinese translation of English Radar, is the abbreviation of English Radio Detection and Ranging, that is: radio detection and ranging. Radar makes use of the basic physical phenomenon that objects will reflect electromagnetic waves, and the basic working principle is that radar radiates electromagnetic signals into space, and detects the electromagnetic waves reflected back from the target, so as to realise the measurement of the target's distance, bearing, elevation and other positional information. With the continuous progress of technology, other relevant information such as target speed can also be obtained.

Radar is an important tool in the field of target information acquisition, and there are many books detailing the working mechanism of radar. This book firstly describes the basic concepts of radar from the perspective of the three basic philosophical issues (readers who are interested in knowing more about professional knowledge are recommended to read "Radar Principles" by Mr. Ding Lufei, "Radar System Analysis and Modeling" by David Barton, "Radar System Analysis and Modeling" by Bassem, and "Radar System Analysis and Modeling" by David Barton. Barton's Radar System Analysis and Modeling, Bassem R. Mahafza's Radar Systems Analysis and Design Using MATLAB and other radar professional works), the three questions that: What is radar? Where does radar come from? Where does radar go? The focus will be to combine the special needs of some marine unmanned platforms, radar design related problems, put forward some technical difficulties and solutions.

12.1 What Is Radar?

12.1.1 Functions of Radar

The basic function of radar has been embodied in the abbreviated name, that is, detection and distance measurement, with the advancement of technology, human beings can control only to a specific direction to launch electromagnetic waves or only to receive a specific direction of the electromagnetic waves reflected back, according to the direction of the beam, you can determine the spatial orientation of the radar, so that the radar has the ability to measure the angle. Along with the improvement of digital technology, using the Doppler effect between electromagnetic wave and target movement, radar has the ability to measure radial velocity. In addition, in specific application scenarios, special system design methods can be used to measure the target's micro-motion characteristics, polarisation scattering characteristics, surface roughness, dielectric properties and other elements.

The basic principle of distance measurement: the radar transmits the electric measuring wave to the space and starts to calculate the time, the electromagnetic wave encounters the target in the space and is reflected, the radar receives the reflected back electromagnetic wave and stops the timekeeping, which obtains the time interval that the electromagnetic wave propagates from the radar to the target and is reflected back. A priori knowledge is known electromagnetic wave propagation speed approximate to the speed of light, know electromagnetic wave propagation speed and electromagnetic wave transmission time, you can calculate the distance between the radar and the target (Fig. 12.1). This method is technically known as Pulse-Delay Ranging.

Let the one-way distance between the target and the radar be R, in m, the time for the electromagnetic wave to travel back and forth between the radar and the target be t, in s, and the propagation speed of the electromagnetic wave be c. Generally, take the speed of light $c = 3 \times 10^8$ m/s, then:

$$R = \frac{c \times t}{2} \tag{12.1}$$

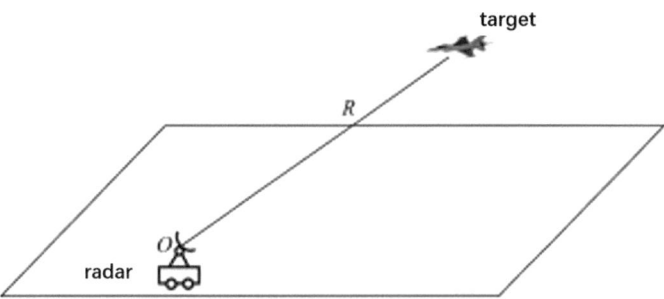

Fig. 12.1 Schematic of target distance and angle measurement

12.1 What Is Radar?

For example: If the round-trip transmission time of electromagnetic wave from radar to target is calculated to be 10us, then the corresponding distance between target and radar is 1.5 km.

Note: If the rigorous calculation, electromagnetic transmission speed in different media is a difference, but this difference in value and absolute value compared to the relative proportion is very small, so in general in the calculation of radar ranging, take the speed of light as the speed of electromagnetic wave propagation.

Angle measurement of the basic principle: Angle measurement that is to obtain the target in three-dimensional space in the azimuth and pitch angle, radar through the antenna will be emitted (or received) electromagnetic energy pooled into a narrow beam, then only when the electromagnetic beam is aligned with the target, the radar's strongest echo, at this time the record of the radar antenna will be directed to the beam of the position, you can get the target's angle information. However, in order to obtain a narrower radar beam, a larger radar antenna size is required, which often has a lot of limitations in engineering, and the radar beam is too narrow, but also makes the coverage of an airspace needs to be lined up with more beams, which is not conducive to the use of engineering. In order to obtain more accurate angle measurement information under a certain band width, the radar also introduces an improved angle measurement method, and in typical cases, the angle measurement accuracy can reach 1/10 of the radar beam.

Fundamentals of velocity measurement: Velocity measurement by radar is usually considered in two dimensions. One dimension is that by measuring the position of the target in three-dimensional space several times and recording the time point corresponding to each position, the speed of the target can be obtained by calculation. This kind of velocity measurement is a secondary algorithm based on the primary measurement information (distance, angle, time), and the measurement error of the primary information will directly affect the accuracy of the velocity measurement information. Another dimension is the use of the Doppler shift phenomenon to directly obtain radial velocity information of the target relative to the radar. The Doppler shift phenomenon, also known as the Doppler effect, refers to the fact that when there is relative radial motion between the transmitting source and the receiver, the frequency of the received signal will change. Physicist Christon Doppler discovered this phenomenon in acoustics in 1842, and the law was used in the field of electromagnetic waves around 1930. When there is relative radial motion between the target and the radar, the carrier frequency of the electromagnetic wave changes when it is reflected by the target, and there is a relationship between this change and the relative radial velocity. Detecting this carrier frequency change amount, the radial relative velocity between the target and the radar can be obtained.

Let v be the relative radial velocity between the target and the radar in m/s, f_d be the Doppler shift in Hz, and λ be the wavelength of the carrier frequency of the electromagnetic wave in m. Then there are:

$$v = \frac{f_d \lambda}{2} \tag{12.2}$$

When the relative distance between the radar and the target is shortening, the echo carrier frequency will increase, and when the relative distance between the radar and the target is expanding, the echo carrier frequency will decrease.

This direct use of primary physical information for velocity measurement is relatively high in accuracy, but produces measurement ambiguity (i.e., a frequency-shifted measurement will correspond to multiple velocity values), which can be achieved by changing the frequency of the radar transmitting pulses, defuzzification algorithms, and so on, to achieve an accurate measurement of velocity.

Doppler shift phenomenon, also widely used in moving target detection and moving target display, etc. Especially when shipboard radar fights against sea clutter, it is one of the important technological paths to differentiate between clutter and target through velocity dimension information.

The basic principle of target profile measurement: the measurement of target profile dimension is not a direct measurement information, it is a calculation information. The most basic principle is that if a target is large enough and the radar beam is narrow enough, the target can form multiple echo points, and the distribution of each echo point in space can form the target's contour information, so as to realise target profile measurement. In order to get more accurate shape measurement information, it is necessary for the radar to have better resolution in distance and orientation dimensions. The distance dimension can be achieved by pulse compression and other techniques, while the azimuth dimension requires a larger antenna aperture to obtain a narrower waveform. However, an unlimited increase in the actual aperture is unrealistic and can be achieved by synthesising the aperture. Here the Synthetic Aperture Radar (SAR), which is commonly used in unmanned maritime applications, is introduced.

The concept of SAR radar originated in the early 1950s, when American scientists came up with the idea of a long antenna, where each unit simultaneously transmits a phase-referenced waveform and then receives it at the same time, forming a narrow beam through back-end processing. If in some scenarios, do not need to transmit and receive at the same time, but the use of an antenna unit, smooth movement along the antenna, time-sharing transmission, reception, and storage of the lower echo information, through the back-end processing, the formation of similar to the detection of a long array of narrow beams of detection effects, thereby moving through the formation of a virtual synthetic aperture, that is, synthetic aperture antenna. The radar using this synthetic aperture antenna technology is called synthetic aperture radar (SAR), which is widely used in the field of ocean exploration because of its ability to provide high-resolution imaging of the target, and is described in more detail in the later chapters of this book.

12.1.2 Radar Operating Frequency and Usage Characteristics

Radar achieves target information acquisition through electromagnetic wave, then the characteristics of electromagnetic wave itself will affect the detection ability of radar. The frequency of electromagnetic wave radiated by the radar is one of the important parameters affecting the radar detection capability, also known as the radar operating frequency. Commonly used radar operating frequency is generally 220 MHz~35 GHz, but modern radar has gradually broken through this frequency limitation, for example, terahertz radar has been used to enhance the frequency to more than 200 GHz, and some ground wave over-the-horizon radar, with a frequency of 2 MHz or so, in order to get a good transmission of the ground wave, to achieve over-the-horizon detection.

At present, there are two common working frequency bands for radar, one is during the Second World War when the radar was first applied, the military used letters such as L, S, C, X and so on to indicate the radar's working frequency for the sake of confidentiality, which was later widely used and is still in use today. Nowadays, most of the typical frequency descriptions used in the domestic radar industry also use this mechanism. In addition, a simpler letter system: A, B, C ......, etc. are commonly used internationally to indicate radar operating frequencies, and will also appear in some reference documents. The radar operating frequency bands described in this book use the first representation, and their corresponding specific frequency relationships are shown in Table 12.1.

Table 12.1 Standard radar letter-frequency correspondence table [1]

Band name	Standard frequency range	Radar frequencies allocated by the International Telecommunication Union
HF	3~30 MHz	
VHF	30~300 MHz	138~144 MHz, 216~225 MHz
UHF	300~1000 MHz	420 MHz~450 MHz, 850~942 MHz
L	1~2 GHz	1215~1400 MHz
S	2~4 GHz	2300~2500 MHz, 2700~3700 MHz
C	4~8 GHz	5250~5925 MHz
X	8~12 GHz	8500~10680 MHz
Ku	12~18 GHz	13.4~14 GHz, 15.7~17.7 GHz
K	18~27 GHz	24.05~24.25 GHz
Ka	27~40 GHz	33.4~36 GHz
v	40~75 GHz	59~64 GHz
w	75~110 GHz	76~81 GHz, 92~100 GHz
mm	110~300 GHz	126~142 GHz, 144~149 GHz, 231~235 GHz, 238~248 GHz

Analysed from the perspective of electromagnetic wave propagation characteristics, the lower the radar operating frequency, the smaller its atmospheric attenuation, but the worse its ability to measure accurately, especially the angular accuracy is poor. Therefore, low-frequency band radar is mostly used in remote and less demanding detection accuracy scenarios. For example, most of the remote early warning radars work below the L band.

S-band radar, has been able to obtain better measurement accuracy, while the atmospheric attenuation is also within the acceptable range, so most of the long-range early warning radar working in the S-band, the more famous S-band radar, including the U.S. Aegis multi-functional phased-array radar (SPY-1 series, SPY-6, etc.), a lot of large-scale national early warning aircraft are also working in the S-band.

C-band can obtain better measurement accuracy, at the same time, the bad weather environment, its attenuation has not deteriorated to seriously affect the system performance, in the same range of action, its consumption of engineering costs are often acceptable. The world's more famous C-band radars include the guidance radar of the U.S. Patriot Weapon System, etc. In the field of shipboard radar, Germany's TRS series and Japan's FCS series are more successful shipboard C-band radars.

Above X-band radar, it has been possible to obtain high measurement accuracy, but at the same time the atmospheric attenuation is relatively large, in the field of precision tracking, measurement, imaging, etc., the use rate is high. In the field of ocean engineering, many navigation radars work in X-band. most of the SAR imaging radars also work in X and higher bands. Worldwide, the more famous X-band radars include the guidance radar in the THAAD weapon system of the United States, and there is also an X-band radar in the AMDR, the newly developed shipborne air defence and anti-missile detection system of the United States [2–4].

The most direct tool for grasping the fundamentals of radar is the radar equation. The radar equation is not only a mathematical model, but also provides a visual understanding of the physical implications of radar operation. The radar equation is a mathematical tool that characterises the radar's range of detection, while linking radar transmission, antenna, space propagation, target characteristics, environmental losses, reception and other factors in the equation. It is very intuitive to show the radar working process.

The following is a simple derivation of the radar equation for radar in free space. This mathematical derivation process can help you understand the basic working principle of radar more rationally.

If the radar transmitter as a point source of radiation, its energy is uniformly radiated in all directions in space, which can be understood as an expanding sphere. If the emitted power of the radiation source is denoted as P_t, the power density at any point on the sphere at a distance R from the radiation source is the total power divided by the area of the sphere $4\pi R^2$, denoted as:

$$S\prime = \frac{P_t}{4\pi R^2} \qquad (12.3)$$

12.1 What Is Radar?

In engineering, radar usually constrains the radiated energy to a particular direction by means of an antenna so that an energy gain, denoted G, is formed and the power density at a point at a distance R is:

$$S_1 = \frac{P_t G}{4\pi R^2} \tag{12.4}$$

The electric measured wave irradiated to the target is re-radiated to different directions by the target, the target size, shape, material and other factors, will affect the characteristics of the re-radiated electromagnetic wave, here these concepts are unified into a parameter σ to characterise the target re-radiated in all directions to scatter the received electromagnetic wave ability. Here again, the radiated target can be taken as a new radiation source, and the energy radiated by this source, similar to the calculation process in the previous step, the power density arriving at the radar can be expressed as:

$$S_2 = S_1 \times \frac{\sigma}{4\pi R^2} = \frac{P_t G}{4\pi R^2} \times \frac{\sigma}{4\pi R^2} \tag{12.5}$$

The radar antenna can receive a portion of the scattered energy, the effective receiving area A_e of the antenna, can receive the energy reflected back from the target, then the radar receives the reflected power of the target is:

$$P_r = A_e S_2 = A_e \times \frac{P_t G}{4\pi R^2} \times \frac{\sigma}{4\pi R^2} \tag{12.6}$$

A target can be detected by a radar when the received power is greater than the radar minimum detectable power $S_{\min}$. The radar minimum detectable power $S_{\min}$ is a system metric that is usually directly related to the design and detection guidelines of the receiving link and to the level of control of system noise (both internal and external).

A simplest view of radar operation is shown in Fig. 12.2.

There are many things to think about by the formula.

If you want to detect a more distant target, you need to get a larger echo energy or reduce the radar minimum detectable signal. Get a greater return energy engineering to achieve the way including increased transmit power, increase transmit gain, increase the effective receiving area of the antenna, select an operating frequency, so that the target can be more reflected electromagnetic wave energy and so on. Reduce the minimum detectable signal, involves more complex radar signal detection knowledge, here will not be expanded in detail, one of the ideas is to amplify the echo signal only in the detection link, while suppressing the noise signal, in the digital signal processing introduced into the radar system, there are some more complex detection algorithms to support the system in the echo energy is very small under the premise of the target, but also can be detected, for example, detection of tracking technology before (Track-Before-Detect, TBD), etc.

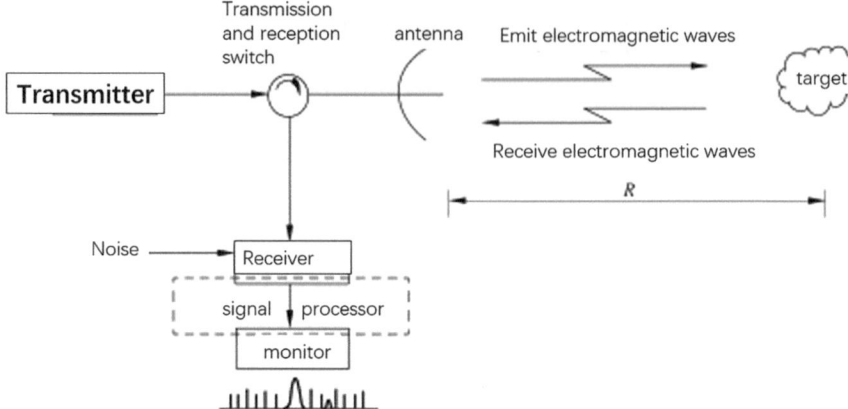

Fig. 12.2 Schematic diagram of radar operating principle

If as a target, do not want to be "seen" by the radar, we must find ways to reduce their own, that is, not to the radar direction of the reflected electromagnetic waves. We usually say that the stealth aircraft (such as the United States F-22), is the use of this principle, the use of some engineering means or new materials, so that the aircraft will not be irradiated in the direction of maximum reflection of electromagnetic waves, although the actual physical size of their own is very large, but the radar on the role of the distance is very short, to achieve the radar "stealth". The radar equations above do not reflect the electromagnetic waves in the direction in which they are directed.

The radar equation above does not reflect the electromagnetic wave propagation of various types of attenuation, in the actual engineering applications, but also need to consider the attenuation factors, such as rain, snow, will produce different degrees of attenuation of electromagnetic waves, and attenuation value and radar frequency band related to the work of the radar, in general, the higher the frequency in the rain and snow in the same distance under the weather, the more serious the attenuation.

And here implies a condition that electromagnetic waves along the straight line in space free propagation, when applied to marine engineering, due to the influence of the curvature of the earth, usually the detection distance of the radar but also consider the sight distance impact, especially for the detection of surface ship targets, often the radar rack height is also an important consideration. In addition, some special atmospheric conditions, will make the electromagnetic wave is not transmitted along the straight line, but the refraction phenomenon, then the use of radar equations to consider more factors, and can no longer use the distance delay method for accurate distance measurement of the target [5].

12.2 Development History of Radar Technology

12.2.1 The Infancy of Radar Technology

The name radar first originated in the 1920s as a secret programme of the US military, but the development of radar technology itself has a much longer history. Radar is an application of radio technology, so to understand the origins of radar, it is important to talk about the development of radio.

The basic principles of radio originated in Europe. Early contributors to the technology were three scientists, Michael Faraday (1791–1867), James Clerk Maxwell (1831–1879) and Heinrich Hertz (1857–1894). The Englishman Faraday, who was entirely self-taught and had little mathematical ability, discovered electromagnetic induction through experimentation and hypothesised that the electromagnetic field could be extended into the vacuum around a conductor, an idea that was the theoretical beginnings of radio. Maxwell was a Scotsman who studied at Edinburgh and Cambridge Universities, a genius in the field of physics "god-like existence", in 1864, he proposed a mathematical model of electromagnetic waves, with a few equations can fully express the electromagnetic radiation, which is the famous Maxwell's system of equations in physics.

The basic principles of radar work were laid down by the German scientist Heinrich Hertz (1857–1894), after whom the unit for describing frequency, the hertz, was named. 1888 or so, in a corner of the physics classroom at the Karlsruhe Institute of Technology in Berlin, Heinrich Hertz used a machine that contained a spark to create the electromagnetic wave. -Heinrich Hertz used a circuit containing a spark gap to generate electromagnetic waves and detected them in a similar circuit not far away. In the process, he not only proved the existence of these electromagnetic waves, but also that, like light waves, they could be reflected by metal surfaces. Thus, his electromagnetic waves "detected" the presence of metal. This seemingly simple experiment laid the physical foundation for radar, verifying that electromagnetic waves are emitted, scattered by the target, and received, the most essential process of radar operation. Unfortunately, however, Heinrich Hertz saw no engineering applications for this discovery.

In 1904, the German scientist Christian Hulsmeyer, shocked by the death of a friend in a ship collision, founded a company and invented a device that used Heinrich Hertz's principle of electromagnetic wave detection for ship collision avoidance, using a 50 cm wavelength spark-gap transmitter and a coherent detector, with the radiated signals emitted by a funnel-shaped reflector and a tube that could be The radiating signal is emitted by a funnel-shaped reflector and a tube that can be aimed at, and the receiver is a free-standing vertical antenna with a semi-cylindrical removable reflector screen. A German patent was obtained for this collision avoidance device. This can be considered the earliest prototype of an engineering application for radar and was the first patent in the field of radar in the world. Unfortunately, however, the technology was still quite early and was not widely recognised and used by society.

Today, radar is widely used in various fields and shines in daily livelihood and military applications, with its earliest engineering applications still in marine engineering. The ocean and radar have long been tied together.

Although Germany first verified the basic principle of radar, and realised the early engineering applications. However, at that time, radio technology was mainly used in the field of communications, and the development and application of the technology was strictly controlled by a few forces. In the following decades, the development of radar technology in Germany basically stagnated [6–8].

12.2.2 Maturity of Radar Technology

The United States was a major contributor to the development of military radio technology from the late 1800s and early 1900s to the end of World War I. In the 1920s, the radio industry was almost entirely devoted to consumer electronics. The pull of the market facilitated the rapid updating and iteration of a large number of basic technologies, which in turn provided a good basis for the rapid development of radar.

With the onset of the Great Depression in 1929, the U.S. Army and Navy also suffered financially, tending to seize scarce funds to sustain what was then still a fragile development force. With product innovation readily available to industrial industries, the focus within government military organisations naturally shifted to developing projects of particular benefit to themselves.

One such activity was called Radio-Echo Detection—the use of radio signals to detect the presence of targets. This activity was soon expanded to include distance or range detection of targets, sometimes referred to as Radio Direction-Finding. This is closely related to undersea acoustic technology, which began before the First World War and is still being developed. The Government realised that applications for this type of activity were of great military relevance and should be kept secret, so these projects became more tightly controlled. Eventually, the government adopted the name Radio Detection and Ranging and used the acronym "RADAR" as a cover, and the name radar was popularised and is still in use today.

In the United States radio to radar development process, there are two research laboratories play a central role: Naval Research Laboratory (Naval Research Laboratory, NRL) and the Army Communications Corps Laboratories (Signal Corps Laboratories, SCL). Both organisations were fully committed to the advancement of military technology and carried out a wide range of activities. Even today, the work of the "U.S. Naval Research Laboratory" remains a respected "calling card" in radar academia.

On 24 June 1930, the Naval Research Laboratory was determining the characteristics of aircraft receiving antennas. The situation was that the U.S. Naval Research Laboratory had a 9.14 m transmitter and an aircraft with a receiving antenna located at Bollinger Field, two miles away. The antenna was a 15-foot wire strung between the cockpit and the tail of the aircraft, which was rotated to obtain a receiving

12.2 Development History of Radar Technology

pattern. When an aircraft passes overhead, the strength of the signal transmitted to the receiver fluctuates—creating a kind of beat-frequency effect. The discovery of this phenomenon kicked off the development of U.S. Navy radar. It was also the beginning of the formal development sequence of the current radar engineering form.

From the 1930s to the outbreak of World War II, the U.S. Naval Research Laboratory and the Army Communications Corps Laboratory were developed similar to the modern radar engineering prototype detection equipment. In January 1939, an early radar system (XAF) was installed on the battleship USS New York and began sea trials, making it the first operational radio detection and ranging device in the U.S. fleet. Over a three-month period, the device was routinely able to detect ships within 10 miles and aircraft at a range of 48 miles. Two U.S. Navy officers, Lieutenant Commander S. M. Tucker and Captain F. R. Felt, came up with the term RADAR, a combination of the initials Radio Detection And Ranging, and in November 1940 the Secretary of Naval Operations directed the use of the term "unclassified" RADAR to refer to the then-secret project. The acronym soon became synonymous with "radar". As you can see, the term radar is inextricably linked to the sea.

At the same time, early warning radars developed by the Army Corps of Communications Laboratory and American Industries were beginning to be used in successive applications. On the morning of 7 December 1941, Privates Joseph Lockard and George Elliott were operating one of these radars when, at 7:02 a.m., a string of dots of light appeared on the screen 136 miles due north. They watched for 18 min, first believing that something was wrong with the radar and then submitting their observations to the aircraft warning system just established by the Army and Navy at Fort Shafter. The duty officer thought it was "nothing special", just a US bomber departing from the mainland. The alert did not get the attention it deserved. Rockard and Elliott followed the plane until 7:39, when it was only 20 miles away. 16 min later, the Japanese attacked Pearl Harbor and World War II was in full swing. History always likes to play such a joke on us, that string of light did not save Pearl Harbor, but did make radar development and engineering applications greatly accelerated. After the attack on Pearl Harbor, in order to protect the Panama Canal Zone from similar attacks, the United States launched an emergency programme to develop radar systems. In the military urgent needs of the traction, radar technology into a period of rapid development.

On the other side of the globe at the same time in history, after the end of World War I, Britain was always on the lookout for another conflict with Germany, and even hoped to achieve early warning by detecting the noise of German bombers, constructing noise amplification devices along the coastline, and searching for people with long hearing as early warning personnel. But it was clear that this type of early warning was not reassuring. Pulled by the urgent need, Britain also went down the route of radio detection technology. After continuous technical attempts and trials, the first demonstration of radio-based detection and ranging was achieved on 17 June 1935, with RDF as the acronym for this work. Throughout the Second World War. The British and German air raid game contributed greatly to the development of radar technology. It also introduced jamming and counter-jamming tactics, into the realm of information contests, and to this day, the jamming and interfering confrontation of radar detection remains an endless and eternal topic.

Before the Second World War, Germany also developed an excellent radar detection system, but Hitler considered it a defensive technology that could not win or lose the war, and used the resources in the development of offensive technology. This decision also influenced the direction of the Second World War to a certain extent. As for if Hitler paid attention to the development of radar technology, whether it will change the course of the war, perhaps this is history for us to leave a technical maze.

Throughout World War II, radar played an important role in air, land and sea combat. The US atomic bomb development project in Manhattan cost about $2 billion a year (by that year's standards), but the development and deployment of radar was a much larger undertaking, costing about $3 billion. It is often said that radar won the war and the atomic bomb brought peace. The massive financial investment also contributed to the great progress of radar technology. After the baptism of the Second World War, the engineering form of modern radar was also basically formed [6–8].

12.2.3 Period of Development of Radar Technology

After the end of the Second World War, radar, as an important information acquisition device, began to be comprehensively applied to various fields. With the progress of the entire industrial technology base, especially the power amplification devices and the rapid progress of digital technology, a new technical way was created for the engineering practice of radar, and some new concepts of radar detection were also pushed forward along with the richness of the means of engineering realisation.

Radar transmitting system from the initial passive antenna plus centralised transmitter form, gradually to passive phased array, active phased array, digital phased array evolution. For the radar system, it brings more flexible spatial beam control capability. Especially the active phased array technology is booming, the original power of the centralised launch, changed to distributed launch, multiple small power launch components together to generate energy gain, even if a certain number of launch components fail, still does not affect the radar to complete the detection task, greatly improving the system reliability indicators, so that the radar can be used for a long time to maintain the state.

With the low noise devices, power amplification devices, filters, analogue-to-digital conversion (Analog to Digital, AD) sampling device technology advances and engineering applications, so that the radar receiving system can be targeted to amplify only some specific signals, while suppressing the other useless signals, improve the radar can be detected by the smallest signal energy, improve the radar detection distance. In particular, the massive use of digital signal processing technology and the continuous enhancement of computing power, radar fundamentally changed the engineering form, building the signal processing field of various types of algorithms in the field of radar to achieve the application of the engineering bridge. Like a mobile phone from the key era into the intelligent era, many algorithms (Application, APP) began to shine on the radar.

The advancement of hardware technology only provides the foundation for the improvement of radar detection capability, and the radar performance is ultimately driven by a large number of new algorithmic concepts. For example, the introduction of pulse compression technology solves the contradiction between radar detection power and distance resolution, the introduction of Moving Target Detection (MTD) and Moving Target Indication (MTI) solves the problem of how to detect the target in the complex environmental background when the target and the radar have relative motion. The introduction of MTD and Moving Target Indication (MTI) solves the problem of how to detect a target in a complex environment when there is relative motion between the target and the radar; and a large number of tracking filter algorithms improve the automation of radar detecting and tracking the target. …… Currently, there are millions of radar industry workers in the world who are constantly working on some specific scenarios to optimise or propose new algorithms for solving the practical problems.

Whether it is high-speed speed radar, car collision avoidance radar, or guide the airport aircraft landing and takeoff air traffic control radar, detecting meteorological elements of the weather radar, radar has been all over the corners of our lives. And in the field of marine engineering, where radar first emerged, radar technology is also booming [6–9].

12.3 Current Status and Trend of Radar Technology Development in Domestic and Overseas Marine Field

After more than 100 years of development, radar has been widely used in all walks of life at present, the development of a multitude of professional branches, broadly speaking, the current state of technology and development trend of radar, will be very messy and huge. So here to introduce the radar in the ocean engineering technology status and development trend, describes the several angles relatively macroscopic introduction to the industry dynamics, only on behalf of the author's personal views.

12.3.1 Digitalisation of Radar Hardware Platforms

In order to better describe the digitalisation of radar, here the author uses more contact with communication equipment in our daily life to do an analogy description. The original communication device is a device that converts sound into current through a device that senses vibration, and through a transmission device, the current is transmitted to the receiving end, and then through a vibration generating device, the current is converted into vibration, which results in the emission of sound. This process has been the transmission of analogue signals, although the system is relatively simple, but the analogue signal transmission process is susceptible to external influences, the slightest interference with it will eventually be reflected in the back end of the transmission, and it is not easy to carry out signal processing and changes. Corresponding to the modern communication device, is still through the perception of vibration device, the sound into electrical signals, but not directly the analogue electrical signals transmitted away, but through a digital sampling device, the current sampling, subsequent processing and transmission, are digital quantities, the transmission of digital quantities is relatively not susceptible to external influences, and there is a complete set of signal processing theory to support the implementation of a variety of processing, flexibility Higher flexibility. At the receiving end, through the digital-to-analogue conversion device and then convert digital quantities into analogue quantities, and finally achieve to the sound changes.

With this process, it is easier to understand the digitalisation of radar. The initial radar, which is the processing and detection of analogue signals, has relatively high limitations and limited flexibility. In modern radar, due to the increasing level of computing hardware, digitisation plays an increasingly important role in radar engineering realisation. In the transmitting link, the traditional radar is a waveform generating device that generates the transmitted waveform, transmits it to the power amplifier, and finally radiates it from the antenna, and in this process, the transmitting link brings about system losses and the introduction of interference signals, and the whole waveform control is not flexible enough. In digital radar, the antenna consists of countless distributed transmitting units, and the radar control system sends the digitised code corresponding to the transmitting waveform to the distributed transmitting units, and the transmitting is finally completed by the digital-to-analogue converter in each unit. As each transmitting unit can be controlled separately, it greatly improves the control flexibility of the whole radar array, and a large array can be launched by different parts of different waveforms, or in the space to flexibly control the launch waveforms. Just like modern smart phones, customised requirements can be achieved through different APPs.

In the receiving link, modern radar is moving the digital sampling node forward, the earlier the digital sampling, the more you can increase the processing flexibility of the system. A simple example: if an analogue input signal, want to be divided into two separate processing, analogue power division device is its energy output separately, so that the energy of each way is only one half of the original, so that they can be processed separately using different processing methods to achieve different purposes. But in the digital domain, the same process can be realised with a single digital copy and paste, without any energy loss. This is just a simple example of the benefits brought about by digital processing links. Because of the introduction of digitisation, a lot of processing can be carried out that is difficult to do in the analogue domain. A large number of the algorithms underlying the functionality of modern radar signal processing are predicated on digitisation. It is also because of the digitisation of radar that it is possible to take full advantage of today's increasing arithmetic power to introduce ideas such as deep learning and big data to enhance the radar's own capabilities.

Note 1: This is not a blind emphasis on the digital front, the radar analogue information also has its inherent benefits, specifically in the engineering design of how to make comprehensive use of their respective advantages in order to achieve the optimal system, which is the value of the radar engineers.

Note 2: In China's radar digital process, China's radar technology, a famous scientist, educator, academician of the Chinese Academy of Sciences, the former president of Xi'an University of Electronic Science and Technology, represented by the old generation of radar workers, did a lot of groundbreaking work to promote China's radar engineering continues to progress. During the preparation of this book, Academician Bao passed away in October 2020 due to illness, and here, on behalf of individuals, I would like to express my deep remembrance of Mr Bao.

Internationally, the degree of digitisation of advanced radar products is approaching its limit. In the whole chain, can achieve digital links, have been the basis of engineering implementation, in the low-frequency band radar products, the degree of its digitalisation is mainly limited to the current balance between the cost of engineering implementation and the market price of the product itself. High-frequency radar is also subject to the current technical level of digital-to-analogue conversion devices, I believe that in the near future, these bottlenecks will gradually break through.

The direct impact of the digitalisation of the hardware platform is that its ultimate capability is to be realised through software algorithms, which is also an important reflection of the flexibility brought about by the digitalisation of the radar.

12.3.2 Softwareisation of Radar Capability Implementation

Early radar processing was realised by hardware circuits, and once a radar was made, its functionality was essentially finalised. Similar to our early acquisition of non-smart phones, after the purchase of the mobile phone, its ability to achieve the function is the mobile phone has been set up in the function of the phone, call, receive text messages, mobile phone alarm. However, in the smartphone era, the mobile phone is only a hardware platform, and the functions it can achieve can be customised through software. Different functions can be realised by downloading different APPs. After the radar is digitised, the capabilities that can be realised are also gradually software-enabled. Without changing the radar hardware, the radar can be continuously redefined and its performance enhanced through software upgrades and updates.

This abstract description makes it difficult to understand the connotation of radar softwareisation. To give a concrete example: the same a radar hardware platform, in marine engineering, may need to complete the detection of airborne targets, detection of surface targets, atmospheric waveguide (related theory in the subsequent chapters) of the inversion of the different functions for the radar hardware platform, are the launch of electromagnetic waves, receiving electromagnetic waves, and then sampling, transformed into a digital quantity, and signal processing, data processing, and finally produce results. Whether the echo signal is finally used for detecting air targets or for atmospheric waveguide inversion is realised by different software. This is the software implementation of the radar capability.

Radar functions are ultimately realised by software, then it can not only be limited to target detection, if the software processing idea is rich enough, its function boundary will also be gradually diversified.

The software of radar capability can also be understood from another level, since the radar function is ultimately realised by software, under the premise of comparable hardware level, the software algorithm and its processing concepts will largely determine the radar performance. Modern radar performance is high or low, most of the decision on the development of the manufacturer's algorithm accumulation and software implementation capabilities, for different application scenarios of the processing details, is often manifested in the final product performance show. However, the domestic radar industry for radar algorithms to improve the importance of the software processing algorithms of the in-depth study is still to be strengthened. We use a lot of processing concepts, still comes from foreign radar industry scholars, for example: the most commonly used navigation radar in marine engineering, the hardware level of domestic products and the world's advanced level is basically equivalent to the software algorithms for foreign dependence is still high, in the sea clutter suppression, target detection in complex scenarios, and other key technologies that require long-term research and accumulation, the core R & D capabilities of the domestic still have a lot of room for progress. There is still much room for progress in domestic core R&D capability.

12.3.3 Blurring of Radar Functional Boundaries

Early radar, inherited a large number of communication technology base, with the independent development of its industry and the deepening of the demand traction, the pursuit of the ultimate performance of the engineering spirit, opened up the radar industry unique ecological environment and state of the art, the formation of a unique form of engineering realisation. But essentially, radar is still the use of the physical characteristics of electromagnetic waves, target detection and measurement of a tool. In principle, as long as the use of electromagnetic waves to achieve the relevant functions of the product are radar "close relatives".

When a direction towards the extreme, will inevitably then towards integration. Radar from the basis of communication technology, and constantly open up their own characteristics of the road, and now has been gradually towards functional integration. Especially in the field of marine engineering, a relatively small engineering carrier, to integrate a lot of functional electronic products, its space is limited, energy is limited under the premise of multifunctionality has gradually become the development trend.

Marine engineering in the most typical shipboard radar, has appeared its functional boundaries of the blurring of the radar is not only radar, but into a radio frequency carrier, because the hardware digital and functional software support. This RF carrier

12.3 Current Status and Trend of Radar Technology Development …

can be a radar, or a communication device (RF carrier transmits an agreed special waveform in a specific direction, another RF carrier receives it and parses the relevant information, which enables communication), or a listening device (RF carrier does not transmit, but only receives indiscriminately, and analyses the received signals after digital sampling, and ultimately senses the electromagnetic signals in the environment), and the same can be done in the field of electromagnetism. characteristics), and likewise can emit noise or spoof signals in the electromagnetic domain as a jamming device. Under the same physical platform, multiple functions can be realised through software, as long as they are not working simultaneously in a microscopic sense, through the continuous rotation of time slices, the integration and integration of multiple functions can be realised in a macroscopic sense, which greatly broadens the boundary of radar use.

At this stage, the multi-functionality of radar is still mainly applied in the military field, due to the high degree of integration of multi-functions of military ships and space limitations, as well as the electromagnetic compatibility control problems arising from the simultaneous work of multiple functions. In the same RF platform, through the software to achieve the demand for different functions is very urgent, the United States and other marine military powers, has long been carried out the relevant technical research, and began to gradually move towards engineering applications.

The future development of unmanned marine equipment, for this highly integrated RF complex products will have more urgent needs.

12.3.4 *Diversification of Radar Information Acquisition Domains*

Early radar, due to the limited refinement of engineering realisation, mainly acquired the distance information of the target through the time domain. In more general terms, it can only achieve the measurement and distance perception of the target through the time interval between the transmitting pulse and the receiving pulse. Subsequently, with the improvement of engineering realisation capability. Radar is expanding its own information acquisition perception domain.

Radar is through the electromagnetic wave to detect the target, that electromagnetic wave inherent characteristics, can be used as a radar to obtain information perception domain. Electromagnetic waves have operating frequency, so the radar can obtain frequency domain information, through the frequency domain of the filter to filter out the antenna to receive the clutter, improve the target detection ability, a certain frequency domain information acquisition, but also can get the target of the fine contour information, combined with the characteristics of the movement with the realization of the target imaging; phased array antenna technology introduction, but also improve the radar airspace information perception ability, through the digital beam formation technology to improve the target spatial resolution, the radar is expanding its own information perception domain. technology, improve the target spatial resolution, in a specific direction to form the antenna's detection gain control, inhibit the interference of the specified direction in space; through the introduction of electromagnetic wave polarisation information, sensing the target of different polarisation scattering characteristics, to obtain the target information in the polarisation domain, for example: in the field of ocean engineering, polarisation information can be used as a kind of sea clutter suppression technology path, by distinguishing between the target and sea clutter of different polarisation scattering characteristics to achieve clutter suppression. By distinguishing the different polarisation characteristics of the target and the sea clutter, we can achieve the purpose of clutter suppression and better target detection under the background of clutter.

Here to do a little expansion of polarisation information, relative to the time domain, frequency domain, air domain information that has been maturely applied in radar engineering, polarisation domain information is not now applied in a large number of radars. However, as an inherent property of electromagnetic waves, with the continuous deepening of radar engineering research, polarisation dimension information mining will definitely become an important development direction in the future.

The concept of polarisation of electromagnetic waves and the concept of polarisation in optics is similar, its English words are described with "polarization", secondary school physics on the concept of polarisation of light has done a popular description of the concept of light as a vector wave, its motion characteristics in addition to the propagation direction of the movement, in the plane perpendicular to the direction of propagation, but also will have motion component. Do a thought experiment, now we and the light in the direction of propagation with the same speed in the forward direction, so that in the direction of propagation dimension, the light for us is stationary, then we see the light is still in the vertical ground up and down swing, is vertically polarised light, if he is in the horizontal ground swinging left and right, is the horizontal polarisation of light. The same concept is the same in the electromagnetic waves of radar. Only the description becomes horizontally polarised electromagnetic waves, vertically polarised electromagnetic waves, or circularly polarised electromagnetic waves.

Note: the concept of polarisation of light is not far from our lives, we watch 3D movies in the cinema, the principle of polarisation of light used to achieve the left eye and the right eye to receive the picture is not the same, so as to form a 3D effect, is a commonly used technology pathway. These perceptions are not the focus of this book, and interested readers can refer to related information for a detailed understanding.

This inherent property of electromagnetic waves, when propagated or reflected, will show different qualities depending on the target, for example: horizontally polarised and vertically polarised electromagnetic waves, both irradiated at the same angle to the same target, the reflected radar echo may be different in the polarisation domain, and if this difference is measured, the target characteristics can be perceived more finely. This is the most fundamental starting point for introducing the polarisation domain into radar detection. In the engineering use, there can be a lot of changes according to the actual situation, such as the aforementioned, sea clutter as an irregular motion medium reflected electromagnetic waves, which in the polarisation domain and ships such as a more stable metal body has a difference, the use of good use of this property, you can improve the probability of detection of ships in the sea clutter background.

The more dimensions of information obtained, the greater the choice of information available for radar processing, the more accurate the radar can sense the target. With the continuous reduction of the cost of engineering realisation and the continuous enhancement of information processing capability. Radar will be more widely used in marine engineering in the future. If there is a prediction of a radar development trend that will surely come true, it is that radar will surely continue to evolve [10–12].

References

1. Ding, Lufei, Fulu Geng, and Jianchun Chen. 2013. *Radar Principles*, 4th ed. Beijing: Electronic Industry Press.
2. Guangyi, Zhang. 1994, 1996, 2000, 2001. *Phased Array Radar System*. Beijing: National Defence Industry Press.
3. Lufei, Ding, and Geng Fulu. 2002. *Radar Principles*, 3rd ed. Xi'an: Xi'an Electronic Industry Press.
4. Jaguo, Lu. 2017. *Synthetic Aperture Radar Design Technology*. Beijing: National Defence Industry Press.
5. Guangyi, Zhang, and Zhao Yujie. 2010. *Phased Array Radar Technology*. Beijing: Electronic Industry Press.
6. Jian, Yang, and Yin Junjun. 2020. *Polarisation Radar Theory and Remote Sensing Applications*. Beijing: Science Press.
7. Wang, S.-Y. 2014. *Marine Navigation Radar*. Dalian: Dalian Maritime University Press.
8. Lufei, Ding, Geng Fulu, and Chen Jianchun. 2014. *Radar Principles*, 5th ed. Beijing: Electronic Industry Press.
9. Yiyao, Dai, Wang Xuesong, Xie Hong, and Xiao Shunping. 2015. *Airspace Polarisation Characteristics of Radar Antenna and Its Application*. Beijing: National Defence Industry Press.
10. Stimson, George W. 2005. *Introduction to Airborne Radar*. Translated by H.P. Wu et al. Beijing: Electronic Industry Press.
11. Mahafza, Bassem R. 2016. *Radar System Analysis and Design (MATLAB Edition)*, 3rd ed. Translated by Wanxing Zhou, Mingchun Hu, Mingya Wu, Jun Sun, et al. Beijing: Electronic Industry Press.
12. Barton, David K. 2012. *Radar System Analysis and Modelling*. Translated by Nanjing Institute of Electronic Technology. Beijing: Electronic Industry Press.

Chapter 13
Radar Technology Principles

Chapter 12 provides a popular science overview of the basic concepts of radar. Today, radar forms have changed dramatically, but no matter what physical form radar has changed to, this chapter provides a simple introduction to the technical principles of radar from the perspective of its detection principles, and combines a typical radar composition with a more relevant and graphic description of the principles.

Some of the basic elements of radar for detection include: energy generation, radiation, reception, and processing of electromagnetic energy, identifying which of the received energy is reflected back from the target, and recording the time interval to complete target detection. The following is an abstract engineering description of the technical principles of radar operation from the perspective of electromagnetic waves, so that it is easier for the reader to understand the relevant content of the later chapters. In order to facilitate understanding, this engineering case does not show all the details of the system design, and is not an accurate modelling of the uniform form of all radars, but carries out the basic elements of the general radar engineering implementation, which can be interpreted as a model room.

First of all, the signal generation, general radar system through the frequency synthesiser (later referred to as frequency synthesis system) to generate radar signals, the figure below is a typical phase reference radar frequency synthesis system. The first step is to generate the transmit signal waveform, such as linear FM signal or phase coded signal, etc., which is adjusted to the radar operating frequency through frequency conversion and output to the antenna feeder system for amplification and transmission [1–3] (Fig. 13.1).

After the excitation signal enters the antenna from the frequency synthesis system, it undergoes power amplification and is radiated through the antenna array to space. In order to facilitate the understanding of the principle of sum and difference goniometry, which is commonly used in radar, we abstract the antenna array into four transmitting units (this is a conceptual abstraction, which is not realised as such in engineering). After spatial transmission and target reflection, the radar antenna in turn receives the electromagnetic waves and forms the various types of signals to

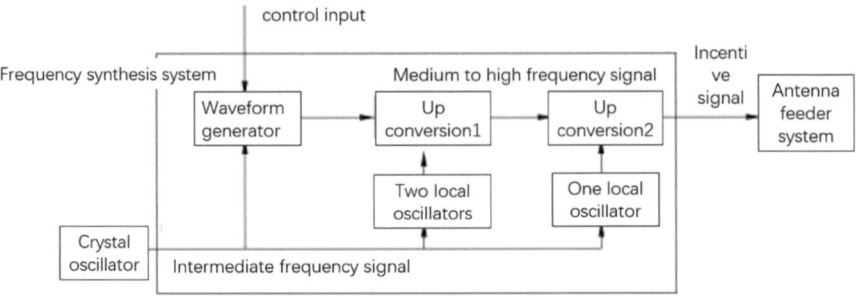

Fig. 13.1 Example of frequency synthesis system engineering practice

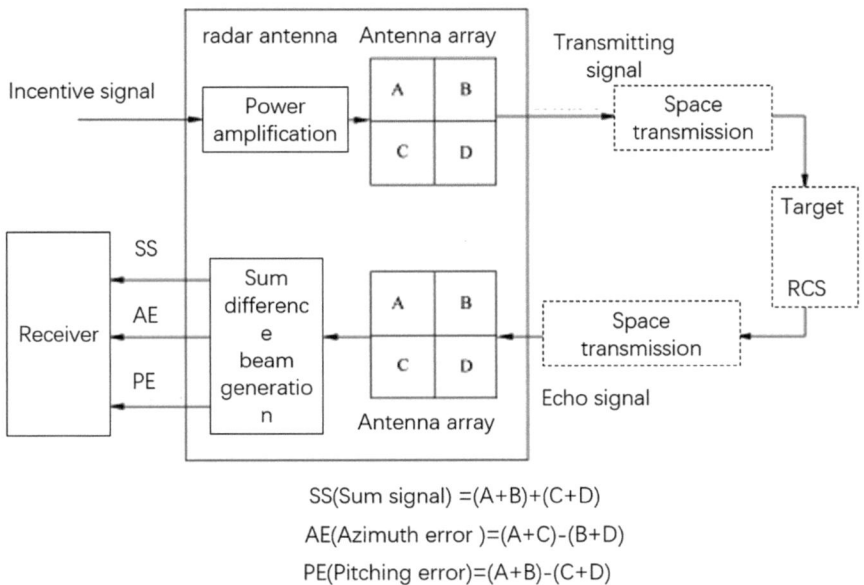

SS(Sum signal) =(A+B)+(C+D)
AE(Azimuth error)=(A+C)-(B+D)
PE(Pitching error)=(A+B)-(C+D)

Fig. 13.2 Example of engineering practice of radar antenna

the receiver through the energy control of the antenna. About and differential beam goniometry related content, and not in this book to expand, interested readers can consult the above mentioned professional books. However, the basic model depicted in Fig. 13.2 is what enables a basic understanding of sum-difference goniometry [2].

The target echo signal received by the antenna is transmitted to the receiver, which is filtered to amplify the useful signal power and filter out the useless noise as much as possible, and after frequency conversion to complete the digital sampling, the subsequent signal processing system can only be achieved through the digital signal processing method to achieve the detection of the target (Fig. 13.3).

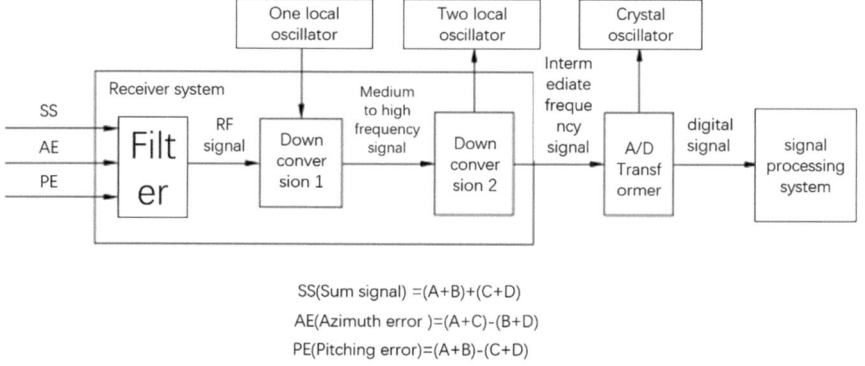

Fig. 13.3 Example of radar receiver engineering practice

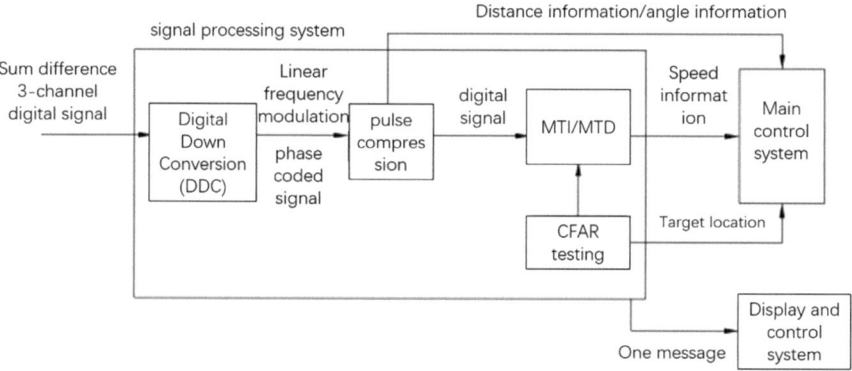

Fig. 13.4 Example of signal processing engineering practice

The signal processing system finally completes the detection of the target, and extracts the target distance, speed, angle and other information. The author believes that the radar signal processing system is the core of the radar, is the key sub-system of its final formation capability. It is also the key to the radar engineering practice and really open the technical level of the engineering implementation unit. Signal processing system involved in the pulse compression, MTI and other concepts, it is difficult to describe in a simple way, there will be a detailed expansion later [1] (Fig. 13.4).

By this point, the target has been detected by the radar and the relevant detection information has been extracted. With this relatively simple and highly abstract model, it helps the reader to understand the subsequent related contents more comprehensively.

13.1 Basic Principles of Radar Transmitter

The radar transmitter does not appear as a sub-system in the model above, and the reader can understand it as the antenna power amplification module in the model above. The main task of the radar transmitter is to generate high power electromagnetic radiation signals. The nature of radar is to detect with electromagnetic waves, where do they come from? Transmitter to solve this problem. Speaking clearly about the transmitter, we must first start from the electromagnetic start to say.

Maxwell first constructed the electromagnetic wave model in mathematics, electromagnetic waves have not been discovered by mankind (this is the greatness of Maxwell). Thirteen years later, Hertz was first in the lab, confirming the existence of electromagnetic waves and revealing the most basic principles of radar detection. A varying electric field can produce a magnetic field, and a varying magnetic field can produce an electric field, which in turn produces electromagnetic waves. That is to say, as long as a changing electric or magnetic field is generated and energy is radiated outwards, electromagnetic waves are produced, and the basic starting point of a radar transmitter is the regular generation of a large number of changing electric or magnetic fields.

So how do you generate changing electric or magnetic fields? In classical radar, the basic principle used is also very simple: the accelerated motion of charged particles. For example: an electron's direction of motion or speed of movement changes, it will bring changes in the electric field, and changes in the electric field will be slightly away from the formation of changes in the magnetic field, changes in the magnetic field is slightly away from the formation of changes in the electric field, so the electromagnetic wave generated.

In everyday life, electrons at room temperature are always in thermal motion. Due to the thermal movement of electrons is the direction and speed of the irregular changes in motion, electromagnetic wave radiation occurs all the time, but most of the electromagnetic wave radiation in the form of thermal radiation (radiation wavelengths longer), there are a small number of radio waves in the form of form or the form of light waves to show. That is to say, thermal radiation, optical radiation, radio radiation, its essence is electromagnetic radiation, the difference lies only in the wavelength of the radiation is different.

The mission of the transmitter is to constrain the phenomenon of electromagnetic radiation by some means, so that it radiates the same wavelength of electromagnetic waves, and the intensity of radiation is much greater than the energy of natural radiation. One common engineering solution idea is to excite a tuned circuit with a strong current. Although the principle is not complex, a long process of engineering optimisation has taken place from the discovery of electromagnetic waves to their engineering application in radar.

In the broad classification of engineering realisation, radar transmitters can be divided into two main categories: continuous wave transmitters and pulse modulated transmitters. Continuous wave transmitter is only used in continuous wave radar, its application surface is relatively narrow compared to pulse radar, but also has

13.1 Basic Principles of Radar Transmitter

its limitations of use, the biggest engineering advantage is that there is no ranging blind spot, a more successful application is our car collision avoidance radar, a lot of continuous wave radar. More radars are using pulsed transmitters. That is, they need to emit a pulsed waveform.

Pulsed transmitters are subdivided into single-stage oscillating transmitters and main oscillator amplifying transmitters. Single-stage oscillation easy to understand description is to use a "knife" to cut a transmit waveform, to achieve the pulse emission, each time to cut to where not to do control, so the initial phase of each pulse emission is not necessarily. Its engineering implementation is relatively simple, but the launch of multiple pulses do not refer to each other, the frequency stability is relatively bad, in some simple, low performance requirements of the application scenarios are still in use. The main amplifier transmitter uses a frequency source to control the whole transmitter link, which can achieve the same initial phase of the waveform transmitted in each cycle, so as to facilitate the radar to accumulate the phase parameter (the basic concept of radar signal processing, which will not be expanded in detail here). Its engineering implementation is relatively complex, but the performance is superior. Especially in some precision tracking radar, this transmitter is generally used.

From the perspective of transmitter materials, early radar, mainly using microwave transistors and microwave ware four-stage tube oscillator transmitter, this technology was first applied in the colour television, its operating frequency is limited to VHF to UHF. subsequently, along with the radar in the military, civil field of demand continues to increase, more and more industrial system for radar work needs to carry out the study of the transmitter system, the speed control tube, Magnetron, travelling wave tube and other transmitter devices technology continues to mature. Especially the magnetron, because of its simple composition, low cost, although the performance and frequency stability is poor, but still today is used in a large number of marine navigation radar.

After the 1960s, solid-state transmitter technology began to mature. Accompanied by the increasing popularity of phased array radar, distributed solid-state transmitter solutions, bringing a qualitative leap in system reliability. In recent years, gallium arsenide (GaAs) field effect transistor technology has been very mature, a large number of applications in phased array radar. Gallium nitride (GaN) devices have also made technological breakthroughs, the radar launch system can provide higher operational reliability, greater radiant energy, better energy conversion efficiency, greater operational duty cycle .... The radar industry, which has solved the basic problem of energy radiation, has also ushered in a faster development opportunity [1].

13.2 Fundamentals of Radar Antennas

In the simplest radar model, a radar antenna is a paraboloid for constraining the electromagnetic wave emission direction to the desired detection direction, which is a device for energy concentration. With the development of radar technology, the form and definition of radar antenna have been gradually blurred, especially in phased-array radar, where the radar antenna has been vaguely defined as a comprehensive engineering entity for waveform generation, power amplification, beam control, echo reception, signal sampling, etc., which has become the core radar hardware and the main cost share. This section provides an overview of the working principles and physical nature of three radar antennas through their physical forms; the detailed technical details cover a very wide range of technologies and are not the focus of this book's presentation. The three antennas are the parabolic reflector antenna, the slit waveguide antenna, and the phased array antenna.

The parabolic antenna is one of the easiest antenna forms to understand how a radar works. Radar wants to transmit electromagnetic waves through the transmitter, after the conductor to the antenna feed, parabolic antenna is the role of the parabolic surface of the reflection, so that the electromagnetic wave energy as far as possible to a direction of propagation, this process in the physical principle, that is, the feed emitted by the spherical wave corrected for the plane wave. This physical process can be more graphically associated with the previously mentioned torch. The working diagram is shown in Fig. 13.5.

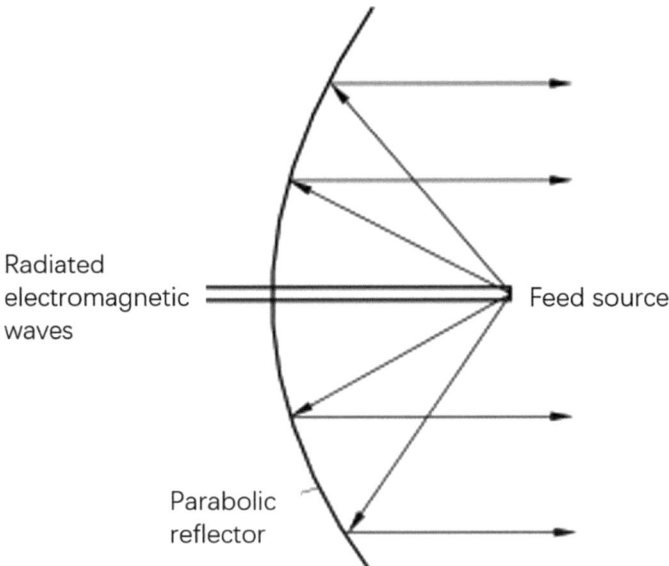

Fig. 13.5 Schematic diagram of parabolic antenna

Slit waveguide antenna is a form of antenna often used in marine navigation radar, and its basic design idea is to open holes on the waveguide, so that the electromagnetic wave transmitted along the waveguide leaks outward from the holes, and according to the spacing and tilt direction of the open holes, it can control the direction of the electromagnetic wave radiated externally to form a beam control function similar to that of a parabolic antenna. Its cost is relatively low, the production is simple, but the production process has certain requirements.

Phased array antenna basic physics principle involves more knowledge, the core idea is that the radar radiates energy through different radiation small unit, if the electromagnetic wave arriving at the target phase is consistent, then it can realise the maximum superposition of energy, it can be simply analogous to more than one person to a direction of force, in order to form the maximum combined force. Engineering implementation, the antenna through the computer control of each small radiation unit of the initial launch phase, so that the electromagnetic wave radiation to reach the target, just when the electromagnetic wave radiation of each radiation unit phase consistency, can be the maximum superposition. Thereby this direction is identified as the beam pointing of the radar, while other directions cannot be maximally superimposed due to phase inconsistency. By changing the initial phase control amount, it is possible to achieve phased array antenna pointing in different directions. From the principle can be found, phased array antenna does not need to mechanically rotate the physical structure of the antenna, you can achieve rapid control of the radar beam, greatly improving the convenience of engineering use. But at the same time, its complexity is also increased accordingly, the engineering implementation and cost to pay more, in some of the cost is not sensitive but the use of the higher requirements of the application field, large-scale application.

The above content describes the process of electromagnetic wave emission, in the process of electromagnetic wave reception antenna assumes a similar role, receiving electromagnetic waves in a specific direction, and sent to the receiver for radar detection target.

Different radar antennas have been endowed with more connotations in the receiving link, such as assisting angle measurement with sum-difference difference, assisting interference detection with auxiliary antennas, and so on. However, the most essential task of antennas in radar reception is still to receive electromagnetic wave energy in a specific direction [3–6].

13.3 Fundamentals of Radar Receivers

The target echo signals received by the radar antenna are often very weak, and the basic function of the receiver is to amplify these weak echo signals while suppressing other signals that we don't want. Ultimately, it helps the signal processing system to achieve the detection of the target.

Radar receivers generally use the super-aberrant system. The basic idea is to use a locally generated signal of a certain frequency to mix with the input signal, inverting the input signal to a predetermined frequency to facilitate amplification, filtering, gain control, sampling and other subsequent processing. The principle of super-exception was first proposed in 1918 as a response to the problem of weak signal reception for remote communications. Initially, the remote signal is directly mixed to the audio, and then the relevant processing, the use of its effect is not good, and the super-exception is generally frequency conversion to a higher frequency than the audio, after filtering, amplification in the detector, its performance is better than the high-frequency direct amplification and the difference in the receiver, so in the early communication system in a large number of use, the initial radar receiver uses a lot of communication technology, so also follows the super-exception The working mechanism of the receiver. And in engineering applications, achieved better results.

With the development of technology to date, the radar receiver has not been a strict concept of ultra-aberratic receiver, many complex systems also use the technology of the second frequency conversion to ensure that the receiver operating bandwidth, but the name of the ultra-aberratic receiver, has been used until now.

Different radar receiver composition is far from each other, modern radar, receiver engineering implementation level, basically determines a radar performance of the lower limit (radar performance of the upper limit is often determined by the signal processing and data processing software). But a variety of receiver implementation programme, the most core links often have three: filtering, mixing, gain control.

Filtering is well understood, the radar only cares about the information brought back by their own transmitted electromagnetic waves, but the antenna will receive a variety of electromagnetic waves without discrimination, and the antenna itself will produce thermal noise (in fact, the antenna's receiving beam control, is a spatial filtering technology). These signals, other than those of interest to the radar, need to be rejected in the receiving link, a process called filtering. The concept of filtering can be visualised by a game in current variety shows, often called "Here comes the wall", where a moving wall is dug out with a specific shaped hole, and the guest needs to perform the same action as the hole in order to pass through the wall, or else be pushed by the wall into the swimming pool behind him. Filtering is generally in the frequency domain "opened a hole", only a specific signal can pass, the other signal energy is suppressed, so as to ensure that the maximum number of only want to retain the signal detected.

Frequency mixing is the frequency of the received signal, transformed to a specific frequency, in order to facilitate radar processing. Frequency mixing is to radar what language is to humans. Language transforms human thinking into a dimension that can be communicated, understood, and processed for a more graphic presentation of human thinking. Frequency conversion transforms the electromagnetic waves received by the radar to a frequency that is more conducive to processing, handling, detection and display, in order to increase the readability of the information detected by the radar. A typical example is easier to understand this process, if the radar operates in the X-band (e.g. 9 GHz), the radar transforms its signal carrier frequency to an intermediate frequency (e.g. 60 MHz) by mixing in the receiver, so that the radar can be designed with a relatively fixed and complex intermediate frequency signal processing circuit to process the intermediate frequency signal. Radar in operation

13.3 Fundamentals of Radar Receivers

will generally change the operating frequency within a certain range, the mixing process can ensure that no matter what the actual operating frequency of the radar, the IF is fixed to transform to 60 MHz, so that the engineering design improves the realisability of the system. Modern radars also usually use digital signal processing, and the conversion to a lower frequency also facilitates digital sampling of the signal by the radar.

Gain control is the amplification or attenuation of signal energy for better detection by the radar. It can be simply understood that the radar's detection system can only better detect signals within a certain energy range. Gain control is to attenuate the large signal and amplify the small signal to ensure that the energy level is controlled within the detection range. After the introduction of digital sampling in modern radars, digital sampling AD chips can often only sample signals within a specific energy range. Gain control is more important in the sense that the signal energy of the radar receiver link is ultimately controlled to within the operating range of the digital sampling chip.

In the radar receiver engineering practice, often cross use different targeted above links, and ultimately the pursuit of systematic optimal reception link. In practice, many more precise issues will be considered, but the basic idea is no more than "exclude differences, the original source".

Radar applications in marine engineering often involve some special processes in the receiver. Here is a brief introduction to the sensitivity time control (Sensitivity Time Control, STC) and automatic gain control (Automatic Gain Control, AGC). STC core idea is to control the amplification gain of the receiver link with the distance. The radar starts to receive after transmitting the pulse, but the echoes reflected back from objects closer to the radar are generally stronger in energy, and if a link gain that does not change over time is used, the reflected echoes at these close distances may cause the energy range to exceed the energy that can be processed by the back-end (e.g., exceeding the operating range of the back-end AD sampling, the linear operating range of the amplifier, etc.), thus resulting in difficulties in detection. STC combines the gain control of the link with the distance by some engineering means, which simply means that the closer the distance, the smaller the amplification gain, and then gradually becomes a fixed gain as the distance increases. This can control the entire receiving link device are working in the better interval, but sacrificed the detection sensitivity of small targets in close proximity. In marine engineering, the radar will face the problem of sea clutter (subsequent chapters will expand the introduction), STC control method can effectively control the close-range sea clutter caused by the receiver link saturation, so the general surface search radar will be used to this technology.

AGC control technology is a closed-loop gain control technology. Its core idea is to use the results of the receiver link amplification to control the size of the gain, so that the output of the link is always maintained in a relatively stable interval, this interval in modern radar is often the AD using the best working state of the interval. AGC control technology seems to be a very simple principle, but in the engineering implementation of a lot of technical points that need to be accumulated for a long time, due to its role in the analogue signal domain before digital sampling, the AGC control will cause the whole radar to be damaged. Improper AGC control will lead to the whole radar signal processing abnormality. In marine engineering, due to the complexity of its working scene, the existence of sea clutter, atmospheric waveguide and other phenomena, to the design of the AGC brings more uncertainty, but also more serious will affect the system performance [7, 8].

13.4 Fundamentals of Radar Signal Processing

Radar signal processing is not like transmitters, receivers and antennas which have a clear physical model and a relatively clear engineering architecture. Usually the concept of signal processing is very broad, many professional books that introduce the working principle of radar, we can obviously feel that the transmitter, receiver, antenna, display have special chapters, but there is no chapter called radar signal processor. But the signal processing in the radar final performance embodiment, I think it is the most important one.

Radar signal processing concept is broad, involving the connotation and discipline is also very rich, if you use a sentence to describe what is the radar signal processing, I think it is: from the echo to find the target.

Radar is composed of a series of hardware, which ultimately serves radar signal processing. Finding the target from the echo is the most fundamental soul of radar. Especially after entering the digital signal processing, due to the introduction of digital sampling technology, greatly enriching the connotation of radar signal processing. Through digital signal processing technology, pulse compression, constant false alarm detection and other processing algorithms are implemented, and the final confirmation of the target is generally achieved by means of threshold detection. This will also include anti-jamming processing, clutter suppression, channel correction and so on.

Here in general indirectly some of the conventional radar signal processing algorithms, for a radar, signal processing is the most core and critical content, belonging to the core of each radar manufacturer's trade secrets, so this is not a complete introduction to the signal processing process, respectively, to introduce some of the radar signal processing commonly used core algorithms.

13.4.1 Pulse Compression

Pulse compression technology is a kind of signal processing technology commonly used in modern radar, and the core problem it solves is to resolve the contradiction between radar detection distance and radar distance resolution.

According to the radar equation described in the previous section, it can be equivalently understood as follows: the larger the average power emitted by the radar, the further its detection distance is, especially for some multi-functional radars and

13.4 Fundamentals of Radar Signal Processing

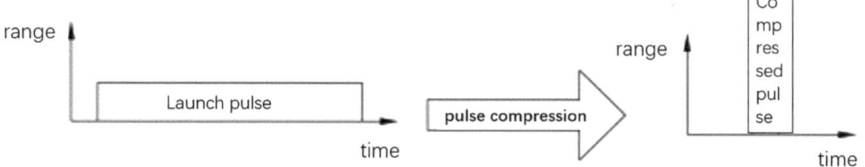

Fig. 13.6 Schematic diagram of pulse compression technology

surveillance radars, which often require the radar to emit a large pulse width. At the same time, according to the basic principle of radar ranging, the distance resolution is inversely proportional to the radar pulse width, that is, the narrower the radar pulse width, the higher the distance resolution. This creates a contradiction between the need to obtain a longer detection distance and the need to obtain a higher distance resolution. To solve this contradiction requires a high peak power narrow pulse, which is difficult to achieve in engineering, the technical bottleneck is difficult to break through. Pulse compression technology is to solve this contradiction through another idea.

The basic idea of pulse compression is that the radar transmits a wider pulse to ensure that the radar can obtain enough average power under a certain peak power. In the signal processing stage, it is then processed into narrower pulses to ensure that the radar has a sufficiently high range resolution. (Fig. 13.6).

Let's not do the tedious formula derivation here first, which is not conducive to the image understanding of pulse compression. Firstly, we will introduce the origin of pulse compression, which is the most important signal processing technology in modern radar.

Pulse compression was first invented on Linear Frequency Modulation (LFM) signals. The so-called linear FM is the frequency of the signal within the pulse, at a fixed rate of increasing or decreasing. This type of tuning is similar to the sound of a bird, so it was called by the inventor: Chirp. Since pulse compression technology was first invented on Linear FM signals, the industry also uses Chirp as a synonym for pulse compression. Scholars familiar with the radar industry who have read simulation programs will often see this nomenclature. The physical model of the linear FM signal pulse compression technique is actually very simple, as the frequency of the radar transmit waveform is linearly varying, then the frequency of the reflected echo is also linearly varying. If we find a filter, different frequencies of the signal through this filter is not the same time, for example: now need to do pulse compression of a pulse within the frequency of the signal increases linearly, as long as we find a filter, the lower the frequency of the signal, the longer the time needed to pass through the filter, and just after the filter delay, all frequencies of the signal in the same time from the filter output. In the time domain, the pulse is compressed into a relatively narrow time, and pulse compression is achieved. If this description is still too abstract, let's simplify the model a little more.

Fig. 13.7 Schematic diagram of the transmit signal

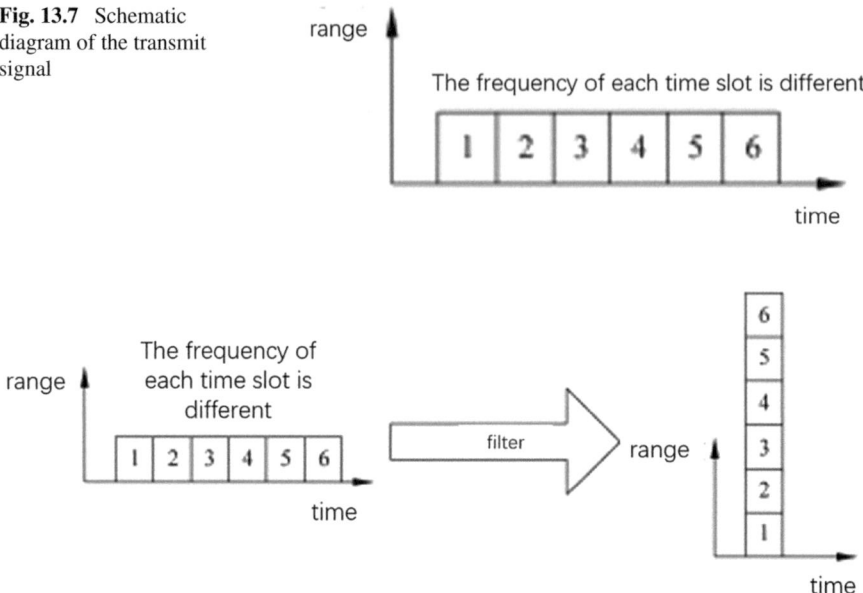

Fig. 13.8 Simplified understanding model of pulse compression

Imagine that the radar transmits a pulse a time-domain signal consisting of different frequency segments. This is shown in Fig. 13.7.

Different time slices pass through the filter with different time, time slice 1 enters first, but its time through the filter is the longest, time slice 2 is the second, time slice 6 enters last, but it takes the shortest time, and the six time slices go through different delays, and just happen to be output to the filter at the same time as shown in Fig. 13.8.

Such filters can be implemented by analogue devices, which is where radar pulse compression technology was originally found. A delay line, for example, enables pulse compression of linear FM signals.

In terms of mathematical modelling, pulse compression has a complete set of theoretical foundations. Modern radar signal processing has entered the digital era. Considering from the mathematical model, pulse compression is a kind of matched filter. Take the radar transmitting signal s(t) as an example, the impulse response of the matched filter is h(t) = s*(T − t). For the radar echo x(t), the pulse compression results in:

$$y(t) = \int_{-\infty}^{+\infty} x(t - \tau) s^*(T - \tau) d\tau \tag{13.1}$$

Digital signal processing after sampling into digital quantities, its pulse compression can be expressed as follows: if the matched filter impulse response is h(k), and the radar echo sampling data is x(k), k = 0, 1, ..., N − 1, then its output of pulse compression is

$$y(n) = \sum_{K=0}^{N-1} x(n-k)h(k) \qquad (13.2)$$

where h(k) is the sampled output of h(t).

Frequency domain pulse compression methods are often used in engineering implementations. The forward and inverse fast Fourier transform algorithm is used to implement the discrete Fourier transform and inverse transform operations, which is fast and has high processing efficiency. The spectrum of the radar transmit signal s(t) is S(ω), then the transfer function of the matched filter is H(ω) = S*(ω), then the output of pulse compression is

$$y(n) = \text{IFFT}\{\text{FFT}[x(k)] \cdot H(\omega)\} \qquad (13.3)$$

Another concept that will be frequently approached in pulse compression technology is distance subflap, also called distance sidelobe. The essence of this is that the spectrum of the pulse signal is not ideally clean. From a frequency domain perspective, the pulse signal introduces other spectral components in the full frequency band, which after passing through the filter will form small output peaks, the sub-flaps of pulse compression. A larger amplitude echo signal may have a higher sub-flap than the pulse compression main flap of some smaller signals, which has an impact on target detection. In radar signal processing, it is also common to use a variety of windowing to suppress the flap. The so-called windowing can actually be understood as a kind of mismatch, by sacrificing part of the width of the main flap at the expense of the suppression of the amplitude of the sub-flap. We often use the weighting methods include: Dolph-Chebyshev, Hemming weighting, cosine squared, etc. Here conclusively given, modern radar signal processing, often use Hemming weighting method. It can effectively suppress the secondary flap, and at the same time, the main flap also spreads relatively undramatic [9–12].

13.4.2 Moving Target Indication

Moving Target Indication (MTI), the core contradiction it solves is the problem of how to detect moving targets in fixed or slow clutter. The basic idea is to make use of the different energy distribution characteristics between the moving target echo and the fixed or slow moving clutter in the Doppler frequency dimension to suppress the clutter, so that the target in the clutter can be separated. The physical model can be simplified to be understood as a filter through which the echoes of a moving target can pass, but the echoes of a stationary or slow-moving target will not pass, or most

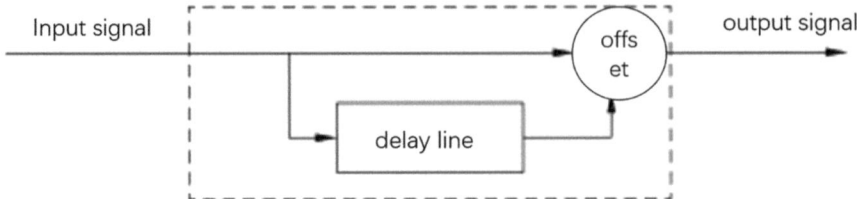

Fig. 13.9 Schematic of the engineering implementation of a single delay line pair of eliminator

of the energy will not pass through the filter, so that only the moving target can be detected.

Its early engineering implementation models make it easier to understand the plain idea of moving target display technology. Before the advent of digital signal processing, delay lines were used in engineering to implement moving target displays, and the simplest model is often referred to as a single delay line canceller. The schematic block diagram is shown in Fig. 13.9.

The simple understanding is that the signal of the current cycle is subtracted from the signal of the previous cycle, if it is the echo generated by the fixed clutter, then the signal of the two cycles is the same, it can be subtracted and cannot be output from the filter, if it is the echo generated by the moving target, then there is a difference between the echo signals generated by the two cycles, it can be output from the filter, and so the moving target display is achieved.

From a mathematical point of view, a typical MTI filter is realised using a non-recursive structure, as shown in the following equation, where a(k) are the coefficients of the filter and N is the number of orders of the filter.

$$H(z) = \sum_{k}^{N} a(k) * z^{-k} \qquad (13.4)$$

For example, the response of the previously mentioned single delay line to the canceller in the Z domain is:

$$H(z) = 1 - z^{-1} \qquad (13.5)$$

As can be seen, it is a simple form of a typical MTI filter.

In radar engineering nowadays, MTI is much more complicated in design, especially after entering the era of digital signal processing, some theoretically more complicated models can be realised through digital signal processing to achieve MTI with better performance. For example, when suppressing slow motion clutter, it is often necessary to move the suppression frequency, or the suppression frequency can be formed adaptively according to the characteristics of the clutter, and this kind of MTI filters are often called adaptive MTI, or AMTI. these techniques are mature technologies in modern radar engineering.

13.4 Fundamentals of Radar Signal Processing

MTI processing can well eliminate the Doppler frequency zero clutter interference generated by fixed targets such as ground objects, and it also has a certain suppression effect on the weak meteorological and other dynamic clutter interference which is not too strong and does not move fast. In ocean engineering, sea clutter is a kind of slow motion clutter, and the process of detecting sea surface targets requires the use of some more advanced moving target display techniques [12].

13.4.3 Moving Target Detection

Moving Target Detection (MTD) addresses the core contradiction of how to better detect moving targets in complex backgrounds.

The MTI technique mentioned in 13.4.2 usually only brings an improvement factor of 20 dB to the system, and its limitations are highlighted in some scenarios with higher clutter intensity. A moving target detection system is an improved system with better performance.

Many professional acquaintances in the radar industry have a misunderstanding of the MTD concept and often confuse the concept of MTD with that of a Doppler filter bank. In fact, MTD is not a single technology, but a detection system. Since it is a system, the definition of MTD in the industry is not completely uniform. Here we quote the viewpoints from Mr Ding Lufei's "Radar Principles" to make a description of MTD system. It is usually considered that on the basis of MTI technology, the following improvements are made to the system to improve the performance of moving target detection, which include:

(1) Increase the linear dynamic range of signal processing;
(2) Adding a Doppler filter to bring it closer to the optimal filter and improve the improvement factor;
(3) The ability to suppress ground clutter (whose average Doppler shift is usually zero) and to simultaneously suppress motion clutter (e.g., weather, bird flocks, etc.);
(4) The addition of one or more clutter maps can serve to aid in the detection of large targets in tangential flight.

Some viewpoints also suggest that Constant False Alarm Rate (CFAR) needs to be included in the MTD system. In short, after these technical means to deal with the composition of the moving target detection system, known as the moving target detection system, MTD system is relatively more central part of the Doppler filter group, so many radar industry personnel are often mixed with these two concepts, strictly speaking, this reference is not accurate.

In order to better understand the concept of MTD system, an early engineering implementation model is also used to illustrate its basic principles. This is the 1970s, the Massachusetts Institute of Technology Lincoln Laboratory (MIT Lincoln Laboratory) designed a set of MTD system, its simple block diagram shown in Fig. 13.10.

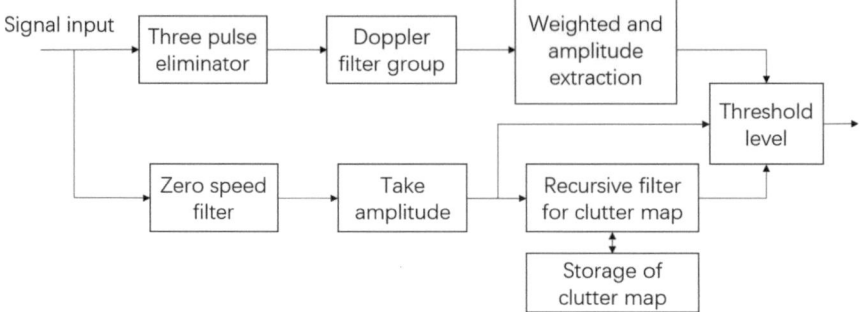

Fig. 13.10 Simple block diagram of the MTD signal processing system

This system uses several techniques to improve the detection of moving targets against a clutter background, firstly a three-pulse pair of cancellers is used to filter out the strongest ground clutter to alleviate the dynamic range requirements of the later processing links, then a Doppler filter bank is connected (the basic principle and role of which is described subsequently), which achieves the signal filtering against the clutter background by means of a narrow-band filter bank, and at the same time, the clutter map detection branch is introduced to achieve the detection of large targets in tangential flight (in tangential flight, the target has a Doppler frequency of 0, which is usually suppressed by the canceller and suppressed as ground clutter in the Doppler filter bank, making it difficult to be detected).

From the introduction of the system, we need to focus on expanding the core concept of MTD: the Doppler filter bank. In fact, the point of view of this technique is very plain. As we mentioned before, the core idea of MTI technique is to design a filter so that signals at a specific frequency point (e.g., zero frequency) can't pass through, and other signals can pass through, so as to achieve the suppression of clutter signals. However, when the clutter is also in motion, and the speed of motion has different patterns, the performance of this filter that can only suppress a specific frequency point is limited. Doppler filter can then play a role, in order to facilitate the understanding of the ideal physical model can be abstracted as follows: there are many filters, each filter only allows a certain frequency signal through, other frequencies are suppressed, these filters pass band superimposed in turn, forming a comb-like filter band. Ideally, for a target with a certain speed, it would pass through only one of these filter banks, and all types of clutter that do not have the same speed as itself would not be able to pass through this filter, so better performance clutter suppression is achieved. Figure 13.11 can help to better understand what a filter bank is.

In engineering implementation, this can be achieved by weighting and summing the N output transverse filters by different pulses. The frequency range covered by the filter bank is 0 to fr, where fr is the pulse repetition frequency of the radar operation. Why the frequency fr is introduced here, we can take this question to understand the

13.4 Fundamentals of Radar Signal Processing

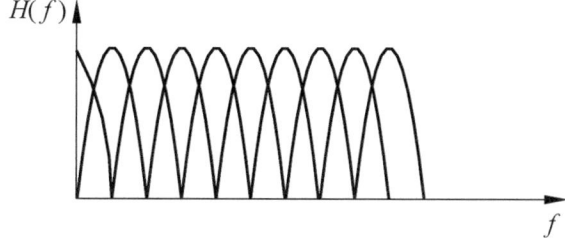

Fig. 13.11 Schematic diagram of the Doppler filter bank

content later. The block diagram of a typical Doppler filter bank implementation is shown in Fig. 13.12.

The delay time ΔT in the figure is usually the pulse repetition period of the radar, and the reciprocal of this pulse repetition period, which is the radar operating frequency fr. Without doing a rigorous formulaic derivation, we can understand how the frequency coverage of the filter bank is related to the radar operating frequency. How many narrowband filters are covered by 0 to fr with this structure? Conclusively, it is given that this number is determined by the number of outputs of the transverse filters. The concept becomes very clear.

In real engineering applications, there are many more technical details of the Doppler filter bank. Often the control of these technical details determines the final use of a radar. In marine engineering, it is especially important to note that the Doppler filter bank is not the ideal filter shape in engineering, and there is still a sub-flap in addition to the main flap. Unweighted filter flap is often still relatively high (may reach − 13 dB or so), Doppler filter bank of the flap, means that other speed of the signal can also pass through the filter, some of the stronger dynamic clutter may have an impact on the target detection. For example, sea clutter is a kind of strong dynamic clutter. In this case, there are generally two ideas to solve this problem: one is to add a pair of cancellers before the narrowband filter bank to suppress most of the energy of the clutter, so that the clutter entering from the sub-flaps will be reduced and the improvement factor will be improved, but this needs to increase the complexity of the

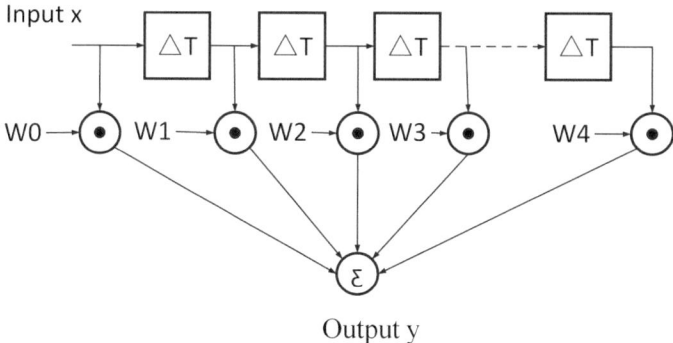

Fig. 13.12 Typical Doppler filter bank

system, and there is a priori knowledge of the characteristics of the clutter motion; the other is to reduce the filter sub-flaps with a weighting method, especially for the main frequency near the clutter in the sub-flaps and refine the design of the sub-flap rejection, but this causes the main flap to spread out [12].

13.4.4 Pulse Detection

After a series of processing, one always ends up with the question, what kind of signal is output from the system as a target? Here the issue of pulse detection comes into play. A common method of pulse detection in radar is threshold detection, which means that the energy of a pulse in the time or frequency domain, exceeding a threshold, is recognised as this being a target and is output from the system. There are also some more complex detection algorithms in modern radar, such as Track-Before-Detect, TBD, etc., which are used to solve some specific problems such as complex scenes, weak target detection, etc., and will not be expanded in detail here.

In conventional pulse detection techniques, the threshold setting is very important due to the idea of threshold determination. This leads to the focus of this section: Constant False Alarm Rate (CFAR), the core contradiction solved by the CFAR technique is how to select a detection threshold, so that the probability of false alarms in the system's output results is controllable.

This concept is still slightly abstract, from the logic of the discussion in this paper, jumped a few elements, here to expand for a brief description. In the radar echo, the real target generated by the echo and all kinds of interference generated by the echo is always accompanied by the method of signal detection through the threshold, there will be some is not the target generated by the echo is detected, this kind of false target detection, it is called false alarm. The probability of these false alarms appearing in all the detected targets is called the false alarm rate, and constant false alarm detection is a method of generating a detection threshold that keeps the false alarm probability constant.

The mathematical model of constant false alarm detection uses a great deal of statistical knowledge, and some relatively clear concepts of the mathematical model can be obtained through rigorous mathematical reasoning. However, this process does not help us understand the basic concept of constant false alarm detection. Here is still through a relatively simple constant false alarm engineering implementation method, to more intuitive experience of this threshold detection method. The method introduced here is called Cell Averaging Constant False Alarm Detection (CA-CFAR), which is a relatively simple engineering method of Constant False Alarm Detection.

The basic idea of Cell Averaging CFAR is to select a certain number of echo signals before and after the detected unit, take the average value of them, and use this average value as the detection threshold. In engineering, in order to avoid energy leakage near the target echo, it is usually also to select a few protection units around the detected unit and not to do the averaging statistics. This simple threshold determination method

13.4 Fundamentals of Radar Signal Processing

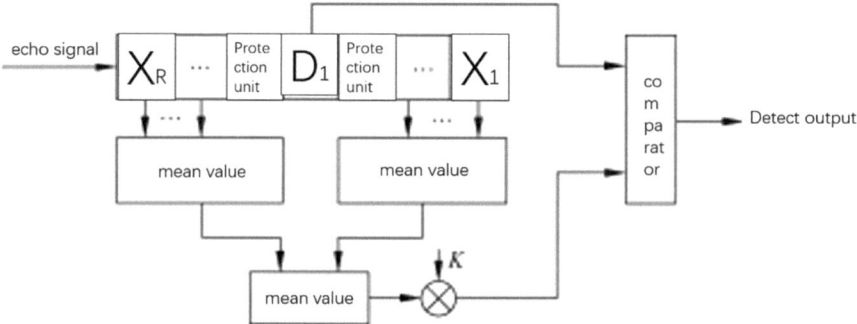

Fig. 13.13 Schematic implementation of the unit-averaged constant false alarm detection project

is a relatively simple constant false alarm detection model. The process is more easily understood through Fig. 13.13.

In this realisation model, the final mean value is also weighted by the coefficients as the final threshold, which is for the flexible control of the system, and through the adjustment of the coefficients, it is easy for the system to be applied to different over detection scenarios.

In marine engineering, there are still some situations need to be considered, so according to the use of the scene, the introduction of some variant algorithms of unit-averaged CFAR, for example, when the marine radar sweeps through the coastal clutter, the clutter intensity will suddenly become larger, if this time is still the selection of the unit-averaged CFAR, envisage that in the radar waveforms just swept to the coast of the clutter, half of the threshold sampling value is still relatively low, which will result in a lower detection threshold. This produces a cascade of false targets. Only when the entire interference level fills the entire reference cell will this clutter be suppressed to some extent. In this case a variant of CFAR, often called Cell Averaging Greatest Of CFAR (CAGO-CFAR), is used, and its engineering implementation flowchart is shown in Fig. 13.14.

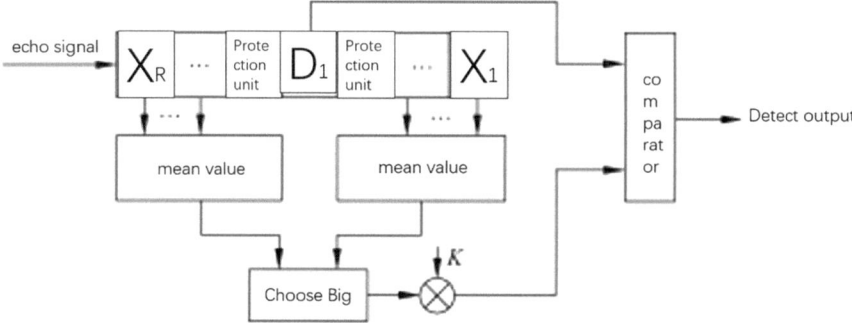

Fig. 13.14 Cell mean selection of larger CFARs

The main change in this model is to select the larger side of the mean of the left and right reference units in the detection unit as the threshold value, and this method can somewhat change the problem of the sudden increase of the interference level shown in the previous section.

There is another situation, for example, the radar in marine engineering needs to detect the formation, so that there may be some larger signals in the reference cells, which raise the threshold threshold, resulting in the target not being detected. So some CFARs also eliminate the largest signal in the reference cell, and then select the larger one in the mean value.

These two algorithms, as well as some similar variants, are also collectively referred to as fast-threshold CFARs, which mainly deal with target detection in some cluttered scenes. In the sea surface target detection of ocean engineering, it is necessary to achieve target detection under the background of strong sea clutter, and this CFAR idea will be used. In some no clutter environment, for target detection in the noise environment, there is a slow threshold CFAR, also known as: noise statistical threshold processing or noise constant false alarm, slow threshold CFAR processing is in the radar rest area of the noise samples sampling, through the large-sample statistical method of estimating the noise mean value, the mean value according to the Gaussian noise standard deviation normalised to get the threshold value, the threshold as the judgment threshold, the criterion for detection is that when the inspected unit is in the strong sea clutter background, it is necessary to achieve target detection, it will use this CFAR idea. The criterion for detection is that when the amplitude value of the inspected unit is greater than this value, it is judged as "target", otherwise it is judged as "no target".

Of course, pulse detection is not constant false alarm detection of this detection method, in order to avoid causing pulse detection is CFAR this wrong view, here to introduce another type of detector, such as: CLEAN detector. CLEAN detector first reference samples for the initial detection, eliminating the large value of the processed data to do the CFAR detection background for the sliding window statistics and detection. The purpose of doing so is to eliminate the influence of large signals on the detection background. A block diagram of one of its typical implementations is shown in Figs. 13.15.

The detection rate and false alarm rate of this kind of detector would not be able to simply use statistics to achieve the model construction, and need to use some more complex methods, such as Monte Carlo simulation.

These basic theories of radar signal processing, in the last century, after the basic determination, there is no revolutionary change, but in the process of specific engineering practice, the understanding of different scenarios, targeted selection of different processing methods and processing details, is the key to determine the final performance of the radar. It is also a radar engineering to achieve the main embodiment of the gap [10–12].

Synthetic aperture radar is not a subsystem technology of radar, but a kind of radar using synthetic aperture technology. Because of the large number of applications in marine engineering, and its technical characteristics and traditional radar has some

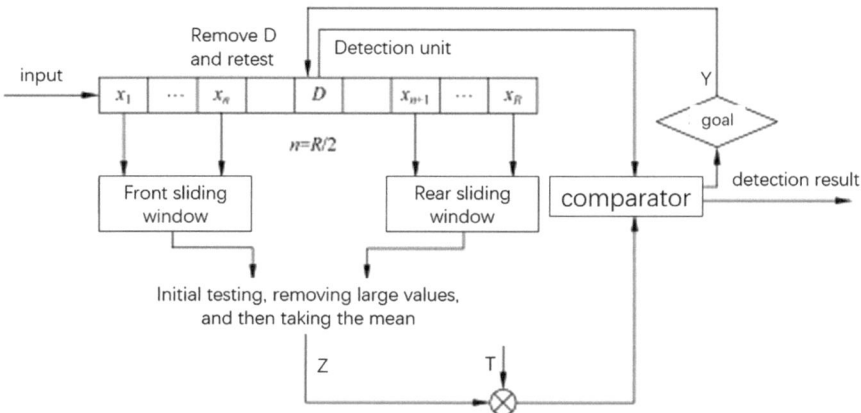

Fig. 13.15 Block diagram of a CLEAN detector implementation

difference, is an important technical branch of radar engineering, here to do a simple principle description.

Synthetic Aperture Radar (Synthetic Aperture Radar, SAR) concept is proposed, mainly to solve the problem of radar detection azimuth resolution. Since the large number of engineering applications of radar technology, researchers have been committed to continuously improve the radar resolution. Resolution can be divided into distance resolution and azimuthal resolution. As mentioned earlier, in modern radar, engineers improve radar distance resolution by means of pulse compression technology for large time-width bandwidth product signals. The intuitive way to improve the azimuthal resolution is to reduce the beamwidth of the radar, which requires increasing the physical length of the radar antenna. But it is clear that it is not possible to increase the physical size of a radar antenna without limit in engineering.

In the early 1950s, a group of scientists at the University of Michigan in the United States proposed that conventional radar produces a narrow beam that is achieved by increasing the physical size of the antenna. For example, a long line array, each unit simultaneously transmits a phase reference signal, and then does phase reference synthesis processing when receiving, thus forming a narrow beam. However, if instead of transmitting and receiving simultaneously with multiple units, a unit is sequentially time-shared along a line, and is not processed immediately after reception, but is stored, compensated, and then processed in unison, so that the signals received by this unit time-shared are also capable of phase-referencing. This method can obtain a resolution similar to that of a longer line array. This is the basic idea of a synthetic aperture antenna. Radars that use this theory are called synthetic aperture radars. In a way, the synthetic aperture technique trades time for azimuthal resolution [4, 12].

References

1. Ding, Lufei, Fulu Geng, and Jianchun Chen. 2013. *Radar Principles*, 4th ed. Beijing: Electronic Industry Press.
2. Guangyi, Zhang. 1994, 1996, 2000, 2001. *Phased Array Radar System*. Beijing: National Defence Industry Press.
3. Lufei, Ding, and Geng Fulu. 2002. *Radar Principles*, 3rd ed. Xi'an: Xi'an Electronic Industry Press.
4. Jaguo, Lu. 2017. *Synthetic Aperture Radar Design Technology*. Beijing: National Defence Industry Press.
5. Guangyi, Zhang, and Zhao Yujie. 2010. *Phased Array Radar Technology*. Beijing: Electronic Industry Press.
6. Jian, Yang, and Yin Junjun. 2020. *Polarisation Radar Theory and Remote Sensing Applications*. Beijing: Science Press.
7. Wang, S.-Y. 2014. *Marine Navigation Radar*. Dalian: Dalian Maritime University Press.
8. Lufei, Ding, Geng Fulu, and Chen Jianchun. 2014. *Radar Principles*, 5th ed. Beijing: Electronic Industry Press.
9. Yiyao, Dai, Wang Xuesong, Xie Hong, and Xiao Shunping. 2015. *Airspace Polarisation Characteristics of Radar Antenna and Its Application*. Beijing: National Defence Industry Press.
10. Stimson, George W. 2005. *Introduction to Airborne Radar*. Translated by H. P. Wu et al. Beijing: Electronic Industry Press.
11. Mahafza, Bassem R. 2016. *Radar System Analysis and Design (MATLAB Edition)*, 3rd ed. Translated by Wanxing Zhou, Mingchun Hu, Mingya Wu, Jun Sun, et al. Beijing: Electronic Industry Press.
12. Barton, David K. 2012. *Radar System Analysis and Modelling*. Translated by Nanjing Institute of Electronic Technology. Beijing: Electronic Industry Press.

Chapter 14
Radar Key Technologies

There are many key technologies in radar itself, and each subsystem and processing link has technical points that need to be breached. A comprehensive introduction to radar key technologies is not the focus of this book. The focus here is on the application of radar in marine engineering, to develop some common key technology introduction.

14.1 Sea Clutter Processing Technology

Radar is through the target reflected electromagnetic waves for detection, but in nature, there are many other targets that we do not want to detect, but reflect a lot of electromagnetic waves and be received by the radar, such as: ground, trees, buildings, clouds and rain (the concept of clutter is relative, such as meteorological radar is used to measure the clouds, clouds and rain are no longer clutter, but to be detected by the target) and so on. These clutter for the detection of the target to bring a lot of trouble, sometimes its energy is even higher than the target itself echo energy, radar wants to detect the target, you need to filter out these clutter. In marine engineering, sea clutter is a special kind of clutter that is difficult to deal with. Especially when radar is used to detect sea targets, how to deal with the interference of sea clutter has become one of the most important key technologies.

Sea clutter is the clutter formed by the radar beam irradiated to the sea surface and backscattered by the sea surface. From the definition of the radar equation, the sea clutter model is discussed:

$$P_r = \frac{P_t G^2 \lambda^2 \sigma}{(4\pi)^3 r^4 L_a} \tag{14.1}$$

where P_r is the reflected power of the sea clutter received by the radar, P_t is the transmitted power, G is the transceiver gain, λ is the radar wavelength, σ is the radar cross-sectional area, r is the slant distance from the radar to the centre region of the irradiated sea surface and L_a is the atmospheric loss.

It is obvious to observe from Eq. (14.1) that the key to estimating the modelling of sea clutter is the estimation of the radar cross-sectional area. The concept of backward scattering coefficient σ^0 is usually introduced for conveniently describing the radar cross-sectional area for sea clutter estimation.

The backward scattering coefficient σ^0 is the normalisation or averaging of the backward scattering intensity on the surface of the scatterer by spatial extent. Since σ^0 is used to describe the backward scattering coefficient of the whole sea surface, its radar cross section area should be in an average sense rather than an exact descriptive value. Therefore, in general terms, σ^0 is the average radar cross-sectional area per unit area.

If it is assumed that the radar irradiated area is the radar reflected area, and the radar irradiated area as a whole is set to be A_c, then the radar cross-sectional area is described by the backward scattering coefficient as σ as:

$$\sigma = A_c \times \sigma^0 \tag{14.2}$$

Bringing Eq. (14.2) into the sea clutter estimation model yields the sea clutter received power calculation formula as:

$$P_r = \frac{P_t G^2 \lambda^2 \sigma^0 A_c}{(4\pi)^3 r^4 L_a} \tag{14.3}$$

Usually radar irradiated area estimation is relatively simple, thus the key to model the sea clutter is to estimate the backward scattering coefficient of the sea surface, σ^0. In this paper, the backward scattering coefficient is estimated using the constant γ model.

The constant γ model is a simple model for estimating surface clutter, and the backward scattering coefficient σ^0 is estimated by the formula:

$$\sigma^0 = \gamma \times \sin \psi \tag{14.4}$$

where γ is a parameter describing the scattering characteristics and ψ is the swept angle with respect to the surface in rad. the model is applied to sea clutter estimation with the value of γ related to the sea state SS (or Beaufort wind level) and radar wavelength. The specific relationship formula is:

$$10 \lg \gamma = 6SS - 10 \lg \lambda - 58 = 6K_B - 10 \lg \lambda - 64 \tag{14.5}$$

where the parameter SS or K_B is used to describe the sea surface conditions and λ is the radar operating wavelength in m. The correlation between the sea surface

14.1 Sea Clutter Processing Technology

state parameters and the wind level, etc., can be obtained from the correspondence in Table 14.1.

Taking a certain type of radar as an example, the typical sea conditions of the radar operating environment are shown in Table 14.2.

Against the wind speed item obtained, the radar operating environment of the sea state potential typical take the value of 4 to 8, selected three typical frequency for simulation: 5300, 5500, 5850 MHz. get the value γ as shown in Table 14.3.

The continuous trend of γ with sea state and frequency is given in Fig. 14.1.

From the above table and trend graph, it is obvious that in the C-band, the effect of frequency change on γ is much smaller than the effect of sea state change on γ.

The change of γ value directly determines the change of backward scattering coefficient under the same grazing angle, which in turn determines the change of

Table 14.1 Sea surface parameters [1]

Sea state (SS)	Beaufort scale for wind speed (K_B)	Air velocity (V_w) (m/s)	Root mean square height deviation (σ_h (m))	Slope (β_0 (rad))
0	1	1.5	0.01	0.055
1	2	2.6	0.03	0.063
2	3	4.6	0.10	0.073
3	4	6.7	0.24	0.08
4	5	8.2	0.38	0.085
5	6	10.8	0.57	0.091
6	7	13.9	0.91	0.097
7	8	19.0	1.65	0.104
8	9	28.8	2.5	0.116

Table 14.2 Working sea state of a certain type of radar

Sea state level	4	5	6	9
Trinity wave height (m)	1.25–2.5	2.5–4.0	4.0–6.0	>11
Wind level	5~7	7~8	8~9	12
Average wind speed (m/s)	8.0~17.1	13.9~20.7	17.2~24.4	>32.7

Table 14.3 Values of γ at typical sea state and frequency points

	5300 MHz	5500 MHz	5850 MHz
SS = 4	7.033×10^{-3}	7.299×10^{-3}	7.763×10^{-3}
SS = 5	0.028	0.029	0.031
SS = 6	0.111	0.116	0.123
SS = 7	0.444	0.461	0.49
SS = 8	1.767	1.833	1.95

Fig. 14.1 Plot of continuous trend of γ with sea state and frequency

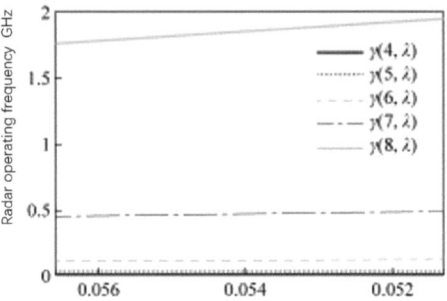

Table 14.4 Backward scattering coefficients at typical sea state and frequency points σ^0

	5300 MHz	5500 MHz	5850 MHz
SS = 4	−39.224	−39.063	−38.795
SS = 5	−33.224	−33.063	−32.795
SS = 6	−27.224	−27.063	−26.795
SS = 7	−21.224	−21.063	−20.795
SS = 8	−15.224	−15.063	−14.795

the intensity of the sea clutter, thus it can be seen that in the C-band, the change of frequency will not bring obvious change to the intensity of the sea clutter, while the change of the sea state plays a decisive role. Table 14.4 gives the typical frequency-to-backscattering coefficients, converted to dB, for three C-bands with the sea state from 4 to 8 at a grazing angle of 1°.

From Table 14.4, it can be clearly observed that the effect of frequency on the backward scattering coefficient is not obvious in the C-band, so only the backward scattering coefficient on the intermediate frequency of 5500 MHz is simulated, which can be a more representative illustration of the effect of the various influencing factors on the σ^0 in the C-band.

When the backward scattering coefficient is estimated using the constant model γ, compensation is required when the swept angle is very low and the swept angle is close to vertical incidence. The compensation is divided into two cases: one is the change in reflectance caused by the directional map-propagation because the F_c is no longer approximately equal to 1 and needs to be re-estimated for the F_c when the grazing angle is less than the critical grazing angle ψ_c, and the other is when the reflectance of the small surface feature is charged by the mirror-like reflectance complementary to the σ_f^0 from the randomly tilted surfaces on the surface when approaching the perpendicular incidence. Each of the two compensations is discussed below.

When the constant θ_R model estimates the backward scattering coefficient, the default state of the derived model is that the direction map-propagation factor φ is approximately equal to 1. However, when the swept angle is less than the critical swept angle ϕ, the direction map-propagation factor φ is no longer approximately equal to 1, and the FC needs to be re-estimated. The critical grazing angle ϕ is defined as:

$$\psi_c = \frac{\lambda}{4\pi \sigma_h} \tag{14.6}$$

where λ is the working wavelength in m; σ_h is the parameter used to describe the sea surface condition—the root mean square height deviation, and the correspondence with the sea state is shown in Table 14.1; for example, in sea state 4, when the working wavelength is 0.055 m, the critical swept angle ψ_c is 0.63 degrees.

When the swept angle is less than the critical swept angle ψ_c, the estimation formula for F_c is:

$$F_c \approx \frac{\psi}{\psi_c} = \psi \frac{4\pi \sigma_h}{\lambda} \tag{14.7}$$

1. Compensation method for near-vertical incidence

For near vertical incidence, a compensation factor σ_f^0 is added to compensate for the reflectivity due to randomly inclined surfaces on the surface resembling mirrors. σ_f^0 is defined as:

$$\sigma_f^0 = \left(\frac{\rho}{\beta_0}\right) \exp\left(-\frac{\beta^2}{\beta_0^2}\right) \tag{14.8}$$

The value of σ_f^0 is negligible when the grazing angle is less than 80 degrees, and its simulation at an incident wavelength of 0.055 m is shown in Fig. 14.2.

At the operating frequency of 5500 MHz (wavelength 0.055 m), the reflectivity simulation results for different sea states and different grazing angles are shown in the Fig. 14.3 (in dB).

In this paper, only from the operating frequency, grazing angle and other elements to do a simple analysis, sea clutter suppression is precisely the need to use these

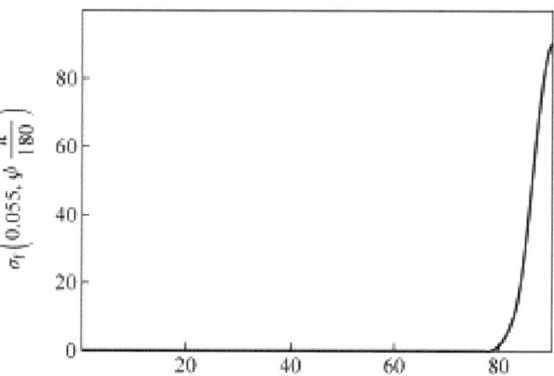

Fig. 14.2 Plot of compensation factor versus swept angle

Fig. 14.3 Simulation of reflectivity for different sea states and different grazing angles

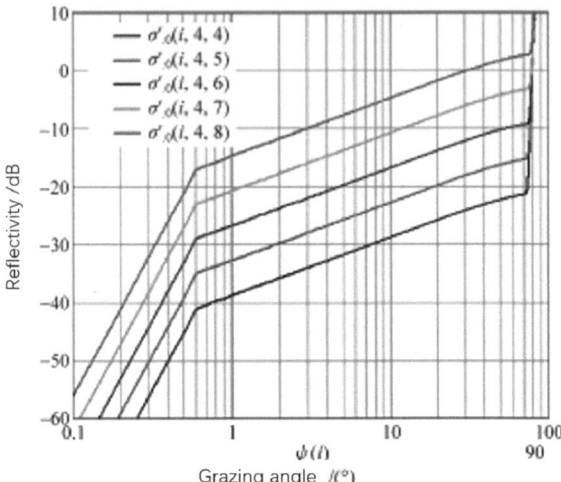

essential elements, in the system designed to refine the filtering model, so as to better detect the target. The suppression algorithm itself and the technical approach is a systematic project, there are more professional works and articles[1-12] in the industry, interested readers can learn more about it, the author is not convenient to go into detail here.

14.2 Dynamic Platform Beam Control Technology

The moving platform beam control here is mainly for shipboard radars, and for airborne radars, shore-based radars, etc., used in marine engineering, they are not in the focus of the discussion in this section.

First think about a simple application scenario to make it easier to understand what moving platform beam control technology is now a radar working on a ship is required to keep an eye on a target and detect its trajectory and current position in space. This requires the radar to keep the transmit and receive beams aligned with this target at all times. In the simplest scenario, where both the radar and the target are stationary, then the radar antenna just needs to be solved for a beam pointing and placed there all the time. Now the situation has changed, the target is still not moving, but the boat is moving, the antenna beam still need to always point to the target, we need to adjust their own beam pointing, as shown in the following Fig. 14.4.

This simple example is just to better understand the beam control of a moving platform, the real situation is more complicated. The target itself is in motion, the radar is also in motion at the same time, in order to achieve real-time precision tracking of the target, you need to cancel out the effect of the two motions in the control link at the same time. Here can be simply the antenna pointing understood

14.2 Dynamic Platform Beam Control Technology

Fig. 14.4 Radar at different moments requires different antenna pointing

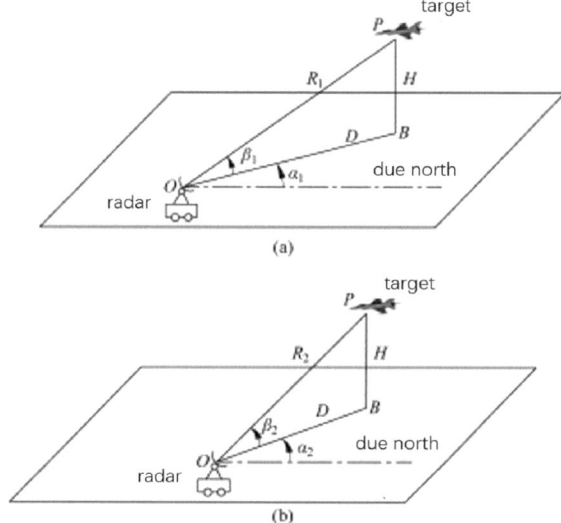

as a torch, light can only be straight to a direction, to be in a moving car, has been using a torch to shine a deer also in the running, you need to constantly adjust your wrist, here the brain automatically complete this adjustment of the control of the closed-loop, while in the radar, this is much more difficult.

For the target's movement, the radar can be based on detection data, do the target trajectory prediction, as long as the time interval relative to the target's movement speed is appropriate, modern mathematical filtering model can be relatively accurate prediction of a certain point in time on the target's spatial location, for the dynamic control of the beam to do the basis.

For the radar itself motion, in addition to the radar itself control, but also need the platform attitude sensing equipment, relatively accurate to get the spatial motion characteristics of the platform. For example, if the platform is a ship, it is necessary to get the ship's transverse rocking, longitudinal rocking, pitching, movement speed and other information, it can be solved to control the radar antenna beam pointing to a point in space, and how to control the antenna itself beam.

This series of solving process needs to be completed in quasi real time and perfectly aligned in time and space before accurate tracking of the target can be realised under a moving platform[2-4].

14.3 Sea Environment Adaptability

The problems of high temperature, high humidity and high salt spray faced by radar equipment are basically similar to those described in the Optoelectronics chapter. Same as electronic equipment, the three defence design of radar is an important issue in the application of ocean engineering.

In addition to the problems similar to those of optical equipment, some radar equipment for ocean engineering has a large volume and weight. The vibration and impact in the working environment will also have a greater impact on the radar performance. The design process of large radar equipment should focus on the impact of these physical quantities on radar performance, especially the impact of long-term service life.

In ocean engineering, as a kind of electronic equipment, the design of radar corrosion resistance in the sea environment has become an important design index. This point is the radar in marine engineering than other environment using radar is more special point. Radar design and manufacturing process, a large number of metal materials, and metal materials in the ocean's high temperature, high humidity, high salt spray environment, prone to corrosion phenomenon, especially some welding, grooves and other parts of the three anti-measures is difficult to deal with, is often prone to corrosion of the weak links. Metal corrosion will cause a decline in various aspects of radar performance, for example: the corrosion of the main structural components will affect the overall structural strength of the radar, which in turn affects the detection accuracy, the corrosion of the key parts of the circuit board may cause a decline in electrical performance, or even cause a serious short-circuit, triggering damage to the device or even serious damage to the entire machine, and so on. Rubber, plastic, synthetic materials and other non-metallic materials, but also due to the impact of the environment, the accelerated 'aging', resulting in less strength, brittle and fragile, such as the outer layer of the strong power cable is usually a rubber material, rubber aging may result in leakage and other security issues, the radar operator caused by the problem of life safety. So the design of three defences for radar in marine engineering is often as important as the design of equipment indicators and functions.

The marine environment is prone to cause radar equipment quality problems, mainly reflected in the following aspects.

High temperature, high humidity environment leads to changes in the basic parameters of electronic components, especially some of the core components that are more sensitive to the operating temperature, such as electronic chips, solid-state array of power amplification devices, etc., some polymer materials viscosity is reduced, sealing weakened, insulation weakened, etc., for different periods of time in the high temperature and high humidity environment, amplified physical expansion coefficients, resulting in chip detachment, welded joints disengagement, active parts structure detuning, etc., causing irreversible functional damage;

14.3 Sea Environment Adaptability

Electronic equipment high humidity environment and high salt spray environment together, in the period of the surface condensation formation of electrolyte, accelerate the electrochemical corrosion of metal materials, insulating materials in the electrolyte attachment, insulation performance decline, and even cause the period of short circuit;

Salt spray and air action, may form on the metal protective film has a destructive effect of ions (for example: chloride ions), these ions can accelerate the corrosion of metal, directional destruction of electroplated parts;

High temperature, high humidity environment also provides a good growth environment for the growth of mould, mould can also accelerate the corrosion of metal, reduce insulation performance, resulting in a significant reduction in the reliability of electronic devices;

Radar's sea environment adaptability is the fundamental application in marine engineering, especially on the radar reliability has a substantial impact on the service life and functional performance has a strong correlation. Therefore, this chapter slightly expands the adaptation of the sea environment, combing the factors that need to be considered in the design of radar sea environment adaptability. The content of this paper guided overview of the design of the sea environment adaptability of the main points, each direction also need more detailed information to complete the design[13-16].

14.3.1 Basic Concepts of Material Corrosion

Corrosion is a material in the environmental medium of chemical, electrochemical action and physical factors under the synergistic effect of the phenomenon of damage. Corrosion of materials should have the following conditions: (1) the material and the environment constitute the same system; (2) the material interacts with the environment; (3) the material undergoes chemical or electrochemical damage.

In metal materials, corrosion is positioned as the destruction or deterioration caused by chemical or electrochemical interaction between metal and the surrounding environment. In the marine environment, high temperature, high humidity and high salt spray accelerate the electrochemical action, thus the metal corrosion becomes faster.

In addition to metals, other materials also corrode. For example, the rubber material of the outer layer of cables in radar, corrosion can also occur, but this kind of corrosion is more accustomed to use another word to describe: 'aging'. The ageing of rubber is also accelerated under conditions of high salt spray and strong light.

In the classification of metal corrosion, there is a more special corrosion phenomenon, can be subdivided into galvanic corrosion, also called heterometallic corrosion, refers to the two different electrochemical properties of the material in the environment with the surrounding medium constitutes a circuit, the potential correction of the metal corrosion rate slows down, the potential of the metal corrosion of the more negative speed up. This particular phenomenon is reasonably applied to form a protective measure to reduce the effects of corrosion, which will be described in the subsequent text, where the reader can also pay attention to such phenomena.

14.3.2 Selection of Metal Materials for Radar Design in Ocean Engineering

The anti-corrosion measures of radar is a systematic project, which needs to consider a number of factors from the source of system design, including material selection, processing technology, three-proof treatment and so on.

In the choice of materials, the choice of metal materials is particularly important, radar design process, there will be a large number of metal, the corrosion resistance of the selected metal itself becomes the lower limit of the radar corrosion resistance. Commonly used metal materials for the radar design process include: stainless steel, titanium alloy, aluminium alloy, copper and copper alloy. In the case of slight corrosion, environmental factors, aluminium alloy due to its relatively light weight, moderate cost, in the strength can meet the requirements, but also on the weight and cost of certain constraints, in fact, is widely used by the radar, but in the marine engineering, aluminium alloy is relatively more susceptible to corrosion, and is not recommended to choose again.

When it comes to corrosion resistance, ordinary people will first think of stainless steel. Stainless steel is also true that it is widely used by radar in marine engineering. From the type of points, stainless steel is divided into: martensitic stainless steel, ferritic stainless steel, austenitic stainless steel, duplex stainless steel and so on. 304 stainless steel, which we often hear in our daily life, is a kind of austenitic stainless steel, and its stable properties are often used as food grade use, such as thermos cups, hot water kettles and so on. The following table gives some typical stainless steel and grade and the corresponding properties (Table 14.5).

The basic principle of corrosion resistance of stainless steel is that a surface oxide film will be formed in the atmosphere, which has been the occurrence of corrosion. But the ocean's high temperature, high humidity, high salt spray will destroy the stainless steel surface oxide film, thus inducing the occurrence of corrosion. Different stainless steel in the environment of the corrosion resistance characteristics of the performance is not the same. Materials science monographs have relevant detailed test data, this part of the content is not as the focus of this book, just conclusive given, in general, austenitic stainless steel is often chosen, of which the 316 series of corrosion resistance is better than the 304 series. However, the use of often also need to do in the special treatment of corrosion-resistant coatings. Under the condition of acceptable cost, some advanced duplex stainless steel is also a choice, after passivation treatment its corrosion resistance is better. But the project itself is the cost and technology of the art of compromise, many scenarios can not be one side of the pursuit of the ultimate.

14.3 Sea Environment Adaptability

Table 14.5 Properties and grades of common stainless steel[14-16]

Group	Grades	Properties
Austenitic stainless steels	304	Good corrosion resistance, wide range of uses, stamping, bending and other hot workability, no hardening phenomenon of heat treatment
	301	Compared with 304, Cr, Ni content is less, cold working to make the tensile strength and hardness increase, non-magnetic, but after cold working there is magnetic
	316	Add Mo, so its corrosion resistance and high temperature strength is particularly good, can be used in harsh conditions
	316L	As 316 steel low carbon series, in addition to the same characteristics with 316 steel, its better resistance to intergranular corrosion properties
Ferritic stainless steels	410L	Processability, resistance to welding deformation, resistance to high-temperature oxidation performance is good
Martensitic stainless steel	410	Representatives of martensitic steel, high strength, but not suitable for use in harsh corrosive environments
Duplex Stainless Steel	2205	Good mechanical properties and corrosion resistance

In addition to stainless steel, radar also uses a large number of other metals, such as structural steel, titanium alloy, copper alloy, aluminium alloy and so on. In marine engineering, all these metal materials have to be protected by surface treatment or coating before they can be used in radar. Among them, in cost-insensitive engineering applications, titanium alloys are a preferred material in the marine environment, where they are relatively stable in hot and humid environments, but at the same time, attention should be paid to the specific use of temperature limits.

14.3.3 Selection of Polymer Materials for Radar Design in Ocean Engineering

Radar design cannot be separated from polymer materials, radar design in ocean engineering, more need to focus on the selection of polymer materials. The concept of polymer materials is relatively broad, generally polymer compounds as a matrix, and then through other additives together constitute the material, collectively referred to as polymer materials. Radar commonly used in plastics, rubber, paints, binders, etc. are polymer materials.

Unlike the electrochemical corrosion of metal materials, polymer materials 'corrosion' is generally manifested as aging. The composition of polymer materials, determines its electrochemical inertia, in the high salt spray environment, induced metal corrosion of the electrochemical reaction, the polymer materials basically have no effect, so polymer materials are commonly used for metal surface treatment and coating protection. But polymer materials on solar radiation, temperature, oxidation, special liquids, gases, etc., but relatively sensitive, easy to cause its cracks, brittle, hard, and even lose strength, these phenomena are generally called aging. Polymer materials, due to their adaptive nature, once aging is likely to cause a significant reduction in radar reliability. In particular, it may cause some safety problems (e.g. cable leakage), which requires special attention. In the selection of materials to fully consider its aging characteristics, but also pay attention to the daily maintenance, a reasonable extension of its service life, when the need to advance maintenance, timely replacement of spare parts and so on.

In radar design, plastics may be used for some auxiliary devices shell (such as temperature and humidity sensors, random instrumentation), some structural parts of the outsourcing (to prevent friction), cable bundling line, etc. Plastics will not be used in high-end radar in a large area, but if necessary, should try to select the type of sea environment with good adaptability. In particular, some plastic products may be in contact with metal corrosion, the selection needs special attention.

Radar will also use polymer composite materials in the design, the most common is the radome. Especially in marine engineering, radome can play a role in windproof, rainproof, improve the radar working environment and so on. Spherical radome can take advantage of the shape, reduce wind resistance, is the radar significantly improve the ability to work under strong winds. Some radomes also have frequency selection effect, which can improve the electromagnetic spectrum control ability of the radar as a whole. For example: glass fibre reinforced plastic (FRP) is light and hard, non-conductive, with high mechanical strength and corrosion resistance, and is generally chosen as the material for radomes. In order to improve the aging-resistant service life of FRP, it can be coated with another layer of paint or covered with a layer of aging-resistant polyester film. As the aging-resistant film is attached to the FRP forming mould, the film combines excellently with the FRP resin to form an integral whole, and this type of radomes are conveniently assembled, weather-resistant, and are heavily used by the radar in marine engineering.

Radar will also be used in the rubber, adhesives, sealants and other polymer materials, the basic principle of choice is to adapt to high temperature, high humidity, high salt spray environment, try to avoid interaction with metal materials. If the outdoor use, need to consider more strong radiation and humidity and other factors. Polymer materials are relatively better in terms of stability, is a large number of auxiliary materials used in marine engineering. Its aging characteristics in addition to the characteristics of the material itself, but also with the use of maintenance has a greater relationship, in the design stage, these factors need to be taken into account as far as possible.

14.3.4 Some Considerations for Radar Design in Offshore Engineering

In addition to material selection, there are some other considerations for radar design in ocean engineering. This is not a systematic design basis, but only lists some aspects that the author believes should be considered.

First of all, the thermal design of the system in the marine environment should be given special consideration, because of the specificity of the environment, should try to use simple and reliable cooling technology, because the cooling system itself may be the weak link of the radar. The system should be designed with air filtration device as much as possible, and the air entering the cooling system can greatly improve the reliability of the system after dry treatment. System thermal design should also give full consideration to the problem of condensation, in the overall structural design, it is necessary to give full consideration to the drainage of the various structural links, to avoid the formation of water accumulation in the local weak points.

Radar design in marine engineering components selection, to give more consideration to the derating design, for some failure, may cause a significant decline in the overall performance of the system or fatal failure of the key components, try to select the highest level of derating design principles. To a certain extent, this kind of derating design will cause the increase of system cost, but a lot of radar equipment with similar performance in marine engineering, its price is much higher than the land equipment, which is a very important aspect of the reason.

One of the most important aspects of the radar in the design process to do a good job in the three anti-design. The so-called three-proof, is moisture-proof, salt spray, mould-proof short. High-performance radar will do three-proof design, but the three-proof design of radar in marine engineering needs to be more refined. The first principle of three-proof design in marine engineering is that there should be no dead space for three-proof. In the radar system at all levels of the three anti-protection process, there will be some processing blind spots, such as some very difficult to touch the structure of the seam, after the installation of the structure caused by the internal dead space, etc., these three anti-blind spots in the ordinary radar may not cause a fatal impact, but in the marine engineering radar, often become a corrosive cut point, and ultimately lead to a significant decline in overall performance. In the process of moisture-proof design, it is necessary to minimize the direct exposure of the functional material itself, and avoid the direct contact of different metals; the basic design principle of anti-salt spray is sealing, and the commonly used means is to coat the surface of the components with organic coating; the main measure of anti-mould is to choose the materials that are not easy to grow mould, and some particularly sensitive external exposed parts can be selected to be coated with anti-mould paint. Three-proof process, the most commonly used is the use of three-proof paint, but marine engineering should also be especially according to the specific use of the environment to choose different varieties, focusing on consideration of the use of the occasion of light and temperature and other factors, which are induced by the three-proof paint aging of the main external triggers. Here conclusively given acrylic, polyurethane and poly-p-xylene class three-proof paint performance is relatively stable.

Note: polyurethane class three-proof paint: long-term dielectric properties, superior moisture resistance and solvent resistance, and low-temperature environment, stable performance; poly p-xylene class three-proof paint: is currently known as high-frequency components, high-density components, high-insulation components of the most effective protective coating materials, electrical insulation and good protection.

There are two other special links in radar engineering that may be weak links in corrosion resistance, namely, radar electrical or structural welding points and electrical series connections. These post-processing links should pay more attention to their three anti-protection treatment, if the welding is a different metal, more through the three anti-protection process to prevent its galvanic coupling corrosion. Some of the strong electrical connection points because of its connection is not good for three anti-protection treatment, more likely to cause environmental exposure, especially in these places, focus on sealing treatment to avoid it becoming a weak link.

14.4 Electromagnetic Compatibility Control Technology

The problem of electromagnetic compatibility in radar applications has two levels of issues. One is the problem of electromagnetic compatibility design of the radar itself. The second is the problem of mutual interference between the radar, as an electronic information device, and other electronic devices in the compact space.

14.4.1 Significance of Electromagnetic Compatibility Design of Radar in Ocean Engineering

Radar as a kind of detection equipment through electromagnetic waves, its ultimate is to detect the weak induction current generated by electromagnetic echo, and then get all kinds of information about the target. The electromagnetic compatibility design of radar in marine engineering is very important, which can be understood from two levels: one is that the electromagnetic echo of the target or its formation of the induction current can not be contaminated, otherwise it will interfere with the detection, for example: if the system will be mixed with some noise signals into the analogue circuit, it will cause difficulties in the detection of the signals themselves; the second is that some electromagnetic interference will affect the radar's own performance of the components, for example: some strong electromagnetic pulses will affect the performance of digital chips. For example, some strong electromagnetic pulse will have an impact on the performance of the digital chip, resulting in the system crash, can not run the programme normally, etc. If the circuit design is not considered comprehensively enough, it may even cause permanent damage to the chip.

So the radar system electromagnetic compatibility design has high requirements, with the digital development of radar technology, radar itself that will contain analogue circuits, but also will contain digital circuits, crosstalk control between the two types of circuits, but also need to be special technical means. For example: through the scientific grounding system design, try to avoid the impact on the stability of the power supply of other equipment, but also to avoid the voltage changes generated by other equipment on the radar itself. Especially in the marine engineering shipboard radar, equipment installation is compact, are in a common platform, can't avoid mutual electromagnetic compatibility influence by pulling the physical space. Therefore, shipboard radar has strict electromagnetic compatibility requirements and design specifications to ensure the stability of the electromagnetic environment throughout the ship. The corresponding industry standards have detailed descriptions, and there are domestic organisations specialising in electromagnetic compatibility testing and verification.

At the same time, radar as a kind of electromagnetic wave radiation equipment, its radiation of electromagnetic waves may also have an impact on other equipment. General marine engineering, the ship is the basic carrier of detection equipment, a ship often more than one kind of use of electromagnetic wave work equipment, such as radar and communication will use electromagnetic waves, which requires different equipment control do not affect each other. The basic idea of EMC control is interleaving. The purpose of EMC control is achieved through spatial masking, time-sharing in the time domain, and spacing in the frequency domain. Ensure a more compact carrier, all kinds of equipment can be at the same time or functionally do not affect each other at the same time work.

14.4.2 Design of Electromagnetic Compatibility of Radar in Offshore Engineering

The electromagnetic compatibility design of radar in ocean engineering is mainly the electromagnetic compatibility design of shipboard radar. From the design point of view, the electromagnetic compatibility design of shipboard radar is mainly to analyse, predict, control and evaluate the electromagnetic compatibility between or within systems, to limit the emission level of electromagnetic harassment of electrical, electronic devices, equipment or systems to the permissible level range, so as to achieve the purpose of protecting the electromagnetic environment, and meanwhile in the environment where there is electromagnetic harassment the electrical, electronic devices, equipment or systems have the ability of not At the same time, under the environment of electromagnetic interference, electrical, electronic devices, equipment or systems have the ability to not reduce the operational performance, to achieve electromagnetic compatibility and the best cost-effective ratio.

Electromagnetic compatibility design itself is a professional, and closely linked with the system electrical design, radar itself within the system of electromagnetic compatibility design, the main concern is the deterioration of performance caused by their own interference, but also consider the system generated by the conductive, radiated emissions on the neighbouring system of the harmful effects of the conductive, radiated emissions from the outside of the system caused by the harmful effects of the harmful effects of the conductive, radiated emissions. Electromagnetic compatibility design is a systematic project, can not be achieved through a number of technical means of cut-off, this book briefly introduces some commonly used design methods, but these methods need to be used in an integrated manner, and to be closely coupled with the system circuit design. Here is a brief introduction to three more important electromagnetic compatibility design methods: filtering, grounding, shielding.

Filtering is the use of filtering technology to deal with electromagnetic noise in the frequency domain, providing a low impedance bypass channel for the source of electromagnetic nuisance, in order to achieve the purpose of suppressing electromagnetic nuisance. In marine engineering, the more important three filtering directions include: power filtering, receiving filtering, transmitting filtering.

Power filtering

Shipboard radar in marine engineering, in general, the ship uses a unified power supply, various systems on the power quality requirements are different, of which the radar is a relatively high power quality requirements of electronic equipment. This requires the power supply filtering, that is, to reduce the instability of the power supply system on its own, but also to reduce the radar system itself produces interference crosstalk to the power supply system, the impact on other electronic equipment. The industry has a clear standard specification for the power filter design of shipboard radar equipment, and there are mature filter products available in the industry. However, the filter parameters should be adjusted appropriately according to the actual situation in order to achieve the purpose of filtering out specific interference.

Receiving Filter

Radar design, the entire receiving link and digital signal processing, in fact, is a filtering plus detection process, described here from the point of view of electromagnetic compatibility, just a side of the radar system design. The filtering here is from the point of view of eliminating electromagnetic crosstalk within the system and system construction. General radar system, the demand in this regard is not very strong, if not spatially tightly coupled, can be achieved through the physical space between the mutual control of electromagnetic interference, but in marine engineering, the physical space between the interval is often difficult to achieve, it is necessary to filter out the interference of other electromagnetic radiation signals through the filter. For example: the same ship on the two radars, the working frequency band is similar, one of the radar emitted by the electromagnetic wave of the spurious signal, it may be through the receiving link into another radar, due to the spatial distance is very close, this spurious signal may be much larger than the target echo signal, resulting in the reception of the link of the front section of the saturation, in this case, it is necessary

14.4 Electromagnetic Compatibility Control Technology

to target the system to select the filter for a specific frequency of the signal for power Suppression, while the system's operating band of signals can pass through smoothly. This electromagnetic compatibility design is the shipboard radar this space tightly coupled electronic system is relatively special.

Transmit Filtering

Following on from the logic of the previous section, it is good to understand transmit filtering. In a relatively compact platform, if everyone thinks only of themselves and disregards the feelings of 'others', then the end result is that no one is well off. For the sake of the overall interests of the maximum control of their own electromagnetic wave emission, to avoid their own work outside the frequency band of energy radiation. Modern radar engineering system, radar harmonics, spurious suppression is the main object of emission filtering. In particular, the harmonics of the radar's main operating frequency, whose energy is generally larger, need to be in the various links of the signal generation link (e.g.: clock reference, intrinsic oscillator, DDS signal generation output) to design filters, systematic control of the purity of the final emitted signals, and to ensure that the system does not have an impact on the other electronic equipment to the maximum extent possible.

Grounding design is to provide a common pathway for all types of signals (including but not limited to all types of useful signals, noise signals, etc.) in the system, including protective grounding, signal grounding, etc. The design of the grounding body, the arrangement of the ground wire, the impedance of the earth wire in a variety of different frequencies are the key to electromagnetic compatibility design. It is also the basic yardstick to reflect a system engineer's design skills. Especially for the radar such a complex electronic information system, its system-level grounding design capability, to a large extent, will affect the radar machine reliability.

In marine engineering, the grounding design of radar is still a difficult point in the grounding design of shipboard radar. The entire ship tightly coupled, various types of electronic equipment must be strictly enforced uniform design specifications, any one device can not be executed in accordance with the requirements of grounding, may lead to the entire ship's electronic equipment grounding design disorder. In general, the grounding of radar equipment in marine work is divided into three types: signal ground, safety ground, noise ground. Signal ground is the ground for low-level sensitive circuits, usually the reference reference for signal circuits and DC power supplies, and is divided into digital and analogue ground when necessary. Digital ground and analogue ground should also be avoided between each other crosstalk, in general, the digital ground for digital circuits, its large number of 0, 1 level switching will be formed in the ground signal components are more complex, compared to its possible impact on analogue circuits. Safety ground, as the name suggests, in order to ensure the safety of personnel, so that the equipment shell and the formation of low impedance connection between the ground, used to prevent the induction of equipment charged and insulation damage to play a protective role, usually for the machine and chassis shell, household appliances connected to the shell is the safety of the ground. Noise ground, also known as interference ground, for non-sensitive circuits, power transformers, electrostatic shielding layer, as well as generating inrush currents and large power components of the ground. The overall design requirements of these three kinds of ground independent of each other, and ultimately in the whole ship at a point to complete a single point of grounding, this is the radar in the ocean engineering and other radar is different from one of the design concept.

Shielding design is the use of a variety of conductive and electromagnetic materials, manufactured into a variety of shells and connected to the earth, in order to cut off through the space of electrostatic coupling, inductive coupling or cross-EMF coupling formation of electromagnetic noise propagation pathway. In marine engineering, in order to ensure that the working environment of each device, generally will have uniform requirements for the shielding performance of radar equipment, according to the different levels of requirements, the need to carry out different levels of shielding design. In general, shielding is a technical means of controlling electromagnetic noise generated by its own non-active radiation units, perceptual interpretation can be illustrated by an example, such as radar a combination of electronic equipment with electronic equipment, in the absence of shielding, its electronic equipment in the work of all kinds of different frequency bands of electromagnetic noise (from the most fundamental physical phenomenon to understand the phenomenon, any induced current back to the formation of induced Electromagnetic waves, and radiation to space), the purpose of shielding is to make these electromagnetic noise maximum do not radiate into free space, so as not to have an impact on other sensitive electromagnetic induction equipment. In general, high-power or high-voltage devices need to do shielding; some sensitive circuits, such as frequency sources, clock sources, etc., need to do shielding. Shielding materials can be selected from high conductivity, high permeability of metal materials, increase the shielding shell in the design.

The detailed unfolding of EMC design specifications is not the focus of this book, which is more inclined to throw light on the important role of EMC design in radar engineering. The ability to control such details is often the key point of the current level of radar technology, is the radar from 'have' to 'fine' must cross the important links.

The use of radar technology in marine engineering confronts a love-hate physical phenomenon: atmospheric waveguide. Commonly known as electromagnetic wave propagation in the atmosphere is not strictly straight line propagation, but will be absorbed, refraction, reflection, scattering and other phenomena. Atmospheric waveguide is the electromagnetic wave in the transmission process is 'special refraction' of a phenomenon.

In certain meteorological conditions, in the atmospheric boundary layer, especially in the atmosphere near the sea surface propagation of electromagnetic waves, its propagation trajectory downward curved. When the curvature of the trajectory exceeds the curvature of the Earth's surface, electromagnetic waves will be partially trapped in a certain thickness of the atmosphere within a thin layer, as in the propagation of the metal waveguide, the formation of waveguide propagation of the atmosphere in a thin layer is called the atmospheric waveguide layer cause. We call the refraction of rays with curvature smaller than the critical refraction the trapped refraction, and electromagnetic waves propagating in such an atmosphere usually have the possibility of waveguide propagation.

When electromagnetic waves propagate in the atmosphere near the surface of the sea, the atmospheric waveguide on its impact is mainly manifested in two aspects one is to increase the propagation distance, and the other is to increase the electric field strength. Electromagnetic waves in the waveguide layer back and forth constantly reflected, the attenuation of its energy greatly slowed down, so electromagnetic waves in the waveguide layer can be long-distance propagation, usually in the atmospheric waveguide propagation distance can be several times its normal value. The target detected by the radar is a target far away in the horizontal direction. If the atmosphere is viewed as a layered spherical body, the radar's ranging error is generally not greater. However, when there is an atmospheric waveguide and the radar wave propagates into the waveguide, the apparent distance of the detected target is very different from the actual distance, sometimes up to several kilometres to one or two hundred kilometres. Considering the influence of the curvature of the earth, the actual elevation angle of the target at this time should be a negative value.

It can be seen that, on the one hand, radar can make use of the atmospheric waveguide to carry out over-the-horizon detection, breaking through the limitations of the range of vision caused by the curvature of the earth, and detecting targets at longer distances. Many countries are equipped with microwave over-the-horizon radar, the detection distance of the sea target reaches more than 120 km, far beyond the radar range caused by the curvature of the earth, is the use of atmospheric waveguide related principles. On the other hand, the existence of atmospheric waveguide also seriously affects the detection performance of conventional radar, in a certain detection angle, electromagnetic waves will fall into the atmospheric waveguide, resulting in the radar detection blind spot, and due to the existence of atmospheric waveguide, to the radar itself ranging, angle measurement has brought about interference, the need for specialised data processing algorithms, to circumvent the impact of atmospheric waveguide.

14.5 Atmospheric Waveguide Processing Technology

The use of radar technology in marine engineering confronts a love-hate physical phenomenon: atmospheric waveguide. Commonly known as electromagnetic wave propagation in the atmosphere is not strictly straight line propagation, but will be absorbed, refraction, reflection, scattering and other phenomena. Atmospheric waveguide is the electromagnetic wave in the transmission process is 'special refraction' of a phenomenon.

In certain meteorological conditions, in the atmospheric boundary layer, especially in the atmosphere near the sea surface propagation of electromagnetic waves, its propagation trajectory downward curved. When the curvature of the trajectory exceeds the curvature of the Earth's surface, electromagnetic waves will be partially trapped in a certain thickness of the atmosphere within a thin layer, as in the propagation of the metal waveguide, the formation of waveguide propagation of the atmosphere in a thin layer is called the atmospheric waveguide layer cause. We call the refraction of rays with curvature smaller than the critical refraction the trapped refraction, and electromagnetic waves propagating in such an atmosphere usually have the possibility of waveguide propagation.

When electromagnetic waves propagate in the atmosphere near the surface of the sea, the atmospheric waveguide on its impact is mainly manifested in two aspects one is to increase the propagation distance, and the other is to increase the electric field strength. Electromagnetic waves in the waveguide layer back and forth constantly reflected, the attenuation of its energy greatly slowed down, so electromagnetic waves in the waveguide layer can be long-distance propagation, usually in the atmospheric waveguide propagation distance can be several times its normal value. The target detected by the radar is a target far away in the horizontal direction. If the atmosphere is viewed as a layered spherical body, the radar's ranging error is generally not greater. However, when there is an atmospheric waveguide and the radar wave propagates into the waveguide, the apparent distance of the detected target is very different from the actual distance, sometimes up to several kilometres to one or two hundred kilometres. Considering the influence of the curvature of the earth, the actual elevation angle of the target at this time should be a negative value.

It can be seen that, on the one hand, radar can make use of the atmospheric waveguide to carry out over-the-horizon detection, breaking through the limitations of the range of vision caused by the curvature of the earth, and detecting targets at longer distances. Many countries are equipped with microwave over-the-horizon radar, the detection distance of the sea target reaches more than 120 km, far beyond the radar range caused by the curvature of the earth, is the use of atmospheric waveguide related principles. On the other hand, the existence of atmospheric waveguide also seriously affects the detection performance of conventional radar, in a certain detection angle, electromagnetic waves will fall into the atmospheric waveguide, resulting in the radar detection blind spot, and due to the existence of atmospheric waveguide, to the radar itself ranging, angle measurement has brought about interference, the need for specialised data processing algorithms, to circumvent the impact of atmospheric waveguide (Fig. 14.5).

Through the description of the definition of atmospheric waveguide, it can be found that atmospheric waveguide is not a description of a physical medium, but a description of a physical phenomenon. There are several classifications that can form the cause and nature of this physical phenomenon, and one of the more recognised classifications in the industry is to classify it into: ground-hugging atmospheric waveguide and overhanging atmospheric waveguide.

There are three specific subdivisions of grounded atmospheric waveguides: the first is called a surface waveguide, which is a waveguide layer close to the ground; the second is called a surface-based atmospheric waveguide, which is characterised by a base layer underneath the waveguide layer. There are also cases where both types of atmospheric waveguides are collectively referred to as surface waveguides. These two types of waveguides generally occur in the boundary layer atmosphere

14.5 Atmospheric Waveguide Processing Technology

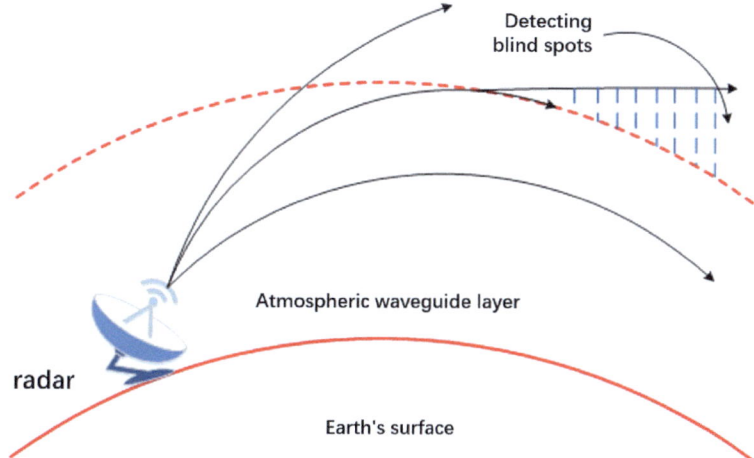

Fig. 14.5 Schematic diagram of detection blindness caused by atmospheric waveguide

below 300 m altitude, and their significant feature is that the atmospheric corrected refractive index at the top of the waveguide layer is smaller than the atmospheric corrected refractive index at the ground. This type of waveguide occurs when there is a strong inversion layer near the sea surface or when warm, dry air moves over a cold, wet sea surface. The third type of waveguide is called evaporating waveguide, which is generally found in the near-surface atmosphere at a height of 40 m or less, and it consists of a thin trap layer. It consists of a thin trap layer, which is formed when the evaporation of water vapour from the sea surface causes the atmospheric humidity to decrease sharply with height over a small altitude range above the sea surface. The height of the evaporating waveguide varies with geographic latitude, season, time of day, etc. The height of the evaporating waveguide is usually higher during the summer and daytime at low latitudes. Evaporative waveguides occur frequently in the marine atmospheric environment and may occur in any sea area.

Suspended atmospheric waveguide generally occurs in the bottom of the troposphere at a height of 3000 m or less in the atmosphere, also known as lifting waveguide. This waveguide bottom lifts above the Earth's surface and usually consists of a suspended trap layer superimposed on top of a suspended base layer. It is characterised by an atmospheric corrected refractive index at the top of the waveguide layer that is greater than the atmospheric corrected refractive index at the surface. The height of the lower boundary of the overhanging waveguide is typically tens or hundreds of metres above the ground, above which an inverse temperature layer typically occurs. The following figure gives a perceptual illustration of several atmospheric waveguides through the relationship between waveguide strength and height (Fig. 14.6).

In the above figure, h is the height of the top of the waveguide, h1 is the height of the top of the trapped layer, h2 is the height of the bottom of the base layer, d

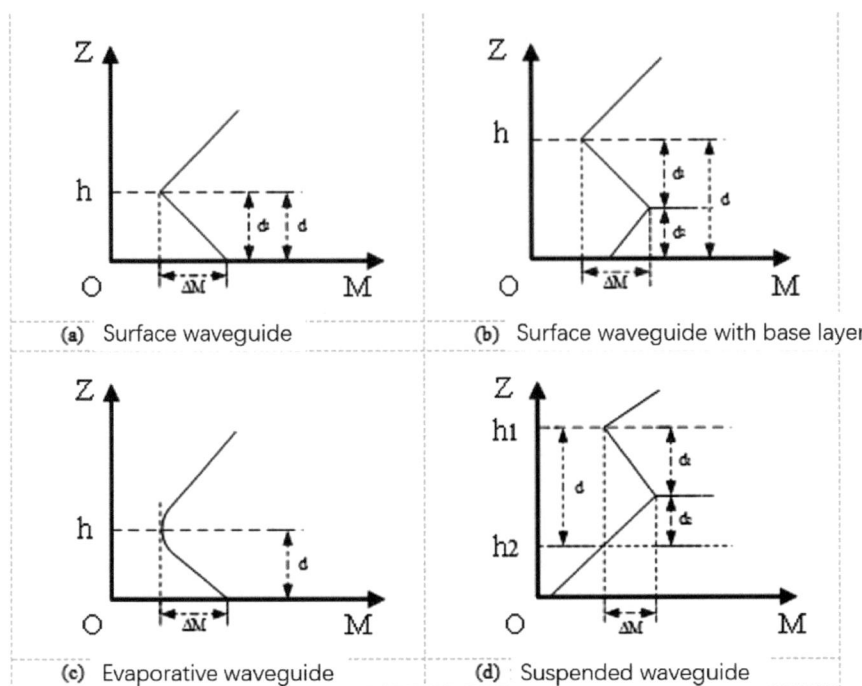

Fig. 14.6 Schematic diagram of atmospheric waveguide and its characteristic parameters

is the thickness of the waveguide, d1 is the thickness of the trapped layer, d2 is the thickness of the base layer, and ΔM is the waveguide strength.

If analysed quantitatively from a mathematical point of view, a quantitative description is given in the book 'Radar System Analysis and modelling' (David K. Barton). The base atmosphere has a refractive index gradient at ground level of units per metre. When this gradient drops below $-0.157N$ units per metre, an electromagnetic wave leaving the antenna may be captured in a waveguide and propagate a long distance with low loss. The maximum elevation angle at which an electromagnetic wave can be captured is

$$\theta_d = \sqrt{2(N_s - N_d - 0.157h_d)} \, (mard) \tag{14.9}$$

The maximum wavelength that can be captured is

$$\lambda_{max} = 0.0025 h_d \sqrt{N_s - N_d - 0.157h_d} \, (m) \tag{14.10}$$

In the above equation, N is the refractive index, which can be expressed as

$$N = \frac{77.6}{P}\left(P + \frac{4810p}{T}\right) \tag{14.11}$$

Electromagnetic waves have an important propagation property in addition to frequency: polarisation. The effect of the form of polarisation on electromagnetic waves is given here only conclusively. Under the same other conditions, the maximum free-space trapped wavelength of a horizontally polarised electromagnetic wave is one-third of the maximum trapped wavelength of a vertically polarised wave. That is, at the same wavelength, vertical polarisation is more likely to be trapped by atmospheric waveguides, and thus has an effect on detection. Similarly, vertical polarisation is a more suitable choice for shipboard microwave over-the-horizon radars from the point of view of using atmospheric waveguides[1,17-18].

References

1. Ding, Lufei, Fulu Geng, and Jianchun Chen. 2013. *Radar Principles*, 4th ed. Beijing: Electronic Industry Press.
2. Guangyi, Zhang. 1994, 1996, 2000, 2001. *Phased Array Radar System*. Beijing: National Defence Industry Press.
3. Lufei, Ding, and Geng Fulu. 2002. *Radar Principles*, 3rd ed. Xi'an: Xi'an Electronic Industry Press.
4. Jaguo, Lu. 2017. *Synthetic Aperture Radar Design Technology*. Beijing: National Defence Industry Press.
5. Guangyi, Zhang, and Zhao Yujie. 2010. *Phased Array Radar Technology*. Beijing: Electronic Industry Press.
6. Jian, Yang, and Yin Junjun. 2020. *Polarisation Radar Theory and Remote Sensing Applications*. Beijing: Science Press.
7. Wang, Shiyuan. 2014. *Marine Navigation Radar*. Dalian: Dalian Maritime University Press.
8. Lufei, Ding, Geng Fulu, and Chen Jianchun. 2014. *Radar Principles*, 5th ed. Beijing: Electronic Industry Press.
9. Yiyao, Dai, Wang Xuesong, Xie Hong, and Xiao Shunping. 2015. *Airspace Polarisation Characteristics of Radar Antenna and Its Application*. Beijing: National Defence Industry Press.
10. George, W.S. 2005. *Introduction to Airborne Radar*. Translated by H. P. Wu et al. Beijing: Electronic Industry Press.
11. Bassem, R.M. 2016. *Radar System Analysis and Design (MATLAB Edition)*, 3rd ed. Translated by Wanxing Zhou, Mingchun Hu, Mingya Wu, Jun Sun, et al. Beijing: Electronic Industry Press.
12. David, K.B. 2012. *Radar System Analysis and Modelling*. Translated by Nanjing Institute of Electronic Technology. Beijing: Electronic Industry Press.
13. Merrill, I. S. 2003. *Radar Handbook*, 2nd ed. Beijing: Electronic Industry Press.
14. Merrill, I. S. 1987. *Radar Handbook. Ship Navigation Radar: Book I*. Beijing: People's Traffic Press.
15. Yunze, Zhang. 1987. *Ship Navigation Radar: The Second Volume*, 1987. Beijing: People's Transportation Press.
16. Yunze, Zhang. 1990. *Ship Navigation Radar: The Third Book*. Beijing: People's Transportation Press.
17. Lina, Li. 2000. *Navigation Automation*. Beijing: People's Transportation Press.
18. Xinggu, Zhang. 2010. *Marine Radar*. Dalian: Dalian Maritime University Press.

Chapter 15
Application of Radar Technology in Marine Engineering

Radar, as an information acquisition sensor, is widely used in marine engineering due to its physical property of working all day and all night. And from the beginning of target detection, more application areas have been derived. For example, environmental sensing, hydro-meteorological conditions observation and so on. It is because of the invention of radar, greatly enriched the means of human exploration of the secrets of the sea, for the development and utilization of marine resources to provide assistance.

15.1 Military Application of Radar in Ocean Engineering

Although the Second World War has ended, quite a long time, the world is in a state of relative peace, but traction radar technology level continues to develop in marine engineering, or radar in the military application.

Radar as a kind of information acquisition equipment, is an important part of the modern ship air defence operations, radar performance is often a modern combat ship's key combat indicators, assuming the ship's own protection and formation protection of the important combat mission. This part of the content mainly refers to the Jane's Defence and the World Sea Radar Handbook. For more details, please refer to other sources.

15.1.1 US AN/SPY-1 Series Shipboard Radar

When it comes to the air defence radar of modern combat ships, it must not bypass the famous American AN/SPY-1 series shipboard radar, which cooperates with the formation of the weapon system, there is a more notable name: Aegis (Aegis).AN/

SPY-1 series shipboard radar was initially equipped with Ticonderoga missile cruisers, and then successively equipped with the Ticonderoga missile cruisers. Ticonderoga' guided-missile cruiser, and then equipped aircraft carriers, guided-missile destroyers, etc., and now the largest number of shipboard radar is installed in the "Arleigh Burke" class guided-missile destroyer's main battle shipboard radar. The U.S. Navy not only continues to upgrade the AN/SPY-1 series, but also in the development of more advanced AN/SPY-6 radar, forming the current U.S. shipborne air defence radar of the main equipment map.

Pulling the development of AN/SPY-1 radar is the U.S. Navy for a new surface missile system needs. This missile system needs to search the air defence area, and the key targets (such as aircraft, missiles) to track, guide the weapon system to complete the strike. In the face of this demand, the traditional naval single-sided array rotating radar obviously can not meet, and phased array radar flexible beam control capabilities, a good fit for the weapon system's operational needs, gave birth to the development of the AN/SPY-1 radar.

AN/SPY-1 series radar with full airspace search, automatic target detection, multi-target tracking; for the command and decision-making system to provide detected target data; for the weapon control system to provide target and interceptor missile tracking data, under the control of the weapon control system for the missile to provide mid-range guidance instructions, and to the end of the irradiation radar to deliver pointing data, etc. AN/SPY-1 series radar AN/SPY-1 series radars have AN/SPY AN/SPY-1A, AN/SPY-1B, AN/SPY-1C, AN/SPY-1D and AN/SPY-1E series of products. Which AN/SPY-1 for the test equipment, not installed warships, the subsequent series of products in the framework of the basic products, constantly upgraded. And installed in various types of ships.

AN/SPY-1 series radar is a fixed S-band electronically scanned phased array radar, was a very advanced passive phased array system radar (now look at its technical system has been relatively backward, the latest development of the AN/SPY-6 has been an active phased array system, the reliability is greatly improved).

In the early stage of equipping troops, AN/SPY-1 series radar enjoys world-wide fame. AN/SPY-1A and its improved radar is the most equipped shipboard passive phased array radar, and over the years, through continuous improvement, it has achieved remarkable results in cost reduction, weight reduction, size reduction, improved reliability and availability, and anti-jamming performance, which greatly improves the technical performance and tactical performance of the radar system. The technical and tactical performance of the radar system has been greatly improved.

15.1.2 New SMART Three-Coordinate Radar from the Netherlands

The SMART radar is also well known around the world. SMART (Signaal Muhibeam Acquisition Radar for Targeting) is an acronym for the Multibeam Target Acquisition

15.1 Military Application of Radar in Ocean Engineering

Radar of the Dutch Telecommunications Equipment Company. The radar is an all-weather, shipborne, three-coordinate radar, which is the main detector for information processing and weapon control systems, and is used to equip ships of all classes from small frigates upwards.

The SMART radar is available in three versions. SMART-S (basic) for air-to-air area defence; SMART-X (auxiliary) for point defence systems; and SMART-L (long-range) for long-range search.

In 1981, the development programme for the SMART-S radar system began, and in 1983 a formal agreement was reached between the Dutch telecommunication equipment company (Signaal) and the UK-based MEL to jointly bid to meet the requirements needed to equip the Royal Navy's Type 23 frigates with the new Search and Target Indication Radar system. Until 1985, the SMART-S radar was under development, and between April and June 1990, the radar was fitted to the S-class frigates for sea trials, which included a full functional trial and a test of the system's effectiveness against multiple targets in a marine environment, and in the same year, the SMART radar derivative, the MW.08 C-band medium- and short-range multibeam three-coordinate radar, entered service in Portugal. In 1991, the Netherlands Telecommunications Equipment Company (TELECOM) developed a long-range version of the SMART radar, the SMART-L long-range alert radar system. On 24 July 1991, a contract was signed with TELECOM for the development of the SMART-L radar in the amount of US$25 million. The contract calls for the first of eight SMART-L radars to be delivered in mid-1995 to equip Royal Netherlands Navy test ships. Telecommunications Equipment plans to produce an average of two SMART-L radars per year between 1995 and 2005. In November 1991, Telecommunications Equipment signed a $5.8 million development contract with FEL-TNO (Physics and Electronics Laboratory) for the ARTIST (Advanced Radar Technology for Improved Surveillance and Tracking) research programme. The air defence system resulting from the ARTIST programme will be combined with the SMART-L and SMART-S radars, the STING fire control radar, infra-red sensors and anti-aircraft missiles. 1992 saw the entry into service of the SMART-S radar, and in February 1995 the antenna arrays and various components of the SMART-L radar, which had been completed in pilot production, were transferred to the SMART-S. L radar's antenna array and various electronic equipment and software were assembled into the radar's video processing cabinet. From the second quarter to the end of 1995, the SMART-L radar was subjected to far-field antenna measurement tests and outdoor system performance tests.

The cost of the SMART-S radar was reported in 1994 to be about $10 million per unit; the SMART-L radar probably cost about $12 million.

Since 1985, the Dutch Navy has ordered two SMABT-S radars for the Helmskirk frigate, six SMART-S radars for the Cotton Eyre frigate, and two SMART-S radars for the Karel Dorman frigate. Karel Dorman frigate, and eight SMART-S radars, as well as eight SMART-L radars; Spain ordered four SMART-L and four SMART-S radars for the F-100 frigate; in 1989, Germany ordered four SMART-S radars for its F-123 frigates, and in 1998, four SMART-S radars for its F-124 frigates, and in 1998, four SMART-S radars for its F-124 frigates. F-124 frigate in 1989, and in 1998,

Germany ordered three SMART-L radars for its F-124 frigates, and the Dutch Navy ordered four SMART-L radars to equip its frigates. The radar will form part of the long-range radar component of the German, Dutch and Western Tri-national Frigate Cooperation (TFC) programme's air defence warfare system.

The SMART-S radar is expanded here to give the reader a more perceptive understanding of the relevant content from earlier in the book. The SMART-S radar antenna is mounted on the mast top and is hydraulically stabilised to $\pm 30°$ of transverse and $10°$ of longitudinal swings. It consists of a single-array free-tilt broadband transmitter array and a multi-array (16) stripline receiver array. The ultra-low subflap level single-plane phased array enables the system to perform accurate three-coordinate search and tracking.

Pre-processing of the received signal is first done automatically within this radar antenna array element to ensure high sensitivity, and then the receiver output is fed to a digital beamformer to generate 12 independent elevation beams for Doppler FFT (Fast Fourier Transform) processing and automatic tracking. These 12 stacked beams cover the elevation range of $0°$–$90°$.

The transmitter uses a high-power inter-pulse phase-referenced travelling wave tube amplifier (TWT) in the pulse compression regime, with a peak output power of 145 kW. It has three transmission modes: fixed frequency, frequency shortcut during scanning, and inter-pulse frequency hopping.

The radar receiver uses moving target display (MT1), Doppler FFT and linear video output. The noise figure is 3.0 dB, the bandwidth is 2.2 MHz, and the compressed pulse width is 0.6 microseconds. In addition, two microprocessors are equipped, one for target recording and the other for the terminal computer. In order to reduce the processing load, the moving target display by FFT is used only in areas affected by meteorological clutter. In case of interference, the SMART-S radar automatically changes the pulse repetition frequency (PRF).

The SMART-S radar is an engineering success, and many of its design details are a model for marine engineering radars to follow. From the three defence performance of the hardware to the refinement of the algorithmic model, there are very many desirable features.

15.1.3 German TRS Series Radar

There have been debates on which bands are more suitable for use in marine engineering. The development route of the United States, described earlier, was to choose the S-band. While taking into account the power, it also ensures a certain measurement accuracy.

However, some countries, such as Europe and Japan, have chosen the higher C-band for their shipboard radars. the TRS series radars are the best of the C-band radars.

15.1 Military Application of Radar in Ocean Engineering

TRS-3D series has been equipped with many navies, including multi-functional warships, cruisers, frigates, etc. Its main functions include sea-to-air search, sea-skimming detection, and surface fire control. Its latest TRS-4D active phased array sea-to-air surveillance radar has been delivered to the German Navy with the latest F125 frigate.

The TRS-4D radar has two modes of operation, rotating and quad array, of which the one installed on the German Navy's F125 frigate is in quad array status. The radar adopts the T/R component of Gallium Nitride (GaN) semiconductor technology, with a maximum detection range of 250 km, capable of detecting targets with $\sigma < 0.01$ m^2, and possessing three-dimensional tracking capability for more than 1000 air and sea targets. In the equipment test, TRS-4D demonstrates extremely high test accuracy, especially for small targets such as UAVs, missiles and periscopes.

The radar has the following performance characteristics:

(1) Excellent response time and survivability: fast track initiation and confirmation. Strong detection of air and surface targets with low signal signature. Threat assessment based on unique classification strategies.
(2) Advanced targeting: precise tracking of small, fast-moving targets, countering asymmetric threats. Enhanced threat surveillance for high-priority areas.
(3) Networking Capabilities: Provide newer collaborative capabilities such as unprecedented directed search and tracking.
(4) Expanded situational awareness capability: High tracking update rate to ensure effective self-defence and area defence.
(5) High Operational Reliability: Adoption of high reliability technology with slow degradation of performance.
(6) Designed to meet current and future needs: through flexible software-defined radar management, a wide variety of mission requirements can be met (Fig. 15.1).

C-band is a good choice for ships that need to achieve strong early warning and observation capability on a single radar with a certain degree of detection accuracy. C-band is a better choice, especially for some medium-sized ships that have only one main radar, and C-band has a better institutional advantage as the operating frequency of the main detection equipment.

The current worldwide shipboard radar spectrum is very rich, a comprehensive introduction of its content is not the focus of this book. Here we introduce three famous types of radars from the perspectives of development history, equipment process and main performance. In order to make the readers have a basic knowledge of the military application of radar in marine engineering. To understand this important branch of radar application from a more perceptive point of view. Many important technological breakthroughs in modern radar are achieved in the demand for shipboard radar traction, and its important technical status, it is worth highlighting, but limited to the sensitivity of its content, it is inconvenient to expand too much in this book.

Fig. 15.1 F125 frigate with TRS-4D radar

15.2 Navigation Radar

Navigation radar is the most widely used branch of radar applied in marine engineering. Navigation radar is mainly used for navigation and collision avoidance in the process of ship navigation as well as in and out of harbours, narrow waterways and other areas, especially in bad weather with poor visibility, it is the necessary electronic equipment to ensure the safety of ship navigation, and its performance is good or bad, which is closely related to the safety of ship navigation. At the ship operation level, there are two common avoidance methods using the detection results of navigation radar: distance avoidance line (radar safety distance line) and bearing avoidance line (safety bearing line). Due to the continuous development of water transport, the use of navigation radar is gradually expanding the scope of application in harbours, coasts, river monitoring and management, etc., and the navigation radar needs to be continuously enhanced.

In 1904 the German Christian Hulsmeyer first applied for a single-station pulse radar patents, its marketing for the sea ship collision avoidance radar prototype, in fact, is a kind of navigation radar in a broad sense. For more than a century, marine navigation radar continues to make progress (Table 15.1).

By the 1980s, with the rapid development of military radar technology, civil navigation radar new technologies have been applied to some extent, such as waveguide slit array antennas, solid-state modulators, digital terminal display, digital signal processing, solid-state amplifiers, continuous wave regime, etc. Compared with the rapid development of large-scale phased-array radar for military use, navigation radar's positioning in terms of functionality determines that it will not pursue a significant improvement in performance. After the basic functions are available, there is no need for large-scale iterative updating of its technology. The result is that

15.2 Navigation Radar

Table 15.1 Key historical points in the development of navigation radar

Year	Event
1946	Use of radar to prevent ship collisions
1960	The International Regulations for Preventing Collisions at Sea, 1972, set out requirements for the proper use of radar and for marking and mapping
1972	Merchant ships are required to be equipped with radar
1974	The US Coast Guard issued regulations requiring all vessels entering US waters to be equipped with and use collision avoidance systems
1977	The 1981 Amendment to the International Convention for the Safety of Life at Sea, 1974 (SOLAS) provides for the number of radar and automatic radar plotters to be installed on ships of different tonnages and the dates on which they should be installed
1981	Use of radar to prevent ship collisions

the navigation radar does not follow the main radar of large surface ships in terms of technology, and constantly tries to update new technology. Although some minor modifications have been introduced, such as: combined navigation (access to ship automatic identification equipment, running electronic charts), beam sharpening, dual-range display, etc., none of them involve major technological innovation. On the contrary, the fierce competition in the field of civil navigation radar has promoted the continuous reduction of the cost of navigation radar, and the market price has been constantly down, which, to a certain extent, is another dimension of technological innovation. Nonetheless, navigation radar has revealed some unique technological advances in technology, and is closely integrated with the use of demand. Mainly presented in the following aspects:

(1) detection system to phase reference radar development

Magnetron non-phase reference system radar due to its cost advantage, for quite a long time will also occupy a certain market share of navigation radar. But its performance has been a bottleneck. As the cost of electronic devices continues to dip, the navigation radar of phase reference system gradually begins to show its competitive advantage.

At the technical level, the magnetron non-phase reference system radar has high operating voltage, limited transmit pulse width accuracy, and limited magnetron maintenance. It brings a series of problems such as work stability, reliability, and maintainability. Although large-scale use has resulted in a high degree of technological maturity, the inherent characteristics of these devices have made it difficult to bring about transformational changes through linear technological improvements. In contrast, solid-state devices can use the full-phase reference system, bringing more choices for radar signal processing and data processing, which can effectively improve the overall detection performance of the device. The large-scale application of highly integrated electronic devices, such as DSP and FPGA, in the field of radar, provides a technical solution to the solid-state navigation radar's high-integration, high-performance, high-reliability, and low-cost.

In some higher-end navigation radars, continuous wave operating modes are also integrated. Continuous wave regime navigation radar transmitting power is low, for the need to work with the operator of the common exposure of the working environment, to avoid physical harm to the personnel (although now the industry does not have a recognised radar electromagnetic wave on the human body harm scientific analysis). At the same time, the biggest advantage of the continuous wave system is that its detection blind spot is small. For the narrow waterway barrier navigation has a unique system advantages. Especially in the use of low visibility scenarios, it can provide a strong guarantee for the safe navigation of the ship.

(2) Navigation radar multifunctionality and multi-sensor data fusion

With the progress of electronic technology, the navigation radar is not only the initial barrier function, but also evolved into a radio frequency transceiver platform. Any function that relies on radio frequency transceivers to complete, constantly separated from the device, towards multi-functional integration. For example: most of the modern navigation radar with automatic marking (ARPA) function, complete ship navigation collision avoidance, in the large ships with shipboard helicopters, but also be able to complete the helicopter guidance function; show control interface can be superimposed with the preset charts; detection results can be with the ship automatic identification system AIS (Automation Identification System) target Fusion; large ships multi-radar network detection, and data fusion processing, in order to achieve a dead-angle navigation situational awareness; radar detection and optoelectronic equipment detection results fusion in order to achieve a high degree of credibility, high-value information results display ...... navigation radar is also in the functional boundaries of the continuous expansion of the excavation of its own commercial value, and seeking to system contribution in naval electronic equipment.

(3) Frequency band differential competition

In order to seek different market segments, navigation radar also derives more working frequency bands. In general, most of the navigation radars work in centimetre wave, and the working frequency is mainly S, C, X. In some special application occasions, Ka-band navigation radar also appears. Analysed from the principle, S band below the operating frequency, will bring the resolution of the trouble, and more than Ka band operating frequency, will bring greater atmospheric attenuation, especially the need to navigate to play a role in the visibility of the lower rain, foggy weather, often high-frequency electromagnetic wave attenuation is also more serious, so there is no need to go further to the high-frequency band to extend the operating frequency.

In specific use, medium-sized ships are generally equipped with two navigation radars at the same time, many times in order to make up for each other's advantages and disadvantages of the frequency band, the general choice of dual-band, such as the S-band and X-band with X-band radar has a small antenna size, orientation resolution is good, good detection of targets under the sea clutter, the most widely used shipboard radar band. S and C-band in the rain and fog in the attenuation of reduction, the surface of the sea is suitable for detecting targets in bad weather and sea conditions. S and C bands have reduced attenuation in rain and fog and small reflection in sea surface, which are suitable for detecting targets in bad weather and sea conditions. Millimetre wave (Ka band) radar antenna size is small, high resolution, but there are atmospheric, rain and snow attenuation shortcomings, more suitable for rivers, lakes, or near the coast and other narrow channels, dense ship navigation waterway.

15.2.1 Application Form of Navigation Radar

The most basic application of navigation radar in marine engineering is radar positioning and radar navigation.

Radar positioning is the process of using the bearing and distance information detected by radar, combined with the operation of the object marker and the chart to find the ship's position. To accurately get the radar positioning, but also need to have certain basic operation ability, first of all, need to select the object marker used for positioning, this work itself has a certain technical difficulty, should be combined with the ship's actual position and the object marker itself echo reflection characteristics, to ensure that the echo recognition is correct; Secondly, to the radar to detect the distance and bearing data to use the correct, fast, accurate, because the ship itself in the process of movement, its The spatial position of the ship itself is in the process of movement, its spatial position is in constant change; Finally, it also needs the correct nautical chart operation. In fact, in today's comprehensive application of global positioning system, radar positioning through navigation radar, has been a more traditional means, the convenience of its use can not be compared with the Beidou positioning, GPS positioning, etc., but the more ancient means, often the more reliable, as a kind of ship positioning of the last resort, and often the value of its use in the most extreme cases, and will come to the fore.

Radar navigation that is, in the ship in and out of ports, narrow waterways and coastal navigation, in order to avoid the dangerous points near the route, the use of radar detection information, a reasonable way to plan the navigation route. When using radar navigation, depending on the environment, there are relatively mature risk avoidance operation schemes (such as the distance risk avoidance line and bearing risk avoidance line briefly mentioned above), but in actual operation, it is still necessary to match with navigation experience in order to achieve safe navigation. The detection echo display of navigation radar is often complex, and in relatively dangerous shipping lanes, it is also necessary to carefully study charts, master the hydro-meteorological environment, make good use of the main object markers, and correctly identify buoys and false targets.

In addition to the common ship navigation function, with the innovation of technology, navigation radar is also widely used in port traffic management, ecological environmental protection and security monitoring scanning and other fields.

(1) Ship navigation and port traffic management

From the point of view of individual ships, the most widely used radar function is ship navigation. With the help of navigation radar to detect the hydrological conditions and weather conditions in the sea near the ship, the best travelling line of the ship is planned, so as to improve the navigation efficiency of the ship and avoid the ship from hitting the reef, running aground or encountering storms.

Port traffic management is the most traditional functional application of navigation radar. Foreign countries through the upgrading of navigation radar transmit power, strengthen the target tracking function, the use of large-calibre antennas, additional information input and output interfaces, etc., to effectively achieve the port traffic management. At present, the more common applications abroad are phased array and multi-polarisation system radar, and are widely used in large ports for ship navigation and risk avoidance, entry guidance, heading planning and directional tracking and other fields. In addition, some of the hydrological instability of the port, but also the navigation radar for hydrological observation and flood control alarm.

(2) Eco-environmental protection

As airports, coastal and onshore wind farms, landfills, oil and gas fields, and tailings ponds in mines of all kinds are off-limits to many animals (especially birds of prey), and especially wind farms, the much-criticised socio-ecological disadvantage of green wind energy is that it leads to a large number of deaths of birds of prey. It is estimated that each wind turbine on land alone may kill up to 60 birds per year.

In view of this, from the perspective of the concepts of ecological protection and sustainable development, the use of the navigation radar hardware platform, combined with the flexible signal processing and data processing models at the back end, has equipped the navigation radar with the capability of detecting and tracking various organisms near the aforementioned prohibited zones, as well as collecting and researching information on the activities of birds of prey and protecting migratory birds, resident birds, raptors and bats, thus greatly reducing the mortality rate of birds of prey. In addition, by integrating radar with various acoustic, optical and other deterrent equipment, and system control equipment at airports or wind farms, timely warnings can be issued to drive away birds approaching the danger, thus achieving the purpose of protecting birds of prey. For example, in the wind turbine area of a wind power plant, the coverage of navigation radar signals is realised, so that when a flying bird approaches, it can be scared away by releasing noise or, through sensors, stop the rotation of the fan pages of the wind turbine. In addition, it can also take advantage of big data technology, through the long-term monitoring of navigation radar, to analyse the movement patterns of organisms, and formulate targeted protection countermeasures, so as to achieve a win–win situation in terms of ecological environment and economic benefits.

15.2 Navigation Radar

(3) Safety Monitoring and Scanning

In the field of ship safety monitoring and scanning, a broadband linear FM continuous wave radar is used more often. The advantage of this kind of radar is that the resolution is very high, and the volume of a single radar is not big and the weight is light. Through the group network monitoring of multiple radars, it is possible to achieve 360° dead-angle-free monitoring of large ships, effectively preventing attacks by terrorists or pirates.

In the emerging surface unmanned boat equipment, often navigation radar is necessary radio frequency detection equipment, its all-day, all-weather target detection capability, is to ensure that the unmanned boat safe navigation is an important auxiliary means, the typical unmanned boat at home and abroad, most of them are equipped with a variety of specifications of the navigation radar. As a radio frequency detection platform, the navigation radar of unmanned boat is also developing in the direction of multi-functionality. In the foreseeable future, the navigation radar of unmanned boat on the water surface will develop in the direction of integrated radio frequency electronic equipment, and will gradually expand the hydro-meteorological data detection and collection capability, communication capability, electromagnetic environment perception and directional jamming capability, etc., in addition to the basic navigational and navigational functions, and will continue to improve the level of artificial intelligence to lead the way. The level of artificial intelligence will be continuously improved, leading the continuous development of autonomous navigation planning technology in the field of marine use.

15.2.2 Domestic and Foreign Navigation Radar Development Power

As of around 2020, the global ocean-going navigation radar market is about 25,000 sets per year; for other merchant ships, fishing vessels and yachts, only the U.S. and the Far East have a radar market of about 30,000 sets in the S-band (3 GHz) and about 800,000 sets in the X-band (9 GHz). High-end marine radar products in foreign markets are often configured as X, S dual-band, such as the U.S. Northrop Grumman's Bridge-Master E340 radar system, Kevin body of the company's SharpEye series of radar systems, Denmark's Scanter radar, the Japan Wireless Corporation's JMA-9100 and JMA-9900 series, Germany's Atlas JMA-9100 and JMA-9900 series from Japan Wireless, and the 9500–9800 ARPA series from Germany, all of which are heavily fitted to large cruise ships and cargo ships. These manufacturers are firmly in control of the market share, while at the same time in the development of technology has been at the forefront of the world. Navigation radar is not an absolute high-tech electronic equipment platform, but its high reliability, strong environmental adaptability, low cost and other requirements, there is still a certain technical threshold. In particular, the detection of small targets under the background of sea clutter, high sea conditions under the detection of adaptability, etc., the need for a large amount of data accumulation and excellent software processing model, these aspects of the cutting-edge technology, basically by the top of the foreign research and development of the manufacturers to control.

Compared with foreign countries, the domestic navigation radar development started late, before the 1950s, China's navigation radar mainly rely on imports. Subsequent product development, but mainly foreign capital, joint ventures and factories. The industry also has a few adhere to the introduction of technology and independent research and development to the combination of research and development units, including the Shanghai Radio and Television Kaige Communications Radar Equipment Factory, Guangzhou Haihua, Zhisen Navigation Electronic Technology Co. In recent years, there are also emerging forces of navigation radar, such as HELANSON, Beijing Radio Measurement Research Institute, etc. It is believed that the joining of these relatively more technically strong research and development manufacturers can promote the better development of navigation radar industry and technology in China.

15.3 Synthetic Aperture Radar

Synthetic Aperture Radar (SAR) is also heavily used in ocean engineering. In a sense, the landmark event that initiated human research and use of SAR originated from the ocean. In 1987, the Jet Propulsion Laboratory (JPL) of the National Aeronautics and Space Administration (NASA) of the United States of America launched the world's first oceanic satellite carrying SAR (Seasat-1), and SAR became inseparably linked with the ocean. And it has made great progress and technological breakthroughs since then.

In the field of ocean engineering, according to the different loading platforms of SAR, it can be divided into star-carried SAR and airborne SAR, among which airborne SAR can be subdivided into manned SAR and unmanned aircraft SAR.

Star-carried SAR orbit depth is high, can be from hundreds of kilometres to tens of thousands of kilometres, high orbit brings a wide field of view, you can once get a wide range of observation images, relative to optical equipment, SAR penetration is better, can be in the cloud layer is relatively thick scene, get a certain observation ability, and electromagnetic wave some unique properties, can also support the study of some special scenes. At the same time, because the satellite is not subject to the restrictions of airspace sovereignty, can break the national boundaries of the restrictions on the formation of sensitive areas of observation capabilities, access to strategic deep sea field of observation data. This is the advantage of satellite-borne SAR from the point of view of use. From the technical aspect, the satellite platform running orbit is relatively stable, more conducive to the application of SAR, smooth running trajectory can greatly simplify the back-end processing of the amount of

15.3 Synthetic Aperture Radar

computation, simplify the satellite's burden of bearing, but also more conducive to obtain high-quality observation images.

The platform manoeuvrability of airborne SAR is stronger, and it can be used for high-resolution imaging in some specific areas. Since the airborne platform is not as demanding in terms of load constraints compared to the satellite-borne platform, relatively more complex detection systems can be developed. The airborne SAR platform extends new technology branches and application modes such as SAR imaging, SAR/MTI, interferometric SAR, polarised SAR and so on. And due to the airborne SAR flight altitude can be controlled, electromagnetic wave atmospheric attenuation factors can be based on the use of scenarios to consider, airborne SAR working frequency band can be extended to millimeter wave, the country has a mature Ka-band SAR products, the upward expansion of the frequency band brings higher resolution and the extraction of fine features of the target, but also for the expansion of application scenarios to provide the possibility. Among them, unmanned airborne SAR is an important development direction for airborne SAR in the future due to the good economy of the platform, which can be extended on time and perform tasks under high manoeuvrability. Meanwhile, due to the unmanned operation, it puts forward new requirements on the reliability and automation capability of the SAR load itself, and the volume and weight constraints of the load by the UAV itself also put forward the technological development direction of the high integration of the unmanned airborne SAR.

Considering from the application scenario, SAR plays an important role in ocean engineering. From the perspective of military application, SAR is an important means of reconnaissance for air- and space-based platforms due to its all-day, all-weather operating characteristics. Especially because star-carried SAR can break through the limit of national boundaries, it is an important means of military target information acquisition and irreplaceable sensors. SAR images are also very sensitive to changes in the sea surface structure, which can be combined with the corresponding background knowledge of the operational environment of the marine meteorological and hydrological information to perceive. SAR can also make use of the electromagnetic wave on the geometry of the waves, the roughness of the sea surface and other sensitivities, the ocean characteristics of the qualitative SAR can also use electromagnetic wave sensitivity to wave geometry and sea surface roughness to qualitatively identify ocean characteristics, and combine meteorological and hydrological background knowledge to carry out large-scale ocean feature identification. Some high-precision SAR images can even make use of the perturbation of sea surface characteristics by underwater moving targets, which in turn leads to changes in the SAR image to achieve the detection of underwater targets. In the field of civil applications, SAR has been widely used in marine monitoring. Including sea surface oil spill detection, sea ice situation assessment, seaweed distribution situation control. For example, the sea surface oil spill is the use of oil film cover will lead to the sea surface on the electromagnetic wave backward scattering characteristics change, and achieve the effect of detection. In addition to this specific situation detection, SAR is also used in large-scale accumulation of marine data for a long time, continuous monitoring of sea ice changes, large-scale monitoring of marine transport conditions, regular monitoring of the marine climate, and so on, and even a lot of major climate change prediction, but also a large number of use of SAR images to control the information.

References

1. Ding, Lufei, Fulu Geng, and Jianchun Chen. 2013. *Radar Principles*, 4th ed. Beijing: Electronic Industry Press.
2. Zhang, Guangyi. 1994, 1996, 2000, 2001. Phased Array Radar System. Beijing: National Defence Industry Press.
3. Ding, Lufei, and Fulu Geng. 2002. Radar Principles (3rd ed.). Xi'an: Xi'an Electronic Industry Press.
4. Jaguo, Lu. 2017. *Synthetic Aperture Radar Design Technology*. Beijing: National Defence Industry Press.
5. Guangyi, Zhang, and Zhao Yujie. 2010. *Phased Array Radar Technology*. Beijing: Electronic Industry Press.
6. Jian, Yang, and Yin Junjun. 2020. *Polarisation Radar Theory and Remote Sensing Applications*. Beijing: Science Press.
7. Wang, S.-Y. 2014. *Marine Navigation Radar*. Dalian: Dalian Maritime University Press.
8. Lufei, Ding, Geng Fulu, and Chen Jianchun. 2014. *Radar Principles (5th Edition)*. Beijing: Electronic Industry Press.
9. Yiyao, Dai, Wang Xuesong, Xie Hong, and Xiao Shunping. 2015. *Airspace Polarisation Characteristics of Radar Antenna and Its Application*. Beijing: National Defence Industry Press.
10. Stimson, George, W. 2005. *Introduction to Airborne Radar*. Translated by H. P. Wu, et al. Beijing: Electronic Industry Press.
11. Mahafza, Bassem R. 2016. *Radar System Analysis and Design (MATLAB Edition)* 3rd ed. Translated by Wanxing, Zhou, Mingchun Hu, Mingya Wu, Jun Sun, et al. Beijing: Electronic Industry Press, 2016.
12. Barton, David K. 2012. Radar System Analysis and Modelling. Translated by Nanjing Institute of Electronic Technology. Beijing: Electronic Industry Press.
13. Merrill I Skolnki. Radar Handbook, 2nd ed. Beijing: Electronic Industry Press, 2003.
14. Merrill I Skolnki. Ship Navigation Radar: Volume I. Beijing: People's Traffic Press, 1987.
15. Zhang, Runze. 1987. *Ship Navigation Radar: Book two*. Beijing: People's Transportation Press.
16. Zhang, R.-Z. 1990. Ship Navigation Radar: Volume III. Ship navigation radar: book three. Beijing: People's Transportation Press.
17. (U.S.) Jerry L. Ivors et al. 1991. Principles of modern radar. Beijing: Electronic Industry Press.
18. Skolnik, M. I. 2001. *Introduction to Radar Systems*, 3rd ed. New York: McGraw-Hill.
19. Zhang, G. Y., D. C. Wang, H. G. Hua, et al. 2001. *Phased Array Radar for Space Detection*. Beijing: Science Press.
20. Jingcheng, Xiang, and Zhang Mingyou. 2005. *Millimetre Wave Radar and Its Applications*. Beijing: National Defence Industry Press.
21. Zhizhong, Zhang. 2004. *Introduction to Airborne and Satellite-Borne Synthetic Aperture Radar*. Beijing: Electronic Industry Press.
22. Zhang, M.Y., Lv, M. 2004. *Signal Detection and Estimation*. 2nd ed. Beijing: Electronic Industry Press.
23. 2005. World Ground Radar Handbook. Beijing: National Defence Industry Press.
24. 2004. Airborne Radar Handbook. Beijing: National Defence Industry Press.
25. Mingyou, Zhang, and Wang Xuegang. 2006. *Radar Systems*, 2nd ed. Beijing: Electronic Industry Press.

26. You, He, Jian Juan, Zhang Jingwei, et al. 2006. *Radar Data Processor Applications.* Beijing: Electronic Industry Press.
27. You, He. 2010. *Multi-sensor Information Fusion and Applications.* Beijing: Electronic Industry Press.
28. Li, Li.-Na. 2000. *Marine Automation.* Beijing: People's Transportation Press.
29. Zhang, Xinggu. 2010. Marine Radar. Dalian Maritime University Press.
30. Xiaoniu, Yang, et al. 2001. *Principles and Applications of Software Radio.* Beijing: Electronic Industry Press.

Part V
Communications

Chapter 16
Overview of Communications Technology

16.1 Development History of Maritime Communication Technology

Communications technology refers to the methods and measures used to transmit information from one location to another. Communication technology is an extremely important component of electronic technology. According to the order of historical development, communication technology successively from the human body to transfer information communication to simple signal communication, in the development of wired communication and wireless communication [1].

Maritime communication as an important part of communication technology, if the means of communication used to distinguish, can be roughly divided into the following three stages.

1. Communication in ancient naval warfare

Until the nineteenth century, the navy ship communication successively used the hand signal (later develops for the hand flag signal), the wolf smoke, the five-colour flag, the flag tassel and the pyrotechnics and the rocket. In addition, also use some audio tools to transmit a variety of combat orders and coordination information.

2. Ship communication in the radio era

During the Russo-Japanese War, radio waves were used for the first time in fleet communications 10 years after the invention of radio.

3. Computer and satellite communications era of naval communications in order to change the naval fleet communications passive and difficult situation, the late 1950s, the U.S. Navy took the lead in the use of the Naval Tactical Data System (NTDS) for shipboard computers between the data communications. The data chain mainly used short-wave bands because of the long distance required for communication. However, because of the inherent weaknesses of short-wave

circuits, the data chain gradually shifted to the use of UHF satellite channels for a considerable period of time before many new short-wave communication technologies were developed and after the emergence of satellite communication means.

It is difficult to ensure fast, accurate, confidential and uninterrupted military communications by relying only on a single mode and means of communication. After a long period of research and development and use in war. Modern naval communications have included almost all means of communication, the use of electromagnetic waves has covered the entire electromagnetic spectrum from long waves to light waves [2].

16.2 Development Status of Foreign Maritime Communication Technology

Maritime communications can be divided into analogue and digital communications according to the technical system; radio and wireline communications according to the transmission medium; optical fibre communications, satellite communications and mobile communications according to the transmission technology; and long-wave, short-wave, ultra-short-wave and microwave communications according to the operating frequency.

16.2.1 Current Status of Foreign Maritime Shortwave/ultra Shortwave Communications

Short-wave communications in long-distance communications has always occupied an important position, is considered an effective and economical means of long-range military communications, widely used in military strategic and tactical communications.

At present, the widely researched and applied short-wave digital transmission systems and equipment, generally using the channel bandwidth in the 300–3000 Hz band range of 'narrow-band' system. Due to the limitations of the bandwidth and short-wave time-varying fading channel, even if a variety of adaptive techniques to improve the performance of short-wave data transmission, the practical data transmission rate can only be up to 2400 b/s, the highest 4800 b/s, and the ability to resist man-made interference is extremely poor.

In order to improve the anti-jamming ability of the system, the spread spectrum (frequency hopping or direct sequence spread spectrum) technology is usually used. However, narrowband frequency hopping is difficult to escape the enemy frequency tracking jammer interference. And in the frequency hopping operating conditions (especially when the hopping speed is higher), to achieve high-speed data transmission is more difficult, because the frequency hopping radio frequency conversion and synchronisation to take up a considerable part of the time, thus reducing the time to transmit data; and there are always a part of the many frequency hopping channels in poor conditions, or even serious interference can not be used, so it is difficult to ensure the necessary reliability of the transmission.

It can be seen that the current short-wave data transmission system is still difficult to adapt to the requirements of future battlefield electronic countermeasures. Therefore, it is hoped to seek a new technical way that can both improve the data transmission rate and transmission reliability, as well as enhance the anti-jamming ability [2]. Research has shown that the use of broadband fast frequency hopping data transmission technology, can achieve the above purposes.

In recent years, foreign countries have developed a new type of spread spectrum communication system, whose frequency hopping bandwidth is more than 1.5 MHz, and the frequency hopping rate is as high as 2560 hops/s, or even 5000 hops/s. This type of system has unique characteristics of data transmission mode and signal waveform, which organically matches with the fast frequency hopping rate, and obtains a high data rate. Reliable communication capability can be provided on skywave channels affected by fading and multipath effects.

Since the 1980s, foreign armies have also developed a series of airborne short-wave adaptive frequency hopping radios. Frequency hopping rate of 5–50 jump/s, frequency hopping bandwidth is generally 50–200 kHz. the United States ARC-171 (V), ARC-182 (V) and so on are the new type of airborne short-wave radio with frequency adaptive and frequency hopping anti-jamming function.

It can be seen that the broadband fast frequency hopping data transmission technology can improve the transmission rate, transmission reliability, but also has a strong anti-jamming ability; can be used for military short-wave data transmission, can better support the data voice, fax, static images and computer data security communication services, so that short-wave communications on a new level.

16.2.2 Current Status of Foreign Digital Microwave Communications

Microwave communications usually have terrestrial microwave relay communications, one-point-to-multipoint microwave communications, satellite communications, microwave scattering communications, and meteor aftermath communications.

Figure 16.1 is a schematic diagram of a microwave relay communication line, the trunk line can be up to several hundred kilometres or even thousands of kilometres long, and there can be more than one branch line. In addition to setting up microwave terminal stations at the end of the line, a number of microwave relay stations and microwave splitter stations are also set up in the middle of the line at certain intervals. The main features of microwave communication: wide frequency band, large communication capacity; high transmission quality, stable and reliable communication; high antenna gain, good confidentiality; long-distance communication on the

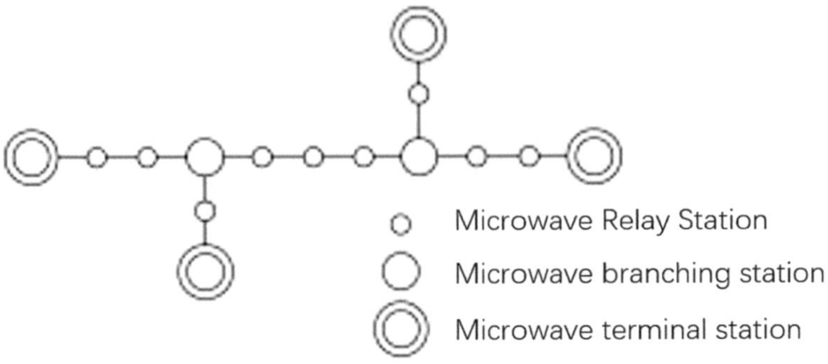

Fig. 16.1 Schematic diagram of a microwave relay communication line

ground can be used in the way of 'relay'; convenient and flexible, low cost, and easy to form a comprehensive service digital network.

In addition, SDH microwave communication is a new generation of digital microwave transmission system, which has the advantages of both SDH digital communication and microwave communication, and has significant advantages compared with the traditional PDH microwave communication, which adopts a variety of key technologies, improves the system's anti-jamming and anti-decaying performance, meets the needs of the new communication services, and the system has strong management capability, which is the new development direction of the transmission network.

At present, digital microwave in the communication system is mainly used for trunk fibre-optic transmission system in the event of natural disasters in the emergency repair, as well as for various reasons is not suitable for the use of fibre-optic lots and occasions. Short-distance branch line connection within the city, such as mobile communication base station, base station controller and base station interconnection, wireless networking between local area networks, etc., both small and medium-capacity point-to-point microwave can be used, but also without the need to apply for the frequency of the microwave digital spread spectrum system; the future of broadband service access (such as LMDS); wireless microwave access technology.

16.2.3 Status of Foreign Maritime Satellite Communications

Inherent weaknesses of short-wave circuits: the ionosphere is a time-varying dispersion channel, its transmission characteristics vary randomly with different seasons and day and night, fading is serious; its passband is much narrower than the microwave and ultrashort-wave, and can not transmit television or high-speed data; susceptible to ionospheric harassment and the impact of nuclear explosions at high altitude; and due to the transmission of directionality of the weak and easy to be intercepted by the enemy eavesdropping and interception, etc. [3]. Therefore, before the development of many new short-wave communication technologies and after the emergence of satellite communication means, the data chain gradually shifted to the use of UHF satellite communications for a long period of time.

16.3 Domestic Maritime Communication Technology Development Status

Since the twentieth century, inland urban communications have undergone rapid changes from cable to fibre optics and analogue to digital. However, due to the complexity and diversity of the water environment and the difficulty of building base stations on the water, the development of ship communication is significantly lagging behind that of inland urban communication. Relative to inland city communications, ship communications mainly have the following characteristics: ships are a long-term high-speed movement of the carrier, and at the same time, due to the water surface of high temperature, humidity, salt spray and other harsh conditions, can not be and inland cities as large-scale laying of communication cables. Therefore, ships often can only keep in touch with the shore base by means of wireless communication; the environment on the water surface is more complicated compared to inland cities, so the reliability and anti-interference requirements for communication are higher.

At present, China's ship wireless communication methods are mainly maritime satellite, Tiantong 1 satellite, 4G communication system, Beidou satellite navigation system and so on[4].

1. Maritime Satellite

Maritime satellite is a communication satellite used for radio contact between sea and land, which is a practical high-tech product integrating global maritime routine communication, distress and safety communication, special and war preparation communication. Maritime satellite can be customised with different business standards according to the communication needs of users, and it is the most advanced emergency communication system in the world at present. For large civil ocean-going ships, maritime satellite is the preferred communication method. The advantages of maritime satellite are as follows: maritime satellite has no communication blind area, and the coverage of its communication network includes almost all regions on the earth except the poles; the communication is stable and reliable, and since most of the communication medium of maritime satellite is the cosmic vacuum above the earth's atmosphere, it will not be interfered by the complex environment on the ground.

However, the key technology of maritime satellite is in the hands of the developed countries such as Britain, the United States and France, which poses a certain threat to China's information security; moreover, the bandwidth of maritime satellite is relatively small, which can't satisfy the intelligent robots of unmanned boats that have a large amount of data and a high degree of real-time interaction with the information of the shore base, and at the same time, due to the limitation of the volume and load capacity of the unmanned boats and power supply capacity, it's usually difficult to install this kind of large-scale carrier-type satellite communication terminal.

2. Tiantong-1 Satellite

The Tiantong-1 satellite communication system belongs to the geosynchronous geostationary orbit (GEO) satellite communication system, which consists of a space segment, a ground segment and a user segment. The space segment is planned to be composed of several GEO mobile communication satellites, and it is expected to become the second largest global satellite communication system after the maritime system. Its main features are as follows: safe and reliable, the Tiantong-1 satellite communication system adopts China's independently researched and developed satellite networks, system platforms, chip modules and communication terminals, with military-grade confidentiality protection capability; feature-rich, in addition to basic voice text messaging and network access functions, it also has additional functions such as terminal position tracking and data return; rich terminal types, in addition to conventional handheld and portable terminals, there are also terminals specially designed for shipboards and other mobile devices. In addition to conventional handheld and portable terminals, there are also carrier-type terminals specially designed for ships; breaking the monopoly of foreign advanced technology and enriching the means of communication for ships.

However, the highest bandwidth of TDT-1 satellite is only 384kbps, and the data fee of TDT is as high as 16 RMB/MB. For small unmanned boats on the water surface, TDT-1 satellite is not suitable for communication between unmanned boats and the shore base, no matter from the point of view of real-time data or communication cost.

3. 4G Communication System

In comparison with long-distance voyage, ships sailing near coastline and inland waterways are usually covered by the base station signals of the three major operators, and high-quality public mobile communication network can meet the communication requirements of near-shore ships.4G communication, as the mainstream representative of current public mobile communication network, provides faster transmission rate and lower cost compared with the previous 2G and 3G. The use of 4G communication in the navigation process near the shore of the ship has the advantages of low construction cost, wide coverage area and high communication rate. For information interaction of the same data length, 4G communication takes short transmission time, low power consumption and high reliability, and the shore base can better interact with the unmanned boat. It can be seen that the use of 4G communication in the unmanned boat near-shore navigation not only guarantees the data quality and improves the communication efficiency, which greatly reduces the cost of communication. However, the water surface does not have the same large area covered with service base stations of operators as inland cities.

The transmission distance of 4G communication system on the water is limited, usually the navigation area of unmanned boats is wide, and 4G communication system alone cannot meet the needs of unmanned boats when they are sailing far away from the shore.

4. BeiDou Satellite Navigation System

BeiDou short message is a unique function of BeiDou satellite navigation system, which is mainly used for the transmission of ship monitoring data in the field of ship communication. The advantages of BeiDou short message communication are as follows: fast response speed, low communication delay, strong anti-jamming ability; high confidentiality, large coverage area, the BeiDou satellite navigation system has been built to cover the entire Asia–Pacific region, as long as the BeiDou satellites can be covered by the scope of the BeiDou short message can be through the BeiDou short message for data communication. BeiDou short message communication technology can effectively solve the communication problem of unmanned boats when they are sailing on the far shore, realise the remote information interaction between unmanned boats and the shore base, and ensure the safety of unmanned boats' autonomous navigation. With the continuous improvement and perfection of BeiDou satellite navigation system, its communication range and message transmission rate will be able to better meet the communication needs of unmanned boats. However, BeiDou short message communication has its limitations in the application field of unmanned boat: the single communication capacity is limited, and the general civil BeiDou's communication capacity is only about 70 bytes; the communication frequency is limited, and the frequency of civil BeiDou's communication is about 1 min; it is unable to transmit pictures and videos; there is a certain packet-loss rate in the short message; and there is no communication acknowledgement, so that the reliable communication needs to take relevant auxiliary measures.

References

1. Xiaojin, Pu, and Si Zhigang. 2003. *An Overview of E-Commerce*. Beijing: Machinery Industry Press.
2. Wu, Bin, and Liang Wu. 2018. Status Quo and Development of Foreign Maritime Wireless Communication. *Ship Electronic Engineering* 4.
3. Wang, Ling, and Binxiang Zhang. 2016. Ship Communication Navigation Technology and Development Trend. *Ship Electronic Engineering* 36(3):17–21.
4. Fan, Qian. 2020. *Surface Small Unmanned Boat Communication Navigation System*. Hainan: Hainan University.

Chapter 17
Principles of Communications Technology

17.1 Airborne Platform Communications

Airborne platform (or elevated platform) communication refers to the use of various near-Earth spacecraft loaded with radiocommunication equipment or systems to achieve antenna electric communication between ground stations (stations) or between networks. It is a means of over-the-horizon, large-area communications and emergency communications, and actually converts over-the-horizon communications into over-the-horizon communications by means of elevated antennas, generally working in the VHF, UHF and microwave frequency bands. Although the construction and maintenance of airborne communication platforms are more complicated, they are smaller in investment, lower in cost and more flexible than satellite communication systems.

Airborne platform communications can be used at the tactical, operational, and strategic levels to form a three-dimensional communications network that integrates air and land, as shown in Fig. 17.1. Tactical communication layer of the air platform communication system is mainly loaded with small-capacity communication relay equipment, deployed to the brigade below, battalion above the combat team, the communication radius of 50–200 km; battle communication layer loaded with multipurpose communication relay equipment or airborne communication nodes, the radius of coverage of 200–300 km, mainly deployed in the army below the division, brigade combat troops; strategic communication layer assembly of airborne communication nodes, adaptive joint C4IS communication nodes and airborne communication nodes. The strategic communication layer is equipped with airborne communication nodes and adaptive joint C4ISR nodes, covering a radius of 300–500 km, and is mainly deployed in the echelons of the army and above.

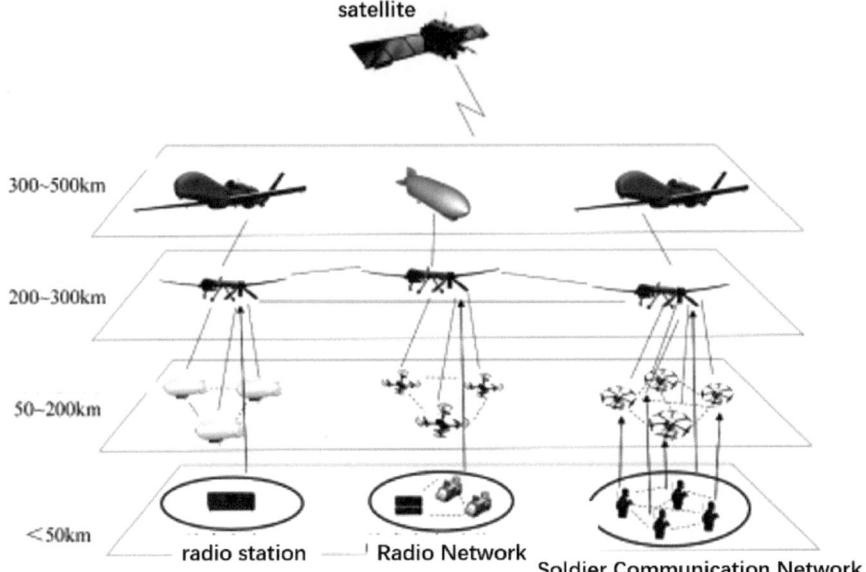

Fig. 17.1 Comprehensive network composition of air platform communication system

17.1.1 Basic Principle of Air Platform Communication

The air platform communication takes the lifting carrier as the platform, and puts the radio communication equipment in the carrier for relaying, forwarding or exchanging radio signals, and carries out communication between multiple ground stations. The higher the air platform rises, it means that the propagation of radio waves by the influence of the terrain is smaller, the larger the communication coverage area, can use the air platform to transfer the letter of the distance between the two stations is also farther. The relationship between platform height and communication distance is approximated by the formula

$$R = 3.57 \quad (17.1)$$

where R is the radius of the coverage area (km); h is the height of the overhead antenna (m).

For example, if the antenna of a relay repeater station is elevated from 15 to 1000 m by using an airborne platform, the radius of the retransmission coverage is extended from 13.8 to 113.8 km.

17.1.2 Main Features of Air Platform Communication

1. Wide week communication coverage.

The air platform greatly increases the effective height of the antenna, reduces the influence of the curvature of the earth on the apparent propagation of the radio wave, makes the transmission path open and close to the propagation conditions of free space. Increase the communication distance, not only to meet the communication requirements of large coverage, but also to facilitate the formation of networks, mobile and flexible; adjust the altitude of the relay repeater platform, the communication distance can reach dozens of kilometres to hundreds of kilometres. In the mountain jungle, scattered islands, desert areas, large areas of water and other special environments, the air platform communication can play its special role.

2. Modular system architecture.

Air platform communication system is a channelised, modular structure of the open platform, according to different uses to change the corresponding module, in the aircraft loaded with different functions of the communication module, it can be composed of different uses of the air platform communication system, in order to achieve the joint coordination of air and ground communication systems.

3. Flexible network configuration.

The air platform communication can adapt to the relay and forwarding of different systems, and can be used as a point-to-point relay station, a base station of the communication network, and also as a broadcasting central forwarding station. As long as the communication equipment or system of the air platform and ground communication equipment or system of the same system, can be relayed, and allows several different systems to work independently at the same time.

4. Complex electromagnetic environment.

Because of the limited space of the air platform, more antennas and the distance between them is close, the transmission power is large and the reception sensitivity is high, the working frequency band is wide and the electromagnetic signal is dense, there are not only all kinds of electronic and electrical equipment inside the platform to produce high and low frequency interference, but also susceptible to the external thunder and lightning, and other natural interference and interception and interference from the enemy.

5. Unattended working mode.

Regardless of whether the aircraft is unmanned or manned, the control and management of communication equipment on the platform requires no human intervention, through the ground command on the platform equipment for remote control operations, unified network management by the ground system.

6. Harsh working environment.

Compared with the ground equipment, the communication equipment mounted on the platform requires higher reliability, vibration-proof, shock-resistant, humidity-resistant, high and low temperature-resistant, lightning-proof; communication equipment should be durable, highly reliable, small size, light weight, low power consumption, good interchangeability, and convenient maintenance.

17.1.3 Airborne Platform Carriers

There are two types of near-Earth space vehicles (NESVs) for air platform carriers: fixed-wing aircraft, helicopters, etc., which are called 'heavier-than-air vehicles'; and airships, balloons, etc., which are called 'lighter-than-air vehicles'. The following are four types of tethered balloons, manned aircraft, drones and airships, which have become important carriers of relay communications for modern air platforms.

1. Tethered balloon platform

Tethered balloon is the use of cables to tie the balloon on the ground device of an aircraft without power drive, the composition of the package balloon, tethered cable, throw shop facilities, working gas, data telemetry telemetry remote control system, logistical supply part of the application of the field is very much, it is a very important carrier of the air communication platform.

(1) Balloon. It mainly includes balloon shell, gas chamber, windshield, balloon auxiliary device, balloon control device. The balloon shell is made of polymer multilayer composite material, which has good strength/quality ratio, anti-ultraviolet radiation, weather resistance, abrasion resistance, high and low temperature resistance, corrosion resistance, not easy to age, good sealing and other excellent properties. The shape is designed as streamlined, with a cross-shaped, Y-shaped or inverted Y-shaped tail to reduce wind resistance. The upper airbag is filled with emotional gas to provide lift; the lower secondary airbag is filled with air to regulate the air in or out to keep the pressure inside the balloon greater than the surrounding air pressure to ensure that the balloon will not be deformed when it is windy. The effective negative nucleus is placed in the lower part of the balloon, and there is a windshield outside to protect the electronic equipment carried.
(2) Tethered cable. Cable tensile strength, light weight, good conductivity of the exterior. Cable in the power supply cable, optical fibre to select information, guide the leakage of lightning metal mesh, the outer layer has a sheath. Tethering cable can be attached to the VHF / UHF communication system antenna.
(3) Tethering and mooring facilities. It includes a mooring tower, the main winch for balloon retrieval and release, three or four small rope winch rotating base, horizontal trusses, and the main winch for balloon retrieval and release.
    rotating base, horizontal truss and operation console.

17.1 Airborne Platform Communications

(4) Gas supply facilities. Provide the balloon with floating gas (usually helium) on board.
(5) Data telemetry telecontrol system.
  The balloon's working status, environmental parameters, such as altitude, wind speed, temperature, pressure, valves and blower working status, as well as the balloon's pitch angle, roll and direction of the data transmitted to the ground control terminal, and control information to the balloon.
(6) Supply guarantee. It includes power supply, gas supply, weather forecast, system maintenance and life supply.

Tethered balloon has all-weather working ability, not easy to be detected by radar and infrared detector, also not easy to be detected visually. Even if the balloon is penetrated by bullets, due to the small pressure difference between the inside and outside of the balloon, the nitrogen escapes very slowly, the rupture will not be torn and expanded, and can be maintained in the air for several hours. If a special adhesive is used, it can be repaired quickly. Balloon in the air can continue to work 15 days-30 days, life will be more than 10 years. Most small tethered balloon systems are vehicle-mounted, usually on two flatbed trailers, and can be moved hundreds of kilometres in a relatively short period of time, and can also be towed by bulldozers or helicopters. Generally speaking, only 5 people are needed to operate the balloon, and the system can be put into work within 6 h, with the speed of 6–240 m/min, and the preparation of inflation is about 2 h.

The main performance indexes of the tethered balloon system (typical values) are: total length 19.74 m; total width 8.236 m; total height 9.658 m; balloon volume 450 m^3; working height 800 m; payload 100 kg; working time 14 days.

2. UAV platform

In recent years, the development and application of military unmanned aerial vehicles (UAVs) have gained momentum. By carrying different equipment UAV can perform communication relay, reconnaissance and surveillance, ground attack, electronic jamming, target positioning, attack damage, effective assessment and other tasks. Drones can be divided into two categories: tactical drones and drones with endurance.

The UAV itself has an on-board ground system to ensure sustained flight through the air at a minimum radius as required.

(1) UAS

1) UAS Composition

It is generally divided into two major parts: the air unit and the ground unit.

The air unit mainly refers to the body of the UAV and the automatic navigation system, and the more advanced UAVs are also equipped with automatic stabilisation system, so that the UAV can fly independently and stably at a fixed altitude, as shown in the typical block diagram in Fig. 17.2 (taking AW-4 as an example).

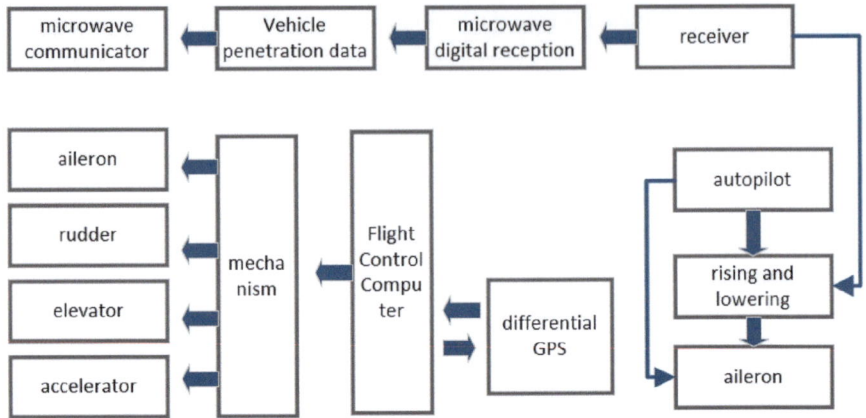

Fig. 17.2 Block diagram of UAV airborne equipment principle

Ground station unit consists of navigation equipment, microwave receiver, wireless MODEM, remote control transmitter, navigation computer, etc. Typical functional block diagram is shown in Fig. 17.3 (taking AW-4 as an example).

The relay communication equipment as the communication load will be introduced in the following section.

2) Usage of UAV

The UAV can be used in a catapult launch mode or in a skidding mode. The catapult launching method, i.e. taking off with solid rocket booster on the launcher of the

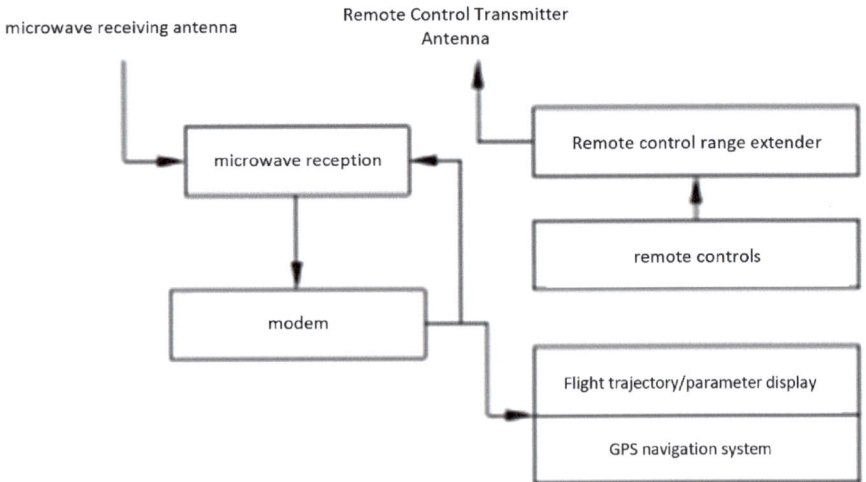

Fig. 17.3 Functional block diagram of the UAV ground subsystem

17.1 Airborne Platform Communications

aircraft launching vehicle and recovering with parachute, does not need a special take-off and landing runway, and is suitable for use in the field conditions. The UAV can be used many times.

(2) Advantages of UAV

1) Low investment, short building time, less maintenance.
2) Low cost and good benefit. There is no pilot on board, eliminating the need for air-conditioning, oxygen, ejection chairs and other facilities for aircrews. The safety factor is higher than that of manned aircraft, glass fibre, plastic and other lightweight materials can be used, low flight resistance, low fuel consumption.
3) Strong survivability. Electromagnetic response area is small, infrared radiation, sound, visual and other characteristics are not obvious.
4) Mobile and flexible. Can take off from trucks or warships and recover with parachutes or recovery nets.
5) Tactical potential. Wide flight territory, can extend the vertical and horizontal distance of the battlefield.

3) Typical performance indicators

Practical ceiling 4000 m Task load 12 kg

Maximum endurance 4 h Launching mode: skidding, catapulting.

Maximum remote control distance 150 km Recovery mode Skidding, parachuting

Maximum take-off mass 30kg.

3. Manned aircraft platform

Manned aircraft platform is mainly helicopters. Helicopters are divided into water, land and amphibious three types. Helicopters are equipped with communication, navigation, need to reach, meteorological and other avionics and instrumentation equipment, the requirements for airports are not high, good mobility.

Typical helicopter performance parameters are as follows.

Practical ceiling 4000–6000 m

Maximum endurance 4 h

Load mass of several hundred kilograms

The helicopter's lifting altitude can reach 6000 m, the equipment mass can reach several hundred kilograms, and the power supply problem can be easily solved. Its limitation lies in the more electromechanical equipment, including compass, airborne radio, altimeter, airborne radar, etc., the lifting load and on-board equipment are more likely to produce intermodulation interference, mirror frequency interference and other electromagnetic compatibility problems. In addition, the interference generated by the on-board antenna coupling, as well as the helicopter fuselage (aluminium alloy material) shielding and other issues need to be considered in a comprehensive manner, unified design.

4. Airship platform

Airship is able to lift the aircraft vertically, it is equipped with a power plant can fly horizontally and steering. According to the airship on the complete bearing frame, can be divided into two types of airship hard and soft. Hard airship frame consists of longitudinal keel, truss, transverse frame and diagonal tension line, and there is a skin to bear the aerodynamic force outside the frame. This kind of airship is large in size, the engine and the passenger and cargo compartments are located in the boat. Soft airship airbag filled with a certain pressure by the gas to maintain the shape of the pod and the engine hanging in the airbag below.

The gases that can generate buoyancy are hydrogen, nitrogen, ammonia and hot air. Hydrogen is easy to burst into flames, unsafe; nitrogen is safe, but expensive; ammonia has a corrosive effect on the structural materials, harmful to the human body; air heated to reduce the density of buoyancy can be generated, but the same buoyancy generated by hot air airships than nitrogen airships double the size. The power unit installed on the airship is generally used to generate horizontal tension and steering, and can also generate vertical tension during take-off.

Long flight time and fuel saving are the biggest advantages of airships, and there is no need to build airports and runways. Unlike helicopters, airships do not require any fuel for take-off or hovering and can therefore be parked at high altitudes for long periods of time. The airbag shell of the blimp consists of thousands of small bags. Tests have shown that even when 40 bullets hit a blimp at the same time, only a tiny amount of nitrogen leaks out.

Because the airship has both floating gas lift, and power drive, thus less energy consumption, strong load capacity, safety, economy than the helicopter is better, is a better air platform communication carrier.

17.2 Surface Platform Communications

Surface communication means use ships, buoys, etc. as maritime platforms to quickly establish communication security and carry out rescue missions in the event of maritime emergencies or natural disasters.

17.2.1 Application of Unmanned Boats in Maritime Communication

Unmanned boat is a kind of surface movement platform capable of autonomous navigation. For different application scenarios, China has continuously launched unmanned boats with professional advantages, such as 'Tianxing No. 1', 'Seafly01' and 'Blue Whale'. The shore-based monitoring sub-system of the unmanned boat can be set up on top of other surface ships to realise the group cruising mode centrally

17.2 Surface Platform Communications

controlled by the mother ship. The communication modes of the unmanned vessel are UHF/VHF frequency digital transmission radio, 4G wireless network, Ad-Hoc self-grouping network and satellite communication. The communication between the unmanned boat and the shore-based control centre adopts digital transmission radio within the line-of-sight range, and if the line-of-sight range is exceeded, a 4G wireless network can be arranged in the near-shore waters within 50 km, and the 4G wireless network is the preferred way of near-shore communication of the unmanned boat because of its high communication rate and low construction cost. In the sea search and rescue mission, the unmanned boat is far away from the shore, and it can choose Inmarsat, Beidou, Tiantong No. 1 and other satellite communication modes to carry out far-sea communication. At the same time, unmanned boats are flexible in networking and can quickly build communication networks, which is suitable for communication between unmanned boats with high mobility. The search and rescue range of unmanned boats is limited, and the search and rescue range can be extended through cooperative operation with UAVs in practical application.

17.2.2 Application of Buoy in Maritime Communication

Emergency buoy is a floating communication platform, and the communication mode is mainly satellite communication, as well as CDMA, GPRS, 4G mobile communication network, digital transmission radio, Wi-Fi and so on. It mainly includes emergency life-saving buoy, communication relay buoy and emergency monitoring buoy, as shown in Table 17.1.

Emergency lifesaving buoy is used when the submarine is in danger underwater, and the cableless form can cooperate with the navigation radar system to respond to the radar scanning signal, so that the rescue ship or aircraft can quickly locate the submarine in danger. The cable form is usually equipped with ultra-short-wave life-saving radio and wired telephone to realise the function of external alarm and communication. The communication relay buoy is generally a towed buoy, which is used as a relay station for submarine communication and enables the submarine to communicate with the shore ship when sailing underwater. The oil spill tracking and monitoring buoy adopts the Beidou satellite positioning platform to achieve real-time

Table 17.1 Emergency float application scenarios and functions

Emergency buoys	Application scenarios	Functions
Emergency buoys	Submarine underwater distress	Alarm and location
Communication relay buoy	Submarine communication relay	Message relay
Emergency monitoring buoy	Offshore oil ship oil spill	Real-time tracking of oil spill trajectory

tracking and monitoring functions for different sea conditions and different oil films, so as to reduce the economic losses and environmental hazards caused by oil spill accidents of oil tankers at sea.

17.2.3 Application of Mesh/Ad-Hoc Network in Maritime Communication

Each node of Mesh/Ad-Hoc network is equipped with data transmission and routing relay function, which can quickly build a high-quality network, and is widely used in ships, unmanned boats, and unmanned aerial vehicle networking. In the near-shore situation, ships can use Mesh network communication, which is composed of Mesh routers to form a backbone network, providing multi-hop network connection for Mesh clients through wired Internet network. Ships can communicate with RAS directly within the coverage area of Radio Access Station (RAS), and if it is beyond the coverage area of RAS, the mesh network can be co-constructed with other ships/buoys. The biggest difference between the Ad-Hoc network and the mesh network is that there is no base station in the Ad-Hoc network. In the case that base stations cannot be installed far away from the coast, Ad-Hoc network is more suitable for randomly moving communication terminals compared to the weak mobility structure of Mesh network. The maritime mobile communication system based on Ad-Hoc network usually selects the AODV (Ad-HocOn-demand Distance Vector Routing) protocol, which sends routing data packets (RREQ) in the form of broadcasting, and the intermediate nodes will update their own route caching information when receiving RREQ packets, and continue to forward the RREQ packet when they judge that they have never seen the RREQ When it determines that it has never seen the RREQ packet, it continues to forward the packet and sends a Route Response (RREP) packet to the data source vessel. The AODV protocol has low complexity and avoids a large amount of routing information congestion, and therefore, it is well adapted to maritime mobile grouping networks where the topology is constantly changing.

Bibliography

1. Zheng, Zhou, Zhou Huilin, et al. 2002. *Practical Handbook of New Technology in Communication Engineering: Mobile Communication Technology*. Beijing: Beijing University of Posts and Telecommunications Press.
2. (E) Hall MPM. 1984. Convection Propagation and Radio Communications. Convection Propagation and Radio Communication. Beijing: National Defence Industry Publishing House.
3. Baowei, Lv, and Wang Zhensong. 2003. *Theory of Radio Wave Propagation and Its Applications*. Beijing: Science Press.
4. Ba, Yong, Zhongzhao Zhang, and Naitong Zhang. 1999. *Feasibility Study of Stratospheric Communication System in Military Applications*. System Engineering and Electronic Technology.

5. He, Chen, and Hongwen Zhu. 1999. New Technology of Broadband Wireless Relay Stratospheric Communication. *Computer and Network*
6. Wu, Youshou. 2000. Stratospheric Communication System in Development. *Engineering Physics* 4: 1–8.
7. Air Force Preparatory Academy and Beijing Institute of Aeronautics and Astronautics. In *Proceedings of the Colloquium on the Development and Application of Floating Vehicles*. Beijing: Air Force Preparatory Academy and Beijing Institute of Aeronautics and Astronautics.
8. Zhang, Dongchen, and Ji Zhou (eds.). 2015. *Military Communications* (2nd ed.). Beijing: National Defence Industry.
9. Lin, Bin, Zhiqiang Zhang, et al. 2020. A Review of Maritime Emergency Communication Network Technologies Integrated with 'Air, Space and Haiti'. *Mobile Communication.*

Chapter 18
Communication Key Technology

18.1 Channel Receiver Realisation Technology

In information theory, symbols are the main object of study. Firstly, the source symbols are mapped into channel symbol sequences X = (x1, x2, …, xn, …), and then the corresponding channel output sequences Y = (y1, y2, …, yn, …) are generated from these sequences. In a digital communication system, what the transmitter sends is not the symbol sequence X itself, but the corresponding continuous time waveform s(t,x). Since the distribution of the symbol sequence to the channel waveform is done by the modulator, the distribution of the output sequence depends not only on the input sequence X, but also on the set of parameters θ = {θT, θC}, where the subset θT is the transmitter parameters, and the subset θC is the channel parameters, mainly including the phase θ or the time delay ε. These parameters are not known to the receiver, and the essential role of the receiver is to accurately recover the transmitted symbols according to the output sequence. These parameters are unknown to the receiver, so the essential role of the receiver is to accurately recover the transmitted symbol messages according to the output sequence [1].

As shown in Fig. 18.1, the physical communication model of a digital receiver is divided into two parts: the inner receiver and the outer receiver. In order to recover the transmitted symbol sequence, firstly, the inner receiver has to accurately estimate the required unknown parameters from the received signals, and then, these estimation results are used as real values and sent to the outer receiver so that the performance of the outer receiver is as close as possible to the ideal channel conditions, and next, the outer receiver further completes the optimal decoding of the transmitted sequence.

This section firstly focuses on the two parts of channel synchronisation and channel estimation and equalisation in the inner receiver for single-carrier and multi-carrier systems respectively, and then takes the DVB-T system and our digital terrestrial TV broadcasting system as examples.

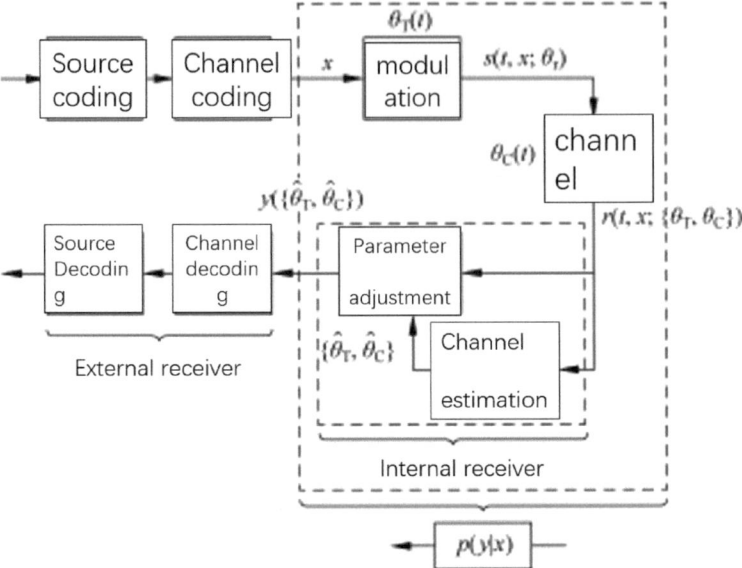

Fig. 18.1 Digital receiver communication model

18.1.1 Overview

Digital terrestrial TV broadcasting channels are mainly located in the VHF and UHF bands, which is a kind of time dispersive wireless channel that is currently used more often and with very poor quality. The frequency bandwidth of TV transmission signals is several megahertz wide, and due to the presence of additive noise, co-frequency interference and linear distortion such as multipath transmission, this results in static and dynamic multipath interference as well as the Doppler effect, which in turn results in inter-symbol crosstalk of the data received at the receiving end of the system. Therefore, to obtain better reception performance in such poor transmission channels, channel characteristic reduction using accurate channel synchronisation, channel estimation and equalisation is necessary.

Channel synchronisation, occupies a very important position in the internal receiver of a digital transmission system. In general, synchronisation includes timing restoration (also known as symbol synchronisation) and carrier synchronisation. The process of acquiring symbol timing is called timing recovery and includes frame synchronisation and sample clock synchronisation. First of all, in the digital transmission system, the digital signal sequence is usually transmitted in a certain frame structure, the receiving end can only accurately identify its frame structure to correctly recover the transmitted data, which requires that the receiving end must generate the timing consistent with the frame structure of the transmitted data; then, in order to correctly detect a series of serial symbols appearing in sequence, the receiving end must know the starting moment of the appearance of each symbol, which requires

18.1 Channel Receiver Realisation Technology

that the sampling clock coincides with the frequency and phase of the received symbol sequence. The process of acquiring the local carrier signal is called carrier synchronisation. In demodulation of digital carrier signals, it is classified as coherent demodulation and incoherent demodulation depending on whether or not known local carrier information is required. Coherent demodulation has superior noise immunity compared to incoherent demodulation, while coherent demodulation requires the receiver to provide a local carrier signal of the same frequency and phase as the received carrier signal.

There are many classification criteria for synchronisation. Depending on whether the synchronisation algorithm requires data information or not, it can be classified as Decision Directed (DD)/Data Aided (DA) and Non-Data Aided (NDA). The known data sequence is called DA synchronisation, while the NDA algorithm averages the various possible sequences without knowing the data sequence. Depending on where the synchronisation error signal is extracted from the received signal, the synchronisation algorithms can be classified as Feedforward (FF) and Feedback (FB). The FF estimation involves the extraction of the error signal before the synchronisation recovery unit, whereas the FB estimation involves the extraction of the error signal from after the synchronisation recovery unit, and then the corrected signal is fed back to the unit at the front end. The FB estimator is also known as an error feedback synchroniser because the feedback structure itself carries the ability to automatically track slow changes in parameters. Depending on whether the estimation of the synchronisation parameters requires information about other synchronisation parameters, the synchronisation algorithms can be classified into two types: correlation estimation with other parameters and irrelevant estimation. As the name suggests, the uncorrelated estimation does not require information about the synchronisation of other parameters, and this type of algorithm makes the synchronisation of each parameter independent of each other, and there is no sequential order. According to the type of synchronised digital signals, synchronisation algorithms can be divided into two types: continuous signal estimation and burst signal estimation. Continuous signals require algorithms to have the ability to track timing changes over a long period of time, while burst signals require algorithms to capture a short period of time, i.e., the synchronisation must be completed in a short period of time.

Channel estimation is another key component of the digital receiver after channel synchronisation. The so-called channel estimation, also known as channel compensation, that is, the channel characteristics of the equalisation, refers to the receiver side of the equalisation of the characteristics of the channel and the experience of the opposite characteristics of the channel, thereby offsetting the time-varying and multipath characteristics of the channel caused by the inter-code interference, that is, through the channel estimation can be eliminated by the channel's frequency and time double selectivity. In addition, because the channel is time-varying, the channel estimation also needs to be able to automatically adapt to the changes in the channel so that the channel is compensated by the equalisation, so it is also known as adaptive equalisation. In addition to channel compensation, equalisation also includes data equalisation, i.e. recovery of source data using channel estimation results.

There are many channel estimation methods currently applied to wireless broadcasting systems. According to the channel compensation domain, channel estimation is mainly divided into time domain equalisation, frequency domain equalisation and the combination of both. On the one hand, in single-carrier systems, time-domain equalisation is the most commonly used method. The so-called time-domain equalisation is to consider the impulse response in the time domain so that the impulse response of the whole system including the time-domain equaliser satisfies the condition of no inter-symbol interference. Time-domain equalisation uses the response it produces to compensate for the distorted signal waveform, and ultimately effectively eliminates the inter-symbol interference on the sampling judgement moment, and thus has been widely used in many areas of digital communications, such as modems, mobile communications, ADSL, etc., and is a more mature technology. On the other hand, in OFDM systems, frequency domain equalisation is more commonly used, with the aim of making the total transfer function satisfy the distortion-free transmission conditions, i.e., correcting the amplitude characteristics and group delay characteristics. The classification criteria for channel estimation vary. According to channel compensation algorithms, channel estimation is mainly classified into least square (LS, also known as forced-zero) and minimum mean square error (MMSE) criterion, and MMSE in the frequency domain has the optimal theoretical performance, but due to the need to know the statistical characteristics of the channel, it is difficult to be widely applied in practical systems because of the high implementation complexity. The frequency-domain MMSE has the best theoretical performance, but it is difficult to be widely used in practical systems due to the high implementation complexity because it requires known channel statistics. For simplicity, one approach is to simplify the MMSE algorithm by using the optimal low-rank theory, which is called low-rank linear MMSE (LMMSE), the second approach is to implement it by singular value decomposition (SVD), and the third approach is to implement it by using transform domains, such as Fourier transform domain and wavelet transform domain wavelet transform domains. Equalisers implementing channel estimation are further classified into linear and nonlinear equalisers according to whether the channel estimation results are used for feedback control or not. If the output is not used for feedback control, the equaliser is linear; conversely, if the output is used to feedback the system and changes the subsequent output of the channel compensation, the equaliser is nonlinear, as in the case of the decision feedback equalizer (DFE) and the maximum likelihood sequence equalizer (MLSE) sequence equalizer (MLSE), etc. The American ATSC/8-VSB standard single-carrier system makes use of the decision feedback equalizer (DFE) for channel compensation: the training sequence carried in the first data segment of each field in the ATSC data frame is used for training, and the two neighbouring training sequences are separated by 24.2 ms, whereas for the fast-varying multipath, the can only be performed using adaptive blind equalisation. In order to eliminate multipath interference and achieve good results, DFE generally requires a huge number of equalisation filter taps, which greatly increases the complexity and cost of the communication system. In addition, in strong multipath environments, DFEs naturally have the disadvantages of being prone to self-excitation and instability due to the fact that DFEs are infinite impulse response structures (IIRs), so the focus of the current ATSC equaliser research is still on how to improve system stability and reduce complexity. European DVB-T/COFDM is a typical OFDM modulated multicarrier system, according to the frequency of OFDM subcarriers, DVB-T divides the frequency selective fading channel into a number of flat fading subchannels, which overcomes the influence of ISI brought about by the multipath effect, and the current DVB-T system adopts the method of frequency-domain channel estimation, which mainly includes judgement-feedback frequency-domain estimation and conduction-frequency-domain estimation frequency domain estimation. China's DTMB/TDS-OFDM system, on the other hand, adopts the time-domain training sequence to complete the channel estimation of the OFDM system, and uses the training sequence instead of the guide frequency for the channel estimation, which further improves the spectrum utilisation of the channel. According to the division of channel compensation implementation method, channel estimation is mainly divided into data-assisted method, judgement pointing method and blind estimation method. DA methods generally perform channel estimation with the help of a certain guide frequency or training sequence. In the case of time-varying channels, the guide frequency or training sequence should be sent over and over again according to a certain pattern. A multipath fading channel can be viewed as a two-dimensional signal in time and frequency, so the spacing of the guide frequency or training sequences in the time and frequency domains depends mainly on the correlation time and correlation bandwidth of the channel. When performing channel estimation, the channel is sampled at different points in the time-frequency two-dimensional space using the guide frequency or the training sequence, and the frequency response value of the whole channel can be obtained using sample interpolation as long as the sampling frequency satisfies the Nyquist sampling criterion in the time and frequency domains. Unlike the DA approach, the DD approach uses the adjudicated estimates as known data for channel estimation. The blind estimation method, on the other hand, completes the channel estimation with complete unknown of the transmitted signal, so that the transmitter does not have to send special guide or training sequences, which greatly improves the spectral efficiency of the system compared to DA and DD, but the blind estimation needs to receive enough data to get a reliable estimation result, coupled with the fact that the broadcasting channel is time-varying, and for the time-varying fading channel, the channel estimation must also be able to track the variations of the time-varying channel, which requires the guide or training sequence to be inserted into the transmit sequence in some consecutive manner, thus limiting the use of blind estimation methods.

18.1.2 Mathematical Foundations

1. Channel synchronisation [1]

Channel synchronisation algorithms have been extensively studied and successively proposed for different types of channel synchronisation starting from the 1960s to the

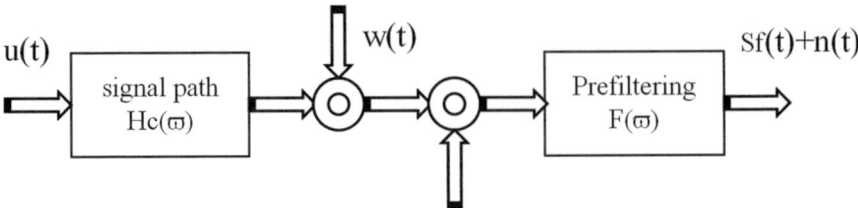

Fig. 18.2 Baseband Linear Model for Carrier-Modulated Digital Communication Systems

present. In principle, the criteria for the estimation of these parameters in channel synchronisation are all maximum likelihood (ML) synchronisation algorithms. In particular, the Maximum posteriori Probability (MAP) criterion is identical to the ML criterion when the input data satisfies an equal probability distribution. Although the synchronisation parameter (θ, ε) is time-varying, it is much slower compared to the data sign interval, and thus, the synchronisation parameter can be considered as a constant over long data segments. Therefore, the next derivation of the synchronisation algorithm usually assumes that the synchronisation parameter to be estimated remains constant over a signal frame.

A typical baseband linear model of a carrier-modulated digital transmission system is shown in Fig. 18.2. Assume that u(t) is any type of linearly modulated signal transmitted within the baseband

$$u(t) = \sum_n a_n g_T(t - nT - \varepsilon_0 T), \qquad (18.1)$$

where the channel symbol {an} is an arbitrary set of signals selected from the complex plane, gT(t) is the impulse response of the pulse-forming filter, T is the symbol period, the frequency response of the linear channel is Hc(ω), and the additive Gaussian white noise is w(t). The phase difference between the transmitter and receiver is expressed as $\theta 0 = \theta T(t) - \theta R(t)$, then the equivalent baseband signal at the receiver is

$$r_f(t) = s_f(t) + n(t) = \sum_n a_n g(t - nT - \varepsilon_0 T) e^{j\theta_0(t)} + n(t), \qquad (18.2)$$

where g(t) is the pulse waveform seen at the receiver's end, which can be expressed as

$$g(t) = g_T(t) \otimes h_C(t) \otimes f(t). \qquad (18.3)$$

Next, the generated $r_f(t)$ baseband signal enters the receiver synchronisation circuit section to estimate the synchronisation parameters such as carrier frequency bias and timing clock. Consider the detection of a sequence a with N symbols, the received signal is further obtained from Eq. (18.3).

18.1 Channel Receiver Realisation Technology

$$r_f(t) = \sum_{n=0}^{N-1} a_n g(t - nT - \varepsilon_0 T) e^{j\theta_0(t)} + n(t). \tag{18.4}$$

Assume that the samples $\{r_f(kT_s)\}$ of $r_f(t)$ are sufficient statistics. It can be shown that when the prefilter $|F(\omega)|2$ is symmetric with respect to $T_s/2$, the noise process can be viewed as a complex white Gaussian process whose power spectral density N0 is flat in the passband of the prefilter analogue filter $F(\omega)$. Meanwhile the matched filter $g\text{MF}(nT + \varepsilon T\text{-}kT_s)$ at the receiver side satisfies the Nyquist sampling theorem, i.e.

$$h_{m,n} = \sum_k g(kT_s - mT)g_{\text{MF}}(nT - kT_s) = \begin{cases} h_{0,0} & (m = n); \\ 0 & (\text{other}). \end{cases} \tag{18.5}$$

The likelihood function can be simplified as

$$L(r_f | a, \varepsilon, \theta) \propto \exp\left\{-\frac{1}{\sigma_n^2}\left[2\text{Re}\left(\sum_{n=0}^{N-1} |h_{0,0}|^2 |a_n|^2 - 2a_n^* z_n(\varepsilon) e^{-j\theta}\right)\right]\right\}, \tag{18.6}$$

where the abbreviated symbol $z_n(\varepsilon) = z(nT + \varepsilon T)$ is the output of the matched filter, denoted as

$$z_n(\varepsilon) = \sum_{k=-\infty}^{\infty} r_f(kT) g_{\text{MF}}(nT + \varepsilon T - kT_s). \tag{18.7}$$

In general, the ML synchronisation algorithm can be obtained by approximating to remove the redundant parameters in the ML function. First, assuming that N is large enough, (18.6) is further approximated by the law of large numbers and $\sum_n |a_n|^2 \to \sum E[|a_n|^2]$ as

$$L(r_f | a, \varepsilon, \theta) = \exp\left[-\frac{2}{\sigma_n^2}\text{Re}\left(\sum_{n=0}^{N-1} a_n^* z_n(\varepsilon) e^{-j\theta}\right)\right]. \tag{18.8}$$

After obtaining the system ML function, the valuation of the corresponding synchronisation parameters is obtained by deriving the maximum value of the ML. There are various existing algorithms for maximum search of the objective function, and the decision of which algorithm to choose is mainly based on the bit rate and the available technology. Specifically there are:

(1) Iterative maximum search method

The iterative maximum search method is to directly solve for the parameter that maximises the objective function by iterating from the initial value. Assuming that the valuation of the data sequence can be obtained or the data sequence a = a0 is known, the necessary but not sufficient conditions for the maximum value of the objective function are

$$\begin{cases} \frac{\partial}{\partial \theta} L(r_f|a, \theta, \varepsilon)|_{\hat{\theta},\hat{\varepsilon}} = 0; \\ \frac{\partial}{\partial \varepsilon} L(r_f|a, \theta, \varepsilon)|_{\hat{\theta},\hat{\varepsilon}} = 0. \end{cases} \quad (18.9)$$

Since the objective function is a concave function of the parameters (θ, ε), if the initialised valuation is within the region of convergence, we can iteratively obtain the zeros of Eq. (18.9) using either the gradient method or the maximum ascent method. The iterative search is useful for obtaining known symbols in the training phase.

The error feedback system uses the error signal to adjust the synchronisation parameters. The error signal is obtained by deriving the objective function and then substituting the nearest $\hat{\theta}_n, \hat{\varepsilon}_n$ estimates.

$$\begin{cases} \frac{\partial}{\partial \varepsilon} L(\hat{a}, \theta = \hat{\theta}_n, \varepsilon = \hat{\varepsilon}_n); \\ \frac{\partial}{\partial \theta} L(\hat{a}, \theta = \hat{\theta}_n, \varepsilon = \hat{\varepsilon}_n). \end{cases} \quad (18.10)$$

From a causal point of view, the error signal may depend only on the sign a_n that has previously been assumed to be obtained further, the error signal can be used to predict the new valuation, i.e.

$$\begin{cases} \hat{\varepsilon}_{n+1} = \hat{\varepsilon}_n + a_\varepsilon \frac{\partial}{\partial \varepsilon} L(a_n, \hat{\theta}_n, \hat{\varepsilon}_n); \\ \hat{\theta}_{n+1} = \hat{\theta}_n + a_\theta \frac{\partial}{\partial \theta} L(a_n, \hat{\theta}_n, \hat{\varepsilon}_n). \end{cases} \quad (18.11)$$

Equation (18.11) gives a first-order discrete error feedback system, where $a\varepsilon$, $a\theta$ determine the bandwidth of the loop, and higher-order tracking systems can be realised by the use of appropriate loop filters. When the error is sufficiently small, we claim to enter tracking mode operation of the error feedback system. The process of changing the system from the initial state to the tracking mode is called locking, and locking is generally a nonlinear process.

It can be seen that there are many similarities between the maximum search and the error feedback system, both obtain the error signal by deriving the objective function. However, there is a fundamental difference between the two, i.e., the maximum search algorithm converges the whole signal to the final estimate through continuous iteration, while the feedback control system only processes the received signal in real time. Currently, since video broadcast data is transmitted as a continuous stream, a search process with less stringent time requirements is required only at the beginning. Used for tracking purposes, the error-feedback structure can be implemented with reasonable complexity, so it is generally used in the synchronisation module of digital TV systems.

2. Channel estimation

(1) Least Squares Criterion [2]

The least squares criterion, also known as the forced-zero criterion, is the most basic and simplest channel estimation algorithm, which will be derived and solved from the time domain perspective in the following.

Assuming that the discrete-time linear channel model has an impulse response $\{h_k\}$, and assuming that a channel equaliser with infinite taps and an impulse response of $\{c_k\}$ is cascaded with it, the equivalent filter can be represented as follows, i.e., the impulse response of this filter is the convolution of $\{c_k\}$ and $\{h_k\}$

$$q_n = \sum_{j-\infty}^{\infty} c_j h_{n-j}. \tag{18.12}$$

Normalising q0 for convenience, the output of the equaliser at sampling moment k is obtained from Eq. (18.12) as

$$\hat{I}_k = I_k + \sum_{n \neq k} I_n q_{k-n} + \sum_{j=-\infty}^{\infty} c_j n_{k-j}. \tag{18.13}$$

wherein the first term is the desired obtained symbol, the second term is an inter-code interference term, and the third term is a noise term. Further, a weighting function of the equaliser taps

$$D = \sum_{\substack{K \neq -\infty \\ K \neq 0}}^{\infty} |q_k| = \sum_{\substack{k \neq -\infty \\ k \neq 0}}^{\infty} \left| \sum_{j=-\infty}^{\infty} c_j h_{k-j} \right|, \tag{18.14}$$

If the inter-code interference is completely eliminated, it is necessary to choose the tapping coefficient $D = 0$, i.e., 0 for all q_n except $n = 0$. The

$$q_n = \sum_{j=-\infty}^{\infty} c_j h_{n-j} = \begin{cases} 1 (n = 0); \\ 0 (n \neq 0). \end{cases} \tag{18.15}$$

It can be seen that the criterion is also known as the forced-zero criterion because of the need to force each interference value to zero. Doing a Z-transform of Eq. (18.15) to transform it into the frequency domain gives us

$$Q(z) = C(z)H(z) = 1, \tag{18.16}$$

Or

$$C(z) = \frac{1}{H(z)}. \quad (18.17)$$

This suggests that complete elimination of inter-code interference would require the use of a channel inverse filter, i.e., a channel equaliser with transfer function C(z) is an inverse filter of the channel model H(z).

(2) Minimum mean square error criterion [2]

The filter that follows the minimum mean square error criterion is the 2D Wiener filter, which can be solved using the orthogonality principle. In the following, the frequency domain MMSE criterion is mainly illustrated using the example of the pilot/training sequence, while the time domain MMSE criterion will not be repeated.

Assuming that the discrete-time linear channel model has a frequency-domain response $\{H_k\}$ and the transmit data is $\{X_k\}$, the data at the receiving end are

$$Y_k = X_k \cdot H_k + W_k, \quad 0 \le K < N, \quad (18.18)$$

where W_k is additive Gaussian noise and σ_n^2 is the noise variance. Using the MMSE criterion, the channel estimation error is

$$J(N) = E|\varepsilon_k|^2 = E|H_k - \hat{H}_k|^2, \quad (18.19)$$

thus

$$J(N) = E|\varepsilon_k|^2 = \sum_{k=0}^{N-1} \frac{E(|W_k|^2)}{|X_k|^2} = \sigma_n^2 \sum_{k=0}^{N-1} \frac{1}{|X_k|^2}. \quad (18.20)$$

The channel estimation error minimisation equivalent under the condition that the pilot/training sequence power is constrained is

$$\begin{cases} \min \sum_{k=0}^{N-1} \frac{1}{|X_k|^2} \\ \text{s.t.} \sum_{k=0}^{N-1} |X_k|^2 = Const. \end{cases} \quad (18.21)$$

Then the error minimisation condition is obtained as

$$|X_0| = |X_1| = \ldots = |X_{N-1}|, \quad (18.22)$$

i.e., constant mode in the frequency domain of the pilot/training sequence. In summary.

(1) The LS estimation algorithm is simple, but the estimation accuracy is greatly affected by Gaussian noise and inter-subcarrier interference (ICI). In contrast, the MMSE estimation algorithm has good suppression of ICI and Gaussian noise, and its algorithm is better than the LS algorithm. Under the same MSE error requirement, the signal-to-noise ratio of MMSE estimation will have a gain of up to 10 to 15 dB over LS.

(2) The MMSE estimation algorithm is mean-square optimal, so the estimation requires the use of the channel's delay power spectrum (subcarrier frequency dependence) and Doppler power spectrum (symbol time dependence). In practice, it is difficult for the receiver to know such statistical information, which in turn increases the complexity, which results in the main disadvantage of MMSE estimation is the high algorithmic complexity, which increases exponentially with the increase in the number of arithmetic points. Coupled with the presence of estimation errors, the performance of the final MMSE estimator deteriorates slightly.

According to the sampling theorem, if the sampling frequency satisfies the Nyquist frequency requirement, the signal can be accurately recovered from the sampling value. If the noise is additive at this point, the use of a Wiener filter allows an approximate recovery of the original signal in the sense of minimum mean square error. Considering the great complexity of the two-dimensional signal processing methods, in practice, the general approach is to decompose a two-dimensional signal into two one-dimensional signals for processing, but even so, the complexity of the Wiener filter is still very high. Accordingly, a variety of simplified methods have been proposed in many applications, and although the performance is somewhat worse than the MMSE estimation, the computational complexity is greatly reduced, and the following examples will be described in detail.

(3) Simplified Multi-dimensional Linear Interpolation [2]

Taking two-dimensional interpolation as an example, two-dimensional linear interpolation performs linear interpolation in both time and frequency directions, similar to the decomposition of a two-dimensional Wiener filter into two one-dimensional Wiener filters. Two-dimensional linear interpolation can be decomposed into two independent interpolation processes: firstly, interpolating the filter in the time direction, and then using the result of interpolating in the time direction and then interpolating the filter in the frequency direction, which is implemented as shown in Fig. 18.3.

Taking the DVB-T system as an example, its dispersive guide frequency is spaced at 4 on the time axis, i.e., every 3 OFDM symbols, the guide frequency is repeated 1 time in the same subcarrier position cycle, as shown in Fig. 18.3a, then the time direction is estimated as

$$\hat{H}^t_{i,n+m} = \left(1 - \frac{m}{4}\right)\hat{H}^p_{i,n} + \frac{m}{4}\hat{H}^p_{i+4,n} (1 \leq m \leq 3), \quad (18.23)$$

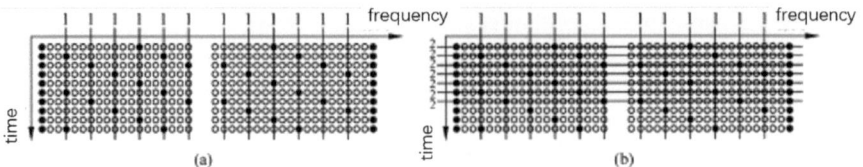

Fig. 18.3 Two-dimensional linear interpolation

The frequency direction is estimated to be

$$\hat{H}_{i,n+1} = \left(1 - \frac{l}{3}\right)\hat{H}^t_{i,n} + \frac{L}{3}\hat{H}^t_{i,n+3} (1 \leq l \leq 2). \tag{18.24}$$

After completing the interpolation filtering in the time and frequency directions, the channel estimation at all points can be obtained, as shown in Fig. 18.3b.

From the above, it can be seen that, in principle, the 2D linear interpolation can track the channel response correctly and in time for channels that change faster in the frequency direction, but it is more sensitive to time-varying channels with large Doppler shifts, and at the same time, multiple OFDM symbols are needed to require better synchronisation performance of the receiving system, i.e., the frequency deviation and common phase error between each symbol should be very small. In terms of hardware implementation, 2D linear interpolation is going to require a large amount of memory. Theoretically, the channel response can be estimated more accurately using higher order polynomial interpolation, but at the same time, its computational complexity increases greatly with the increase of polynomial order.

(4) SVD-based Channel Estimation [3]

The simple interpolation algorithm has poor suppression of subcarrier ICI and Gaussian noise and introduces interpolation errors. For this reason, another method, SVD-based channel estimation, can be utilised.

The singular value decomposition of the autocorrelation matrix of the channel transfer function can be expressed as

$$R_{HH} = U\Lambda U^H, \tag{18.25}$$

where U is the You matrix containing the singular vectors and Λ is the diagonal matrix containing the singular values $\lambda 1 \geq \lambda 2 \geq \ldots \geq \lambda N$. The optimal p-order estimator is

$$H_p = U \begin{bmatrix} \Delta_p & 0 \\ 0 & 0 \end{bmatrix} U^H H_{LS}. \tag{18.26}$$

where Δp is the p × p order upper-left corner matrix of Δ and

18.1 Channel Receiver Realisation Technology

$$\Delta = \Lambda\left(\Lambda + \frac{\beta}{SNR}I\right)^{-1} = diag\left(\frac{\lambda_1}{\lambda_1 + \frac{\beta}{SNR}}, \ldots, \frac{\lambda_N}{\lambda_N + \frac{\beta}{SNR}}\right). \quad (18.27)$$

From the above, UH can be regarded as a transformation matrix, then the singular value λ_k of the matrix RHH can be regarded as the corresponding component of the channel power at the kth transformation coefficient. And since U is a You matrix, UH can be regarded as a rotation of HLS, and thus the components of UH are uncorrelated with each other.

In summary, a block diagram of a low-order estimator of order p is shown in Fig. 18.4. First, the LS estimate HLS is obtained by multiplying Y by X − 1 at the receiver end, and then, the UH is obtained by rotating HLS next, Eq. (18.26) yields the final channel estimate. Passing through the low-order estimator can be viewed as mapping the LS estimator to a subspace of order p. If the subspace dimension is small and describes the channel characteristics well, an estimator with low complexity and good performance can be obtained. The performance is good in channels with non-integer point sampling, but it also introduces a floor effect in the estimation error since the estimation is done only in the subspace of the channel. In order to eliminate this error floor at a given SNR, it is required that the order of the estimator is large enough, however, an increase in the order implies an increase in the computational complexity. In particular, when performing OFDM multicarrier system design, the length of the cyclic prefix has to be larger than the maximum delay extension of the channel, so generally good estimation performance can be obtained when the order p is equal to the length of the cyclic prefix, and at the same time, the total subcarrier length of the OFDM symbols has to be much larger than the length of the cyclic prefix, so that the computational complexity can be greatly reduced.

(5) DFT based channel estimation [4]

In order to reduce the complexity of the MMSE algorithm, the channel estimation can also be done using the fast algorithm of IDFT/DFT.

Firstly, the channel estimation of LS algorithm is performed, then it enters into the time domain through IDFT, and then it is linearly transformed in the time domain, filtered by using the feature that the impulse response energy of the channel in the

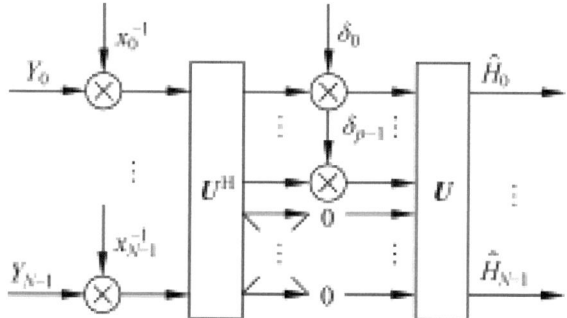

Fig. 18.4 SVD-based p-order channel estimation

time domain is concentrated in fewer sampling points, and finally it enters into the frequency domain through DFT.

The LS estimation within an OFDM symbol can be expressed as

$$H_{LS} = X^{-1}Y = \left[\frac{Y(0)}{X(0)} \frac{Y(1)}{X(1)} \cdots \frac{Y(N-1)}{X(N-1)}\right]^T, \quad (18.28)$$

vectors consisting of the transmit symbol sequence {X(n)} and the receive symbol sequence {Y(n)}, respectively. It is worth noting that the labelling of the time direction is ignored here, and only n is used to denote the serial number of the subcarrier in the frequency domain. The X for performing LS estimation can consist of the guide or training sequence inserted by each OFDM symbol, or it can be the adjudicated data.

The linear minimum mean square error (LMMSE) estimation can be further obtained from Eq. (18.28), namely

$$H_{LMMSE} = R_{HH}\left(R_{HH} + \sigma_n^2(XX^H)^{-1}\right)^{-1} H_{LS};$$
$$R_{HH} = E\{HH^H\}, \quad (18.29)$$

Relevant simulation results have been shown that the performance deterioration due to this approximation is negligible. In the case of probabilistic modulation with equal probability of sending symbols, the

$$E\left\{(XX^H)^{-1}\right\} = E\left\{|\frac{1}{X_K}|^2\right\}, \quad (18.30)$$

Further simplification leads to

$$H_{LMMSE} = WH_{LS};$$
$$W = R_{HH}(R_{HH} + \frac{\beta}{SNR}I)^{-1};$$
$$\beta = \frac{E\{|x_k|^2\}}{E\left\{|\frac{1}{x_k}|^2\right\}}, \quad (18.31)$$

where β is a constant determined by the constellation diagram used for modulation, e.g., β for 16QAM modulation constellation is 17/9. If the autocorrelation matrix and SNR are known in advance, then W has to be computed only once, but in practice W is required to be computed frequently so for simplicity it is necessary to proceed to the DFT channel estimation method. Next, the channel transfer function H obtained by the LS algorithm is Fourier inverse transformed (IDFT) to obtain

$$g_{LS} = IDFT(H_{LS}). \quad (18.32)$$

Further, considering that the auxiliary or judgement data usually oversample the channel frequency response, the channel energy is mainly concentrated in a smaller

18.1 Channel Receiver Realisation Technology

range in the equivalent channel impulse response vector g_{LS} obtained at the receiver, while the noise energy is distributed over the whole range of the vector. The most straightforward approach is to ignore the parameters with small SNR in the g_{LS} and let only the parameters with larger energy go through the DFT into the frequency domain. By using this time domain energy concentration property for time domain filtering, the complexity will be greatly reduced. After obtaining g_{LS}, a linear transformation is performed to obtain $g = Q\, g_{LS}$, and then a discrete Fourier transform (DFT) is performed to obtain $H = DFT(g)$.

It should be noted that both DFT and SVD based methods have to know in advance the frequency domain statistical characteristics of the channel, RHH, and the channel SNR, which are generally unknown in practical applications. Although the DFT algorithm is a simple and efficient interpolation algorithm, when DFT is used to interpolate an N-point negative sequence obtained from uniform sampling, fully accurate interpolation requires not only that the original analogue signal represented by the sampling train is a band-limited signal with a bandwidth less than the Nyquist limit, but also that the spectrum of the original signal be discrete, which implies that the delays of paths in the channel must be distributed in accordance with the OFDM sampling This means that the delay of each path in the channel must be distributed according to the OFDM sampling interval. In practice, although the channel impulse response is of finite length and smaller than the Nyquist limit sampled in the frequency domain, the multipath delay of the channel is generally not distributed according to the OFDM sampling interval, and even the distribution of the power delay of the channel is not discrete but continuous, as shown in Fig. 18.5, so the aliasing phenomenon caused by energy leakage during DFT-based interpolation will bring inevitable errors by rounding off a part of the value when simplifying the algorithm. Bring unavoidable error, which leads to false flat bottoming of the channel estimation.

Both of the above methods are described based on frequency domain correlation. Of course, time-domain correlation can also be used for channel estimator design,

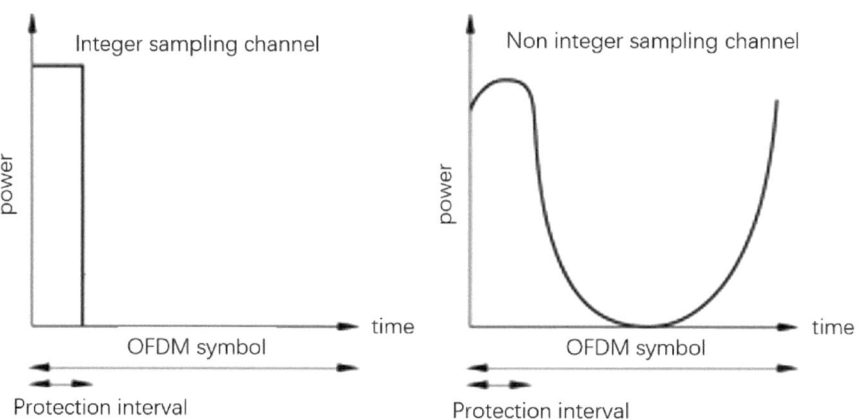

Fig. 18.5 Integer sampling channel and non-integer sampling channel

with a similar principle, using the time-domain correlation matrix instead of the RHH, which will not be described here.

18.1.3 Single-Carrier Systems

1. Timing synchronisation [5, 6]

The problem of timing error includes both frame synchronisation error and sampling clock frequency error. The frame synchronisation error is that when the frame controller intercepts each data frame, the intercept position will have an offset εT with respect to the ideal position. The sampling clock error is that the sampling clock T_s at the receiving end cannot be perfectly aligned with the clock T at the transmitting end.

(1) Typical algorithm for NDA timing estimation

The objective function used to synchronise the parameters (θ, ε) has been given as shown previously. Next, the data- and phase-independent timing estimation is derived first, i.e., the estimate of ε is obtained after removing the redundant parameters a and θ. To remove data dependence, Eq. (18.8) is multiplied by the probability of a symbol, $P(^i a)$, where $^i a$ denotes the ith of M symbols, and then, all M possible scenarios are accumulated. Assuming that the symbols are independent and equally probable, the likelihood function can be obtained

$$L(\theta, \varepsilon) = \prod_{n=0}^{N-1} \sum_{i=1}^{M} \exp\left[-\frac{2}{\sigma_n^2} \text{Re}\left(^i a_n^* z_n(\varepsilon) e^{-j\theta}\right) P(^i a)\right]. \tag{18.33}$$

There are many ways to solve Eq. (18.33), a typical approach is that the phase θ is first considered to be uniformly distributed, which further yields the algorithm associated with the data, i.e.

$$L_2(a, \varepsilon) \approx \prod_{n=0}^{N-1} \int_{-\pi}^{\pi} \exp\left[-\frac{2}{\sigma_n^2} |z_n(\varepsilon)| a_n^* \text{Re}\left(e^{j(-\arg a_n - \theta + \arg z_n(\varepsilon))}\right)\right] d\theta$$

$$= \prod_{n=0}^{N-1} I_0\left(\frac{|z_n(\varepsilon) a_n^*|}{\sigma_n^2/2}\right). \tag{18.34}$$

Since |an| = constant, Eq. (18.34) are the same for all phase modulations (M-PSK). As for M-QAM, in order to obtain an NDA synchronisation algorithm applicable to M-QAM signals, symbol averaging is required. This can be further simplified by expanding the Bessel function into an objective function.

18.1 Channel Receiver Realisation Technology

$$\text{NDA}: \hat{\varepsilon} = \arg\max_{\varepsilon} L_1(\varepsilon) \approx \arg\max_{\varepsilon} \sum_{n=0}^{N-1} |z_n(\varepsilon)|^2;$$

$$\text{DA}: \hat{\varepsilon} = \arg\max_{\varepsilon} L_2(a, \varepsilon) \approx \arg\max_{\varepsilon} \sum_{n=0}^{N-1} |z_n(\varepsilon)|^2 |a_n|^2. \tag{18.35}$$

It can be seen that for M-PSK, the two algorithms are completely equivalent. Next, the logarithmic data-free auxiliary objective function is derived with respect to the parameter ε, which leads to the error-feedback type algorithm

$$\begin{cases} \frac{\partial L(\varepsilon)}{\partial \varepsilon} = \frac{\partial}{\partial \varepsilon} \sum_{n=0}^{N-1} |z(nT + \varepsilon T)|^2 = \sum_{n=0}^{N-1} 2\text{Re}(z(nT + \varepsilon T)\dot{z}^*(nT + \varepsilon T)); \\ \dot{z}(nT + \varepsilon T) = \frac{\partial z(nT + \varepsilon T)}{\partial \varepsilon}. \end{cases} \tag{18.36}$$

Since the summation is performed in a loop filter, we can let ε = and get the error signal

$$x(nT) = \text{Re}\big[z(nT + \hat{\varepsilon}T)\dot{z}^*(nT + \hat{\varepsilon}T)\big]. \tag{18.37}$$

Finally, an approximate differential result is obtained

$$x(nT) = \text{Re}\bigg[z(nT + \hat{\varepsilon}T)\bigg(z^*\bigg(nT + \frac{T}{2} + \hat{\varepsilon}T\bigg) - z^*\bigg(nT - \frac{T}{2} + \hat{\varepsilon}T\bigg)\bigg)\bigg]. \tag{18.38}$$

The detailed derivation process can be found in the related literature. In addition, according to Gardner's algorithm [7], it is required to have 2 sampling values for each data symbol, one near the symbol judgement point and the other near the middle of the two symbol judgement points, and it is independent of the carrier phase deviation, so that the timing adjustment can be completed prior to the carrier recovery, and the timing recovery loop and the carrier recovery loop are independent of each other, which brings convenience to the design and debugging of the demodulator, and it is the best method for the design and debugging of DVB-S and DVB-S and DVB-C systems use the timing error extraction method. However, this algorithm suffers from one major problem, the self-noise is large and a loop filter is generally required to suppress it.

Another commonly used NDA timing estimation algorithm is performed using spectral estimation. Think of $|z_n(\varepsilon)|2$ as a 1/T periodic signal that can be found without using maximal search. Using the 1st coefficient c1 of the $|z_n(\varepsilon)|2$ Fourier series expansion gives an unbiased estimate of the

$$\hat{\varepsilon} = -\frac{1}{2\pi} \arg c_1. \tag{18.39}$$

Since the NDA algorithm provides a unique estimation method for ε and is simpler, it is also more commonly used in practice.

(2) Typical algorithm for DD/DA timing estimation

When DD/DA timing estimation is performed, the estimated value of parameter a is already available or is a known value. Substituting the estimated value of a into the objective function Eq. (18.8) yields

$$L(\hat{a}, \theta, \varepsilon) = \exp\left[-\frac{2}{\sigma_n^2}\mathrm{Re}\left(\sum_{n=0}^{N-1}\hat{a}_n^* z_n(\varepsilon)e^{-j\theta}\right)\right]. \quad (18.40)$$

Since the phase information θ may be unknown, a joint estimate of (θ, ε) is considered.

$$\max_{\varepsilon,\theta}\mathrm{Re}\left[\mu(\varepsilon)e^{-j\theta}\right] = \max_{\varepsilon,\theta}|\mu(\varepsilon)|\mathrm{Re}\left[e^{-j(\theta-\arg\mu(\varepsilon))}\right]. \quad (18.41)$$

Timing estimates can be obtained by first maximising the absolute value of $\mu(\varepsilon)$

$$\hat{\varepsilon} = \arg\max_{\varepsilon}|\mu(\varepsilon)|, \quad (18.42)$$

$$\hat{\theta} = \arg\mu(\hat{\varepsilon}). \quad (18.43)$$

We can use a timing error feedback system for timing recovery. Derive the likelihood function with respect to the parameter ε:

$$\frac{\partial}{\partial \varepsilon}L(\hat{a}, \theta, \varepsilon) \propto -\frac{2}{\sigma_n^2}\mathrm{Re}\left[\sum_{n=0}^{N-1}\hat{a}_n^*\frac{\partial}{\partial \varepsilon}z_n(nT+\varepsilon T)e^{-j\theta}\right], \quad (18.44)$$

The error signal is

$$x(nT) = \mathrm{Re}\left[\sum_{n=0}^{N-1}\hat{a}_n^*\frac{\partial}{\partial \varepsilon}z_n(nT+\varepsilon T)\bigg|_{\varepsilon=\varepsilon'}e^{-j\theta}\right]. \quad (18.45)$$

The sequence to be derived (nT + εT) is passed through a digital filter consisting of hd(kT$_s$) to complete the sequence derivation, and the output of the filter is the resulting error signal, which is fed back to the front-end for timing adjustment.

It is worth explaining that, in the all-digital receiver, the value of the signal at the optimal sampling point is not directly sampled, but is obtained by interpolation using a sequence of sample values of the sampled signal. Thus the interpolation filter is based on the timing adjustment of the signal, rather than on the local oscillating clock or timing waveform. The receiver receives a time-continuous signal r$_f$(t) with data spaced at intervals of the transmit symbol period T. Since r$_f$(t) is band-limited,

18.1 Channel Receiver Realisation Technology

the input signal can be sampled at a fixed clock rate $1/T_s$, yielding a sampling value of r_f (mT_s). Since the transmit symbol period T and the local sampling clock are mutually independent sources and there is always a small difference between the clock frequencies at which each operates, the ratio T/T_s is generally irrational in practical systems. The function of the interpolation filter hI(t) is to interpolate and filter the digital sequence r_f (mT_s) obtained from equally spaced sampling, and to obtain a new sample point, i.e., the interpolation point, at the moment $nhI + \varepsilon TI$. Here, $TI = T/M$ is the interpolation interval; M is an integer indicating the over-sampling rate; ε is the time delay with respect to TI. In digital TV broadcasting, timing loops are using error feedback structures, so only the interpolation control of the error feedback system is described here. In the actual loop operation, the T/T_s ratio is unknown, so the loop is used to generate the extraction coefficients mk and interpolation coefficients μk. The value of the sequence $\{r_f(kT_s)\}$ is calculated from the sequence $\{r_f(kT_s)\}$ at the sampling point, which is known as interpolation, and then the interpolated data is only required for further processing during the time gap mnT_s, which is known as extraction. Extraction is easily implemented in digital circuits by discarding unwanted interpolation points. Interpolation can be implemented in the form of various filters including FIR filters, polynomial FIR filters, Lagrange interpolation, etc.

2. Carrier synchronisation [8, 9]

(1) Carrier phase estimation

1) Carrier phase estimation in DD mode

Assuming that the timing synchronisation has been completed, i.e., ε is known, the objective function is obtained

$$L(\hat{a}, \hat{\varepsilon}, \theta) = \text{Re}\left[\sum_{n=0}^{N-1} \hat{a}_n^* z_n(\hat{\varepsilon}) e^{-j\theta}\right], \quad (18.46)$$

The objective function gets the maximum value

$$e^{j\hat{\theta}} = e^{j \arg \sum_n \hat{a}_n^* z_n(\hat{\varepsilon})}. \quad (18.47)$$

The phase error signal is easily obtained by differentiating θ when a feedback system is used. The following is an example of a first order digital phase-locked loop (PLL) to illustrate the phase error feedback system. Let there be no noise generation, accurate timing and correct sign judgement, i.e., $a_n = \hat{a}_n$ (at this point the DD approach is consistent with the DA approach). In this case, the matched filter output is equal to

$$z_n = a_n e^{j\theta_0}. \quad (18.48)$$

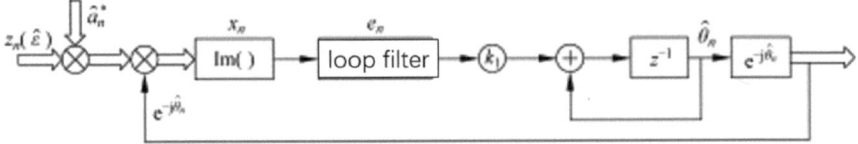

Fig. 18.6 Phase error feedback system

The error signal xn can be obtained as:

$$x_n = \text{Im}\left[|a_n|^2 e^{j(\theta_0 - \hat{\theta}_n)}\right] = |a_n|^2 \sin \psi_n, \quad (18.49)$$

The error signal will be further processed in the loop filter and next the phase valuation is updated in the integrator according to the following equation, namely

$$\hat{\theta}_{n+1} = \hat{\theta}_n + k_1 e_n. \quad (18.50)$$

Substituting gives us that

$$\hat{\theta}_{n+1} = \hat{\theta}_n + k_1 \sin \psi_n, \quad (18.51)$$

where k1 is the loop constant. The entire feedback system is shown in Fig. 18.6.

2) NDA carrier phase valuation

The use of the NDA method makes the performance of the synchronisation unit degraded, so for QAM signals, which are commonly used for high SNR, DD is generally used for phase recovery, unlike for M-PSK signals. The NDA algorithm can be used when reliable data valuation is not available, e.g., in the case of low signal-to-noise ratios. For analogue implementations, the NDA method is used almost exclusively, which reduces complexity, but is not applicable to digital circuit implementations. The NDA algorithm can be summarised as follows

$$F(|z_n(\hat{\varepsilon})|) e^{j \arg z_n(\hat{\varepsilon}) M}, \quad (18.52)$$

where F(|x|) is an arbitrary function. This algorithm is called the Viterbi algorithm as shown in Fig. 18.7.

(2) Carrier frequency estimation

When there is a frequency offset Ω, the linear model presented above has to be modified. When dealing with the frequency offset Ω, the transmitted phase θT is defined as $\theta T = \Omega t + \theta$, where θ is the fixed phase deviation.

We require that the frequency response $C(\omega)$ of the channel and the frequency response $F(\omega)$ of the pre-filter are flat in the frequency range $|\omega| \leq 2\pi B + |\Omega \max|$

18.1 Channel Receiver Realisation Technology

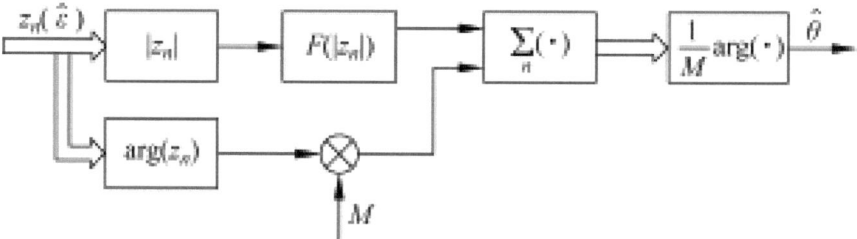

Fig. 18.7 NDA phase recovery algorithm proposed by Viterbi

where B is the single-sideband bandwidth of the signal u(t); Ω_{max} is the maximum frequency error. Under this condition, the signal $S_f(t, \Omega)$ can be written as

$$S_f(t, \Omega) = \sum_{n=0}^{N-1} a_n g(t - nT - \varepsilon T) e^{j(\Omega t + \theta)}. \tag{18.53}$$

At this point the output $z_n(\varepsilon, \Omega)$ of the matched filter is

$$z_n(\varepsilon, \Omega) = \sum_{k=-\infty}^{\infty} r_f(kT_s) e^{-j\Omega kT_s} g_{MF}(nT + \varepsilon T - kT_s). \tag{18.54}$$

Depending on whether timing information is used or not, the frequency estimation algorithms can be classified into the timing-assisted algorithm Dε and the timeless-assisted algorithm NDε. The Dε frequency estimation is carried out before timing recovery, i.e., the frequency estimation is carried out without the information of timing and data, and the derived frequency estimation algorithms must work independently of the other parameters. In the absence of timing information, the signal-to-noise ratio is lower than that of the matched filter output in the presence of timing information, and the corresponding NDε algorithm is susceptible to white noise. The problem can be solved by using a long averaging interval or by using a two-stage approximation method, i.e., a frequency capture stage and a frequency locking stage. When two-stage approximation is used, each stage has an algorithm to fulfil a special purpose. The purpose of the frequency capture phase is to quickly obtain an approximate frequency estimate. Thus, algorithms suitable for large ranges and short times are needed for the capture process, but tracking performance is not considered. In the frequency locking phase, the aim is to optimise tracking performance as a large capture range is no longer required. The algorithms for the frequency-locked phase, which will be described below, have in common the need for correct timing, and therefore require the timing circuitry to work properly even with a certain amount of frequency offset remaining. Again, the corresponding frequency feedback algorithm can be obtained by differentiating the likelihood function with respect to Ω.

1) NDε frequency estimation

This type of algorithm is mostly used in the frequency capture stage, when there is no timing information. That is, the frequency offset Ω has to be estimated first, and by compensating Ω, the other synchronisation parameters are then estimated.

We can obtain the NDε algorithm through spectral analysis by making a Fourier expansion of the time domain waveform $|z(lT + \varepsilon T)|^2$

$$\sum_{l=-L}^{L} |z(lT + \varepsilon T, \Omega)|^2 = c_0 + 2\mathrm{Re}[c_1 e^{j2\pi\varepsilon}] + \sum_{|n|\geq 2} c_n e^{j2\pi n\varepsilon}, \tag{18.55}$$

The analysis shows that the values of c0 are related and not related. The mean value of c0 achieves its maximum value when taken to the true value 0. Therefore, the unbiased estimate that can be obtained by maximising the coefficient $c_0()$, i.e.

$$\hat{\Omega} = \arg\max_{\Omega} c_0(\Omega), \tag{18.56}$$

where the sampling rate is to be satisfied that $1/T_s > 2(1+)/T$ and the data $\{a_n\}$ are to be independently and identically distributed (Fig. 18.8).

$$\hat{\Omega}T_s = \arg\left[\sum_{L_s} r_f(kT_s)r_f^*[(k-1)T_s]\right]. \tag{18.57}$$

2) Dε frequency estimation

In this section, it is assumed that timing synchronisation is established before frequency synchronisation. Since the allowable frequency offset range is not only determined by the capture range of the frequency estimation algorithm itself, but also depends on the ability of the timing algorithm to recover timing when there is a frequency offset, the frequency offset must not be too large at this time. The algorithms in this section are generally applicable when the frequency offset approximately satisfies $|\Omega T/2\pi| \leq 0.15$. This restriction is not strict because these algorithms are generally in the frequency locking phase. If the frequency offset does not satisfy

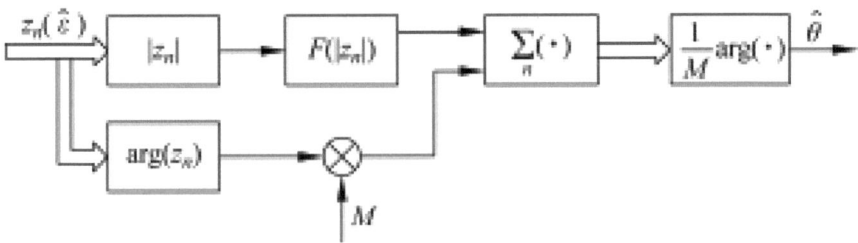

Fig. 18.8 Frequency estimates obtained using incremental phases

18.1 Channel Receiver Realisation Technology

the requirement, then it is first necessary to go through a frequency capture phase to roughly estimate the frequency offset and pull it into the required range.

First we discuss the case with auxiliary data. The auxiliary data indicates that we already know the sign, denoted by $\{a_{0,n}\}$. The ML algorithm requires a joint estimate of $\{\Omega, \theta\}$:

$$\{\hat{\Omega}, \hat{\theta}\} = \arg\max_{\Omega, \theta} \sum_{n=-(N-1)/2}^{(N-1)/2} a_{0,n}^* e^{-j\theta} e^{-jnT\Omega} z_n. \tag{18.58}$$

The two-dimensional maximisation problem of the above equation with respect to $\{\Omega, \theta\}$ can be reduced to a one-dimensional one:

$$\arg\max_{\Omega, \theta} \mathrm{Re} \left[\sum_{n=-(N-1)/2}^{(N-1)/2} a_{0,n}^* e^{-j\theta} e^{-jnT\Omega} z_n \right]$$

$$= \arg\max_{\Omega, \theta} |Y(\Omega)| \mathrm{Re} \left[e^{-j[\theta - \arg(Y(\Omega))]} z_n \right], \tag{18.59}$$

3. Example of ATSC/8VSB system [10]

ATSC system is a typical single-carrier system using VSB, which is a traditional common TV transmission method and its technology has been relatively mature. The receiver of VSB mainly includes demodulation, comb filtering, equalisation, phase tracking and channel decoding. When the receiver is turned on or switched to another channel, the receiver is reset, and each part establishes a synchronous working state in the following order: ① channel selection extracts the signal of the desired channel and filters out the out-of-band signals; ② incoherent AGC limits the unlocked signals to the range required for A/D conversion; ③ carrier extraction; ④ data segment synchronisation and clock extraction; ⑤ coherent AGC of the signal; ⑥ data field synchronisation extraction; ⑦ NTSC interference cancellation filter enable judgement; ⑧ equaliser to complete the adjustment of the tap coefficients; ⑨ TCM and RS decoding start. Several mutually independent loops related to synchronisation will be described in detail, which are used for carrier recovery, symbol clock and data segment and field synchronisation extraction respectively.

(1) Channel filtering and VSB carrier recovery

In order to reliably recover the carrier in harsh environments, the ATSC inserts a frequency guide signal. The inserted guide frequency only increases the signal power by 0.3 dB, which basically has no effect on the system BER performance. The VSB modulator filters out the lower sideband of the signal spectrum and performs appropriate spectrum shaping to minimise inter-symbol interference. The 3dB bandwidth of the transmitted signal is 5.38 MHz with a roll-off factor of 11.5%. That is, the pilot frequency falls at 0.31 MHz from the lower edge of the channel. This is exactly where the analogue NTSC signal spectrum has a high fading, thus producing less interference to the NTSC signal.

The carrier recovery of ATSC is accomplished by using a frequency-locked loop (FPLL) circuit to act on the small pilot signal, as shown in Fig. 18.9. The FPLL combines the frequency loop and the phase-locked loop into a single circuit, with the frequency loop providing a wide bandwidth of frequency locking range (±100 kHz) while the phase-locked loop has only a very narrow bandwidth (less than 2 kHz), and the two summed frequency rejection low-pass filters in the figure are to filter out the 2 × IF signal. During the locking process, the frequency loop uses both in-phase (I) and quadrature (Q) pilot signals. All other data circuits in the receiver use only I-channel signals (VSB signals use I-channel signals with a symbol rate of 2 times the effective bandwidth, i.e., 5.38 MHz × 2 = 10.76 MHz). Before phase locking, the difference frequency signal between the VCO signal and the input signal, as well as the high frequency signal and noise, enters the automatic frequency control (AFC) low-pass filter. The AFC filter determines the capture band of the FPLL, automatically tracking frequencies within the capture band, and the phase discrimination curve of the FPLL is determined by this filter. In this case, the high frequency data as well as noise and interference are filtered out by the AFC filter, and only the differential frequency signal containing the lead signal is passed through. By amplifying and limiting this derivative signal to a square wave of fixed amplitude (±1) and multiplying it by the quadrature signal, an AFC characteristic curve with a bipolar S-curve is obtained. The polarity of the curve depends on whether the frequency of the VCO is higher or lower than the input IF signal.

After filtering and integration by the automatic phase control (APC) low-pass filter, the resulting DC signal is used to adjust the frequency of the VCO to reduce the frequency deviation. The APC low-pass filter determines the phase-locking characteristics of the FPLL, automatically controlling the phase difference between the input signal and the output signal of the VCO when the input signal and VCO output signals have the same frequency. When the frequency difference approaches zero, the APC loop locks the input IF signal to the VCO signal. This is a conventional

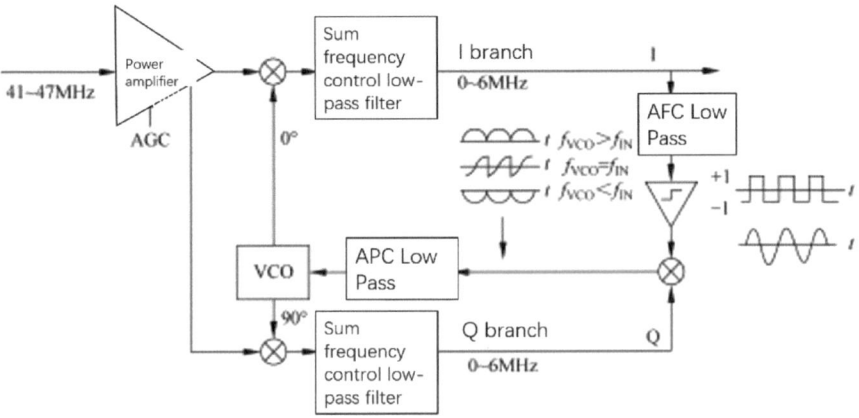

Fig. 18.9 ATSC carrier recovery

phase-locked loop (PLL), which is characterised by two-phase stability, i.e. both 0° and 180° are its stabilisation points. However, the correct phase-locked polarity can be determined by forcing the polarity of the lead-frequency signal to be equal to the known positive transmit polarity. Once locked, the detected lead signal is fixed and the limiter output fed back to the 2nd multiplier is constant. At this point, the phase-locked loop operates as usual and the frequency loop is automatically stopped. The passband of the APC low-pass filter is wide enough to reliably guarantee a lock range of ± 100 kHz rate, but sufficient to resist strong white noise (including data) and PAL co-channel interfering signals. The bandwidth of the PLL is narrow enough to block most of the amplitude-modulated signals and phase-modulated signals generated by the data, but can be tracked over the signals (and therefore also over the lead-frequency signals). (and therefore also on the pilot signal) phase noise in the range up to 2 kHz.

(2) Segment synchronisation and symbol clock recovery

A data-assisted timing synchronisation structure has been obtained above. And in the timing synchronisation of ATSC, segment synchronisation which appears periodically can be used as auxiliary data. The data segment of ATSC has a data format as shown in Fig. 18.10: each data segment is 832 symbols long, including segment synchronisation (4 binary symbols) and data symbols (828 symbols = 187Byte Packt + FEC overhead), in which the data symbols are represented by 3bit The data symbols are expressed in 3 bits (8 levels) and contain 828 × 3 = 2484 bits the ATSC uses the known signals of the 4 symbol segment synchronisation as auxiliary data when performing the timing synchronisation.

Sequence derivation can be accomplished by passing the I-channel digital signal to be derived through a digital filter consisting of h(kT$_s$). Where h(kT$_s$) is a 4-tap digital

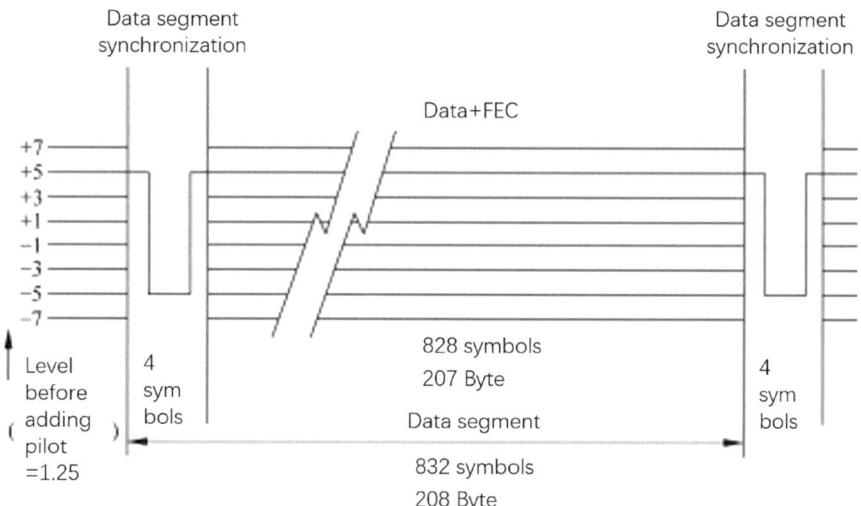

Fig. 18.10 ATSC segment structure

filter, after passing through this filter, the phase identification S-curve is obtained, as shown in Fig. 18.10. The segment synchronisation circuit gives the segment synchronisation moment, and samples the phase-discriminating output at that moment as the loop timing error signal, which is low-pass filtered by the APC and then drives a VCXO (with a centre frequency of 10.762 MHz) for sampling to complete the whole closed loop. It can be seen that the timing loop in the figure uses hybrid timing recovery, which obtains timing information from the receiver samples (rather than from the continuous time signal) and controls the VCXO to sample it. The sampled signal does not require the use of an interpolation filter to complete the timing error recovery. The segment synchronisation detector uses a 4-symbol synchronisation correlator to look for segment synchronisation signals that are inserted repeatedly. The subsequent receiver loop operates only after some predefined confidence level (using a confidence counter) has been reached to confirm that segment synchronisation has been found. Due to the use of specific segment synchronisation signals, segment synchronisation detection and clock recovery works reliably at signal-to-noise ratios of 0 dB or less.

(3) Non-coherent and coherent AGCs

Non-coherent automatic gain control (AGC) is performed whenever a signal (locked or non-locked, or noise/interference) enters the analogue-to-digital converter prior to carrier and clock synchronisation. The AGC adjusts the IF and RF gains accordingly, so that the amplitude of the signal entering the A/D meets the A/D's dynamic range. Coherent AGC is performed by the amplitude of the measured segment synchronisation when data segment synchronisation has been detected. The amplitude of the bipolar synchronisation (discrete level relative to the random data) is determined in the transmitter. Once the synchronisation signals are detected in the receiver, they are compared to a reference value and the difference (error) is integrated. The output of the integrator then controls the IF and 'time-delayed' RF gains, forcing them to provide some value of the correct synchronisation amplitude.

(4) Data Field Synchronisation

ATSC field synchronisation is inserted as shown in Fig. 18.11. The ATSC frame structure is divided into three layers, i.e., data segments, data fields and data frames, where the smallest unit is the data segment.313 data segments constitute a data field, and the first data segment of each data field is the data field synchronisation signal, so that a data field synchronisation is sent every 24.2 ms. Two consecutive data fields #1 and #2 constitute 1 data frame.

The specific structure of the data field synchronisation signal is shown in Fig. 18.12. The polarity of the three alternating 63 bit PN sequences in the figure determines whether field #1 or field #2 is detected.

Detection of data field synchronisation is achieved by comparing each received data segment from the A/D converter (after interference suppression filtering to reduce co-channel interference) with the ideal field #1 and field #2 reference signals in the receiver, without oversampling the field synchronisation. This is because the data segment and the synchronisation clock are established accurately and reliably by the

18.1 Channel Receiver Realisation Technology

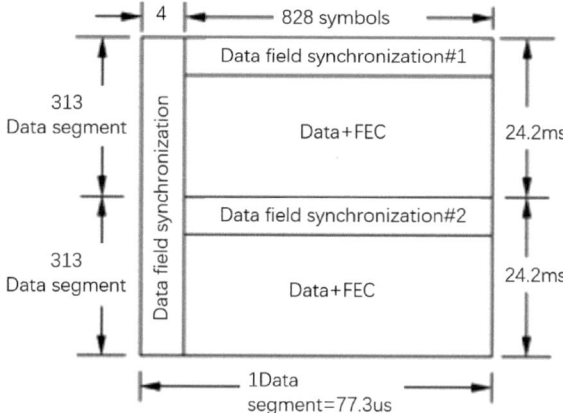

Fig. 18.11 ATSC frame structure

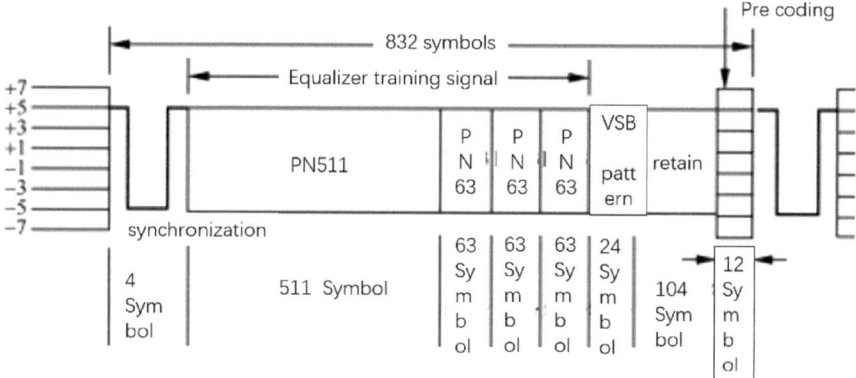

Fig. 18.12 Data field synchronisation signal structure

clock recovery circuit. Therefore, the field synchronisation recovery circuit knows exactly where a valid field synchronisation will occur within each data segment. Simply do the difference operation symbol by symbol, and when a predetermined confidence level is reached (using a confidence counter), the field sync signal is detected on the given data segment, and the data field sync signal can be used by the subsequent circuitry the polarity of the three alternating 63bit PN sequences determines whether the field sync detected is field #1 or field #2 in the data frame such field sync detection is very efficient, even at 0dB SNR. The field synchronisation recovery can be achieved reliably even when the signal-to-noise ratio is 0 dB or lower, and in case of severe interference.

(5) Phase tracking loop

A phase tracking loop is introduced to counter phase noise. The phase tracking loop is an additional judgement feedback loop that further removes the remaining phase noise from the carrier recovery process. In this way, phase noise is removed not by one loop but by two cascaded loops. Since the system is already locked to the conduction frequency rate by the IF PLL (independent of data), the bandwidth of the phase tracking loop can be used as a first-order loop, for which the phase tracking is already maximised. Higher order loops (which are required for frequency tracking) are not as good as first order loops for performing phase tracking, and therefore are not used in VSB systems.

An important feature of VSB modulation is that unlike QAM it does not have equivalent I/Q values and so has an advantage in resisting phase noise. The block diagram of the phase tracking loop principle is shown in Fig. 18.13. Assuming the presence of phase noise, the actual equaliser outputs a sequence of I-way information as I(n), which is first passed through a multiplier for gain control and then fed into a Hilbert filter to approximate the generated Q signal. It is clear from the previous discussion that because of the VSB transmission method used, its I and Q components are interconnected with an almost Hilbert transform filter function. This filter is not complex because it is a class of finite impulse response (FIR) filters with a number of fixed antisymmetric coefficients, while all other coefficients are zero. In addition, many of the filter coefficients are powers of 2, which simplifies the hardware design.

Because the phase tracker operates on 10.76 MS/s (Mega Symbol/s) of data, the bandwidth of the phase tracker is quite wide, about 50 kHz the gain multiplier is also controlled by the judgement feedback, and differs from the phase loop in that it extracts the amplitude information of I'(n) to get the amplitude error signal. Since the loop is a data judgement (DD) based loop, in order for this loop to work correctly, it is required that the data eye diagram has been turned on, i.e., there are few data judgement errors. When the receiver is first started up, the phase tracking circuit can only capture and track the phase noise of a known binary sequence first, as

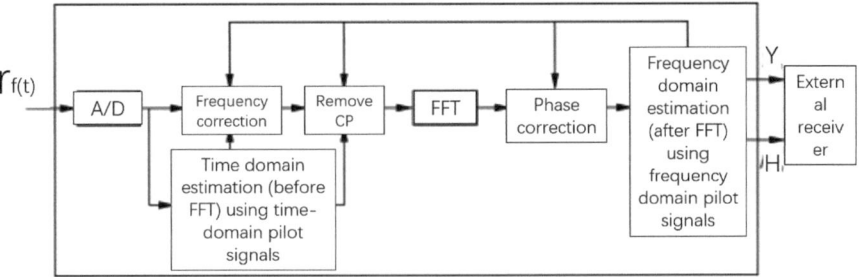

Fig. 18.13 Phase-tracking loop with data judgement feedback

18.1 Channel Receiver Realisation Technology

the data eye diagram is closed at this point and frequency adjustment can only be accomplished by the carrier recovery FPLL circuit first. Once the phase tracking is correctly synchronised on the training sequence, the judgement on the random data is then started and thus the tracking loop also starts tracking the data phase.

4. Channel Estimation and Equalisation

In general, time domain channel estimation and equalisation is more commonly used in single carrier systems. According to whether the output of the time domain equaliser is used for feedback control or not, adaptive equalisers can be classified into linear and non-linear equalisers. If the output is not used for feedback control, the equaliser is linear and vice versa, the equaliser is nonlinear.

(1) Linear equaliser [1]

The most commonly used structure for time domain equalisers is the finite impulse response filter (FIR) with transverse filters. Let the channel have L multipaths, then the input y_k of the equaliser can be expressed as

$$y_k = \sum_{j=0}^{L} h_j I_{k-j} + n_k, \tag{18.60}$$

where $\{n_k\}$ is the Gaussian white noise sequence, $\{h_k\}$ is the channel impulse response tap coefficient, and $\{I_k\}$ is the transmitted data sequence.

As shown in Fig. 18.14, let the number of taps before and after the main path of the equaliser be N. It can be shown that the estimated value of the kth symbol at the output of the equaliser is

$$\hat{I}_k = \sum_{j=-N}^{N} c_j y_{k-j}. \tag{18.61}$$

If the output of the equaliser is different from the sent data I_k, one error in judgement occurs. In fact, whether or not the judgement is wrong depends heavily on the choice of the filter tap coefficients $\{c_k\}$. A large body of literature has been devoted to the optimisation of filter coefficients, and the most commonly used criteria for optimising the equaliser coefficients are the least squares criterion and the minimum mean square error criterion described above.

(2) Judgement Feedback Equaliser [1]

Linear FIR filters eliminate inter-code interference at the expense of signal-to-noise ratio, which has very limited ability to compensate for deep fading, especially for channels with spectral zeros. Therefore, the study of nonlinear equalisers with low computational complexity has become a hot topic of discussion, and the judgement feedback equaliser (DFE) is proved to be an effective solution to this problem.

The DFE equaliser is a typical nonlinear equaliser. Figure 18.15 shows the block diagram of the operating principle of the DFE equaliser, which consists of two filters

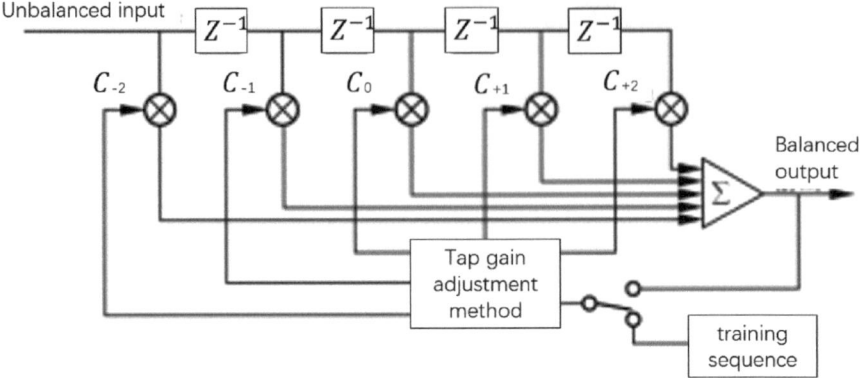

Fig. 18.14 Linear equaliser with lateral filter structure

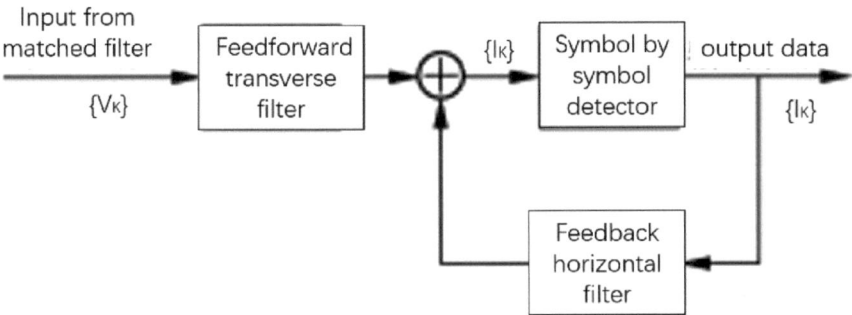

Fig. 18.15 DFE equaliser structure

whose tap intervals are both symbol intervals T, i.e., a feedforward filter (FIR) and a feedback (IIR) filter. The feed-forward part cancels out the inter-code interference that is ahead in time (determined by the position of the reference tap) and converts the ahead inter-code interference into lagging inter-code interference, while the feedback filter, which takes the output of the adjudicator as its input, cancels out the inter-code interference that is lagging in time. In contrast to the linear equaliser, the DFE equaliser operates on the principle that if a transmit symbol has been estimated, its interference (ISI) to the received symbols that follow it can be calculated by the feedback filter, which compensates the adjudicator input signal at the next symbol moment. Linear equalisers have only zeros and should theoretically be used to equalise channels containing only poles; whereas the judgement feedback equaliser can use its feedback feature to eliminate ISI very well, especially when the channel zeros are close to the unit circle. The performance of the DFE is dependent on the accuracy of the judgement, and as a decrease in the signal-to-noise ratio produces more unreliable estimations, the performance of the DFE in a low signal-to-noise ratio environment significantly degradation.

18.1 Channel Receiver Realisation Technology

Equalised output with linear filter

$$\hat{I}_k = \sum_{j=-N}^{M} c_j y_{k-j} = \sum_{j=-N}^{0} c_j y_{k-j} + \sum_{j=1}^{M} c_j y_{k-j}, \qquad (18.62)$$

Unlike, the output of the judgement feedback equaliser is denoted as

$$\hat{I}_k = \sum_{j=-N}^{M} c_j y_{k-j} + \sum_{j=1}^{M} b_j \hat{I}_{k-j}. \qquad (18.63)$$

Its IIR filter in the feedback section operates on M judged values, and no noise is introduced under no misjudgement conditions, i.e., the backward ISI interference can be completely eliminated. And the noise of the equaliser comes only from the feed-forward part (FIR), thus the noise performance of the judgement feedback equaliser is greatly improved compared to the linear equaliser. In addition, the IIR filter in the feedback section cancels the multipath without generating derived multipath, and the length of the IIR section can be determined based on the length of the backward ISI delay that is desired to be cancelled. The performance improvement is more obvious especially under strong channel fading conditions. The IIR can theoretically eliminate the backward interference completely, but it must be required that the feedback volume is judged correctly, and if a misjudgement occurs, it will be fed back into the IIR and enter into the next computation, which affects the later equalisation value, and thus may trigger new misjudgements until the misjudgement value leaves the IIR, and there is the problem of error propagation. Like linear equalisation, judgement feedback equalisation also calculates the tapping coefficients based on the LS criterion and the MMSE criterion.

(3) Adaptive Channel Valuator [1]

The principle of adaptive channel valuer is based on the maximum likelihood criterion to estimate the optimal receiver of the correct symbol sequence from the received signals where symbol interference occurs, hence it is also known as maximum likelihood sequence valuer. The block diagram of the principle of the MLSE valuer implemented with the Viterbi algorithm is shown in Fig. 18.16. It consists of the channel valuer and the Viterbi algorithm, and the structure of the channel valuer is the same as that of the linear lateral equaliser. The channel-induced parameter variations are given by the channel valuer, and the algorithm minimises the error between the actual received sequence and the channel valuer output. When the adaptive channel valuer is activated for equalisation, a known short sequence is used for the initial adjustment of the tap coefficients; when tracking channel variations, the judgement of the signal itself is used directly to form the error signal.

The MLSE valuer is optimal from the point of view of error probability. However, the computational complexity of MLSE in the presence of ISI channels grows exponentially with the channel length, e.g., if the number of symbols is M and the number of interfering symbols causing ISI is L, the Viterbi algorithm has to compute

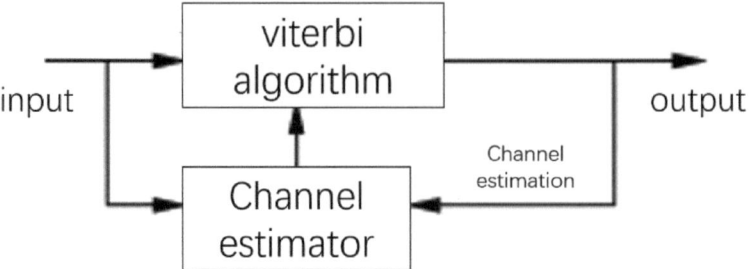

Fig. 18.16 Maximum likelihood sequence valuator based on the Viterbi algorithm

$M(L + 1)$ metrics for each new received symbol. For most practical channels, such a large computational complexity is too expensive to implement. Correspondingly, as mentioned earlier, the computational complexity of the linear filter is a linear function of the channel length L, which is less computationally intensive, but when there is a deep spectral fading in the channel, the linear equaliser will have a large gain on the section of the spectrum where the deep fading occurs, as well as the neighbouring spectra, thus increasing the noise, and it would be better to use a nonlinear process at this point. Currently most digital systems use DFE for equalisation estimation.

(4) Adaptive Equaliser [1]

In the equalisation methods described earlier, it is assumed that the characteristics of the channel are known at the receiver. However, in general, the channel characteristics are unknown a priori and the channel response is time-varying. In such cases, it is often necessary for the equaliser to be able to automatically adjust itself to follow channel changes, called adaptive equalisation.

According to the different adjustment phases, adaptive equalisers are divided into two operating states: training mode operating state and tracking mode equalisation state. In the training mode operating state, a known training sequence is sent to start the equaliser and make it converge quickly so as to complete the initialisation of the weighting of the tap coefficients; in the tracking mode operating state, the equaliser directly uses the judgement of the digital signals transmitted in the communication to form the error signals and adjusts the tap coefficients based on the tracking of adaptive algorithms to maintain the channel equalisation under the optimal equalisation criterion by automatically adapting to the random changes of the channel. In practical systems, special training sequences are often set up in the transmitted digital signals in order to facilitate the convergence and tracking of the equaliser. Based on the criteria such as zero forcing and linear least mean square error mentioned in the previous section, the main adaptive algorithms obtained are zero forcing (ZF), linear least mean square (LMS), recursive least square (RLS) and fast Kalman (fast Kalman), etc., of which the LMS algorithm is one of the most commonly used adaptive equalisation algorithms. When comparing these algorithms, the main considerations are the fast convergence characteristics of the algorithms, the tracking of fast time-varying channel characteristics and the computational complexity, which can be found in the relevant references.

5. Example of an ATSC/8VSB system [11]

The channel estimation and equalisation of the ATSC system mainly relies on its unique adaptive time domain equaliser. The adaptive equalisation algorithm can be used in three ways, i.e., the training sequence (DA) method in field synchronisation, the data symbol (DD) method when the eye diagram is open, and the 'blind' equalisation (NDA) method when the eye diagram is not open.

A typical VSB receiver equalisation structure is shown in Fig. 18.17. The most widely used in ATSC is the LMS adaptive algorithm. When the equaliser is started, the training sequence from the field synchronisation must be used first in order for the equaliser coefficients to converge as quickly as possible. Based on the known training signal, the error signal is obtained by subtracting the locally generated training sequence from the equaliser output. At the beginning when the training signal is used, the eye diagram is closed, and the purpose of using the training sequence is to open the eye diagram. After adaptive equalisation of the training signal, the eye diagram is opened, at which point the judgement data is so correct that the equaliser switches to judgement mode of operation and uses the symbols of the judgement output to generate the error signal. Because the equaliser training sequences are separated by only 24.2 ms, it is not possible for the receiver to achieve equalisation updates at a faster rate, during which rapid changes in multipath cannot be tracked. At this point, blind equalisation techniques that do not require training signals are crucial. In general, the existing blind equalisation algorithms mainly include the constant mode algorithm (CMA) and the reduced constellation diagram (RCA) method, but both methods are applied to QAM signals with two-dimensional constellation diagrams, while the VSB signal is a one-dimensional signal, so it cannot be directly applied in the VSB system, which leads to the proposal of the modified reduced constellation algorithm (MRCA), which determines the VSB by finding a one-dimensional RCA algorithm that judgement region so that the adaptive equaliser converges to complete the blind estimation without the use of training sequences. The basic idea of the MRCA algorithm is that the entire VSB constellation is aggregated into clusters to determine the boundaries of the judgement region. These clusters can be further divided into smaller clusters until each cluster contains only one symbol. Each division of the clusters serves to make more judgements correct and opens the eye diagram further.

In the digital TV terrestrial broadcasting channel, the backward multipath delay may be as long as 20 μs, and in the single-frequency network application the delay even reaches more than 30 μs, and the forward multipath delay is generally required to reach more than 5 μs. Taking the ATSC symbol rate of 10.768MS/s as the calculation, the number of taps of the equaliser is generally selected as 64 levels for forward FIR and 192 levels for backward IIR. At this point, two major difficulties will be encountered: the excessively long number of equaliser stages results in a complex structure and a large scale, which leads to difficulties in implementation and a serious

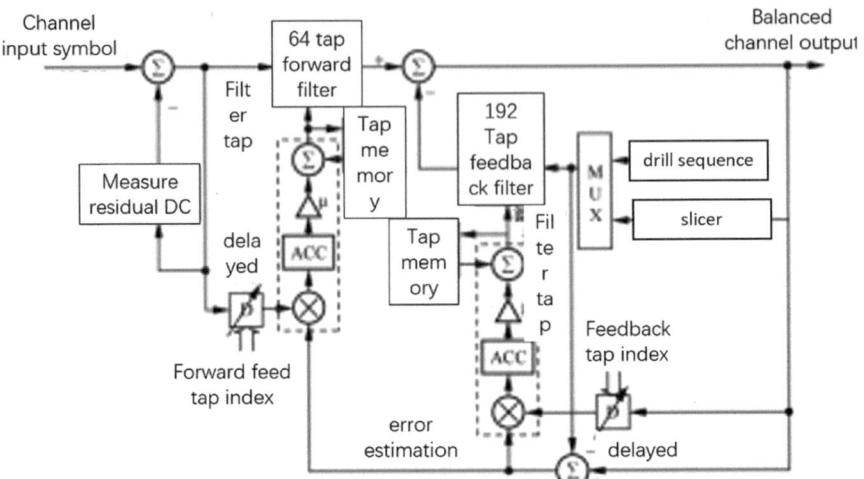

Fig. 18.17 VSB receiver equaliser structure diagram

spread of BERs. Therefore, the current research work on DFE also focuses on the following two points:

(1) Reducing BER spreading

Due to the existence of a large number of equaliser stages and the inherent error propagation phenomenon of DFE, these lead to a rise in the signal-to-noise ratio threshold of the equaliser. To compensate for this, TCM coding is used in the ATSC system, which combines the equaliser with the Viterbi decoding of the back stage to reduce the effect of error propagation, and uses the Viterbi decoded output to feed back to the equaliser as part of the judgement data in the IIR in the DFE. Since Viterbi decoding is time consuming, the judgement data in the IIR part can still be provided by the local hard judgement mechanism.

(2) Reduced complexity

It is important to make the necessary simplifications to the structure of the equaliser containing tens of levels of FIR and hundreds of levels of IIR. Ideally, the tapping coefficients in the IIR are non-zero at locations with multipath and should be equal to zero at all other locations. In practice, since the multipath distribution is very sparse, there are very few taps with large channel energy, and the vast majority of taps are either zero or very close to zero due to system-based noise as well as computational errors and so on. Therefore, the coefficients of the taps at these non-multipath locations can be forced to zero in terms of noise reduction and computational reduction. These zero-coefficient taps can be completely replaced by a beat delayer, at which time the corresponding multiplication and addition structure is also omitted, achieving the purpose of saving hardware resources. For example, the number of taps is compressed by correlating the tap coefficients before and after according to

18.1 Channel Receiver Realisation Technology

the exponential fading property of the channel characteristics, and the number of taps is determined by selecting an appropriate threshold on the tap coefficients using the DFE equaliser with dynamic taps to reduce the size of the FIR and IIR. And from the hardware implementation point of view, the reuse of multipliers is also used to reduce the number of multipliers that account for the heavy head of hardware resources. In particular, it is noted that the adoption of this approach requires the hardware to provide a higher operation clock than the symbol clock of the signal, then the hardware resources can be further utilised by increasing the clock.

Although the ATSC equaliser has been further improved, there are still some problems with the current practical ATSC equalisation system: the pilot signal is severely affected by strong multipath variations in the near future, and the carrier recovery becomes difficult, and the performance of this equaliser drops dramatically when the carrier is not accurately recovered; although the system uses training sequences, the separation between two training sequences is large, and the fast multipath variations in this period cannot be tracked. Although the ATSC system also uses data judgement feedback (DFE), which uses the error signals generated by the data itself for conditioning, to track fast changing multipath, DFE requires the channel to be equalised to a certain degree (less than 10% error judgement) to work properly and is unstable under strong multipath. Therefore, the original design idea, training sequence insertion, and data structure of the ATSC single-carrier system make the system unable to effectively deal with strong multipath and fast-changing dynamic multipath, which results in unstable reception of certain fixes and unsupported mobile reception.

18.1.4 Multi-carrier Systems

1. Timing Synchronisation [5, 6]

As shown earlier, the timing synchronisation problem for multicarrier systems is similar to that for single carriers and will not be repeated here. Only TDS-OFDM systems using time-domain training sequences will be used as an example. Although TDS-OFDM systems use PN sequences as protection intervals to reduce the system overhead, at the same time, they place higher demands on timing deviations. The ISI generated by the presence of timing deviation will destroy the orthogonality between the subcarriers and cause Inter-carrier Interference (ICI). For example, at that time $\varepsilon > 0$, the output of the inner receiver can be expressed as

$$Y_{i,n} = e^{j2\pi \frac{n}{N}\varepsilon} \frac{N-\varepsilon}{N} X_{i,n} \cdot H_{i,n} + n_{i,n} + n_\varepsilon(i,n), \qquad (18.64)$$

2. Carrier synchronisation

As far as carrier frequency synchronisation is concerned, due to the small subcarrier bandwidth of OFDM, it is very sensitive to the carrier frequency deviation, so very accurate carrier synchronisation is required. The frequency deviation in OFDM can be generally decomposed into two parts with respect to the subcarrier spacing: the integer part and the fractional part. The integer part is the part of the frequency deviation that is equal to an integer multiple of the subcarrier spacing, and the fractional part is the part of the frequency deviation that is smaller than the subcarrier spacing. The integer part only causes a cyclic shift in the subcarrier positions and does not destroy the orthogonality between the subcarriers. The fractional part generates ICI, which leads to degradation of signal-to-noise ratio of the signal. OFDM carrier synchronisation generally starts with fine synchronisation in the time domain using cyclic training sequences to estimate the fractional part of the carrier deviation, and then coarse synchronisation in the frequency domain using the frequency guide to estimate the integer multiples of the frequency deviation. This sequence of fine synchronisation followed by coarse synchronisation eliminates the ICI caused by the fractional part of the carrier deviation, so that the subsequent coarse synchronisation of the carrier is not affected by the ICI, and then the fine synchronisation can be performed again using the guide frequency.

(1) Carrier synchronisation using the guide frequency [12]

The tracking phase is introduced first. In the tracking phase it is assumed that the remaining frequency deviation is much less than half of the subcarrier spacing (0.5/T_{sym}). If only 1 sub-channel is considered at this point, the frequency synchronisation problem is analogous to the carrier synchronisation problem in a single-carrier system, and therefore frequency estimation can be performed using, for example, the ML method. This frequency estimation algorithm is to turn the frequency estimation problem into a problem of estimating the frequency offset between the sample values (i.e., $Y_{i,n}$ and $Yi + 1, n$) in two consecutive subcarriers.

When there is a carrier frequency offset Δf, the output of the FFT is expressed as

$$Y_{i,n} = e^{-j\pi i \Delta f T_{sym}} \text{sinc}(\pi \Delta f T_u) X_{i,n} H_{i,n} + n_{i,n} + n_{\Delta f}(i,n). \tag{18.65}$$

Next is the capture phase. In order to avoid using additional capture guides, we use the same frequency guide sequence as described above to implement the capture process, and the capture phase should be completed with $|\Delta f T_{sym}| < 0.5$. The amplitude of $YI_{i,n}$ reaches its maximum value when the remaining frequency deviation is zero. Thus the initial frequency estimate can be searched on the basis of finding the maximum magnitude of $\text{sinc}2(\pi \Delta f T_u)$, i.e., the maximum value can be reached when it coincides with Δf. Thus it is possible to obtain

$$\hat{f} \pi T_{sym} = \max_{f_{trial}} \left[\left| \sum_{n=0}^{N_p} \left(Y_{i,p(n)}(f_{trial}) Y^*_{i+D,p(n)}(f_{trial}) \right) \left(X^*_{i+D,p(n)} X_{i,p(n)} \right) \right| \right], \tag{18.66}$$

f_{trial} is the test frequency used in the capture phase and $YI, p(n)(f_{trial})$ is the output of the FFT cell corrected for the f_{trial} frequency. In order to avoid the maximum value appearing in $f_{trial} - \Delta f = [p(j)-p(j-1)]/T_{sym}$ (j = 1, 2, …… Np), the PN sequence is generally chosen as the frequency guide, which is not only autocorrelation better, but also easy to be systematically implementation, which can be achieved simply by shifting registers. It can be seen that the capture time is directly proportional to the frequency change step used in the search, and when the frequency deviation is relatively large, such as ± 10 subcarrier intervals, the capture time will be longer. However, in the practical application of continuous transmission, it is only necessary to perform the capture process once at the beginning of the transmission, so it is not too much of a problem.

(2) Using time domain training

From the above algorithm derivation, it can be seen that the method used is very similar to the single-carrier method, except that the single-carrier synchronisation is performed in the time domain, while the OFDM system places the processing in the frequency domain. The method of using frequency guides provides better estimation results, but the number of guides inserted is generally less than the data because the insertion of guides reduces the information rate. Also OFDM is affected by frequency bias which causes ICI and can destroy the performance of frequency estimation. Also the algorithm is generally used only after obtaining correctly timed information, thus its synchronisation time is long. In order to overcome the above drawbacks, special training data in the time domain with a repetitive structure is inserted between OFDM signals to complete the carrier synchronisation in the time domain, e.g., using a structure similar to that of Fig. 18.18. In this case, the same algorithm as the single-carrier system can be used completely to complete the synchronisation estimation based on the present signal frame or adjacent signal frames, which will not be described in detail here. This method can complete the synchronisation task much faster, so similar structures are often used in systems requiring fast synchronisation such as IEEE 802.11a.

(3) ML using cyclic prefixes [13]

Almost all multicarrier systems use the insertion of cyclic prefixes to eliminate inter-symbol crosstalk, so the repetitive nature of time-delayed cyclic prefixes can be used to achieve synchronisation as a way of overcoming the wasted system bandwidth from the use of frequency-guided bands.

Consider an OFDM system with N subcarriers and a cyclic prefix length of L. The length of each OFDM symbol is N + L samples, as shown in Fig. 18.19. It is assumed that the channel is non-diffuse and the transmitted signal is affected only by additive Gaussian white noise (AWGN).

Assuming that the starting position of the OFDM symbol is θ and the carrier deviation is Δf, the received signal is therefore

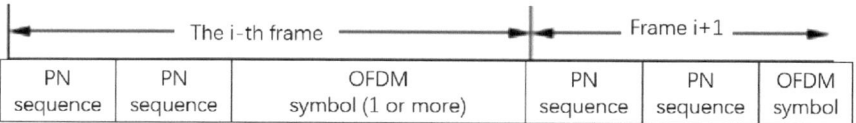

Fig. 18.18 An OFDM frame structure using time-domain training sequences

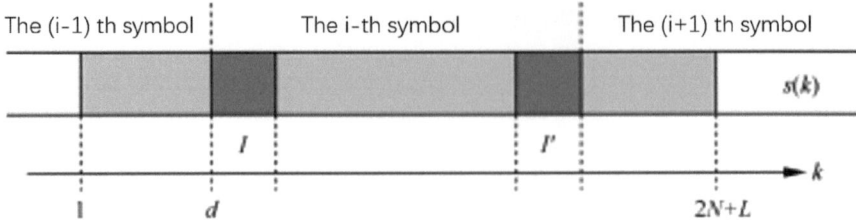

Fig. 18.19 Structure of an OFDM symbol with a cyclic prefix

$$r(k) = s(k - \theta)e^{j2\pi \Delta f k / N} + n(k). \quad (18.67)$$

3. Example of DVB-T/COFDM system

The COFDM used in DVB-T is a standard OFDM system with cyclic prefixes, and the length of the cyclic prefixes can be 1/4, 1/8, 1/16, and 1/32. The number of useful subcarriers is 1705 in the 2K mode, and 6817 in the 8K mode the labelling of the subcarriers is denoted as k ∈ [Kmin; Kmax]. In 2K, Kmin = 0, Kmax = 1704; in 8K mode, Kmax = 6816. In order to obtain stable frame synchronisation, frequency synchronisation, time synchronisation, channel estimation and transmit pattern recognition, and to perform phase noise tracking, not all of these 1705/6817 subcarriers are used for data transmission, and a part of them are used as frequency guides, functionally these Guided frequency is divided into 3 types: dispersed guided frequency, continuous guided frequency and TPS carrier.

(1) Continuous guide frequency is inserted at a fixed position for all OFDM symbols. It is 177 subcarriers in 8K mode and 45 subcarriers in 2K mode. For example, in 2K mode, the carrier positions of consecutive frequency guidance (refers to the number of subcarriers): 10, 48, 54, 87, 141, 156, 192, 201, 255, 279, 282, 333, 432, 450, 483, 525, 531, 618, 636,714, 759, 765, 780, 804, 873, 888, 918, 939, 942, 969, 939, 942, 969, 942, 942, 942, 942, 942, 942, 942, 942, 942, 969 939, 942, 969, 984, 1050, 1101, 1107, 1110, 1137,1140, 1146, 1206, 1269, 1323, 1377, 1491, 1683, 1704.
(2) The dispersal guide is for the Ith symbol in each DVB-T frame (0 ≤ I ≤ 67), for the subset

$$\{k = \text{Kmin} + 3 \times \text{Imod}4 + 12 \text{ pp is an integer, } p \geq 0, \ k \in [\text{Kmin}, \text{Kmax}]\} \quad (18.68)$$

The subcarriers at the location are dispersive guides. In each OFDM symbol, there is a dispersive guide every 12 subcarriers, and the position of the dispersive guide is repeated every 4 OFDM symbols, as shown in Fig. 18.20.

The values of the dispersed guide and continuous guide are generated in a pseudo-random binary sequence (PRBS). The PRBS is initialised to produce an output whose 1st bit w_k corresponds to the 1st useful subcarrier, and the PRBS generates new

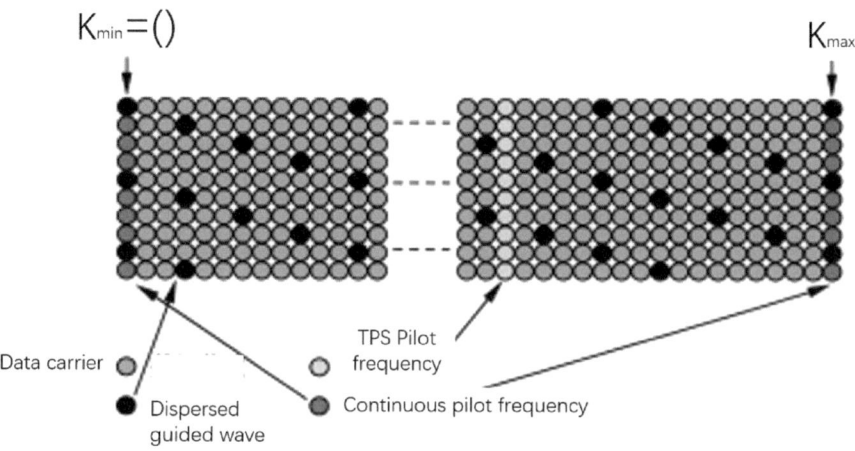

Fig. 18.20 An OFDM Frame Structure Employing a Time-Domain Training Sequence

values for each subcarrier, whether it is a guide or not. The polynomial of the PRBS generator used is: $\times 11 + \times 2 + 1$. Then each guide frequency bit is modulated and the modulated output $C_{i,k}$ is

$$\begin{cases} \mathrm{Re}\{C_{i,k}\} = \frac{4}{3} \times 2(\frac{1}{2} - w_k); \\ \mathrm{Im}\{C_{i,k}\} = 0. \end{cases} \quad (18.69)$$

(3) The TPS guide is used to transmit the adopted signal parameters, such as channel coding and modulation method. TPS guide is inserted at a fixed position in each OFDM symbol as in the case of continuous guide. There are 17 subcarriers in the 2K mode and 68 subcarriers in the 8K mode, and these information bits are modulated by differential coding. For example, in 2K mode, the TPS pilot carrier positions are: 34, 50, 209, 346, 413, 569, 595, 688, 790, 901, 1073, 1219, 1262, 1286, 1469, 1594, 1687.

The autocorrelation is good as we know where the guide signal is inserted, the value of each guide is calculated from pseudo-random binary sequence modulation, and the guides are sent at about 3 dB power above the data. When performing COFDM synchronisation, the autocorrelation of the guide frequency signals can be used to do the synchronisation estimation. Considering the different insertion positions of the dispersed FG in different symbols, if the dispersed FG is used, it is necessary to know the FG position of the current OFDM symbol, which undoubtedly increases the difficulty of implementation. In addition, the spacing of the TPSs is the same, which makes the autocorrelation peaks many, and increases the difficulty of judgement. Due to the small number of TPSs and the difficulty of implementation, the main guide frequency used for synchronisation is the continuous guide frequency. The guide frequency based OFDM synchronisation methods mentioned above can be used and these estimators give good results in channels with large frequency deviations and multipath fading. To take out the guide frequency it is necessary to first take out the OFDM data block for FFT using an FFT window, which requires a roughly accurate frame synchronisation, which is typically done using a cyclic prefix based method.

The entire DVB-T synchronisation process begins with frame synchronisation, using an algorithm that generally correlates the data in an OFDM block with its guard interval. Since the protection interval is the same as the last part of each OFDM block, the correlation between the protection interval and the last part of the OFDM block will have a strong peak. After signal filtering, the resulting signal is used by the coarse timing synchronisation process to control the position of the FFT window.

After completing the frame synchronisation, the N-point data of each OFDM symbol can be taken out for FFT through the FFT window. In the previous discussion, the various synchronisation deviations are considered separately, but in practice these deviations are always present and affect each other at the same time, and it is not possible to separate them to achieve the synchronisation separately as in the above analysis, so it is generally necessary to consider them at the same time for the joint synchronisation Therefore, these deviations are generally considered at the same time and joint synchronisation is carried out. At this time the output of FFT can be expressed as

$$Y_{i,n} = I_{i,n} X_{i,n} H_{i,n} + n_{i,n} + n_{\text{ICI}}(i, n). \tag{18.70}$$

It should be noted that the application environment of DVB-T generally has more serious multipath interference, at this time, it is generally necessary to combine the results of the channel estimation to obtain a more accurate timing synchronisation, so that the FFT window position is set more optimally. By observing the channel estimation, the FFT window is made to contain as much channel energy as possible. A simple method is to generate the main path of the channel impulse response roughly at half the protection interval. The time-domain impulse response can be obtained from the dispersive guides: divide the received dispersive guides by the value they send to get the undersampled channel frequency response, and use IFFT to get the time-domain channel response.

4. Channel Estimation and Multi-Carrier System Examples

(1) Guided frequency domain channel estimation [14]

Most of the frequency domain channel estimation algorithms for OFDM systems are based on the estimation algorithm of the guide frequency. The terrestrial wireless channel is a time-varying frequency-selective fading channel, and the transmitted data is often associated with a certain frame structure, and the channel is usually assumed to be constant over a frame due to the relatively slow change of the broadcast channel. When using OFDM techniques, the entire IDFT block is to be viewed as a linear time-invariant system, and only linear time-invariant systems can be subjected to DFT transformations, which can greatly simplify the complexity of the channel estimator. In order to get a fast and accurate channel estimation, the guide frequency information is inserted into the transmit sequence in a particular way.

18.1 Channel Receiver Realisation Technology

1) Guide frequency insertion style

The two most important parameters for selecting the guide frequency insertion pattern are the maximum Doppler frequency F_d and the maximum multipath delay τ_{max}. The guide frequency symbols are transmitted at specific positions in the time-frequency grid, which can be regarded as two-dimensional samples of the channel transfer function H(f,t), and the samples have to be close enough to satisfy the sampling theorem to avoid distortion. Therefore, the density minimum of the guide frequency symbols is determined by the Nyquist sampling theorem. Let there be N_t OFDM symbols spaced in the time direction and N_f subcarrier spacing in the frequency direction

$$N_t = \frac{1}{2F_d T_{sym}}, N_f = \frac{1}{\tau_{max} \Delta F_c}, \tag{18.71}$$

To make the channel estimation highly robust, the worst case of Doppler frequency and maximum delay of the channel is considered. Also more sampling of the fading channel is required to obtain reliable channel estimates. In order to be able to track the time–frequency variations of the transfer function in time, the guide frequency symbols should be placed close enough; but on the other hand, there should not be too many guide frequency symbols so as not to make the data rate too low, so that a compromise between the data rate and the performance of the channel estimation has to be achieved in the actual design, i.e., the one that is selected according to the application environment and the transmission service, and the compromise between the estimation accuracy and the effectiveness of the transmission should be carried out in the actual system, and it is generally It is recommended to use the sampling theorem in the time–frequency direction 2 times the number of guided-frequency symbols.

The common form of the guide frequency insertion is shown in Fig. 18.21. Figure 18.21a is the comb form of FGD, FGD signals are uniformly distributed in each OFDM symbol, and this form is sensitive to frequency selectivity; Fig. 18.21b is the periodic transmit FGD, and a certain OFDM symbol is full of FGD data, and this form is suitable for slow fading channels because all carriers are FGD, and this form is insensitive to frequency selectivity compared to the comb form; Fig. 18.21c is the periodic transmit FGD, and the comb form is not sensitive to frequency selectivity; and Fig. 18.21d shows the comb form of FGD insertion, and it is suitable for slow fading channels because all carriers are FGD, and this form is not sensitive to frequency selectivity compared to the comb form. In Fig. 18.21c, the data is interpolated in both the time and frequency directions, using fewer destination guides than the previous two methods. The frequency response of the data subcarrier position to be estimated in the figure can be obtained by interpolating the frequency guide positions around it. As shown in Fig. 18.21b, after sending a guide-frequency symbol, no guide-frequency is inserted in some subsequent symbols, which requires the channel to change very little or even remain unchanged for a considerable time, i.e., the

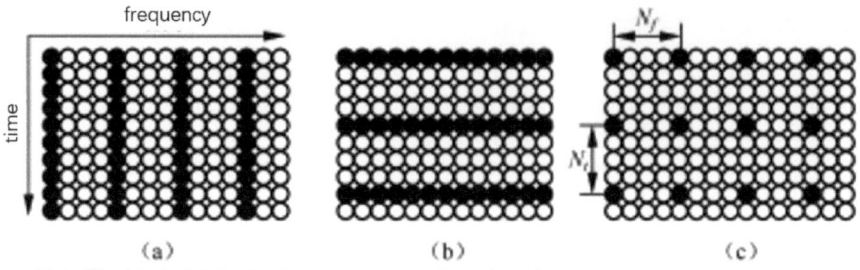

Note: The black dots in the figure represent pilot subcarriers, and the white dots represent data subcarriers

Fig. 18.21 Three common forms of lead frequency

channel is a quasi-quiescent channel, a slow-fading channel. This method is applicable to constant reference channels and can also be applied in WLAN networks. WLAN data is transmitted in bursts. In a burst pulse containing several OFDM symbols, the first few symbols are pilot symbols. The data is sent in the next OFDM symbols. Since the channel is considered to be slow fading, the channel estimation obtained from the first few symbols can be used to do the channel estimation of the data OFDM symbols. In Fig. 18.21c, the time-frequency two-dimensional frequency-guide effectiveness is higher than the first two, and the processing complexity is a little higher, which is applied in broadcasting services such as DVB and DAB.

18.1.5 Introduction to Receivers Within the European DVB-T Standard [15, 16]

The DVB-T receiver is a typical OFDM receiver, as shown in Fig. 18.22 for the DVB-T all-digital receiver system architecture. Except for the analogue front-end, all other parts can be implemented by digital devices, so all-digital demodulation can be achieved.

As shown in Fig. 18.22, the figure includes an RF high-frequency head/analog front-end, followed by an analog-to-digital converter, then the DVB-T channel demodulation and decoding, and the last one is the MPEG-2 source decoding unit. The DVB-T receiver work process is: IF analog signal input, first sampled by the ADC into a digital signal, analog-to-digital conversion before the completion of the coarse AGC control; and then, the baseband Then, the baseband conversion is completed from the IF signal to the baseband signal; next, the baseband signal is under the control of the time synchronisation and carrier synchronisation module for the recovery of time and carrier; further, the cyclic protection interval is removed from the stabilised synchronised signal for the FFT conversion; after the FFT, the fine AGC control, the reference symbols (guide frequency) extraction and the TPS decoding are completed, and the channel estimation and correction, as

18.1 Channel Receiver Realisation Technology

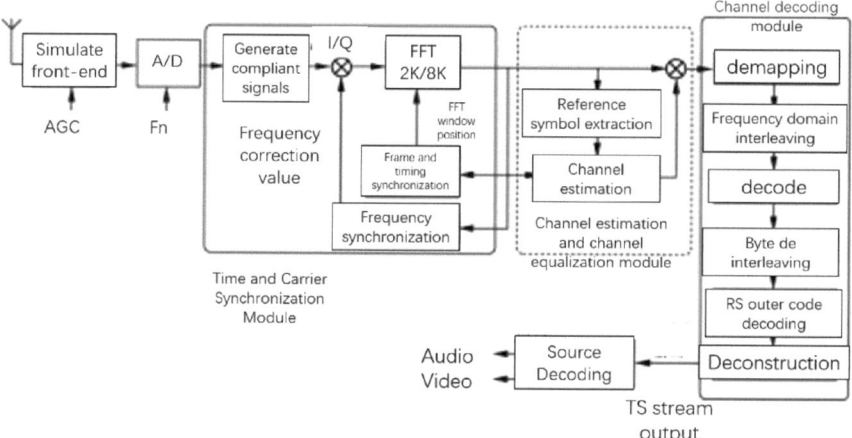

Fig. 18.22 Functional block diagram of DVB-T receiver

well as time fine AGC control are completed by using the extracted guide frequency data. After the FFT, fine AGC control, reference symbol (guide frequency) extraction and TPS decoding are done, then the channel estimation and correction, time fine synchronisation and carrier synchronisation are done with the extracted guide frequency data, then the balanced and corrected channel data are sent to the channel decoder; the channel decoder successively performs demapping, frequency interleaving (inner interleaving), Wirtby (inner coding) decoding, bit interleaving (outer interleaving) decoding, RS decoding (outer coding), and descrambling, and finally the demodulated data are outputted to the source decoder through the MPEG-2 TS interface.

The DVB-T channel demodulation and decoding unit is functionally divided into three main modules: time and carrier synchronisation module, channel estimation and channel equalisation module, and channel decoding module. The first two modules are the main issues of the inner receiver design, which effectively synchronise the OFDM code elements, sampling clock and carrier frequency at the receiving end, and complete the channel estimation, so that the performance of the OFDM outer receiver is close to ideal, while the channel decoding module belongs to the scope of the outer receiver design. The following will focus on the introduction of each module of the inner receiver in order.

1. Timing and carrier synchronisation

Timing and carrier synchronisation are the most critical parts of a DVB-T receiver. Timing synchronisation adjusts the sampling frequency and the position of the FFT window (i.e. frame synchronisation), while carrier synchronisation algorithms overcome the offset caused by the local oscillation. These algorithms use specific consecutive guides inserted into the OFDM spectrum, i.e., a collection of subcarriers modulated with fixed data. Timing recovery and carrier synchronisation with higher accuracy use dispersed pilot frequencies inserted into the OFDM block.

Firstly, timing synchronisation, including frame synchronisation and sample clock adjustment, is done by the following three steps:

(1) Frame header detection

The frame header detection algorithm is to correlate the data in the OFDM block with its protection interval, since the protection interval is the same as the last part of each OFDM block, the correlation between the protection interval and the last part of the OFDM block will appear as a very high peak. After some signal shaping process, the resulting signal is controlled by a coarse timing synchronisation process to control the position of the FFT window. To prevent the very strong co-channel interference generated by analogue TV signals at the same frequency from disrupting the timing synchronisation, the system incorporates a real-valued 11-plug low-pass filter to filter out the analogue TV carrier.

(2) More accurate timing synchronisation

This is to be done after coarse timing synchronisation, carrier synchronisation and guide frequency recovery. The algorithm is obtained by calculating the dispersed pilot frequency after analysing the impulse response of the channel. The undersampled channel frequency response is obtained by dividing the received dispersed guide frequencies by the value they were sent, and the time domain channel response can be calculated using IFFT. The optimal FFT window timing is obtained by delaying the observation window's such that the main path of the channel impulse response is generated at half the protection interval in advance. If the delay is large, the FFT window can be directly offset. When the delay is only one sample point, the position of the FFT window can be corrected by increasing or decreasing the sampling frequency. Because the FFT window drifts very slowly, it is not necessary to calculate the IFFT for every frame. e.g., when the sampling frequency error is 10–6 and the guard interval is 1/4, the FFT window in 2K mode is only offset by one sample point every 100 frames.

(3) Sampling Clock Adjustment

Sampling clock adjustment is accomplished by using changes in the phase of the dispersive lead subcarrier. Sampling frequency errors result in an increasing phase error with respect to the subcarrier frequency. From the increase in phase error a sampling error signal can be calculated, whereupon the sampling clock frequency can be controlled by this error signal using a feedback loop.

Then there is carrier synchronisation. The frequency of the receiver's local oscillator is unlikely to be stable, and Doppler frequency spreading is introduced in time-varying channels, so the receiver needs very accurate carrier synchronisation. Frequency errors cause each QAM signal to have a fixed rate of phase rotation and lead to interference between the subcarriers of each QAM signal. This synchronisation process can be divided into 3 steps. The receiver is powered up with a wide range of initial estimates, and the algorithm reduces the frequency offset to 1/3 of the subcarrier spacing higher precision estimates reduce the frequency error to less

18.1 Channel Receiver Realisation Technology

than 1 Hz. Good carrier synchronisation tracks phase noise, while inaccurate carrier synchronisation produces a CPE that grows linearly with time.

The synchronisation part also needs to complete the Transmission Parameter Signal (TPS) extraction. The TPS extraction is done using specific TPS bits in the OFDM block which are 17 guides in 2K mode and 68 guides in 8K mode and all the guides are encoded using repetition to transmit the same information. To recover the TPS information transmitted in the frame, differential demodulation is used for each TPS subcarrier, followed by averaging the results to obtain a soft judgement.

2. Channel estimation and equalisation

One of the main tasks of the inner receiver section must also provide an estimation of the channel response for each OFDM block in order to correct each received data sample (coherence detection), and it provides confidence information to the soft judgement Viterbi decoder.

In the following description, the subscripts n, k of each descriptor denote the kth subcarrier corresponding to the nth OFDM block. $r_{n,k}$ denotes the observed value (after FFT) of the receiver unit (n,k), while $C_{n,k}$ is the corresponding frequency domain data. Neglecting the phase noise, it can be expressed as:

$$R_{n,k} = H_{n,k} C_{n,k} + N_{n,k}. \tag{18.72}$$

where $H_{n,k}$ is the complex frequency response of the channel; $N_{n,k}$ is a two-dimensional Gaussian noise whose power can vary with subcarrier k.

In order to obtain an estimate of $C_{n,k}$, it is necessary to first obtain an estimate of the channel $H_{n,k}$, $E_{n,k}$ (channel estimation), and divide $R_{n,k}$ by $E_{n,k}$ (channel correction). $D_{n,k}$ is the obtained estimate on $C_{n,k}$:

$$D_{n,k} = H_{n,k}/E_{n,k} C_{n,k} + N_{n,k}/E_{n,k} = C_{n,k} + N_{n,k}/E_{n,k}. \tag{18.73}$$

The signal to noise ratio of $D_{n,k}$ is. This value is to be sent to the channel decoder along with the estimate of sn,k as an estimate of the trustworthiness.

For this purpose the channel estimate $E_{n,k}$ is first obtained. DVB-T receivers generally obtain this estimate by scattering the guided-frequency units, i.e., (k-kmin)mod12 = 3 × (nmod4) of units (n,k), with n set to zero at the beginning of a superframe; that is, each guided-frequency subcarrier [(k-kmin) mod3 = 0] estimates $H_{n,k}$ once in four consecutive OFDM blocks. These estimates are obtained by dividing the received samples by the known guide frequency. Then the estimates of $H_{n,k}$[(k-kmin)mod3 = 0] for each n are obtained by time-domain interpolation or simply by updating every 4 frames. Interpolation gives better performance, but results in a delay of 3 OFDM blocks for the estimates, since we need to store the 3 frames of data and the corresponding lead frequencies. Interpolation also takes into account the estimation of the 'common phase error (CPE)', a random phase error that is the same for all subcarriers and is caused by the phase noise of the local oscillator. The CPE is given in the other part of the estimation, using the continuous guiding frequency inserted in the DVB-T. Finally, in order to obtain an estimate of Hn, k for each k, the receiver uses frequency interpolation (with a ratio of 1:3), which can be done using a FIR digital filter.

18.1.6 Introduction to Receivers Within the Chinese DTMB Standard [17–19]

Since the promulgation of the three international DTTB standards, ATSC, DVB-T and ISDB, many countries and regions have successively formulated their own DTTB standards with independent intellectual property rights. On 30th August 2006, China's National Standardization Administration Committee (NSAC) issued the standard for 'Digital Terrestrial Multimedia Broadcasting (DTTB) Transmission System Frame Structure, Channel Coding and Modulation' (Digital The transmitter and receiver of the DTMB/TDS-OFDM system are shown in Fig. 18.23. At the transmitter side, the TV programme stream firstly undergoes scrambling, Forward Error Correction (FEC) and constellation mapping to obtain the basic frame, and then undergoes time-domain interleaving to form an interleaved frame; next, the interleaved frame is multiplexed with the TPS and then undergoes the frame-body data processing to be multiplexed with the corresponding PN sequence to form a signal frame, and then undergoes the baseband post-processing to be converted into a baseband output signal, and finally the baseband output signal is converted into a baseband output signal. Then, the signal is converted to baseband output signal, and finally, the signal is converted to RF signal (within UHF and Very High Frequency, VHF band) by quadrature up-conversion for transmitting. The receiving end performs the reverse processing.

As shown in Fig. 18.24 is the overall structure of DTMB receiver, which mainly includes inner receiver and outer receiver. Among them, the DTMB inner receiver mainly packages four parts, namely, frame synchronisation, timing recovery, carrier recovery, channel estimation and equalisation. Since DTMB uses PN sequences

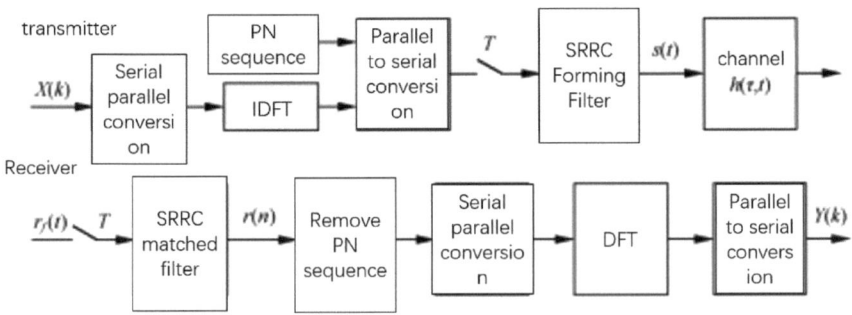

Fig. 18.23 Block diagram of DTMB system transmitter and receiver

18.1 Channel Receiver Realisation Technology

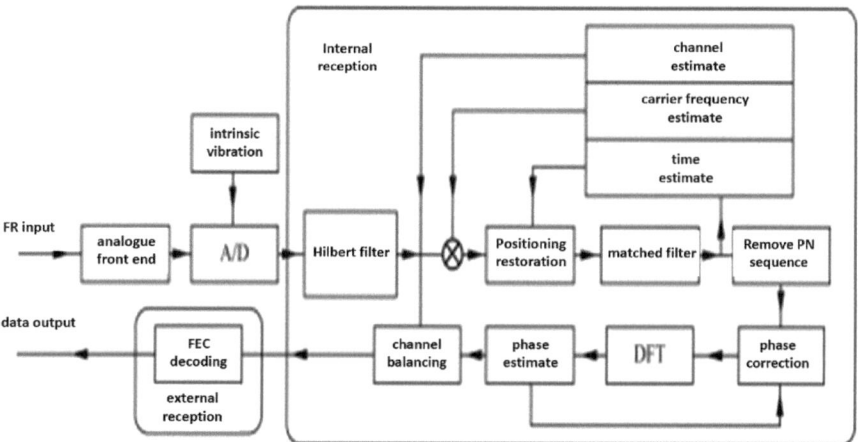

Fig. 18.24 General architecture of the DTMB receiver

to perform frame synchronisation, frequency synchronisation, timing synchronisation, channel transmission characteristic estimation and phase noise tracking in the time domain. DTMB has the superior performance of fast synchronisation speed and accurate channel estimation, which makes the system have the advantages of COFDM and avoid its disadvantages. It has the advantages of high spectrum utilisation, anti-multipath fading, fast synchronisation and accurate channel estimation, which improves the performance of the system. And the external receiver mainly includes time domain deinterleaving, QAM demapping, LDPC and BCH decoding, and data descrambling.

The following will focus on the introduction of each module of the inner receiver in order.

1. Frame Synchronisation

The general timing recovery process is divided into two stages: coarse code element synchronisation first, and then into fine code element synchronisation after completion. Since terrestrial TV broadcasting is a continuous data stream, its fine code element synchronisation algorithm usually adopts the algorithm of feedback structure to obtain better tracking performance. The coarse code element synchronisation for the DTMB system given here is to find a phase for the local PN sequence so that the local PN sequence is phase aligned with the transmit PN sequence. To further make the phase error of the two sequences smaller in the tracking phase and to automatically maintain this high precision phase alignment state under the interference of various external factors, the phase is symbol timing recovery (STR) using closed circuit tracking technique. The external disturbing factors are mainly the sampling clock drift, which leads to the drift of the ICI and timing error of the OFDM system, further deteriorating the frame synchronisation. As shown in Fig. 18.25, which is the implementation block diagram of frame synchronisation, it can be seen that frame synchronisation is mainly divided into the following steps:

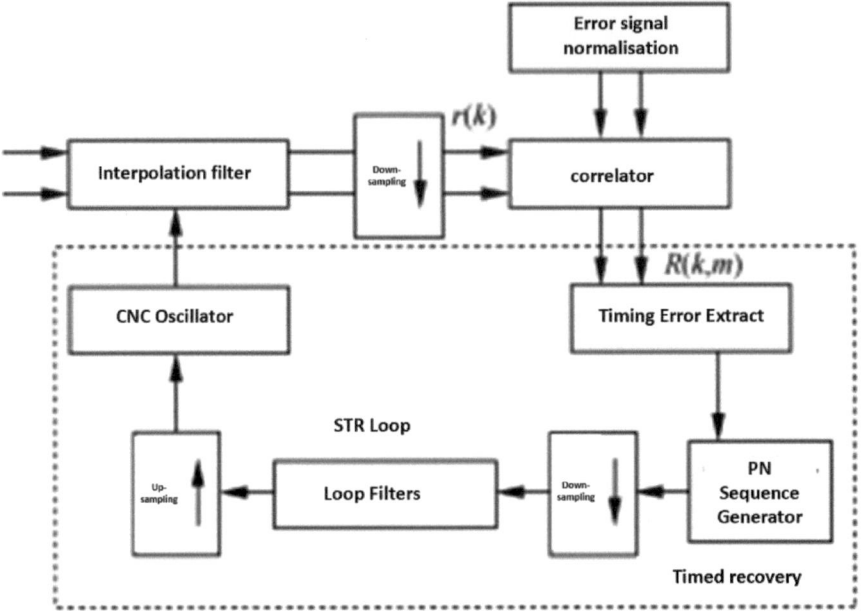

Fig. 18.25 STR system schematic block diagram

(1) Sliding correlation to find correlation peaks

The TDS-OFDM signal frame uses the PN sequence as the synchronisation code the autocorrelation function of the PN sequence is the sequence period K when the phase shift is zero, and the correlation function has a value of -1 when the phase shift is not equal to zero. Accordingly, by detecting the correlation value of the transmitted PN sequence and the local sequence, it can be judged whether the two phases are aligned. In order to reduce the number of correlators for capturing, a sliding search capturing method can be used, which only requires one correlator of length K. Using the sliding correlation technique, the capture time can be greatly reduced and only one correlator is used. In KT time, all possible phases of the PN sequence are searched, which has a high phase search speed. Details can be found in the related literature.

(2) Phase Matching

In the TDS-OFDM system, the PN sequence of each cluster signal frame is determined in advance to ensure that the phase offsets of neighbouring signal frames are unique. Thus, from the obtained ΔPhase, the sequence number of the currently received signal frame in the frame group can be completely determined, and thus the phase of the PN sequence of the subsequent signal frames can be obtained.

In practice, due to noise interference, the correlation peak detector may make a wrong judgement, i.e., there is a false alarm probability and a leakage probability, in particular, when the leakage occurs, the local sequence phase stops searching, and the erroneous synchronisation signals will put the system in an out-of-step state and

18.1 Channel Receiver Realisation Technology   577

prevent it from working properly. To avoid this, a phase matching circuit is required for verification. When the 2nd signal frame correlation peak is obtained, it is not considered to have entered the synchronous state, and it is necessary to compare the correlation peak of the 3rd signal frame with the set threshold value to see if it is greater than the threshold value to verify whether the signal has really entered the synchronous state. If the confirmation circuit judgement is synchronous, indicating that the sequence is truly synchronous, you can start the STR tracking circuit. After entering synchronisation, the reception of subsequent signal frames only needs to update the local PN sequence phase to continue to maintain the synchronisation lock and provide a continuous frame synchronisation signal, which will be used in the extraction of the STR loop timing error, in order to estimate the timing error signal greater than $\pm T_s$. In order to prevent the effect of false alarms in the synchronised state, it is set that the system is considered to be out-of-step only when the correlation peaks are not captured for three consecutive signal frames. If the circuit judgement is confirmed to be out-of-sync, the circuit is restarted.

In particular, it should be noted that the STR loop consists of two main functions: one is to use timing error extraction to obtain an estimate of the timing error; the other is to filter the obtained estimate to drive the CNC oscillator to complete the adjustment of the sampling clock in the time domain, and then through the linear interpolation filter to restore the sampled received data into the data synchronised with the transmitted symbol rate. The STR algorithm uses fourfold sampling in order to ensure timing extraction accuracy and linear interpolation accuracy. The loop improves the accuracy of symbol timing estimation by correlation and has some resistance to multipath, continuing to maintain high accuracy under multipath. As shown in Fig. 18.25. The details can be found in the relevant literature and will not be specifically developed here.

2. Carrier Synchronisation

DTMB proposes to complete the carrier synchronisation in the time domain using the cyclically extended synchronisation header inserted into the PN sequence in the OFDM signal. In this way, a low-complexity time-domain frequency estimation algorithm is proposed based on the inserted synchronisation head. In the initial state of digital terrestrial TV receiver at power-on, there is likely to be a large frequency deviation from the transmitter, in order to ensure that the frequency estimation has a sufficiently large capture range at the same time, to obtain a high estimation accuracy.

In DTMB, the carrier frequency estimation is divided into three stages: in the initial state of power-on, the receiver and the transmitter are likely to have a large frequency deviation Δf, when the receiver is usually unable to carry out accurate timing synchronisation, so the timing synchronisation algorithm generally requires that the Δf can not be greater than a certain threshold. If Δf is larger than this threshold when starting the timing synchronisation, then coarse frequency estimation (CFE) must be performed to capture the frequency deviation within this threshold before timing synchronisation. Since the CFE is performed before the timing synchronisation, there is no timing information at this time, so the CFE algorithm can only be an algorithm without data-assisted approach. Since BPSK is used to modulate the PN sequence, P = 2. The disadvantages of the CFE algorithm due to the dataless assisted approach are that the capture range becomes 1/P of the original and the estimated variance becomes larger due to the introduction of the P-subdivision. After completing the CFE stage, instead of proceeding directly to the coherent AFC (auto frequency control) stage, the incoherent AFC is performed first. The difference between the incoherent AFC and the coherent AFC is that the coherent AFC has to use the timing information whereas the incoherent AFC does not use the timing information. The reason for performing the incoherent AFC first is this: after completing the CFE, the timing synchronisation loop and the fine frequency estimation loop are turned on at the same time, at which point the timing loop needs a few more frames to be accurately synchronised, i.e., to provide accurate timing recovery. Timing within these few frames is not accurate enough, so a non-coherent AFC without timing information is used for synchronisation. The analysis and simulation results show that the method is close to the CRB boundary in the higher SNR region, and with small computational complexity and fast frequency capture time, it is suitable to be applied to all-digital GB receivers.

18.2 Modulation and Coding Techniques

18.2.1 Overview

Compared with analogue television, an important feature of digital television is the use of digital modulation technology. As the design capability of large-scale integrated circuits improves, the processing capability of chips for digital signals is greatly improved, and digital signals can be more easily processed using complex signal processing algorithms. These complex signal processing algorithms include channel estimation, equalisation, error control coding and so on. The wireless channel experienced by terrestrial digital television broadcasting is a bandpass-type channel, and the transmitted baseband digital signals need to be digitally modulated in order to be transmitted in the specified channel, while the terrestrial digital television transmission system is characterised by a large transmission bandwidth, a high data rate, and a high spectral efficiency, which requires a matching digital modulation technology item to meet the above characteristics.

In essence, modulation is a function of the transformation process, which will be transmitted to the binary sequence (usually completed channel coding and the composition of the signal frame) mapped into a carrier waveform with certain properties. Depending on the number of carriers used, digital modulation can be broadly classified into Single-Carrier Modulation and Multi-Carrier Modulation. As the name suggests, single-carrier modulation uses a baseband signal to modulate a carrier, and there is only one carrier signal in a channel, i.e., one modulated signal occupies all the bandwidth of the channel. In single-carrier modulation technique, the attributes that can be changed for a carrier include amplitude, frequency and phase, and hence

18.2 Modulation and Coding Techniques

the basic digital modulation methods also include three types, namely amplitude shift keying (ASK), frequency shift keying (FSK) and phase shift keying (PSK). All existing digital modulation methods can be seen as a distortion and combination of the above 3 basic forms. The term keying comes from the telegraph system, where keying allows the transmission of marks and spaces, or changes in the parameters of the transmission, in this case keying means the use of digital signals to control the carrier waveform. Most of the popular single-carrier communication systems carry data information in both the I and Q paths of the carrier signal, which are digitally modulated using a mixture of ASK and PSK, i.e., the amplitude and phase of the carrier signal are allowed to have a variety of values, and this modulation is known as multicarrier digital modulation, and the commonly used multicarrier modulation methods include Quadrature Amplitude Modulation (QAM), and Residual Sideband Modulation (VSB). Multicarrier modulation is the process of breaking down the high-speed data stream to be transmitted into several low-speed bit streams and modulating several subcarriers in parallel with these bit streams. The most commonly used multicarrier modulation is called Orthogonal Frequency Division Multiplexing (OFDM), which divides a given channel into a number of orthogonal subchannels in the frequency domain, modulates each subchannel with a subcarrier, and transmits the subcarriers in parallel, thus transforming the broadband into narrowband and solving the problem of frequency selectivity into a narrowband, solving the problem of frequency selective fading.

In digital communication, different modulation methods have their own characteristics and are suitable for their applications. For a particular communication system, which modulation method to use requires a compromise between multiple performance metrics.

In the design or evaluation of a digital modulation scheme, the following basic elements usually need to be considered.

(1) Transmission rate

Transmission rate is an important measure of a system's transmission capability [1], and two common types are bit rate and baud rate. Bit rate Rb is the number of binary bits transmitted per unit of time in bits/s; baud rate Rs is the number of modulated symbols transmitted per unit of time in baud/s. For M-round modulation there are

$$R_b = R_s \log_2 M. \tag{18.74}$$

Indeed, the definitions of bit rate and baud rate represent the different forms in which signals are represented at different stages of transmission. For example, at the source and channel compilation code stage at the transceiver end, the information is usually represented in binary form, when the bit rate is used as a unit, while after the modulator mapping and before the demodulator demapping, the information exists in the form of a multiplicity of symbols, when it is more convenient to use the baud rate. In addition, the evaluation of a transmission system usually defines the net payload rate, which is the 'pure' rate of information in transmitted symbols after deducting all the extra overheads due to channel coding and synchronisation fields, and is usually expressed in bits per second (bit/s).

For multivariate modulated signals, since the receiver's judgement is based on symbols, the false symbol rate or false word rate, which is the proportion of symbol errors that occur at the receiver, is more commonly used. The false symbol rate of a linear modulation system is an exact function of the Euclidean distance between the constellation points in its constellation diagram. In general, the denser the constellation points, the higher the probability of incorrect symbol judgement at the receiver side. To more accurately evaluate transmission systems applied to burst channels, the outage probability is also defined as the probability that the number of false codes in a measurement exceeds a specific value, with each outage event representing a failed transmission.

(2) Spectral efficiency

Spectral efficiency ηW, also known as band utilisation, is commonly used as a measure of the effectiveness of a system and is defined as the rate of information transmission per unit bandwidth of the transmission channel in bits/s/Hz. If the bandwidth of the transmission channel is W, then we have

$$\eta_W = \frac{R_b}{W}. \tag{18.75}$$

(3) Signal-to-noise ratio, carrier-to-noise ratio and E_b/N_0

The signal-to-noise ratio (S/N) is the ratio of the average power of the transmitted signal to the average power of the additive noise, while the carrier-to-noise ratio (C/N) is the ratio of the average power of the modulated signal to the average power of the additive noise. They are usually expressed in logarithmic form in dB. The difference between the signal-to-noise ratio and the carrier-to-noise ratio lies in the calculation of the carrier power. The power of the modulated signal in the carrier-to-noise ratio includes the power of the transmitted signal and the power of the modulating carrier, while the signal-to-noise ratio includes only the power of the transmitted signal. Both values are equal for the modulation method of the suppressed carrier. The signal-to-noise ratio and the carrier-to-noise ratio can be obtained directly by measurement at the receiving end.

E_b/N_0 is the bit signal-to-noise ratio, where E_b represents the energy required for each bit of information transmitted and N_0 represents the one-sided power spectral density of a Gaussian white noise channel. Since

$$\frac{E_b}{N_0} = \frac{S/N}{\eta_W}. \tag{18.76}$$

In addition to the above elements, we also have to consider the complexity of implementation, hardware cost, etc. We want a perfect modulation scheme that can transmit data reliably at high speeds with low SNR, and that takes up as little bandwidth as possible. Unfortunately, no single scheme can satisfy all the conditions at the same time. Therefore, the selection of a digital modulation scheme is actually a process of trade-offs between the above constraints according to the actual situation of the system, especially the trade-off between power efficiency and spectral efficiency. For example, error correction coding of information reduces the spectral efficiency, but at the same time the reception power necessary for a given BER is reduced, so the bandwidth efficiency is exchanged for power efficiency; on the other hand, high-degree modulation can increase the spectral efficiency, but the power must be increased in order to ensure reliable reception, so the power efficiency is used to trade-off for spectral efficiency.

18.2.2 Single-Carrier Modulation Technology

Single-carrier modulation techniques are widely used and were in use in most systems such as microwave multipoint distributed systems, wireline, terrestrial and satellite communication systems before the maturity of OFDM systems. Nowadays, despite the development of multi-carrier modulation technology, compared with multi-carrier modulation technology, the advantage of single-carrier modulation technology lies in the lower peak-to-average-ratio (PAPR) of the signal at the same transmit power, and the low performance requirements for amplifiers, so it has been widely used in power-constrained transmission systems. For example, single-carrier modulation is used in satellite communications, cable television networks, and line-of-sight microwave systems (e.g., MMDS). Residual sideband modulation (VSB) is also a typical single-carrier modulation technique, which saves signal energy by filtering the double-sideband signal into a single-sideband signal, and its signal transmission efficiency is high, and the U.S. digital TV standard ATSC adopts the digital TV transmission standard based on VSB technology [2]. Recently, single-carrier frequency division multiple access (FDM) access technology has also been selected for the uplink channel of the 4G-LTE communication system.

One of the major differences between single-carrier modulation and multicarrier modulation is that multicarriers generally use Frequency Domain Equalization (FDE) to combat multipath fading, whereas the classical single-carrier system mainly performs channel equalisation through a time-domain adaptive equaliser, due to the simplicity of multicarrier frequency domain equalisation and the fact that the frequency domain processing is more capable of improving frequency Since multi-carrier frequency domain equalisation is simple, and frequency domain processing can improve the error correction performance under selective fading channels, using multicarrier is more advantageous than single-carrier in environments where multipath fading is more serious. Recently, many scholars have combined single-carrier modulation and multicarrier FDE, and proposed a single-carrier frequency-domain equalisation modulation, generally known as SC-FDE, which is characterised by block-transmission, and the use of cyclic prefixes between blocks and multicarrier modulation to construct the protection interval, so as to use the same frequency-domain equalisation process as multicarrier modulation in the receiver, thus providing the system with the advantages of the traditional single-carrier modulation. The system has the advantages of traditional single-carrier and multicarrier modulation.

Single-carrier modulation mainly includes I- and Q-way baseband signal mapping, band-limiting filtering and quadrature up-conversion circuits, while the commonly used single-carrier modulator band-limiting filtering and up-conversion circuits are identical, the difference is mainly when the I- and Q-way baseband signals are mapped differently, i.e., how to map the data bits to be sent into the I- and Q-way signals. In this section, we first introduce the band-limiting filter, and then introduce the mapping methods for different modulations in turn.

1. Band-limited signals and shaping filters

Digital information is made up of a series of binary logic values 0, 1. Physically represented by a sequence of digital pulses, each 0 and 1 occupies a finite duration T_s. It is known by Fourier analysis that an ideal digital pulse signal with finite duration T_s in the time domain has an infinite spectrum in the frequency domain. The actual physical channel is of finite bandwidth, and a pulse signal with a finite spectrum in the frequency domain has an infinite duration in the time domain, such that consecutive pulse signals with a finite spectrum will overlap in the time domain, resulting in inter-code crosstalk. This means that continuous pulsed signals cannot be directly applied in practical limited bandwidth channels, and engineering generally passes the signals through specific filters for filtering.

2. QPSK digital modulation

In QPSK modulation, the carrier phase value is determined by the value of the transmitted data, which can have four values, each differing by 90 degrees, and each symbol can carry 2 bits of information. Thus the digital information that can be transmitted is 00, 01, 10, 11.

A serial binary code stream at a rate of $2/T_s$ is divided into two data streams (I and Q paths) at a rate of $1/T_s$ by a serial-parallel transform, and the input data is alternately assigned to the I and Q paths. The two data streams are then shaped and filtered by a low-pass filter (square-root rising-cosine filter) to limit the bandwidth of the signals, and then the two data streams are modulated onto the carrier by a dual-band rejection carrier modulation using a local oscillator (LO). The two data streams use the same local oscillator, the only difference being that the local oscillator of one stream is shifted 90 degrees relative to the local oscillator of the other stream, which is how the terms I- and Q-channel are derived. The 90-degree phase shift ensures that the two signals are orthogonal to each other after modulation and use the same bandwidth, which makes them easily separable in the receiver.

3. QAM systems

In a bandwidth efficient transmission system, high volume information must be transmitted by high-progressive modulation. It is intuitively obvious from the signal constellation diagram that if amplitude or phase alone is used to carry the information, the signal constellation points are distributed only on a straight line or a circle, which does not make full use of the signal plane. Based on this consideration, the

18.2 Modulation and Coding Techniques

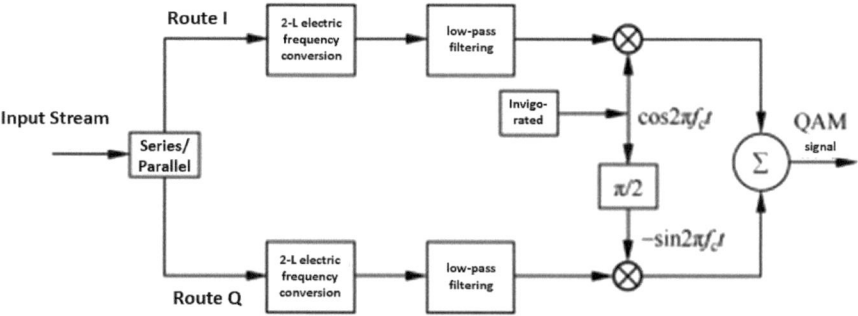

Fig. 18.26 QAM modulator

combination of amplitude and phase modulation—orthogonal amplitude modulation QAM—was born, which can increase the number of constellation points as much as possible under the premise of guaranteeing the minimum Euclidean distance. Currently, multicast QAM modulation schemes are widely used in digital video broadcasting, allowing higher bit rates to be transmitted within a limited bandwidth.

4. QAM Modulator Implementation

Figure 18.26 shows the block diagram of the QAM modulator. The QAM modulator adds only the 2-L level conversion module. Which serves to convert a 2-level sequence to an L-level sequence. The square constellation diagram of the pairs of -binary has. If the serial input stream has a rate of Rb, the level-converted sequence has a rate of Rb/m. Subsequently, each of the two baseband signals is multiplied by the quadrature carrier and synthesised into a QAM signal.

From this modulation process, it can be seen that the QAM signal can be viewed as the sum of two orthogonal multigraded AM signals. On the other hand, since a quadrature QAM signal is exactly equivalent to a QPSK signal, a QAM signal can also be viewed as a linear combination of multiple layers of QPSK signals. For example, a 16QAM constellation diagram can be viewed as consisting of 2 layers of QPSK modulation. The 1st layer of modulation determines which quadrant the constellation point is in, and the 2nd layer of modulation is then mapped to one of the 4 constellation points in that quadrant. This method of constructing a QAM signal is layered modulation. This feature is used in DVB-T systems to provide hierarchical transmission services. The modulated signal is divided into 2 layers (or 2 priorities), i.e., the QPSK signal at layer 1 (high priority HP) and the QPSK signal at layer 2 (low priority LP). The transmitter completes the QPSK mapping first and then performs another QPSK mapping based on the QPSK constellation points. The two layers of mapping usually come from different sources and can use different channel coding to provide different levels of BER protection. The receiver can choose to receive all or only high priority streams according to its own needs and objective reception conditions [4]. We will describe this in more detail in later sections.

5. DVB-based QAM systems

The QAM systems described above are of general interest, and when implemented in a real system, some modulations are usually used jointly to some extent to obtain the desired characteristics. Most of the standards currently used differ little, mainly in the coding techniques used internally and the choice of roll-off factors. However, the main characteristics of single-carrier QAM/QPSK modulators are the same, as described in the above sections. As an example of the implementation characteristics of commonly used modulators, a brief description of DVB-C will be given, and for a detailed description the reader is referred to the DVB-C standard [4].

DVB-C uses differential level QAM modulation, with m being taken as 4, 5, 6, 7 and 8. After channel coding the data is first subjected to a bit-to-symbol transformation. To avoid the phase ambiguity problem at the receiver, the higher two bits of the mapped symbols are differentially encoded before QAM modulation, i.e., Ak and Bk in the figure, where Ak is the highest bit of the symbol (MSB). The encoded data generates I and Q baseband signals according to the constellation diagram mapping relationship. Taking 64QAM as an example, m = 6, which corresponds to q = 4, such that the input bit stream is constellation mapped in groups of 6 bits, denoted as (b5,b4,b3,b2,b1,b0). Thus Ak corresponds to its highest bit Ak = b5, Bk = b4 and q bits correspond to (b3,b2,b1,b0). The I and Q baseband signals are further filtered and shaped using baseband shaping filters (square root ascending cosine roll-off filters with roll-off coefficient of 0.15) (Fig. 18.27).

Table 18.1 gives important characteristic parameters of a QAM system under DVB-C from a network operation point of view, including spectral efficiency (bit/s/Hz), useful information rate (Mbps), and the required carrier-to-noise ratio (C/N) for a standard 8 MHz channel. As can be seen from the table, the 64-QAM system has the highest data transmission capacity, but also the highest carrier-to-noise ratio required to guarantee error-free codes.

6. Residual sideband modulation

The so-called residual sideband modulation (VSB) is a compromise form of bilateral and single sideband modulation [5]. It is the design of an appropriate output filter based on the bilateral sideband modulated signal such that in addition to transmitting one sideband, a portion of the other sideband is retained.

Based on the above analytical formulas of VSB modulated signals in frequency and time domains, it is easy to generate the VSB signals using two methods, i.e.,

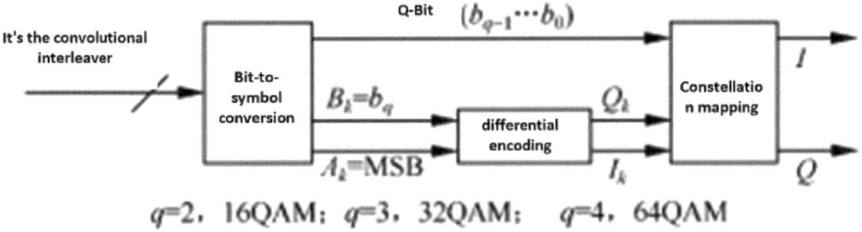

Fig. 18.27 DVB-C baseband signal generation

18.2 Modulation and Coding Techniques

Table 18.1 Characteristic parameters of DVB-C

Modulation type	13-QAM	32-QAM	64-QAM
Spectral efficiency (bit/s/Hz)	4	5	6
Useful information bit rate (Mbps)	25.2	31.9	38.1
Carrier-to-noise ratio in 8MHz channel (dB)	19.7	23.3	25.7
Channel coding	RS(204,188)	RS(204,188)	RS(204,188)

filtering method and phase shifting method. Residual sideband modulation is the modulation method used by the ATSC digital terrestrial television transmission standard in the United States [2]. In order to be compatible with analogue NTSC channels, the transmission bandwidth of the ATSC standard is 6 MHz, and its modulation schematic diagram using the filtering method is shown in Fig. 18.28, where the binary code stream is first divided into one group every 3 bits, and then the eight-level amplitude modulated baseband signal is formed after 2–8 level conversion, followed by up-conversion. Because the amplitude modulation signal is a real signal, its unilateral power spectrum on the carrier frequency is conjugate even symmetry. Engineering is generally higher than the carrier frequency part is called the upper sideband, lower than the carrier frequency part is called the lower sideband, both carry the same information. To generate a VSB signal using the filtering method, a sideband needs to be filtered out of the modulated signal using an analogue filter, which is usually done using a residual sideband filter HVSB(f) with a certain roll-off. The residual partial upper sideband filter, shown in Fig. 18.29, has a roll-off symmetric transfer function within $fc-fv \leq f \leq fc + fv$. In the ATSC standard the VSB modulator transmits symbols at a rate of 10.76 M symbols per second, and by Nyquist's sampling law the required band is 5.38 MHz, so that for the 6 MHz channel bandwidth used for transmission, there is a bandwidth margin of 0.62 MHz. This is filtered with $fv = 0.31$ MHz, which corresponds to a roll-off factor $\alpha = 0.115$ signals.

The filter HVSB(f) required when using the above filtering method is an intermediate-frequency bandpass filter with a bandwidth of about 2 fc and a roll-off band fv of only 0.31 MHz, which makes it extremely difficult to realise such a bandpass filter. It is also very difficult to realise such a near-ideal filter if a high-performance wide-band shift network is required to realise 8-VSB modulation using the phase-shift method, and its phase-frequency characteristics have a step jump. It can be shown that the spectral characteristics of VSB signals are somewhat equivalent

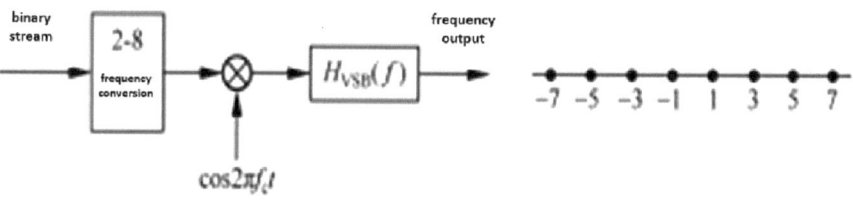

Fig. 18.28 ATSC 8-VSB modulation using the filtering method

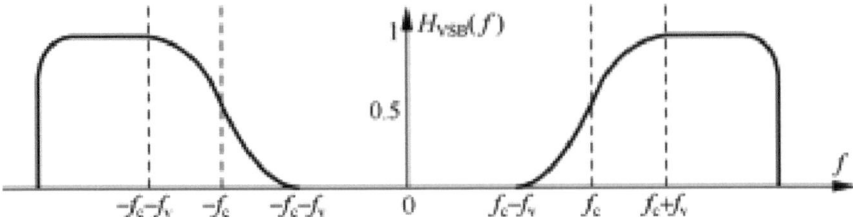

Fig. 18.29 Upper sideband filter for residual section

to 64OQAM signals [5]. The 8-VSB is represented by a real-valued 8-PAM amplitude modulated signal at the beginning, and the 8-VSB signal spectrum is obtained by removing its negative spectrum using the Hilbert transform. The 8-VSB signal is then frequency shifted by FVSB/2 (FVSB is the baseband rate of the 8-VSB) to obtain the equivalent spectrum of the 64OQAM signal.

18.2.3 Multi-carrier Modulation

How to efficiently multiplex the spectrum is a topic that has been explored by scholars since the beginning of wireless communication research. In the past single-carrier era, one way to utilise bandwidth by frequency division was to use several different carrier frequencies and relatively wide bandwidths to transmit different data streams. To make it easier to distinguish the signals at the receiving end, the carrier frequencies were spaced far enough apart so that the signal spectra did not overlap. Another approach is to use carriers of different frequencies to transmit different bits of a single high-rate information stream, rather than using them to transmit separate information streams. In this case, the source should have a parallel output, or the serial source output should be made parallel by passing it through a serial-to-parallel converter. All of these schemes require the use of low-pass filters to process the signals of each carrier in such a way that they do not generate inter-carrier interference but are also able to achieve a certain level of spectral efficiency. However, these algorithms are complex, and because of the use of filters to process the signals, the isolation effect of the filters needs to be taken into account; a similar scheme can be implemented using a method such as that in Fig. 18.30.

The above frequency division multiplexing FDM transmission system can be regarded as an early multicarrier system, but due to the complexity of the implementation, as well as the lack of overlap of the individual frequency bands and the low spectral utilisation, it has not received much attention or been widely used since it was proposed. With the improvement of digital chip processing capability, the use of FFT/IFFT operation to generate orthogonal carriers has become very popular, the use of FFT/IFFT way to generate subcarriers do not need to add filters, the transmitter structure is simple, and through the FFT to get the various subcarriers are orthogonal

18.2 Modulation and Coding Techniques

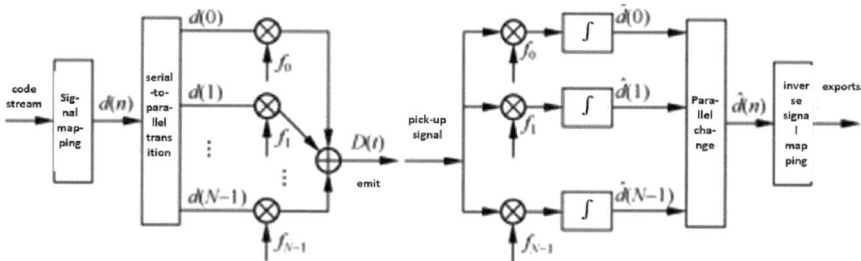

Fig. 18.30 Implementation of FDM multicarrier modulation

to each other, the subcarriers have a part of the overlap between the band utilisation, so it is more than the traditional FDM to improve the frequency band utilisation. It is because of the orthogonality of the subcarriers, so that the FDM modulation method is also known as orthogonal frequency division multiplexing modulation, that is, OFDM. After the use of error correction code to encode each subcarrier of OFDM, it is easy to recover the subcarriers that are affected by the channel fading by error correction, thus reducing the impact of the frequency selective fading of the transmission channel on the signal. This channel-coded OFDM is called COFDM (i.e., Coded Orthogonal Frequency Division Multiplexing).

Since OFDM is robust in multipath transmission and also resists linear distortion of the channel and has high signal spectrum utilisation, it is widely used in various broadband wireless transmission standards. In the past decades, OFDM has been widely used as a modulation method for high-speed data communication in standards such as Digital Audio Broadcasting (DAB), Digital Video Broadcasting Terrestrial (DVB-T), Wireless Local Area Networks (WLANs) 802.11 and 802.16, Asymmetric Digital Subscriber Ring ADSL, Very High-Speed Digital Subscriber Ring VDSL, and 4G solutions (e.g. LTE). For digital broadcasting systems, OFDM modulation is used in the European digital audio broadcasting DAB, the European digital terrestrial TV standard DVB-T, the Japanese ISDB-T digital TV transmission standard and the Chinese digital TV terrestrial transmission standard DTMB. In this section, we will analyse several important issues in the OFDM transmission system and describe the algorithms and principles of OFDM modulation.

18.2.4 SC-FDE

OFDM is a very important modulation method in broadband transmission. Like other modulation methods, it has its own weaknesses.

Firstly, OFDM is sensitive to frequency offsets and phase noise. This is a receiver implementation problem. For OFDM modulation techniques, tuners with better phase noise performance are needed, and the effect of phase noise can be modelled in two parts: one is the common rotating part, which causes phase rotation of all OFDM subcarriers, and is easy to be tracked by the reference signal. The second is the dispersive part, or inter-carrier interference part, which causes noise-like dispersion of the carrier constellation points, which is difficult to compensate and will slightly lower the noise threshold of the OFDM system. Also, in order to reduce the inter-subcarrier interference caused by frequency offsets, OFDM requires better timing and frequency recovery algorithms, which are described in more detail in Chap. 4.

Secondly, OFDM has a high peak-to-average power ratio PAPR. The PAPR of OFDM signals is about 2.5 dB higher than that of single-carrier, which implies the need for a larger transmitter dynamic range, or power fallback, to avoid entering the nonlinear region of the transmitter, and better filtering to reduce the neighbouring-channel interference. Reducing PAPR is one of the research hotspots, and some effective techniques have been proposed in recent years. In digital broadcasting applications, the disadvantage of high PAPR of OFDM only affects a small number of transmitters, not a huge number of receiving users; and when a single-frequency network is used, PAPR will not be a major problem due to the small transmitter power.

Finally, FFT signal processing in OFDM systems is usually performed in blocks of length 4–10 times the channel delay extension. Guide frequency data is also inserted into OFDM symbols for channel tracking and estimation, and in bursty applications it is possible that one or more OFDM symbols may be required to be used as guides, such that OFDM processing using large blocks of data introduces larger delays than in single-carrier systems.

Over the last few years, single carrier modulation techniques combined with frequency domain equalisation, referred to as SC-FDE [5] (Single Carrier Frequency Domain Equalisation), have also found some application in broadband wireless transmission. The peak-to-average ratio of this single-carrier modulation technique is much smaller than that of OFDM, and it can achieve high performance without adaptive modulation. Its most important feature is the equalisation structure which is different from the traditional single carrier and uses the FFT algorithm for equalisation in the frequency domain. This structure will be used for OFDM frequency domain equalisation applied to single-carrier modulation, so that the performance of single-carrier modulation technology in the frequency selective fading of strong channels close to the performance of multi-carrier modulation. Moreover, the data block of the frequency domain equalisation for single-carrier systems is not fixed and can be adjusted according to the real-time nature of the service. This scheme combines the advantages of single-carrier and multicarrier, and has been applied in systems such as 802.16 and more recently in the new generation of Long Term Evolution (LTE) standards for mobile communications.

18.2.5 Channel Coding Techniques

Regardless of the use of single-carrier or multi-carrier modulation, digital TV signals in the channel transmission process will be subject to additive noise, multipath fading, high power nonlinear transmission caused by neighbouring frequencies and the same frequency and other interference factors, of which the terrestrial broadcasting channel in the digital TV three transmission modes face the most interference, but also the most serious, in particular, the multipath of the delay and amplitude of the change in the speed of the complexity is far more than the satellite and cable cable channels, and thus distortion and error codes are also the most serious. As a result, distortion and BER are also the most serious. On the other hand, digital TV puts forward very strict requirements on BER, general communication system BER 7-3 ~ 7-6 can be, while digital TV generally requires to achieve quasi-error-free transmission, transmission BER at least 3×7-6 or less, and high-definition television even requires BER 7-11 or less. In order to make digital TV signals in the channel reliable transmission, minimise the BER, error correction coding must be carried out. As long as the distortion and BER occurring during the transmission of the signal are within the error correction range of the error correction code, the receiving end will be able to demodulate it correctly, thus ensuring the correctness of the information transmission. As introduced in Chap. 1, channel coding is one of the distinguishing marks of digital TV transmission from analogue TV, which, due to the lack of channel coding, will directly affect the received image quality once the transmitted signal is affected by interference and cannot be recovered [7].

Especially for OFDM-based digital TV system, due to the narrow bandwidth of each subcarrier, under the frequency selective channel, there may be some subcarriers in deep fading, at this time, if the channel coding technique is not used, it may be that these subcarriers can not be received normally. Many literatures have pointed out that because in single-carrier system, as long as the overall channel condition is better, then it can perform well; however, for multicarrier system, even if the overall channel condition is better, the deterioration of the individual subcarrier channel will cause the individual symbol error, thus the false symbol rate has been maintained at a relatively high level, so for the Uncoded (Uncoded) system, the single carrier has an advantage over OFDM. When channel coding (Coded) is used, the number of severely deteriorated subcarriers in the frequency-selective channel is generally small compared to that of OFDM, while for OFDM, channel coding is performed for each subcarrier in the frequency domain, so that the receiver can easily use the subcarriers that are not severely deteriorated to recover the severely deteriorated subcarriers through channel error correction, which results in a significant improvement of the performance of the OFDM system, and the equaliser can be used for the reception of OFDM signals during the reception process. In addition, in the process of OFDM signal reception, the equaliser generates the channel information (CSI) of each subcarrier, which can be transformed into the soft information of channel decoding through appropriate calculation, thus achieving better error correction performance. Due to the above reasons, the applied OFDM systems have to use channel coding, at this time OFDM, as mentioned earlier, is called is coded OFDM (COFDM). When the channel frequency selective fading is more severe, COFDM can generally achieve better performance than single carrier with the same coding.

18.3 Channel Characteristics of Wideband Transmission Techniques

18.3.1 Overview

Figure 18.31 gives the location of the channel in a wireless broadcasting system [1]. Wireless broadcasting signals need to use radio waves for transmission, compared with the wired transmission medium, radio propagation characteristics are relatively poor, the radio wave will not only increase with the propagation distance and dispersion loss, but also by the terrain, the building's cover and the 'shadow effect'; the signal through the multi-point reflection, will arrive at the receiving point from multiple paths, this multipath signal amplitude, phase and arrival time are not the same, they are superimposed on each other. This multipath signal amplitude, phase and arrival time are not the same, their superposition will produce level fast fading and delay extension; In addition, wireless broadcasting is often carried out in the fast-moving, which will cause Doppler shift and random FM. Thus, wireless broadcasting systems are far more complex than wired communication systems.

Radio waves in different frequency bands have different propagation methods and characteristics. Terrestrial digital TV system uses VHF/UHF (VeryHighFrequency/UltraHighFrequency) frequency band, and the transmission environment is much more complicated than wired channel: the transmitted signal propagates in the form of radio waves (which may contain direct, reflection, diffraction and scattering, etc.), and is affected by the terrain undulation between the transmitter and the receiver,

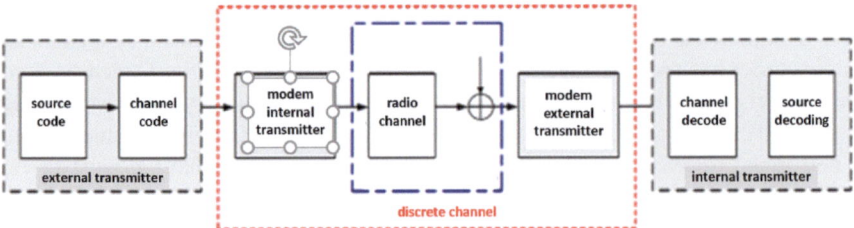

Fig. 18.31 Channels of a radio broadcasting system

18.3 Channel Characteristics of Wideband Transmission Techniques

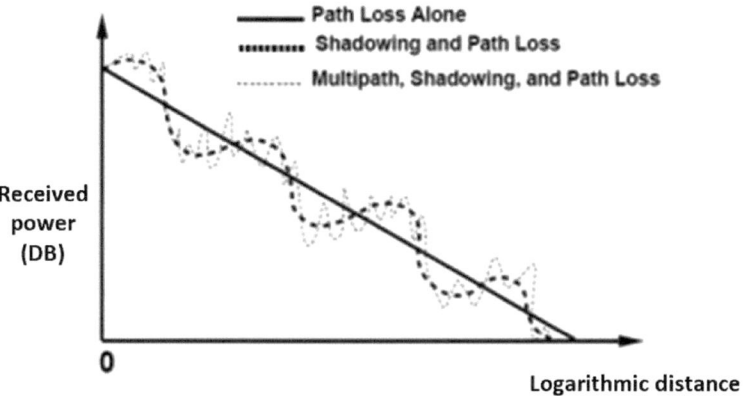

Fig. 18.32 Large-scale and small-scale effects of the wireless broadcast channel

building obstacles, moving objects, etc. The received signal will have many losses, and many factors have time-varying and random characteristics. The received signals are affected by factors such as terrain relief between the transmitter and the receiver, obstacles, moving objects, etc., and the received signals incur many losses, and many of these factors are time-varying and stochastic in nature [2].

For the effect of the wireless channel on the received signal, we can discuss the random time-varying nature of the wireless channel according to the different distances, according to the large-scale effects (Large-scale Effects) and the small-scale effects (Small-scale Effects) from the statistical characteristics to be discussed separately. When the receiver is in a certain location in space, it is received in the vicinity of the position of the signal power of the local average (Local Mean) will be affected by the large-scale effects, these effects include line of sight (Line of Sight, LOS) Path loss (Path loss), Shadowing (Shadowing) fading and other effects. Figure 18.32 shows how received signal power is affected by path loss and fading.

Path loss is the average power value of the received signal that is inversely proportional to the nth power of the signal propagation distance when the distance between the transmitter and receiver varies on a larger scale (hundreds or thousands of metres), where n is the path loss index.

Shadow fading is the signal attenuation caused by electromagnetic waves propagating in space when they are blocked by undulating terrain and tall buildings, behind which a shadow of the electromagnetic field is generated, causing a change in the median field strength. Shadow fading is measured on a large spatial scale and its statistical properties usually conform to a lognormal distribution.

Together, path loss and shadow fading reflect the effect of the wireless channel on the transmitted signal at large scales. Large-scale fading is reflected as a trend in the electrical mean value of the signal over a medium range (hundreds of wavelengths).

The so-called small-scale effect describes the drastic change in the received signal strength over short periods of time (on the order of seconds) or short distances (a

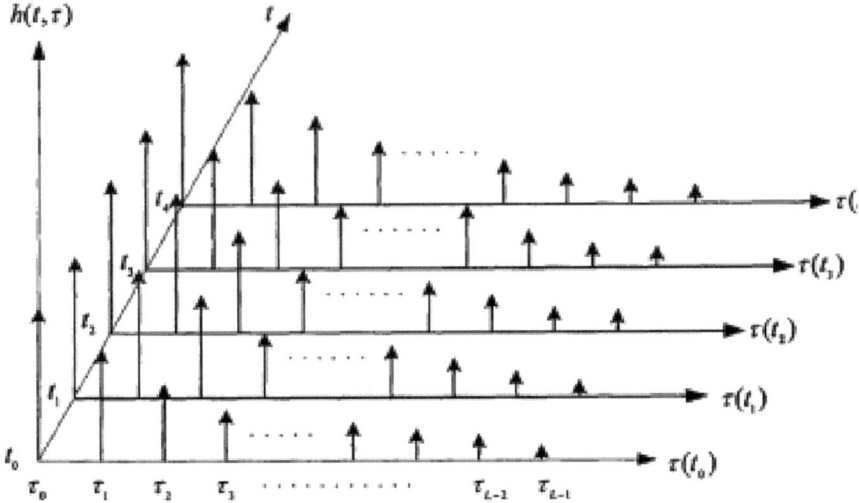

Fig. 18.33 QPSK constellation diagram

few wavelengths). Such drastic variations are mainly caused by the fact that the transmitted signal reaches the receiver via different paths. On a small scale, we are mainly interested in two properties of the wireless channel. The first is the Doppler shift introduced by the relative movement between the receiver and the transmitter, i.e., the wireless channel is time-varying; the second is that the multipath channel will cause the received signal to have a delay extension due to the fact that the paths have different delays in the multipath channel. These two characteristics are shown in Fig. 18.33, which are particularly important for studying the selection of transmission techniques and the design of receivers. Therefore, this book mainly considers small-scale effects.

References

1. Meyr, H., M. Moeneclaey, and S. A. Fechtel. 1997. *Digital Communication Receivers: Synchronization, Channel Estimation and Signal Processing.* New York: Wiley.
2. Proakis, J. G. 2006. *Digital Communications: English Edition*, 4th ed. Beijing: Electronic Industry Press.
3. Edfors, O., M. Sandell, J.-J. Van de Beek, et al. 1998. OFDM Channel Estimation by Singular Value Decomposition. *IEEE Transactions on Communications* 46 (7): 931–939.
4. Edfors, O., M. Sandell, J.-J. Van de Beek, et al. 2000. Analysis of DFT-Based Channel Estimators for OFDM. *Wireless Personal Communications* 12 (1): 55–70.
5. GardnerFM. 1986. A BPSK/QPSK Timing-Error Detector for Sampled Receivers. *IEEE Transactions on Communications* 34(5): 423–429.
6. Deng, Yongjun, Zhixing Yang, Changyong Pan, et al. 2004. A Simple Feedforward Timing Offsets Estimation Algorithm for BPSK/QPSK. In *The 2004 Joint Conference of the 10th Asia-Pacific Conference on Communications, 2004 and the 5th International Symposium on*

Multi-Dimensional Mobile Communications Proceeding: Vol. 1, 214–217. Piscataway, NJ: IEEE.
7. Leng, Weimin, Yu Zhang, and Zhixing Yang. 2008. A Modified Gardner Detector for Multilevel PAM/QAM System. In *ICCCAS 2008: International Conference on Communications, Circuits and Systems*, 891–895. Piscataway, NJ: IEEE.
8. Peng, Kewu, Aolin Xu, and Zhixing Yang. 2008. Optimal Correlation Based Frequency Estimator with Maximal Estimation Range. In *ICCCAS 2008: International Conference on Communications, Circuits and Systems*, 259–263. Piscataway, NJ: IEEE.
9. Liu, Qijia, Zhixing Yang, Jian Song, et al. 2006. A Novel QAM Joint Frequency-phase Carrier Recovery Method. In: *ICACT 2006: The 8th International Conference Advanced Communication Technology*, 1617–1621. Piscataway, NJ: IEEE.
10. Advanced Television Systems Committee. 2006. *A/54AGuide to Use of the ATSC Digital Television Standard, with Corrigendum No. 1*. Washington, DC: ATSC.
11. Whitake, J. 2003. *Digital Television Reception Technology*. Yao Dong Ping (Trans.). Beijing: Electronic Industry Press.
12. Classen, F., and H. Meyr. 1994. Frequency Synchronization Algorithms for OFDM Systems Suitable for Communications over Frequency Selective Fading Channels. In *1994 IEEE 44th Vehicular Technology Conference: Vol. 3*, 1655–1659. Piscataway, NJ: IEEE.
13. Van de Beek, J. J., M. Sandell, and P. O. J. Borjesson. 1997. ML Estimation of Time and Frequency Offset in OFDM Systems. *IEEE Transactions on Signal Processing* 45 (7): 1800–1805.
14. Yin, Changchuan, Tao Luo, and Guangxin Lok. 2004. *Multi-carrier Broadband Wireless Communication Technology*. Beijing: Beijing University of Posts and Telecommunications Press.
15. Combelles, P., C. Del Toso, D. Hepper, et al. 1998. A Receiver Architecture Conforming to the OFDM Based Digital Videobroadcasting Standard for Terrestrial Transmission (DVB-T). In: *ICC 98. Conference Record: IEEE International Communications, 1998: vol. 2*, 780–785. Piscataway, NJ: IEEE.
16. Renzo, P. 2005. *Advanced OFDM Systems for Terrestrial Multimedia Links*. Lausanne, Switzerland: EPFL.
17. National Task Force on Digital Television Terrestrial Transmission. 2006. *GB 20600–2006 Standard for Frame Structure, Channel Coding and Modulation for Digital Television Terrestrial Broadcasting Transmission System*. Beijing: China Standard Press.
18. Song, Jian, Zhixing Yang, Lin Yang, et al. 2007. Technical Review on Chinese Digital Terrestrial Television Broadcasting Standard and Measurements on Some Working Modes. *IEEE Transactions on Broadcasting* 53 (1): 1–7.
19. Wang, Jun. 2003. *Research on Synchronisation and Channel Estimation Algorithms for Terrestrial Digital Television Broadcasting*. Beijing: Department of Electronic Engineering, Tsinghua University.

Chapter 19
Application of Communication Technology in Marine Engineering

19.1 Wide-Band Multichannel Relay Systems

By carrying a relay system on the unmanned aircraft platform, it can provide users with a means of communication that is interconnected over the line of sight. The system can complete medium and long-distance communication air relay; realise medium and long-distance and 'mobile communication' communication; and also realise mutual relay between multiple frequency bands and different communication networks.

1. Typical indicators
   Altitude 300–3000 m.
   Working frequency band Coverage 30–400 MHx.
   Number of forwarding/relaying channels 2.
   The service types are mainly voice and data, also can transmit fax and image.
   Forwarding distance 100–200 km.
   Equipment quality Less than 30 kg.
2. System Composition

Figure 19.1 is a schematic diagram of the air platform wide-band multi-channel relay system. It consists of a wide dynamic range of channel transmission part (ground radio and air platform on the wireless duplex multi-channel equipment); multi-channel controller; and hidden in the transmission equipment antenna common, antenna feed system and so on. Ground vehicle-mounted/backpacked simplex radios work in asynchronous multi-network mode, and the modulation mode, anti-jamming mode and channel bandwidth of the transmitting system can be programmed and controlled to achieve compatibility with the main equipment in service. The multi-channel controller controls the retransmission system according to the signalling, providing services such as registration and deregistration, change of registration, connection management, link establishment, retransmission, link release, etc., for the

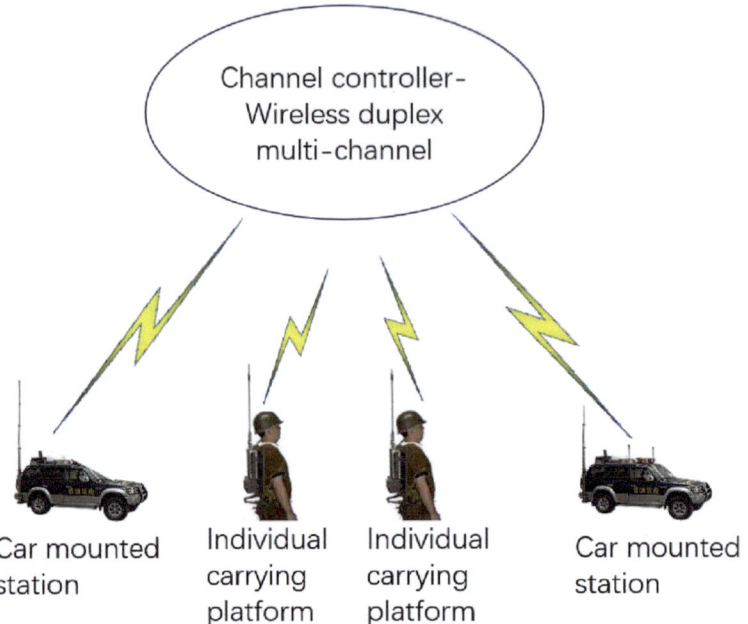

Fig. 19.1 Air platform wide band multi-channel relay system

mobile users on the ground and in the air, with the main functions including signalling processing function, basic service function and channel management function.

3. Main features

In order to achieve the modularity, expandability and interoperability of the payload in the future, the overall structure adopts the open and modular structure, and the main functions are implemented by software to build a variety of wireless transmitting functions of multi-band and multi-channel. Using the VME control bus, a public communication hardware platform for simultaneous multi-channel operation is established on the basis of high-speed A/D, D/A and parallel digital signal processing. Its features are as follows.

(1) Adopting standard bus, on the basis of open public hardware platform, the system has modular programmable signal reorganisation capability through relevant processing software;
(2) Realise the grouping communication of different communication network systems through the communication controller;
(3) Realise various waveform modulation/demodulation, voice coding, anti-jamming communication and final control by digital signal processing;
(4) Implementing the processing of multiple receive/transmit signals simultaneously using parallel processing to achieve simultaneous operation of multiple channels and interconnection of different networks;

(5) New services can be introduced by changing some components, which facilitates the upgrade of system functions.

19.2 Duplex Multichannel Shared Systems

Duplex multi-channel shared system of an air platform centre station is shown in Fig. 19.2, which is equipped with eight duplex frequency hopping (DFH) radios to provide eight duplex channels for transmission of signalling, digital voice and data between the platform centre station and the ground duplex mobile station. Using a dedicated signalling channel control method, the radio channel controller (RCC) can select one of the eight duplex channels as the signalling channel and the remaining seven as the service channel.

RCC provides eight wireless channel interfaces and two group interfaces. It can automatically transfer 8 wireless channels, or multiplex the single signals of 8 wireless channels into 128 kbit/s group signals and send them to the relay machine, or split the group signals from the relay machine into single signals and send them to the duplex channel machine. The RCC manages and controls the secret machine, key management centre and DFH radio of the platform central station, and it also has the group interface with the relay machine. The group road signal is transmitted through digital relay machine to realize group road communication with the ground central station. DFH radio includes three main parts: transceiver, control unit and antenna feeder system, the working frequency is 60–108 MHz, the channel spacing is 50 kHz, the frequency hopping rate is 250 hops/s, the working mode is frequency hopping/time division duplex (FH/TDD), the transmission rate is 16 kbit/

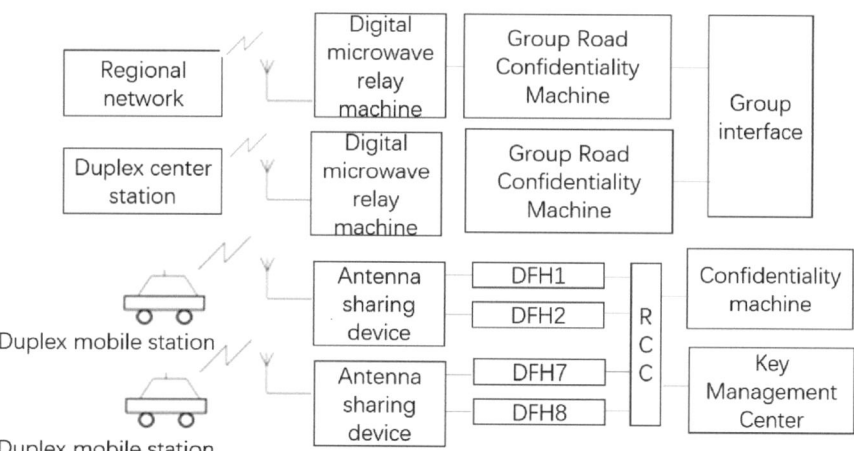

Fig. 19.2 Air platform duplex multi-channel sharing system

s. The system adopts wireless single road digital encryption system, the key management centre and key management centre. The system adopts wireless single-channel digital encryption system to ensure the confidentiality of all mobile user channels; group-channel signals are encrypted by group-channel secrecy machine.

The ground part consists of duplex frequency-hopping radio, mobile terminal and mobile confidential remote control telephone. The simplex radio enters this duplex network through the simplex duty radio and RAP.

19.3 UAV Relay Communication Systems

The trunk transmission system between nodes of the ground communication network or between communication subnets can extend the communication distance through the relay node of the air platform. At present, the representative one is the unmanned aircraft relay and forwarding system, which is composed of an aerial communication platform, a ground master station and user terminals. The total station on the ground consists of a measurement and control station and a communication switch, and the total station exchanges voice or digital information with the terminals on the ground. For example, the sending link from the master station to each terminal can be in TDMA mode and sent out in the form of broadcasting; the uplink from each terminal on the ground to the master station can be sent in FDMA mode; the signals of the two links are relayed through the air platform. Figure 19.3 depicts this kind of relay communication system, the figure shows the frequency points of communication between FA and FB terminals and the air platform, fc is the broadcast sending frequency point of the air platform to each terminal, and $f_1,\ldots f_n$ are the frequency points of frequency-division uplink frequency points of each terminal.

The simplest and easiest way is to use digital multiplexing, such as a group of 30 channels, whose data rate is 2 Mbit/s. As shown in Fig. 19.3, the communication link between the terminus and the air platform as well as the link from the platform to the terminals uses digital multiplexing. The total station will send the signals to each terminal and the remote control signals to the platform according to the digital multiplexing to the platform, the platform receives the demodulation and extracts the remote control signals, and at the same time sends the multiplexed signals to the ground terminals after modulation, each terminal receives the demodulation, extracts the information of the respective time slots, and transmits the signals according to the frequency division mode to the aerial platform, which receives the demodulation of the signals, and then transmits the signals to the aerial platform and then remodulates the signals with its telemetry signals. The platform receives and demodulates various signals, and then sends them to the ground station with their telemetry signals for reconnection and modulation. The total station is equipped with a small switch to exchange signals for end users. Because of the standard digital multiplexing frame format, it is easy to access the local area network (LAN).

The uplink of each terminal can also adopt the CDMA method, applying spread spectrum technology, the platform will demodulate the signals of each channel for

19.3 UAV Relay Communication Systems

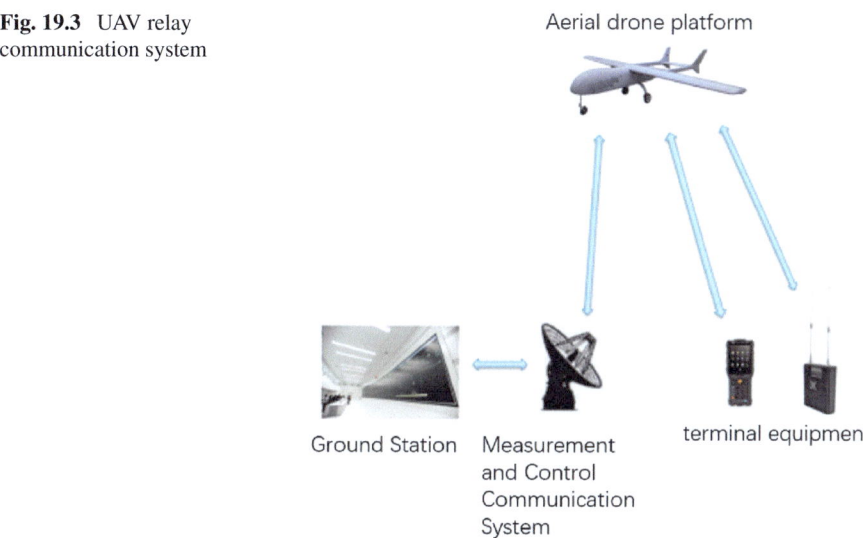

Fig. 19.3 UAV relay communication system

relevant reception, and then rewire them into digital multi-channel signals to be transmitted to the ground master station. This way will increase the amount of equipment and complexity of the platform, but the benefit is to improve the anti-interference performance. The ground master station has two sets of systems for tele-control telemetry and communication network management. The network management system can monitor the terminals, set the parameters and set the working mode, etc. The total station adopts the single pulse tracking system to carry out real-time tracking and measurement and control of the platform. The platform adopts omnidirectional antenna, and the ground master station uses directional tracking antenna. In the actual flight, the control platform should fly according to the smallest possible radius to ensure reliable relay communication. In order to work in a complex electromagnetic environment, the system must have anti-jamming function, and an effective way is to use extended spectrum technology. If the data rate is 2 Mbit/s and the extended gain is 20 dB, the bandwidth after spectrum expansion is 200 MHz, so it is necessary to choose a higher carrier frequency, such as C band or higher.

When the flight of the main platform is close to the endurance time, the backup platform should be used to replace it, and when replacing it, it is required to ensure that the path is switched without any damage, and the ground measurement and control station can measure and control the two UAVs at the same time. If the platform carrier is a circular fixed-wing UAV, it can be as deep as 200 km for long-distance relaying.

19.4 Adaptive Joint C4ISR Nodes

The Adaptive Joint C4ISR Node (AJCN) evolved from the Airborne Communications Node (ACN) programme implemented by the United States Defense Advanced Research Projects Agency (DARPA) in 1998. The concept is to use manned or unmanned aircraft as an airborne communications platform to achieve assured in-theatre communications, out-of-theatre backhaul connectivity, and information intelligence capabilities using the software radio concept. The AJCN implements a modular, scalable, software-defined radio capable of bridging and switching voice or data between two or more radios of different regimes, through the integration of SINCGARS, EPLRS, JTRS broadband networking waveforms, UHF satellite communications, Link-16, intelligence broadcasting and other systems of communication waveforms of the bridge, can be achieved information warfare text and electricity conversion (such as Link-16/variable message format VMF) as the second phase of the AJCN programme, it further develops the concept of software-defined radios in the initial conception, the main goal is to develop a modularity, scalable, multifunctional C4ISR-based architecture. The main goal is to develop a multifunctional C4ISR payload based on a modular, scalable architecture.

This section only briefly describes the communication aspects of AJCN.

AJCN mainly uses manned aircraft or unmanned aircraft as an airborne communication platform to extend the communication distance and realise the seamless link between different radios of different military branches. With the development of technology, the design of AJCN has expanded the scope of application, and in addition to airborne platforms, AJCN is also applicable to ground and sea platforms. The difference is that the altitude of the airborne AJCN gives it an advantage in wireless transmission; while in terms of application, ground and maritime platforms are basically similar to airborne platforms. The carriers for airborne AJCN platform communications (Fig. 19.4) include three classes:

1. Tactical level, which is applied to a small system, such as the UAV called Shadow 200.
2. The theatre level, where the application vehicle is a CH-53 aircraft or a Predator UAV. The load capacity specially designed for the Predator is 45 kg and it can accommodate 12–16 radio channels.
3. The strategic class, which is carried by the Global Hawk UAV. The Global Hawk has a payload mass of more than 400 kg, can accommodate more than 100 channels, and can receive very weak and useful signals by using advanced broadband receivers, and can perform precise geo-location through cross-links to other platforms. The communication area is 120 km deep and 200 km wide, with frequency coverage from 20 MHz to 2 GHz.

Fig. 19.4 ACN for different carriers

Bibliography

1. Zhou, Zheng, Huilin Zhou, et al. 2002. *Practical Handbook of New Technology in Communication Engineering: Mobile Communication Technology*. Beijing: University of Posts and Telecommunications Press.
2. (E) Hall MPM. 1984. *Convection Propagation and Radio Communications*. Beijing: National Defence Industry Publishing House.
3. Baowei, Lv., and Wang Zhensong. 2003. *Theory of Radio Wave Propagation and Its Applications*. Beijing: Science Press.
4. Yong, Ba, Zhang Zhongzhao, and Zhang Naitong. (1999). Feasibility Study of Stratospheric Communication System in Military Applications. *System Engineering and Electronic Technology*, 10
5. Chen, He, and Zhu Hongwen. 1999. New Technology of Broadband Wireless Relay Stratospheric Communication. *Computer and Network*, 12.
6. Youshou, Wu. 2000. Stratospheric Communication System in Development. *Engineering Physics* 4: 1–8.
7. Air Force Preparatory Academy and Beijing Institute of Aeronautics and Astronautics. 2005. *Proceedings of the Colloquium on the Development and Application of Floating Vehicles*. Beijing: Air Force Preparatory Academy and Beijing Institute of Aeronautics and Astronautics.
8. Dongchen, Zhang, and Zhou Ji, eds. 2015. *Military Communications*, 2nd ed. Beijing: National Defence Industry Publishing House.

Part VI
Prospects

Chapter 20
Future Outlook for Unmanned Maritime Systems

Unmanned combat is profoundly changing the face of war and is one of the top choices for future combat equipment. According to Xinhua News Agency, on 23 July 2020, General Secretary Xi Jinping said when inspecting the drone control teaching facilities at the Air Force Aviation University and learning about the training of drone controllers, "Now that various types of unmanned aircraft systems have appeared in large quantities, unmanned combat is profoundly changing the face of war. We need to strengthen the unmanned combat research, strengthen the professional construction of unmanned aircraft, strengthen the actual combat education and training, and accelerate the training of unmanned aircraft application and command personnel."

The future "14th Five-Year" equipment procurement focus on the direction of the important military investment direction, I think, can be from the "consumable equipment" and "future combat equipment" two dimensions of screening The "consumable equipment" and "future combat equipment" two dimensions of screening. "Consumable equipment" the logic of choice is to increase the intensity of combat training and combat readiness resulting in increased demand for weapons and equipment; "future combat equipment" is to comply with the new military changes in the choice of weapons and equipment track.

20.1 An Important Form of Future Warfare is the Unmanned Nature of Warfare

Unmanned equipment refers to unmanned, completely remote-controlled operation or autonomous operation according to pre-programmed procedures, carrying offensive or defensive weapons to carry out combat tasks of a class of weapons platforms, mainly including unmanned (UAS), unmanned ground vehicle (UGV), underwater unmanned underwater vehicle (UUV), surface unmanned vessel (USV) and so on.

With the rapid development of the new military change, human war is transforming into information war, and unmanned war has become one of the important development trends. Various kinds of unmanned combat platforms have begun to emerge in local wars, showing great potential for development, and have been increasingly valued by various countries, with a very strong development momentum. There are many successful cases of unmanned equipment in many modern wars. The author believes that unmanned equipment will show even greater development potential in future wars.

20.2 Characteristics of Future Development of Unmanned Equipment

1. Unmanned equipment growth logic: strong consumption attributes, huge potential for civil-military sharing

    We believe that unmanned equipment, as an emerging type of equipment, future growth logic is mainly reflected in the following two aspects: (1) In the military field, unmanned equipment in the war application scenarios continue to expand, strong consumption attributes.

    (1) In the military field, unmanned equipment in the war application scenarios continue to expand, strong consumption attributes, demand space. As a "consumable" product, unmanned equipment has the attribute of "easy to consume". Since the destruction of unmanned equipment will not bring casualties, and the direct loss compared to the manned combat platform is much smaller; in addition, the modern war missiles, radar and other offensive weapons and combat aids are becoming more and more advanced, greatly increasing the risk of equipment destruction. Take drones as an example, as a "combat tool" is a typical consumable, in recent years, on the battlefield, drones were shot down or destroyed reports continue.
    (2) In the civil field, the future unmanned equipment will play an irreplaceable role in surveying and mapping, inspection, exploration, meteorology, logistics, environmental monitoring, post-disaster rescue, underwater salvage and many other aspects.

2. Unmanned equipment to "nine" development

    The future development of unmanned aerial vehicles (UAVs) will show nine trends: detection and combat integration, long-duration lag time, structural stealth, micro-small size, high degree of intelligence, comprehensive integration, synergistic use, combat network, and serialisation of equipment. Among them, long endurance UAVs, micro-small UAVs, combat UAVs and UAV clusters will be the key direction of development (Fig. 20.1).

    This book explains the significance of the development of unmanned equipment at sea and the history and current situation of the development at home and abroad, and

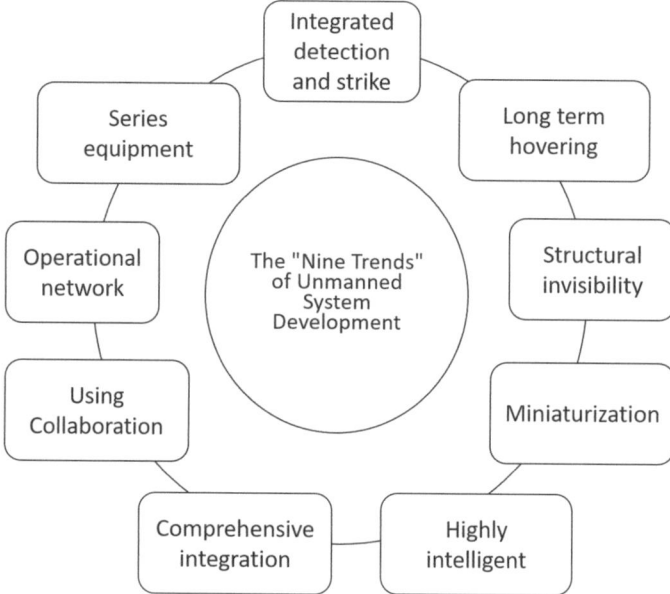

Fig. 20.1 Trends in unmanned systems

analyses the technical characteristics of unmanned equipment at sea from the dimensions of the basic principle, key technology and application in marine engineering by taking unmanned aircraft and unmanned boat equipment platforms as examples. The four necessary technologies to realise the intelligent development of maritime unmanned equipment are introduced in detail: acoustic technology, optoelectronic technology, radar technology and maritime communication technology.

The advantages of maritime unmanned equipment such as low cost, multi-functions and strong mobility make it have a broad application prospect, and the development of artificial intelligence technology further improves its performance. In the future, unmanned equipment at sea will greatly affect the pattern of marine transport and marine resources development, and a large number of unmanned military equipment will also be used in the war, becoming an important part of the marine.

Bibliography

1. *Xingye Military Industry In-Depth Series of Four Military Drones Industry In-Depth Research.* 2017. Beijing: Securities Research Report.
2. *Unmanned Equipment Industry Research.* 2020. Beijing: Guosheng Securities Research Report.
3. *Industry Research on Unmanned Aircraft Series.* 2017. Beijing: GCSC.

MIX
Papier aus verantwortungsvollen Quellen
Paper from responsible sources
FSC® C105338

If you have any concerns about our products,
you can contact us on
ProductSafety@springernature.com

In case Publisher is established outside the EU,
the EU authorized representative is:
Springer Nature Customer Service Center GmbH
Europaplatz 3, 69115 Heidelberg, Germany

Printed by Libri Plureos GmbH
in Hamburg, Germany